Lecture Notes in Control and Information Sciences

Edited by A. V. Balakrishnan and M. Thoma

For information about Vols. 1–21 please contact your bookseller or Springer-Verlag.

Lecture Notes in Control and Information Sciences

Edited by M. Thoma and A. Wyner

96

H. J. Engelbert
W. Schmidt (Eds.)

Stochastic Differential Systems

Proceedings of the IFIP-WG 7/1 Working Conference
Eisenach, GDR, April 6-13, 1986

Springer-Verlag
Berlin Heidelberg GmbH

Editors
Hans Jürgen Engelbert
Wolfgang Schmidt

Department of Mathematics
Friedrich-Schiller-University
6900 Jena
GDR

ISBN 978-3-540-18010-4

Library of Congress Cataloging in Publication Data

IFIP Working Conference on Stochastic Differential Systems (5th : 1986 : Eisenach, Germany)
Stochastic differential systems.
(Lecture notes in control and information sciences ; 96)
"Fifth IFIP Working Conference on Stochastic Differential Systems" — — Pref.
1. Stochastic differential equations — — Congresses.
I. Engelbert, Hans Jürgen. II. Schmidt, W. (Wolfgang)
III. IFIP WG 7.1. IV. Title. V. Series.
QA274.23.I35 1986 519.5 87–16533
ISBN 978-3-540-18010-4 ISBN 978-3-540-47245-2 (eBook)
DOI 10.1007/978-3-540-47245-2

Originally published by Springer-Verlag Berlin Heidelberg New York in 1987

PREFACE

The Fifth IFIP Working Conference on Stochastic Differential Systems took place in Eisenach (GDR), April 6-13, 1986. The conference was organized by IFIP TC 7/1 and the Friedrich-Schiller-University of Jena, in co-operation with the Karl-Weierstraß-Institute of Mathematics of the Academy of Sciences of GDR, the Technical University of Dresden, and the Humboldt-University of Berlin.

The meeting was intended to continue the traditional line of the foregoing conferences in Kyoto (1976), Vilnjus (1978), Visegrad (1980), and Marseille-Luminy (1984) and to focus on topics of present research in the field of stochastic differential systems. As orientation, particular emphasis was put on infinite-dimensional stochastic systems and stochastic partial differential equations, stochastic equations and diffusions and, especially, numerical methods for them, and stochastic control and filtering. The conference was attended by 73 participants from 17 countries: German Democratic Republic (37), USSR (8), France (5), Federal Republic of Germany (4), Bulgaria (3), Switzerland (3), Great Britain (2), Poland (2), Austria (1), Canada (1), Czechoslovakia (1), Denmark (1), Finland (1), Italy (1), Romania (1), USA (1).

The scientific program was very intensive. At the meeting 10 lectures of 50 minutes, 12 of 40 minutes, and 29 talks of 30 minutes were presented.

This volume contains most of the texts of these lectures. They highly reflect the orientation given for the scope of the conference. We found it useful to arrange the contributions in several groups of related subjects. Needless to say that the solution to this classification problem is by no means unique. Finally we decided to divide the volume into the following four parts:

1. Infinite-Dimensional Stochastic Systems and Random Fields. Stochastic Partial Differential Equations.
2. Stochastic Equations and Diffusions. Approximation of Diffusions.
3. Stochastic Control Theory.
4. Special Problems in Martingale Theory and Stochastic Calculus.

We hope that the volume will give a representative insight into the work of the conference and that it would enforce the cooperation of scientists from all over the world working on stochastic differential systems and their applications.

We should like to thank all participants for their contributions not only to the scientific success but also to a very hearty and friendly meeting of people from different countries. Needless to say that the meeting took place not only in the lecture room. Participants enjoyed the rich historical traditions in Eisenach, situated at the foot of the Wartburg, one of the oldest and most famous German castles, and the surroundings, which are closely connected with M. Luther, J.S. Bach, J.W. von Goethe ...

It is a pleasure to express our gratitude to other members of the International Program Committee and its chairman Professor A. V. Balakrishnan (who encouraged us to organize this conference) for valuable assistance. We are indebted to the members of the Local Organizing Committee and many others who helped in the organization of the conference. Finally, our special thanks are due to M. Venth for her immense and careful typewriting in the preparation of the conference and, in particular, of the manuscript of this volume.

H.J. Engelbert

W. Schmidt

LIST OF PARTICIPANTS

L. ARNOLD	Universität Bremen, Fachbereich Mathematik/ Informatik, Postfach 330440, D-2800 BREMEN (FRG)
H. BECHER	TH "Carl Schorlemmer" Geusaer Str., 4200 LEUNA-MERSEBURG (GDR)
R.J. CHITASHVILI	Sector of Probability Theory and Math. Stat., Tbilisi Mathematical Institute, Georgian Academy of Sciences, Plekhanov Ave 150a, TBILISI 12, 380012 (U.S.S.R.)
S.K. CHRISTENSEN	Schønbergsgade 15, DK-1906 FREDERIKSBERG C, (Denmark)
N. CHRISTOPEIT	Institut für Ökonometrie und Operations Research, Universität Bonn, Adenauerallee 24-42, D-5300 BONN 1, (FRG)
J.M.C. CLARK	Imperial College of Science & Technology, Department of Electrical Engineering, Exhibition Road, LONDON SW7 2BT, (Great Britain)
M.H.A. DAVIS	Department of Computing and Control, Imperial College, 180 Queen's Gate, LONDON SW7 2BZ, (Great Britain)
O. ENCHEV	Centre of Mathematics, P.O. Box 325, 7000 ROUSSE, (Bulgaria)
H.J. ENGELBERT	Friedrich-Schiller-Universität Jena, Sektion Mathematik, 6900 JENA (GDR)
N. El KAROUI	Laboratoire de Probabilités, Université de Paris VI, 4, Place de Jussien, 75230 PARIS CEDEX 5, (France)
K.H. FICHTNER	Friedrich-Schiller-Universität Jena, Sektion Mathematik, 6900 JENA (GDR)
K. FLEISCHMANN	Akademie der Wissenschaften der DDR, Karl-Weierstraß-Institut für Mathematik, Mohrenstraße 39, 1086 BERLIN (GDR)
H. FÖLLMER	ETH Zürich, Mathematik ETH-Zentrum, 8092 ZÜRICH (Switzerland)

S.D. GAIDOV	Centre of Mathematics, P.O. Box 325, 7000 ROUSSE (Bulgaria)
J. GÄRTNER	Akademie der Wissenschaften der DDR, Karl-Weierstraß-Institut für Mathematik, Mohrenstraße 39, 1086 BERLIN (GDR)
A. GERMANI	Instituto di Analisi dei Sistemi ed Informatica del CNR, Viale Manzoni, 30, 00185 ROMA (Italy)
L. GIRAITIS	Institute of Mathematics and Cybernetics Academy of Sciences of the Lithuanian SSR, K. Pozelos str., 232 600 VILNIUS 54 (U.S.S.R.)
W. GRECKSCH	TH "Carl Schorlemmer" Leuna-Merseburg, Sektion Mathematik, Otto-Nuschke-Straße 4200 MERSEBURG (GDR)
B. GRIGELIONIS	Institute of Mathematics and Cybernetics, Academy of Sciences of the Lithuanian SSR, K. Pozelos str., 232 600 VILNIUS (U.S.S.R.)
A.A. GUCHTCHIN	Mathematical Institute of the Academy of Sciences of the U.S.S.R., u. Vavilova, 117333 MOSCOW (U.S.S.R.)
U. HAUSSMANN	Department of Mathematics, University of British Columbia, VANCOUVER, B.C. V6T 1Y4 (Canada)
L. HOY	TH "Carl Schorlemmer" Leuna-Merseburg, Sektion Mathematik, Otto-Nuschke-Str., 4200 MERSEBURG (GDR)
K. HELMES	Institut für Angewandte Mathematik der Universität Bonn, Wegelerstraße 6, D-5300 BONN (FRG)
G. JETSCHKE	Friedrich-Schiller-Universität Jena, Sektion Mathematik, 6900 JENA (GDR)
G. KALLIANPUR	Institute for Mathematics and its Applications, 514 Vincent Hall. 206 Church Street, University of Minnesota, MINNEAPOLIS, MN 55455 (U.S.A.)
H. KOREZLIOGLU	E.N.S.T., 46 rue Barrault, 75013 PARIS (France)
P. KOTELENEZ	Universität Bremen, Fachbereich 3 Mathematik & Informatik, Kufsteinerstraße D-2800 BREMEN 33 (FRG)
U. KÜCHLER	Humboldt-Universität Berlin, Sektion Mathematik, Unter den Linden 6, 1086 BERLIN (GDR)

H. LANGER	Technische Universität Dresden, Sektion Mathematik, Mommsenstraße 13, 8027 DRESDEN (GDR)
N.L. LAZRIEVA	Sector of Probability Theory and Math. Stat., Tbilisi Mathematical Institute, Georgian Academy of Sciences, Plekhanov Ave 150 a, TBILISI 12, 380012 (U.S.S.R.)
J.P. LEPELTIER	Département de Mathématiques Université du Maine, Route de Lavel, 72017 LE MANS CEDEX (France)
F. LIESE	Wilhelm-Pieck-Universität Rostock Sektion Mathematik, Universitätsplatz, 2500 ROSTOCK (GDR)
J.-U. LÖBUS	Friedrich-Schiller-Universität Jena Sektion Mathematik, 6900 JENA (GDR)
P. MANDL	Department of Probability and Mathematical Statistics, Charles University, Sokolovska 83, 18600 PRAGUE 8 (Czechoslovakia)
R. MANTHEY	Friedrich-Schiller-Universität Jena, Sektion Mathematik, 6900 JENA (GDR)
R. MIKULEVICIUS	Institute of Mathematics and Cybernetics Academy of Sciences of the Lithuanian SSR, K. Pozelos, 54, 232600 VILNIUS (U.S.S.R.)
S.A. MOLCANOV	Moscow State University, Faculty of Mathematics and Mechanics, 117 234 MOSCOW (U.S.S.R.)
M. NAGASAWA	Seminar für Angewandte Mathematik der Universität Zürich, Freiestraße 36 CH-8032 ZÜRICH (Switzerland)
E. PARDOUX	Université de Provence, 3 Place Vicor-Hugo 13331 MARSEILLE CEDEX 3 (France)
N.I. PORTENKO	Mathematical Institute, Academy of Sciences of the Ukrainian SSR, ul. Repina 3, 252601 KIEV (U.S.S.R.)
E. PLATEN	Akademie der Wissenschaften der DDR, Karl-Weierstraß-Institut für Mathematik, Mohrenstraße 39, 1086 BERLIN (GDR)
M. RICHTER	Technische Hochschule Karl-Marx-Stadt, Sektion Mathematik, PSF 964, 9010 KARL-MARX-STADT (GDR)
W. RÖMISCH	Humboldt-Universität Berlin, Sektion Mathematik, Unter den Linden 6, 1086 BERLIN (GDR)

F. RUSSO	Ecole Polytechnique Fedérale, Départment de Mathematiques, 1015 LAUSANNE (Switzerland)
P. SALMINEN	Abo Akademi, Matematiska Institutionen Sähriksgaten 3, SF-20500 ABO 50 (Finland)
J. vom SCHEIDT	Ingenieurhochschule Zwickau, Dr.-Friedrichs-Ring 2a, 9500 ZWICKAU (GDR)
W. SCHENK	Technische Universität Dresden, Sektion Mathematik, WB WMS, Mommsenstraße 13, 8027 DRESDEN (GDR)
B. SCHMALFUß	Ingenieurhochschule Köthen, Bernburger Straße, 4370 KÖTHEN (DGR)
W. SCHMIDT	Friedrich-Schiller-Universität Jena Sektion Mathematik, 6900 JENA (GDR)
T. SHIGA	Department of Applied Physics, Fac. Sci. Tokyo Institute of Technology, Oh.okayama, TOKYO 152 (Japan)
J. STOYANOV	Bulgarian Academy of Sciences, Inst. of Mathematics, P.O. Box 373, 1090 SOFIA (Bulgaria)
C. STRICKER	Université de Franche-Comté-Besancon Faculté des Sciences et des Techniques, Laboratoire de Mathematiques, Route de Gray, 25030 BESANCON CEDEX (France)
K. TWARDOWSKA	Uniwersitet Jagiellonski, Instytut Informatyki, ul Kopernika 27, 31-501 KRAKOW (Poland)
C. VARSAN	INCREST, Department of Mathematics, Bdul Pacii 220, 79622 BUCHAREST (Romania)
A. WAKOLBINGER	Johannes Kepler Universtiät Linz, Institut für Mathematik, A-4040 LINZ (Austria)
J. ZABCZYK	Institute of Mathematics, Polish Academy of Sciences, Sniadeckich 8, 00-950 WARSAW (Poland)

CONTENTS

1. INFINITE-DIMENSIONAL STOCHASTIC SYSTEMS AND RANDOM FIELDS. STOCHASTIC PARTIAL DIFFERENTIAL EQUATIONS

4. SPECIAL PROBLEMS IN MARTINGALE THEORY AND STOCHASTIC CALCULUS

Infinite-dimensional Stochastic Systems
and Random Fields.
Stochastic Partial Differential Equations

LONG-TIME FLUCTUATIONS OF WEAKLY INTERACTING DIFFUSIONS

D.A. Dawson and J. Gärtner

Department of Mathematics
and Statistics
Carleton University
Ottawa, Ontario
Canada K1S 5B6

Karl-Weierstrass-Institut
für Mathematik
Akademie der Wissenschaften
der DDR
DDR-1086 Berlin, PF 1304

1. INTRODUCTION

It is well-known that the ferromagnetic Ising model on the lattice $\mathbb{Z}^2$ with spins $x_k \in \{-1,+1\}$ and formal Hamiltonian

$$H(\underline{x}) = \sum_{|k-l|=1} |x_k - x_l|^2$$

exhibits a phase transition. In the low temperature region there exist two pure phases (i.e. two ergodic Gibbs distributions) having negative and positive mean magnetization, respectively.

Let us consider the Glauber dynamics [9] in a large but finite volume $V \subset \mathbb{Z}^2$. This is a reversible Markov jump dynamics on the state space $\{-1,+1\}^V$ with the property that its unique invariant probability measure coincides with the Gibbs distribution in V corresponding to free boundary conditions. Suppose that this dynamics starts with a configuration of spins the "greater part" of which has negative magnetization. Then, because of the coexistence of the above two phases, one expects that during a long time period the system will remain in the regime with "typical" negative spins. But from time to time the stochastic system will make attempts to escape from this regime: there will occur relatively large islands of positive magnetization. After a long but finite time one of these islands is expected to become so large that the system undergoes a transition into a new regime with configurations consisting of "mostly" positive spins.

The understanding of such *dynamical phase transitions* (sometimes called *tunnelling*) and their mathematically rigorous investigation is a challenging problem. For models with short-range interaction such as the Ising model this seems to be a difficult task. One therefore tries to understand several aspects of the tunnelling mechanism by investigating simpler models. In [6] W.G. Faris and G. Jona-Lasinio studied related questions for a non-linear heat equation with small noise disturbances. Recently F. Comets [1] obtained analogous results for a model with discrete spins ±1 on a periodic lattice and "weighted" mean-field interaction.

Our research is devoted to the investigation of dynamical phase transitions and related problems for a Curie-Weiss model with continuous spin [5]. The proofs of the results stated below can be found in a somewhat more general setting in [2], [3], [4], and [8].

2. THE EQUILIBRIUM DISTRIBUTION

We consider a system $\underline{x} = (x_1,\ldots,x_N)$ of N real-valued spins with joint probability distribution

$$(1)\quad p_N(d\underline{x}) = Z_N^{-1} \exp\left\{ -\frac{1}{2\sigma^2} N^{-1} \sum_{k,l=1}^{N} |x_k - x_l|^2 \right\} \mu_\sigma(dx_1) \ldots \mu_\sigma(dx_N) ,$$

where

$$\mu_\sigma(dx) = Z^{-1} \exp\left\{ -\frac{2}{\sigma^2} U(x) \right\} dx$$

and Z_N and Z denote normalizing constants. That is, p_N is the equilibrium distribution (Gibbs distribution) associated with the Hamiltonian

$$\mathcal{H}_N(\underline{x}) = \frac{1}{4N} \sum_{k,l=1}^{N} |x_k - x_l|^2 + \sum_{k=1}^{N} U(x_k)$$

which describes a system with mean field interaction in an external field with potential U. The quantity $\beta^{-1} = \sigma^2$ may be interpreted as temperature. For concreteness we will assume in the rest of this paper that U is a symmetric double-well potential of the form

$$U(x) = x^4/4 - x^2/2 .$$

Besides of the particle configuration $\underline{x} = (x_1,\ldots,x_N)$ we will investigate the empirical measure

$$X_N = N^{-1} \sum_{k=1}^{N} \delta_{x_k} ,$$

where δ_x denotes the Dirac measure at x. Let $\mathcal{M}_R$ denote the space of probability measures on $\mathbb{R}$ the fourth moment of which does not exceed R. As space of admissible probability measures we take

$$\mathcal{M}_\infty = \bigcup_{0<R<\infty} \mathcal{M}_R$$

furnished with the strongest topology which induces on $\mathcal{M}_R$ the weak topology for each $R < \infty$. This non-metrizable topology allows a simple characterization of convergence and compactness of subsets. In particular, we have $\nu_n \longrightarrow \nu$ in M_∞ iff $\nu_n \longrightarrow \nu$ weakly and the sequence (ν_n) belongs entirely to M_R for some $R<\infty$.

We define the *free energy functional* $F : M_\infty \longrightarrow (-\infty, \infty]$ of the equilibrium distribution of the empirical measure by

$$F(\nu) = \langle \nu \otimes \nu, |x-y|^2 \rangle + \beta^{-1} \langle \nu, \log \frac{d\nu}{d\mu_\sigma} \rangle$$

if $\nu \in M_\infty$ is absolutely continuous with respect to μ_σ . Otherwise we set $F(\nu) = \infty$. ($\langle \nu, f \rangle$ denotes the integral of f with respect to ν .) This formula exhibits the decomposition of the free energy functional into internal energy minus temperature times relative entropy. The *Gibbs free energy* f is then the minimum of the functional F.

Let π_N denote the law on M_∞ of the empirical measure X_N under p_N. We have the following large deviation result.

<u>THEOREM 1.</u> Let A be a Borel subset of M_∞ . Then

$$- \inf_{\nu \in A^o} I(\nu) \leq \liminf_{N \to \infty} N^{-1} \log \pi_N(A)$$

$$\leq \limsup_{N \to \infty} N^{-1} \log \pi_N(A) \leq - \inf_{\nu \in \overline{A}} I(\nu) ,$$

where $I : M_\infty \longrightarrow [0,\infty]$ is the *equilibrium action functional* defined by

$$I = \beta(F - f)$$

and A^o and $\overline{A}$ are the interior and the closure of A, respectively.

This theorem states in particular that for large N the law of the empirical measure X_N is concentrated in a small neighborhood of the global minima of the free energy functional F. Let us describe the structure of the global minima. There exists a critical temperature $\sigma_{cr}^2 \approx 0.914$ with the following properties. For $\sigma \geq \sigma_{cr}$ the functional F has exactly one global minimum ν_o with symmetric density $d\nu_o/dx$ (Fig. 1). If σ becomes smaller than σ_{cr}, then the measure ν_o turns into a "saddle point" and there appear two new global minima ν_- and ν_+ with negative and positive mean value, respectively (Fig. 2). These properties of the free energy functional reflect the fact that our mean-field model exhibits a ferromagnetic phase transition.

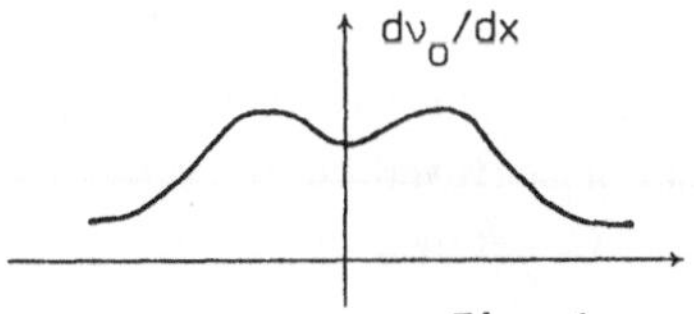

Fig. 1

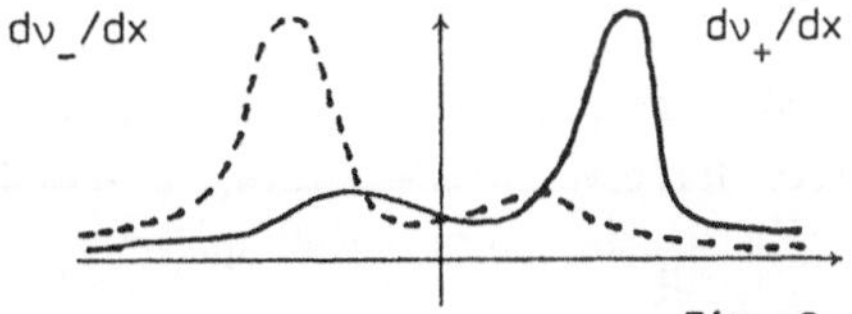

Fig. 2

3. THE STOCHASTIC DYNAMICS

A natural way to introduce a reversible stochastic dynamics having equilibrium distribution (1) consists in considering the Itô equations

$$dx_k(t) = -\frac{\partial H_N}{\partial x_k}(x_1(t),\dots,x_N(t))\,dt + \sigma\,dw_k(t)\,, \qquad k = 1,\dots,N,$$

where $w_1,\dots,w_N$ are independent one-dimensional Wiener processes. This system can be written in the form

$$(2)\quad dx_k = -\Big[\,U'(x_k) + N^{-1}\sum_{l=1}^{N}(x_k-x_l)\,\Big]dt + \sigma\,dw_k\,, \qquad k = 1,\dots,N.$$

Our aim is to study the behavior of the *empirical process*

$$X_N(t) = N^{-1}\sum_{k=1}^{N}\delta_{x_k(t)}$$

as $N \to \infty$. Applying Itô's formula to (2), we get

$$d\,\langle X_N(t),f\rangle = \langle X_N(t),L(X_N(t))\,f\rangle\,dt + N^{-1/2}\,dM_N^f(t)$$

for all test functions $f \in \mathcal{D}$, where $\mathcal{D}$ denotes the space of infinitely differentiable real functions with compact support. The operators $L(\mu)$, $\mu \in M_\infty$, are defined by

$$L(\mu)\,f(x) = \frac{\sigma^2}{2}\,f''(x) - [U'(x)+x-m(\mu)]\,f'(x)\,,$$

where

$$m(\mu) = \int_{\mathbb{R}} x\,\mu(dx)\,.$$

M_N^f is a continuous martingale with quadratic variational process $\langle\!\langle M_N^f\rangle\!\rangle$ given by

$$\frac{d}{dt}\langle\!\langle M_N^f\rangle\!\rangle_t = \sigma^2\,\langle X_N(t),|f'|^2\rangle\,.$$

Thus $X_N(.)$ can be considered as a random perturbation of order $N^{-1/2}$ of the *McKean-Vlasov dynamics*

$$(3)\quad \frac{d}{dt}\langle\mu(t),f\rangle = \langle\mu(t),L(\mu(t))\,f\rangle\,, \qquad t \geq 0,\ f \in \mathcal{D}\,.$$

The empirical process $X_N(.)$ is a Markov diffusion process on the "Riemannian manifold" M_∞ with drift vector $\mu \longmapsto L(\mu)^*\mu$ and diffusion tensor associated to the "Riemannian norm" $\|\cdot\|_\mu$ in the "tangent space" T_μ to M_∞ at μ which is defined by

$$(4)\quad \|\vartheta\|_\mu^2 = \frac{1}{2\sigma^2} \sup_{f \in \mathcal{D}} \frac{|\langle\vartheta, f\rangle|^2}{\langle\mu, |f'|^2\rangle}, \qquad \vartheta \in \mathcal{D}'.$$

(T_μ is a linear subspace of the Schwartz distributions $\mathcal{D}'$.) Therefore the results stated below may be considered as a certain infinite dimensional generalization of the Freidlin-Wentzell theory [7] on small random perturbations of finite dimensional dynamical systems.

Given $0 \le s < t < \infty$, let $C_{s,t} = C([s,t];M_\infty)$ be the space of continuous paths $[s,t] \longrightarrow M_\infty$ furnished with the strongest topology that induces on $C([s,t];M_R)$ the topology of uniform convergence for each $R < \infty$. Let $C_\infty = C([0,\infty);M_\infty)$ furnished with the weakest topology for which the projections onto all $C_{0,T}$, $0 < T < \infty$, are continuous. Let $M^{(N)}$ denote the subspace of M_∞ consisting of empirical measures associated with N-particle systems. We denote by $\{P^{(N)}_{\nu,s};\ (\nu,s) \in M^{(N)} \times \mathbb{R}_+\}$ the family of probability laws on C_∞ induced by the empirical process $X_N(.)$ associated to (2). The projection of $P^{(N)}_{\nu,s}$ onto $C_{s,t}$, $0 \le s < t < \infty$, will also be denoted by $P^{(N)}_{\nu,s}$. We will write $P^{(N)}_\nu$ instead of $P^{(N)}_{\nu,0}$. We have the following law of large numbers.

<u>THEOREM 2.</u> Given $\nu_N \in M^{(N)}$ and $\nu \in M_\infty$, suppose that $\nu_N \longrightarrow \nu$ in M_∞. Then

$$P^{(N)}_{\nu_N} \longrightarrow \delta_{\mu(.;\nu)}$$

in the sense of weak convergence of probability laws on C_∞, where $\mu(.;\nu) \in C_\infty$ is the unique weak solution of the McKean-Vlasov equation (3) with initial datum ν.

The McKean-Vlasov dynamics (3) exhibits the following qualitative behavior. If $\sigma \ge \sigma_{cr}$, then ν_0 is the only equilibrium state and this state is asymptotically stable (in the topology of M_∞). For $0 < \sigma < \sigma_{cr}$ there exist exactly three equilibrium positions ν_-, ν_0, ν_+, where ν_- and ν_+ are stable and ν_0 is unstable. Each path of the McKean-Vlasov dynamics is attracted by one of these equilibria. It turns out that σ_{cr}, ν_-, ν_0, and ν_+ are the same as in Section 2.

Suppose for the moment that $0 < \sigma < \sigma_{cr}$. If N is large, then the empirical process $X_N(.)$ will normally follow the McKean-Vlasov path leading into a small neighborhood of the stable equilibrium state ν_- or ν_+ and then perform small fluctuations around it. But from time to time the system will make attempts to escape from the domain of attraction of this equilibrium via a large deviation. Sooner or later one of these attempts will be successful and the empirical process will undergo a transition into a small neighborhood of the second stable equilibrium state. To describe this dynamical phase transition, we first present the underlying large deviation result.

THEOREM 3. Given $\nu_N \in M^{(N)}$ and $\nu \in M_\infty$, suppose that $\nu_N \longrightarrow \nu$ in M_∞. Let A be a Borel subset of $C_{s,t}$, $0 \le s < t < \infty$. Then

$$- \inf_{\mu(.) \in A^0, \mu(s)=\nu} S_{s,t}(\mu(.)) \le \liminf_{N \to \infty} N^{-1} \log P^{(N)}_{\nu_N,s}(A)$$

$$\le \limsup_{N \to \infty} N^{-1} \log P^{(N)}_{\nu_N,s}(A) \le - \inf_{\mu(.) \in \bar{A}, \mu(s)=\nu} S_{s,t}(\mu(.)).$$

The *action functional* $S_{s,t}: C_{s,t} \longrightarrow [0,\infty]$ is given by

$$S_{s,t}(\mu(.)) = \int_s^t \|\dot{\mu}(u) - L(\mu(u))^* \mu(u)\|^2_{\mu(u)} \, du$$

if $\mu(.) \in C_{s,t}$ is absolutely continuous in the distribution sense. Otherwise $S_{s,t}(\mu(.)) = \infty$. In the above $\|.\|_\mu$ denotes the norm defined by (4) and $L(\mu)^*$ is the formal adjoint of $L(\mu)$ acting in D'.

4. QUASIPOTENTIAL AND FREE ENERGY FUNCTIONAL

An important quantity for describing the tunnelling behavior of the empirical process is the so-called quasipotential. Let ρ be an arbitrary equilibrium of the McKean-Vlasov dynamics (3). Then the *quasipotential* Q_ρ is defined by:

$$Q_\rho(\nu) = \inf \{ S_{s,t}(\mu(.)): -\infty < s < t < \infty, \ \mu(s) = \rho \ , \ \mu(t) = \nu \} .$$

We next establish a close relationship between the quasipotential which is a dynamical notion and the free energy functional F of the equilibrium distribution of the empirical process.

THEOREM 4. Suppose that $\nu \in M_\infty$ belongs to the domain of attraction of ρ (with respect to the McKean-Vlasov dynamics). Then

$$Q_\rho(\nu) = \beta(F(\nu) - F(\rho)) .$$

Furthermore, if $F(\nu) < \infty$, then the infimum in the definition of Q_ρ is attained at $\bar{\mu}(u) = \mu(-u;\nu)$, $u \in (-\infty,0]$, that is, the time reverse of the McKean-Vlasov trajectory. Up to a time shift, this is the only path at which the infimum is attained.

5. DYNAMICAL PHASE TRANSITIONS

In this section we assume that $0 < \sigma < \sigma_{cr}$. Then the McKean-Vlasov dynamics has exactly three equilibrium states, namely ν_-, ν_0, and ν_+. Their domains of attraction will be denoted by D_-, D_0, and D_+, respectively. Let V_+ be an open neighborhood of ν_+ such that $\overline{V}_+ \subset D_+$ and let

$$\tau_+(\mu(.)) = \inf\{\, t \geq 0: \mu(t) \in V_+ \,\},$$

$$\sigma_-(\mu(.)) = \inf\{\, t \geq 0: \mu(t) \notin D_- \,\}.$$

Abbreviate $\Delta F = F(\nu_0) - F(\nu_-)$. In the next theorem we describe the asymptotic behavior of the tunnelling time from ν_- to ν_+ as $N \longrightarrow \infty$.

THEOREM 5. Given $\nu_N \in M^{(N)}$ and $\nu \in D_-$, suppose that $\nu_N \longrightarrow \nu$ in M_∞. Then

$$\lim_{N\to\infty} P^{(N)}_{\nu_N}\{\, e^{N(\beta\Delta F-\delta)} < \sigma_- \leq \tau_+ < e^{N(\beta\Delta F+\delta)} \,\} = 1$$

for each $\delta > 0$. Moreover,

$$\lim_{N\to\infty} N^{-1} \log E^{(N)}_{\nu_N} \sigma_- = \lim_{N\to\infty} N^{-1} \log E^{(N)}_{\nu_N} \tau_+ = \beta\Delta F ,$$

where $E^{(N)}_{\nu_N}$ denotes expectation with respect to $P^{(N)}_{\nu_N}$.

We remark that

$$\beta\Delta F \sim \frac{\sigma_{cr}^2/2 - 1/3}{\sigma_{cr}^6}\, m_\pm^4(\sigma) \qquad \text{as } \sigma \uparrow \sigma_{cr},$$

where the mean magnetization $m_\pm(\sigma) = \int x\, \nu_\pm(dx)$ behaves like

$$m_\pm(\sigma) \sim \pm \frac{\sigma_{cr}^2}{2} \sqrt{\frac{\sigma_{cr}^2}{\sigma_{cr}^2/2 - 1/3}} \left(\frac{1}{\sigma^2} - \frac{1}{\sigma_{cr}^2}\right)^{1/2} \qquad \text{as } \sigma \uparrow \sigma_{cr}.$$

Let $\mu_-(.)$ and $\mu_+(.)$ denote the paths of the McKean-Vlasov dynamics leading from ν_0 (at time $-\infty$) to ν_- and ν_+ (at time $+\infty$), respectively. Up to a time shift, these paths are unique. Let V be an open neighborhood of ν_0 and let T_- and T_+ be such moments that $\mu_-(T_-)$ and $\mu_+(T_+)$ belong to V. Set

$$\hat{\mu}_-(s) = \mu_-(T_-+s), \quad \hat{\mu}_+(s) = \mu_+(T_++s), \qquad s \geq 0.$$

Let V_+ and τ_+ be the same as above.

The next theorem states that there exists a typical *tunnelling path* along which the empirical process performs the transition from a small neighborhood of the stable equilibrium ν_- into a small neighborhood of the stable equilibrium ν_+.

THEOREM 6. There exist random times $\sigma, \tau : C_\infty \longrightarrow [0,\infty]$ such that for all $\nu_N \in M^{(N)}$ and $\nu \in D_-$ with $\nu_N \longrightarrow \nu$ in M_∞, all $T > 0$, and all open neighborhoods $\mathcal{V}_-$ of $\hat{\mu}_-(.)$ and $\mathcal{V}_+$ of $\hat{\mu}_+(.)$ in $C_{0,T}$:

$$P_{\nu_N}^{(N)}(A) \longrightarrow 1 \qquad \text{as } N \longrightarrow \infty ,$$

where

$$A = \{ T < \sigma < \tau < \tau_+, \ \mu(\sigma-.) \in \mathcal{V}_-, \ \mu(t) \in V \text{ for } t \in [\sigma,\tau], \ \mu(\tau+.) \in \mathcal{V}_+ \} .$$

REFERENCES

[1] F. Comets, Tunnelling and nucleation for a local mean field model, preprint, 1985.

[2] D.A. Dawson and J. Gärtner, Large deviations from the McKean-Vlasov limit for weakly interacting diffusions, Stochastics, to appear.

[3] D.A. Dawson and J. Gärtner, Large deviations, free energy functional, and quasipotential for a mean field model of interacting diffusions, Technical Report Series of the Laboratory for Research in Statistics and Probability, Carleton Univ. Ottawa, 1986.

[4] D.A. Dawson and J. Gärtner, Dynamical phase transitions and metastability for a mean field model of interacting diffusions, in preparation.

[5] R.S. Ellis and C.M. Newman, The statistics of Curie-Weiss models, J. Statist. Phys. 19 (1978), 149-161.

[6] W.G. Faris and G. Jona-Lasinio, Large fluctuations for a nonlinear heat equation with noise, J. Phys. A: Math. Gen. 15 (1982), 3025-3055.

[7] M.I. Freidlin and A.D. Wentzell, Random Perturbations of Dynamical Systems, Springer-Verlag, New York, 1984.

[8] J. Gärtner, On the McKean-Vlasov limit for interacting diffusions, I, II, AdW der DDR, Karl-Weierstrass-Institut für Mathematik, preprint, 1986.

[9] R.J. Glauber, Time-dependent statistics of the Ising model, J. Math. Phys. 4 (1963), 2, 294-307.

AN ESTIMATION PROBLEM FOR GENERALIZED GAUSSIAN PROCESSES

Ognian Enchev

Centre of Mathematics,
P.O.Box 325,
7000 Rousse
BULGARIA

Introduction

This paper was originated by the work of Hitsuda [4]. It treats the problem

(1) $X(t)=B(t)+Y(t)$, $0 \leq t \leq 1$,

where $B=(B(t) : 0 \leq t \leq 1)$ is a Brownian motion (i.e. a "noise") and $Y=(Y(t) : 0 \leq t \leq 1)$ is a Gaussian process which is interpretted as "signal". It is assumed to be independent of the noise B. It is also assumed in [4] that Y and B have one and the same RKHS. The last assumption allows to treat Y and B as two independent Gaussian random functionals $\phi_1(.)$ and $\phi_2(.)$ on the Hilbert space $L^2[0,1]$ which can be written in the form

(2) $\phi_1(f)= \int_0^1 f(t)dY(t)$, $\phi_2(f)= \int_0^1 f(t)dB(t)$, $f \in L^2[0,1]$.

Our goal is to generalize the problem (1) and respectively to extend the results of [4] in the following directions:

1) Instead of Y and B we shall consider two arbitrary generalized Gaussian processes $(\phi_1(x) : x \in H)$ and $(\phi_2(x) : x \in H)$ indexed by an <u>arbitrary</u> Hilbert space H, i.e. $\phi_1(.)$ and $\phi_2(.)$ are continuous linear transformations of H into some Gaussian subspace $\mathcal{G} \subset L^2(\Omega,\Sigma,\mathbb{P})$ (cf. [8], [9]). Thus the problem (1) we transform into the problem

(3) $\phi(x)=\phi_1(x)+\phi_2(x)$, $x \in H$.

2) In general we shall not impose any assumptions for $\phi_1(.)$ and $\phi_2(.)$ which can make them similar to the Brownian motion, martingale process or process with bounded variation. So, in the general formulation of the problem (3) $\phi_1(.)$ and $\phi_2(.)$ appear just as two arbitrary processes (the words "generalized" and "Gaussian" will be omitted sometimes for convenience).

3) We shall not assume that $\phi_1(.)$ and $\phi_2(.)$ are independent.

4) Nothing will be assumed about the RKHS's of $\phi_1(.)$ and $\phi_2(.)$.

The transformation of the problem (1) into the problem (3) is motivated by some Quantum-Mechanical interpretations of Gaussian processes, connected with the Fock-Cook quantization and the Weyl-von Neumann commutation relations (cf.[7],[8],[9]). This allows to formulate some classical results also for the noncommutative case, i.e. when the observables Y(t) and B(t) are elements of some noncommutative C*-algebra with a given state. From the other hand such a reformulation of the problem (1) makes possible the greater use of operator-theoretic methods which, as we shall see, arise quite naturally in the treatment of the problems (1) and (3).

1. Preliminaries About Generalized Gaussian Processes and the Linear estimation

Let H be a real Hilbert space and let $(\phi(x) : x \in H)$ be a generalized Gaussian process indexed by H and given on the probability space $(\Omega,\Sigma,\mathbb{P})$, i.e. $\phi(.):H \to L^2(\Omega,\Sigma,\mathbb{P})$ is a linear operator (we treat $(L^2)=L^2(\Omega,\Sigma,\mathbb{P})$ as a Hilbert space, provided with the standard L^2-norm) with norm $\|\phi\| < +\infty$ such that $(\phi(x) \in (L^2) : x \in H)$ is a Gaussian family. Note that everywhere by "Gaussian distribution" we mean "Gaussian distribution with vanishing mean". The (real) algebra of all bounded linear operators on H we shall denote by $\mathcal{L}(H)$ and the inner product in H we shall denote by $(.,.)_H$ or simply by $(.,.)$. Let $A=\mathrm{Cov}(\phi)$ be the covariance operator of the process ϕ, i.e. the only operator from $\mathcal{L}(H)$ defined by $\mathbb{E}(\phi(x)\phi(y))=(Ax,y)$, $x,y \in H$, ($\mathbb{E}(.)$ denotes the integration w.r.t. $\mathbb{P}$) which is automatically positive and selfadjoint. We call $\phi(.)$ "unit process" if $\mathrm{Cov}(\phi)=\mathbb{1}$ (the identity operator). For any $A \in \mathcal{L}(H)$, $A=A^*$, $A \geq 0$, one can construct with the methods of [8, Chap.I] a probability space $(\Omega,\Sigma,\mathbb{P})$ and an unit process $\phi^0(.)$ indexed by $\overline{\mathrm{Ran}(A^{\frac{1}{2}})}$ ($\overline{\cdot}$ means closure). Then the process

$$\phi(x)=\phi^0(A^{\frac{1}{2}}x), \ x \in H,$$

obviously has covariance operator $\mathrm{Cov}(\phi)=A$. From the other hand for any process $(\phi(x) : x \in H)$ with $\mathrm{Cov}(\phi)=A$ one can define for $y \in \mathrm{Ran}(A^{\frac{1}{2}})$

$$\phi^0(y)=\phi(x) \ ,$$

where $x \in H$ is such that $A^{\frac{1}{2}}x=y$. Then $\phi^0(.)$ is correctly defined on $\mathrm{Ran}(A^{\frac{1}{2}})$ and by continuity it can be extended to an unit process indexed by $\overline{\mathrm{Ran}(A^{\frac{1}{2}})}$. Thus any Gaussian process $\phi(.)$ indexed by H is equivalent to an unit Gaussian process $\phi^{o}(.)$ indexed by $\overline{\mathrm{Ran}(\mathrm{Cov}(\phi)^{\frac{1}{2}})}$ and this equivalency is carried out by the relation

$$\phi(x)=\phi^{o}(A^{\frac{1}{2}}x) \quad , \ x\in H.$$

For this reason, in what follows, the Hilbert space $\overline{\mathrm{Ran}(\mathrm{Cov}(\phi)^{\frac{1}{2}})}$ will play the role of the RKHS for the process $\phi(.)$. Obviously $\overline{\mathrm{Ran}(\mathrm{Cov}(\phi)^{\frac{1}{2}})}$ is unitary equivalent to the Hilbert subspace of (L^2) (actually it is a Gaussian space)

$$\mathcal{G}(\phi)=\overline{\mathrm{span}(\phi(x) \ : \ x\in H)}\subset L^2(\Omega,\Sigma,\mathbb{P}) \ .$$

<u>Definition 1.1</u> We shall say that $(\phi(x) \ : \ x\in H)$ is nondegenerated process, or equivalently that $\mathrm{Cov}(\phi)$ is nondegenerated operator, if $0\notin \mathrm{spectrum}(\mathrm{Cov}(\phi))$.#

Thus nondegeneracy of $\phi(.)$ means that $\phi(.)$ is continuous linear isomorphism between H and $\mathcal{G}(\phi)$.

Let $(\phi_1(x) \ : \ x\in H_1)$ and $(\phi_2(y) \ : \ y\in H_2)$ be two processes indexed by the Hilbert spaces H_1 and H_2, which are generally different, but (!) given on one and the same probability space $(\Omega,\Sigma,\mathbb{P})$. Then there exists unique linear operator $T:H_1 \to H_2$ such that

$$\mathbb{E}(\phi_1(x)\phi_2(y))=(Tx,y)_{H_2} \quad , \ x\in H_1 \quad , \quad y\in H_2 \ .$$

We call T joint covariance operator between ϕ_2 and ϕ_1 and we shall denote it by $\phi_2\&\phi_1$. We have $\phi_1\&\phi_2=(\phi_2\&\phi_1)^*$ and $\mathrm{Cov}(\phi)=\phi\&\phi$.

<u>Lemma 1.1</u> $\|\phi_2\&\phi_1\|\leq\|(\phi_1\&\phi_1)^{\frac{1}{2}}\|.\|(\phi_2\&\phi_2)^{\frac{1}{2}}\|$, where $\|.\|$ denotes the uniform operator norm.#

Let $A_1=\mathrm{Cov}(\phi_1)=\phi_1\&\phi_1$ and $A_2=\mathrm{Cov}(\phi_2)=\phi_2\&\phi_2$. Then

$$\phi_1(x)=\phi_1^{o}(A_1^{\frac{1}{2}} \ ), \quad x\in H_1; \quad \phi_2(y)=\phi_2^{o}(A_2^{\frac{1}{2}}y), \quad y\in H_2,$$

where $\phi_1^{o}(.)$ and $\phi_2^{o}(.)$ are unit processes indexed by $\overline{\mathrm{Ran}(A_1^{\frac{1}{2}})}$ and $\overline{\mathrm{Ran}(A_2^{\frac{1}{2}})}$. Let $S:\overline{\mathrm{Ran}(A_1^{\frac{1}{2}})} \to \overline{\mathrm{Ran}(A_2^{\frac{1}{2}})}$ be such that $S=\phi_2^{o}\&\phi_1^{o}$. Then according to Lemma 1.1 S is a contraction and we see that $A=\phi_2\&\phi_1$ can be written in the form

$$A=A_2^{\frac{1}{2}}.S.A_1^{\frac{1}{2}} \quad .$$

Then $A^*=\phi_1\&\phi_2=A_1^{\frac{1}{2}}.S^*.A_2^{\frac{1}{2}}$. We call S joint correlation operator between ϕ_2 and ϕ_1 and we shall denote it by $\phi_2:\phi_1$. Thus

$$\phi_2\&\phi_1=(\phi_2\&\phi_2)^{\frac{1}{2}}.(\phi_2:\phi_1).(\phi_1\&\phi_1)^{\frac{1}{2}} \ ;$$

$$\phi_1\&\phi_2=(\phi_2\&\phi_1)^*=(\phi_1\&\phi_1)^{\frac{1}{2}}.(\phi_1:\phi_2).(\phi_2\&\phi_2)^{\frac{1}{2}} .$$

Theorem 1.1 Let H_1 and H_2 be Hilbert spaces and let $A_1\in\mathcal{L}(H_1)$, $A_1\geq 0$, $A_1^*=A_1$, $A_2\in\mathcal{L}(H_2)$, $A_2\geq 0$, $A_2^*=A_2$. Let $S:\overline{Ran(A_1^{\frac{1}{2}})}\to\overline{Ran(A_2^{\frac{1}{2}})}$ be an arbitrary contraction. Then one can construct a probability space $(\Omega,\Sigma,\mathbb{P})$ and two processes $(\phi_1(x) : x\in H_1)$ and $(\phi_2(y) : y\in H_2)$ given on $(\Omega,\Sigma,\mathbb{P})$ such that: $A_1=Cov(\phi_1)$, $A_2=Cov(\phi_2)$ and $\phi_2\&\phi_1=A_2^{\frac{1}{2}}.S.A_1^{\frac{1}{2}}$.#

The last theorem can be derived from the general results for contractions [6, Chap.1] or it can be proved directly [2]. Geometrically it means that for any two Hilbert spaces H and K and for any contraction $S:H\to K$ one can imbed H and K in a larger space in such a way that S will be transformed into the orthogonal projector of the image of H onto the image of K.

From now on, till the end of the section, we shall assume that $\phi_2(.)$ is a nondegenerate process. Then for every $x\in H_1$ one can correctly define

$$\hat{x}=A_2^{-\frac{1}{2}}.S.A_1^{\frac{1}{2}}x=A_2^{-1}.A_2^{\frac{1}{2}}.S.A_1^{\frac{1}{2}}x=(\phi_2\&\phi_2)^{-1}.(\phi_2\&\phi_1)x\in H_2 .$$

We also define the following process indexed by H_1

$$(\phi_1/\phi_2)(x)=\phi_2(\hat{x}),\ x\in H_1 .$$

Proposition 1.1 (a) $\phi_2\&(\phi_1/\phi_2)=\phi_2\&\phi_1=A_2^{\frac{1}{2}}.S.A_1^{\frac{1}{2}}$;

(b) $\phi_1\&(\phi_1/\phi_2)=A_1^{\frac{1}{2}}.S^*.S.A_1^{\frac{1}{2}}$;

(c) $Cov(\phi_1/\phi_2)=(\phi_1/\phi_2)\&(\phi_1/\phi_2)=\phi_1\&(\phi_1/\phi_2)=A_1^{\frac{1}{2}}.S^*.S.A_1^{\frac{1}{2}}$.#

Definition 1.2 For the processes $(\phi_1(x) : x\in H_1)$ and $(\phi_2(y) : y\in H_2)$ given on one and the same probability space $(\Omega,\Sigma,\mathbb{P})$ we shall say that:

(a) $\phi_1(.)$ and $\phi_2(.)$ are orthogonal (notation $\phi_1\perp\!\!\!\perp\phi_2$) if $\mathcal{G}(\phi_1)$ and $\mathcal{G}(\phi_2)$ are orthogonal subspaces of $L^2(\Omega,\Sigma,\mathbb{P})$;

(b) $\phi_1(.)$ is submitted to $\phi_2(.)$ (notation $\phi_1\subset\phi_2$) if $\mathcal{G}(\phi_1)\subseteq\mathcal{G}(\phi_2)$;

(c) $\phi_1(.)$ and $\phi_2(.)$ are equivalent (notation $\phi_1\cong\phi_2$) if $\phi_1\subset\phi_2$ and $\phi_2\subset\phi_1$.#

Proposition 1.2 (a) $\phi_1\perp\!\!\!\perp\phi_2$ iff $\phi_2\&\phi_1=0$;

(b) $\phi_1\subset\phi_2$ iff $S^*.S$ is the identity operator on $\overline{Ran(A_1^{\frac{1}{2}})}$ or which amounts to the same iff S is an isometry.#

Consider now the following process indexed by H_1

$$\tilde{\phi}(x)=\phi_1(x)-(\phi_1/\phi_2)(x) ,\quad x\in H_1 .$$

From Proposition 1.1 (a) and Proposition 1.2 (a) it follows that $\tilde{\phi} \perp\!\!\!\perp \phi_2$. Since $(\phi_1/\phi_2) \subset \phi_2$ this means that for each $x \in H_1$ $(\phi_1/\phi_2)(x)$ coincides with the orthogonal projection in (L^2) of $\phi_1(x)$ on the Hilbert subspace $\mathcal{G}(\phi_2)$. For this reason we call the process (ϕ_1/ϕ_2) "ϕ_2-filter of ϕ_1" or "ϕ_2-estimator of ϕ_1".

Thus we see that $\phi_1 \subset \phi_2$ iff $\phi_1(x)=(\phi_1/\phi_2)(x)$, $x \in H_1$, and it is now very easy to obtain the general type of processes submitted to ϕ_2.

Proposition 1.3 $\phi_1 \subset \phi_2$ (remind that ϕ_2 is assumed nondegenerated) iff there exists a bounded linear operator $L: H_1 \to H_2$ such that $\phi_1(x)=\phi_2(Lx)$, $x \in H_1$.#

Let $H_1' \subset H_1$ and $H_2' \subset H_2$ be Hilbert subspaces and let P_1 and P_2 be respectively the orthoprojectors in H_1 and H_2 with $\mathrm{Ran}(P_1)=H_1'$ and $\mathrm{Ran}(P_2)=H_2'$. Consider the following processes

$$\phi_1'(x)=\phi_1(P_1x),\ x \in H_1; \quad \phi_2'(y)=\phi_2(P_2y),\ y \in H_2 .$$

Then

$$\phi_1'\&\phi_1'=P_1.(\phi_1\&\phi_1).P_1 \ ; \quad \phi_2'\&\phi_1'=P_2.(\phi_2\&\phi_1).P_1 \ ;$$

$$\phi_2'\&\phi_2'=P_2.(\phi_2\&\phi_2).P_2 .$$

Let $(H_i^t \subset H_i : 0 \le t \le 1)$ for $i=1,2$ be nondecreasing families of Hilbert subspaces of H_1 and H_2 such that $H_i^0=\{0\}$, $H_i^1=H_i$, $i=1,2$ and let $(P_i^t : 0 \le t \le 1)$, $i=1,2$ be the corresponding chains of orthoprojectors. Define for every $t \in [0,1]$ the processes

$$\phi_i^t(x)=\phi_i(P_i^tx),\ x \in H_i,\quad i=1,2 .$$

Definition 1.3 We shall say that $\phi_1(.)$ is ϕ_2-adapted process if $\phi_1^t \subset \phi_2^t$, $0 \le t \le 1$.#

As it is easy to verify $\phi_1(.)$ is ϕ_2-adapted process iff $\phi_1 \subset \phi_2$ and $L.P_1^t=P_2^t.L.P_1^t$, $0 \le t \le 1$, where L is the operator obtained in Proposition 1.3. Thus when $\phi_1(.)$ and $\phi_2(.)$ are indexed by one and the same Hilbert space $H=H_1=H_2$ and $P^t=P_1^t=P_2^t$, $0 \le t \le 1$ then the ϕ_2-adaptivity of $\phi_1(.)$ means that $\phi_1(x)=\phi_2(Lx)$, $x \in H$, for some $L \in \mathcal{L}(H)$ for which $(P^t : t \in [0,1])$ is an eigenchain (we refer to [5, Sec.9.1] and mainly to [3, Chap.1] for this terminology).

Definition 1.4 (a) Let $m:[0,1] \to L^2(\Omega,\Sigma,\mathbb{P})$ be any function. We shall say that $m(.)$ is ϕ_2-martingale if

$$\mathbb{E}(\phi_2^s(y)[m(t)-m(s)])=0,\ 0 \le s \le t \le 1,\ y \in H_2.$$

(b) Let $(\varphi^t : 0 \le t \le 1)$ be any family of processes indexed by H_1.

We shall say that (φ^t : $0 \le t \le 1$) form a ϕ_2-martingale if for any fixed $x \in H_1$ the function $t \to \varphi^t(x)$ is a ϕ_2-martingale in the sense of (a).

(c) For a fixed reference family (H_1^t : $0 \le t \le 1$) in H_1 we shall say that process ($\phi_1(x)$: $x \in H$) is ϕ_2-martingale if $(\phi_1^t(.)=\phi_1(P_1^t.)$: $0 \le t \le 1$) forms a ϕ_2-martingale in the sense of (b).#

Proposition 1.4 . $\phi_1(.)$ is ϕ_2-martingale iff

$$(\phi_1^t/\phi_2^s)=(\phi_1^s/\phi_2^s),\ 0 \le s \le t \le 1,$$

or equivalently iff

$$\phi_2^s \& \phi_1^t=\phi_2^s \& \phi_1^s,\ 0 \le s \le t \le 1.\ \#$$

Corollary 1.1 $\phi_1(.)$ is ϕ_2-martingale iff $(\phi_1/\phi_2)(.)$ is ϕ_2-martingale or, which amounts to the same, iff $\phi_1 \& \phi_2 . P_2^t=P_1^t . \phi_1 \& \phi_2 . P_2^t$, $0 \le t \le 1$.#

In particular if $\phi_1(.)$ and $\phi_2(.)$ are indexed by one and the same Hilbert space $H=H_1=H_2$ and $P^t=P_1^t=P_2^t$, $0 \le t \le 1$, then $\phi_1(.)$ is ϕ_2-martingale iff (P^t : $0 \le t \le 1$) is an eigenchain for $\phi_1 \& \phi_2$.

Proposition 1.5 If (φ^t : $0 \le t \le 1$) forms a ϕ_2-martingale in the sense of Definition 1.4 (b) then $((\varphi^t/\phi_2^t)$: $0 \le t \le 1$) also forms a ϕ_2-martingale.#

2. The Linear Estimation Problem

In this section we treat the problem

(2.1) $\phi(x)=\phi_1(x)+\phi_2(x),\ x \in H.$

We assume that processes $\phi(.)$, $\phi_1(.)$ and $\phi_2(.)$ are indexed by one and the same Hilbert space H with some fixed reference family ($H^t \subset H$: $0 \le t \le 1$), i.e. $\{0\}=H^0 \subseteq H^s \subseteq H^t \subseteq H^1=H$, $0 \le s \le t \le 1$. Let (P^t : $0 \le t \le 1$) be the corresponding chain of orthoprojectors. For each $t \in [0,1]$ define the following processes, indexed by H

$$\phi^t(x)=\phi(P^t x),\ x \in H;\ \phi_i^t(x)=\phi_i(P^t x),\ x \in H,\ i=1,2\ .$$

Let $A_1=Cov(\phi_1)=\phi_1 \& \phi_1$, $A_2=Cov(\phi_2)=\phi_2 \& \phi_2$ and $S:\overline{Ran(A_1^{\frac{1}{2}})} \to \overline{Ran(A_2^{\frac{1}{2}})}$, $S=\phi_2 : \phi_1$. Then we have

$$\phi_2 \& \phi_1=A_2^{\frac{1}{2}}.S.A_1^{\frac{1}{2}}\ ;\quad \phi_1 \& \phi_2=A_1^{\frac{1}{2}}.S^*.A_2^{\frac{1}{2}}=(\phi_2 \& \phi_1)^*\ ;$$

$$A=Cov(\phi)=A_1+A_2+A_1^{\frac{1}{2}}.S^*.A_2^{\frac{1}{2}}+A_2^{\frac{1}{2}}.S.A_1^{\frac{1}{2}}\ .$$

To reduce the problem (2.1) to those considered in [4] one needs aditionally to assume that: $H=L^2[0,1]$; $H^t=L^2[0,t]$, $0 \le t \le 1$; ϕ_1 corresponds to the Brownian motion, i.e. $A_1=\mathbb{1}$; and finaly that $\phi_1 \perp\!\!\!\perp \phi_2$, i.e. S=0. Especially assumptions $A_1=\mathbb{1}$ and S=0 can greatly simplify the computations in this section.

Theorem 2.1 If $\|S\|<1$ and if one of the processes ϕ_1 and ϕ_2 is nondegenerated then ϕ is also nondegenerated.#

Everywhere below, while considering problem (2.1), we shall assume that $\phi^t(.)$ is nondegenerated process indexed by H^t. It holds, by virtue of Theorem 2.1, if for example $P^t.A_1.P^t$ or $P^t.A_2.P^t$ is a nondegenerated operator from $\mathcal{L}(H^t)$ for each $t\in[0,1]$.

Now observing that

$$(\phi\&\phi_1)=A_1+A_2^{\frac{1}{2}}.S.A_1^{\frac{1}{2}} \ ; \quad (\phi\&\phi_2)=A_2+A_1^{\frac{1}{2}}.S^*.A_2^{\frac{1}{2}}$$

we can construct with the technique of the previous section the "ϕ-filters" of $\phi_1(.)$ and $\phi_2(.)$. Namely we have

$$(2.2) \qquad (\phi_1/\phi)(x)=\phi(A^{-1}.[A_1+A_2^{\frac{1}{2}}.S.A_1^{\frac{1}{2}}]x), \ x\in H \ ;$$

$$(2.3) \qquad (\phi_2/\phi)(x)=\phi(A^{-1}.[A_2+A_1^{\frac{1}{2}}.S^*.A_2^{\frac{1}{2}}]x), \ x\in H \ .$$

For each $x\in H$ the r.h.s.'s of the above equalities are the orthogonal projections of $\phi_1(x)$ and $\phi_2(x)$ on the Hilbert subspace $\mathcal{G}(\phi)$, i.e. (2.2) and (2.3) are the best least square estimates of $\phi_1(x)$ and $\phi_2(x)$ given ϕ. It is then clear that

$$\phi(x)=(\phi_1/\phi)(x)+(\phi_2/\phi)(x), \ x\in H \ .$$

Our main result is the following.

Theorem 2.2 Let $x,y\in H$ and let $\mathfrak{z}+\eta$ be the decomposition of the random variable $\phi_1(x)+\phi_2(y)$ into the summ of two random variables $\mathfrak{z}\in\mathcal{G}(\phi)$ and $\eta \perp\!\!\!\perp \mathcal{G}(\phi)$. Then $\mathfrak{z}$ and η are given respectively by

$$\mathfrak{z}=\phi(A^{-1}([A_1+A_2^{\frac{1}{2}}.S.A_1^{\frac{1}{2}}]x+[A_2+A_1^{\frac{1}{2}}.S^*.A_2^{\frac{1}{2}}]y))$$

$$\eta=\phi_1(x-A^{-1}[A_1+A_2^{\frac{1}{2}}.S.A_1^{\frac{1}{2}}]x - A^{-1}[A_2+A_1^{\frac{1}{2}}.S^*.A_2^{\frac{1}{2}}]y) +$$

$$+\phi_2(y-A^{-1}[A_1+A_2^{\frac{1}{2}}.S.A_1^{\frac{1}{2}}]x - A^{-1}[A_2+A_1^{\frac{1}{2}}.S^*.A_2^{\frac{1}{2}}]y) \quad .\#$$

Observing that

$$\phi^t\&\phi^s=P^t.A.P^s; \ \phi_1^t\&\phi_1^s=P^t.A_1.P^s; \ \phi_2^t\&\phi_2^s=P^t.A_2.P^s;$$

$$\phi_2^t\&\phi_1^s=P^t.A_2^{\frac{1}{2}}.S.A_1^{\frac{1}{2}}.P^s;$$

$$\phi^t\&\phi_1^s=P^t.[A_1+A_2^{\frac{1}{2}}.S.A_1^{\frac{1}{2}}]P^s;$$

$$\phi^t\&\phi_2^s=P^t.[A_2+A_1^{\frac{1}{2}}.S^*.A_2^{\frac{1}{2}}]P^s,$$

we can easily extend the above results also for the processes ϕ^t,

ϕ_1^t and ϕ_2^t. Namely for $0 \leq s \leq t \leq 1$ we have

$$(\phi_1^t/\phi^s)(x) = \phi((P^s.A.P^s)^{-1}.P^s[A_1 + A_2^{\frac{1}{2}}.S.A_1^{\frac{1}{2}}]P^t x), \quad x \in H; \tag{2.4}$$

$$(\phi_2^t/\phi^s)(x) = \phi((P^s.A.P^s)^{-1}.P^s[A_2 + A_1^{\frac{1}{2}}.S^*.A_2^{\frac{1}{2}}]P^t x), \quad x \in H\ . \tag{2.5}$$

Respectively for $x,y \in H$ $\phi_1^t(x)+\phi_2^t(y)$ can be decomposed as $\zeta^s+\eta^s$, $\zeta^s \in \mathcal{G}(\phi^s)$, $\eta^s \perp\!\!\!\perp (\phi^s)$ in the following way

$$\zeta^s = \phi((P^s.A.P^s)^{-1}.P^s([A_1 + A_2^{\frac{1}{2}}.S.A_1^{\frac{1}{2}}]P^t x + [A_2 + A_1^{\frac{1}{2}}.S^*.A_2^{\frac{1}{2}}]P^t y)) \tag{2.6}$$

$$\begin{aligned} \eta^s = {} & \phi_1(x - (P^s.A.P^s)^{-1}.P^s([A_1 + A_2^{\frac{1}{2}}.S.A_1^{\frac{1}{2}}]P^t x + [A_2 + A_1^{\frac{1}{2}}.S^*.A_2^{\frac{1}{2}}]P^t y)) + \\ & + \phi_2(y - (P^s.A.P^s)^{-1}.P^s([A_1 + A_2^{\frac{1}{2}}.S.A_1^{\frac{1}{2}}]P^t x + [A_2 + A_1^{\frac{1}{2}}.S^*.A_2^{\frac{1}{2}}]P^t y))\ . \end{aligned} \tag{2.7}$$

The results of [4, Sec.3] can be derived from (2.4) - (2.7) by letting $A_1=\mathbb{1}$ and $S=0$.

3. Martingale Problems

In this section we shall extend the results of [4], concerning the martingale problems. All processes $\phi(.)$, $\phi_1(.)$, $\phi_2(.)$ are assumed to be as in the previous section.

Theorem 3.1 Let $\phi_1(.)$ be ϕ_2-martingale and let $(P^t : 0 \leq t \leq 1)$ be eigenchain for $A_1=\mathrm{Cov}(\phi_1)$ (i.e. $P^t.A_1.P^t=A_1.P^t$, $0 \leq t \leq 1$). Then $\phi_1(.)$ is ϕ-martingale.#

Remark Actually in the above theorem it is assumed that $\phi_1(.)$ is both ϕ_1-martingale and ϕ_2-martingale. We preserve this assumption till the end of the section.#

Corollary 3.1 The family of processes (namely "filters") $((\phi_1^t/\phi^t) : 0 \leq t \leq 1)$ form a ϕ-martingale in the sense of Definition 1.4 (b), i.e. for $0 \leq s \leq t \leq 1$ we have

$$(\phi_1^t/\phi^t)(x) - (\phi_1^s/\phi^s)(x) \perp\!\!\!\perp \mathcal{G}(\phi^s), \quad x \in H\ .\#$$

The last result is similar to those of [4, Lemma 3.2]. Note that in the problem treated in [4] the assumptions of Theorem 3.1 above are automatically fulfilled since $\phi_1(.)$ and $\phi_2(.)$ are independent and $A_1=\mathbb{1}$.

We shall describe next all possible types of ϕ-martingales for a given process $(\phi(x) : x \in H)$ and some fixed reference family $(H^t \subset H : 0 \leq t \leq 1)$.

Proposition 3.1 Let $m:[0,1] \to L^2(\Omega,\Sigma,\mathbb{P})$ be a ϕ-martingale, adapted to $(\mathcal{G}(\phi^t) : 0 \leq t \leq 1)$ meaning that $m(t) \in \mathcal{G}(\phi^t)$ for each t. Then m is expressible in the form

$$m(t)=(\phi/\phi^t)(x_0) \ , \ 0\leq t\leq 1,$$

for some fixed $x_0\in H$.#

Thus any martingale $(m(t) : 0\leq t\leq 1)$ in the sense of Definition 1.4(a) can be written in the form

$$m(t)=\phi((P^t.A.P^t)^{-1}.P^t.Ax_0),\ 0\leq t\leq 1,$$

for some $x_0\in H$. Actually x_0 is given by $\phi(x_0)=m(1)$.

Theorem 3.2 Let $\tilde{\phi}(.)$ be a process indexed by the same Hilbert space H as $\phi(.)$ is and let, as above, $A=Cov(\phi)$. Then $\tilde{\phi}(.)$ is ϕ-adapted process which is ϕ-martingale iff there exists $L\in\mathcal{L}(H)$ such that $(P^t : 0\leq t\leq 1)$ is an eigenchain for the operators L and L*A and $\tilde{\phi}(x)=\phi(Lx)$, $x\in H$.#

4. The Innovation Problem

Let $(H^t : 0\leq t\leq 1)$ be any reference family in the Hilbert space H and let $(\phi(x) : 0\leq t\leq 1)$ be any Gaussian process indexed by H with $A=Cov(\phi)$. Assume that $P^t.A.P^t$ is a nondegenerate operator for each t (as in the previous sections by P^t we denote the orthoprojector in H with $Ran(P^t)=H^t$) and preserve this assumption till the end.

The main question we are discussing in this section is the following:

When there exists a ϕ-adapted ϕ-martingale $m:[0,1]\to(L^2)$ such that for each t

$$\mathcal{G}(\phi^t)=\text{span}(\phi(P^s x) : 0\leq s\leq t,\ x\in H)= \tag{4.1}$$

$$=\overline{\text{span}(m(s) : 0\leq s\leq t)} \quad ?$$

The answer follows directly from Proposition 3.1 and it is the same as those given in [4, Theor.4.1]. We have that (4.1) holds if and only if there exists $x_0\in H$ such that for every fixed $t\in[0,1]$ the following two relations are equivalent:

i) $((P_s.A.P_s)^{-1}.P_s.Ax_0,\ y)=0,\ 0\leq s\leq t$;

ii) $P^t y=0$.

However we can reformulate the innovation problem according to our terminology in the following manner. Let $(\tilde{\phi}(x) : x\in H)$ be another process indexed by H and given on the same probability space. Assume that $\tilde{\phi}\subset\phi$. Then according to Proposition 1.3 $\tilde{\phi}(x)=\phi(Lx)$, $x\in H$, for some $L\in\mathcal{L}(H)$ and it is obvious that $\tilde{\phi}\cong\phi$ iff Ran(L) is dense in H, or, which amounts to the same, iff $Ker(L^*)=\{0\}$. Note that $Cov(\phi)= L^*.A.L$.

The standard Brownian motion process $B(t)$, $0 \le t \le 1$, via the relation

$$\phi(f)= \int_0^1 f(t)dB(t), \quad f \in L^2[0,1],$$

might be regarded as an unit Gaussian process indexed by the Hilbert space $L^2[0,1]$. In that sense every unit Gaussian process indexed by an arbitrary Hilbert space can be considered as an analog of the Brownian motion.

Consider again the process $\phi(.)$ introduced at the beginning of this section. When there exists an unit process $(\tilde{\phi}(x) : x \in H)$ which is ϕ-adapted ϕ-martingale and such that $\tilde{\phi} \cong \phi$ (i.e. $\mathcal{G}(\tilde{\phi}) \equiv \mathcal{G}(\phi)$)? According to the above remarks this question is equivalent to the following one. When there exists $L \in \mathcal{L}(H)$ such that: i) $Ker(L^*)=\{0\}$; ii) $(P^t : 0 \le t \le 1)$ is an eigenchain for L and $L^*.A$; and iii) $L^*.A.L=\mathbb{1}$? One possible answer can be given under the additional assumption that $(P^t : 0 \le t \le 1)$ is a maximal chain of orthoprojectors and that $tr((\mathbb{1}-A)^2) < +\infty$ (A is apriory assumed invertible!). Then according to [5, Theor.9.1.2] A and A^{-1} admits the following factorizations along the chain $(P^t : o \le t \le 1)$ (cf. also [3]):

$$A=(\mathbb{1}+W_-).D^{-1}.(\mathbb{1}+W_+) \; ; \quad A^{-1}=(\mathbb{1}+X_+).D.(\mathbb{1}+X_-) \; .$$

Then directly from [5, Theor.9.1.2] we come to the following conclusion.

<u>Proposition 4.1</u> If $tr((\mathbb{1}-A)^2) < +\infty$ then the operator L with the properties i), ii) and iii) above does exist and it can be taken to be $L=(\mathbb{1}+X_+).D^{\frac{1}{2}}$.#

It should be pointed out that if $tr((\mathbb{1}-A)^2) < +\infty$ then there exists another probability measure $\tilde{\mathbb{P}}$ equivalent to $\mathbb{P}$ under which $(\phi(x) : x \in H)$ is an unit process. (cf. [7],[8],[9])

However applied to the standard case when $H=L^2[0,1]$ and $H^t=L^2[0,t]$ the last result is saying nothing new. It only means that every Gaussian stochastic process $X(t)$, $0 \le t \le 1$, for which

$$\mathbb{E}(X(s)X(t))=\min(s,t)+\int_0^s dx \int_0^t dy.K(x,y), \quad K \in L^2([0,1]^2),$$

is equivalent to the standard Brownian motion (equivalent in the sense of the probability measures) and has multiplicity 1. Nevertheless the last reasonings show how the innovation problem can be reduced to a factorization problem. The similarity between these two problems was pointed out in [4].

References

1. O. Enchev, 'Doob-Meyer decomposition for processes which might not be semimartingales', (to appear).
2. O. Enchev, 'Gaussian random functionals', a technical report.
3. I. Gohberg and M. Krein, "Theory and applications of Volterra operators in Hilbert space", Am. Math. Soc. Transl., Vol. 24, Providence, RI, 1970.
4. M. Hitsuda, 'Wiener-like integrals for Gaussian processes and the linear estimation problem', Proc. of a Conf. of Stochastic Analysis, Evanstone, 1983, (M.Pynsky edt.), 167-177.
5. G. Kallianpur, "Stochastic Filtering Theory", Springer-Verlag, N.Y., 1980.
6. B. Sz.-Nagy and C. Foias, "Harmonic analysis of operators on Hilbert space", North-Holland & Kiado, London and Budapest, 1970.
7. D. Shale, 'Linear symmetrics of the free boson field', Trans. Amer. Math.Soc., 103 (1962), 149-167.
8. B.Simon, "The $P(\varphi)_2$ Euclidian (Quantum) field theory", Princeton Univ. Press, Princeton, NJ, 1974.
9. B.Simon, " Functional integration in Quantum Physics", Academic Press, N.Y., 1979.

A CRITICAL MEASURE-VALUED BRANCHING PROCESS WITH INFINITE ASYMPTOTIC DENSITY

Klaus Fleischmann
Academy of Sciences of G.D.R.
Karl Weierstrass Institute of Mathematics
Box 1304, Berlin, DDR-1086

The purpose of the paper is to report on some critical effect arising in a space-time branching model. Critical effect should mean here that in a particular dimension of space, the so-called critical dimension, we have a quite different behavior of the system.

1. Model

Consider locally finite measures μ defined on Euclidean space R^d of dimension d, i.e. measures μ such that $0 \leq \mu(B) < \infty$ for all balls B in R^d. Each μ is interpreted as a population. If $\{X(t); t \geq 0\}$ is a process with such states μ, then $X(t,B)$ is the mass of the population in B at time t.

The process under consideration can at least formally be described by a stochastic differential equation of the following type:

$$dX(t) = \Delta X(t)dt + (X(t))^{1/2}dW(t) , \quad t \geq 0 . \tag{1}$$

Here Δ is the Laplacian and $\{W(t); t \geq 0\}$ a Brownian motion whose incrementsare Euclideanwhite noise on R^d. For such an approach we refer to Dawson [1] and Shiga [9].

Instead of this we use a construction made by Iscoe [7]. Here X is defined as a Markov process with the following Laplace transition functional

$$\mathbb{E}\{\exp[-\langle X(s+t), \varphi\rangle] \mid X(s)=\mu\} = \exp[-\langle\mu, V_t\varphi\rangle] \tag{2}$$

where $s,t \geq 0$ are time points, μ the introduced states, and φ belongs to some class F of non-negative test functions.Here $\langle\mu,f\rangle$ denotes the integral $\int\mu(dx)f(x)$ and $V_t\varphi(x) = v(t,x)$ is the solution to

$$\frac{\partial}{\partial t} v(t,x) = \Delta v(t,x) - v^{1+\beta}(t,x) , \quad t \geq 0 , \quad x \in R^d \tag{3}$$

withinitial condition $v(0,x) = \varphi(x)$. The constant $0 < \beta \leq 1$ is fixed (The equation (1) corresponds to the value $\beta=1$).

To interprete that process, let us first neglect the non-linear term in (3). Then we have the heat equation with solution $V\varphi = H\varphi$, say. Hence $\langle \mu, V_t\varphi \rangle = \langle H_t\mu, \varphi \rangle$ and from (2) we conclude $X(t) = H_tX(0)$. In other words, without the non-linear term in (3), X is nothing else than the deterministic heat flow. Consequently, one feature of the model is a <u>transport of mass</u> according to the heat flow.

Consider now the opposite term on the right-hand side of (3). Then the equation can be solved explicitely:

$$v(t,x) = \varphi(x)\,[\,1 + \beta t\, \varphi^{\beta}(x)\,]^{-1/\beta}, \quad x \in R^d, \quad t \geq 0 .$$

Setting $\mu = \delta_x$, equation (2) defines the Laplace transition function of a critical continuous state Galton-Watson process with parameter β which evolves at position x. Thus, the second feature of the system is that at each position x we have a <u>fluctuation of mass</u> as in critical continuous state Galton-Watson processes.

In the real model we have a superposition of both effects: deterministic transport and random branching. That different parts of the population develop independently (the typical <u>branching property</u>) can be seen immediately: dividing $\mu = \mu_1 + \mu_2$, the right-hand side of (2) leads to a product of Laplace functionals.

We mention that the process X can also be considered as the so-called <u>"diffusion" approximation</u> for particle models, for instance for the branching Brownian motion, if we have a high density of particles with small mass and short lifespan, cf. Dawson [1] . In this case, the critical offspring distribution of the splitting particles is assumed to belong to the normal domain of attraction of a <u>stable</u> distribution with exponent $1+\beta$ (Clearly, for $\beta=1$ the offspring variance is finite).

On the model under consideration there are a lot of results, see Dawson [2] for a recent survey.

<u>2. An Infinite Mean System</u>

We are interested in the behavior of the process as time increases if we start at time $t=0$ with an <u>infinite asymptotic density of mass</u>: The mass $X(0,KB)$ in a large ball divided by the volume $|KB|$ of that ball should a.s. converge to infinity as $K \to \infty$. In other words, although the initial population $X(0)$ has locally finite mass, the large scale density of mass is infinite caused by some large clumps.

One possible choice of such an infinite mean system is the following. Assume that $X(0)$ has Laplace functional

$$\mathbb{E} \exp[\,-\langle X(0), \varphi \rangle\,] = \exp[\,-\int dx\ \varphi^{\gamma}(x)\,] , \quad \varphi \in F$$

where $0<\gamma<1$ is a given constant. This means that $X(0)$ is a translation invariant random measure which is independent in each point (i.e. has independent increments) and is stable with exponent $\gamma<1$. In particular, $\mathbb{E}X(0,B)=\infty$ for all balls B.

To discuss the model as $t\to\infty$ on a <u>heuristic</u> level assume for a moment for the dimension of space that $d=0$. Then we simply have a critical continuous state Galton-Watson process which dies out:

$$X(t) \to 0 \quad \text{as} \quad t\to\infty \quad \text{a.s.}$$

If $d\neq 0$ is "small" we could speculate that the branching component dominates the transport component. Although the overall density of mass is infinite at time $t=0$ and does not change as t develops, we can imagine that the critical branching leads to the effect that as time increases the clumps originating from $X(0)$ become larger and larger and more and more seldom: locally the mass will disappear.

Conversely, if d is "large" enough the motion component will dominate because of the high degree of freedom. This will cause a strong averaging of the infinite mean system: the process will locally explode.

However, these intuitive considerations do not explain the behavior in "intermediate" dimensions.

An exact answer to these problems is given in Dawson, Fleischmann, Foley, and Peletier [3]:

<u>Theorem</u>. <u>As</u> $t\to\infty$, <u>the Laplace function</u>

$$\mathbb{E}\exp[-\theta X(t,B)], \quad \theta>0$$

<u>converges to some constant</u> c <u>which is independent of the ball</u> B <u>in</u> R^d. <u>This constant</u> c <u>equals one or zero if the dimension</u> d <u>is smaller or greater than</u> $2\gamma/\beta$, <u>respectively</u>. <u>However, in the case</u> $d=2\gamma/\beta$ <u>we have</u> $0<c<1$.

This implies that in low dimensions the process becomes extinct, in large dimensions it explodes, and in the <u>critical dimension</u> (if it exists) the process has a mixed behavior. Here both extinction and explosion are involved. More precisely, c is the probability that the random mass in B converges to zero, and $1-c$ is the probability that this mass becomes unbounded. Thus, we conclude that in this critical dimension for arbitrarily large times a positive fraction of space is empty (or at least has arbitrarily small mass) and its complement is unboundedly filled up with mass. In the limit, the huge clumps are met in a fixed bounded region with probability $1-c$.

Note that the critical dimension is $d=1$ for the mentioned diffusion approximation to the critical branching Brownian motion with finite offspring variance ($\beta=1$) and with stable initial random measure $X(0)$ with exponent $\gamma = 1/2$. This example also shows that the infinite mean system may explode in cases where the corresponding finite mean system goes to extinction (look at $d=2$).

The study of these measure-valued branching processes with infinite asymptotic density was motivated by the open problem of existence of (non-trivial) invariant states for systems that do not have invariant states with finite mean; see for instance Liggett [8], Fleischmann [5], Holley and Liggett [6], Durrett and Liggett [4]. Unfortunately, the present set-up does not lead to a positive answer to that problem.

To attack that problem by another choice of $X(0)$ we are led to the following open analytic problem. Consider the differential equation (3) in the case $\beta=1$ and $d=2$. Assume

$$\varphi(x) = \varepsilon \exp(-|x|^2), \quad x \in R^2.$$

Then we know (joint result with L. A. Peletier) that the solution v to (3) has the property that

$$(\log t) \int dx\, v(t,x) \tag{4}$$

is bounded away from zero and infinity as $t \to \infty$. However, will the limit behavior of (4) depend on the initial value ε?

References

1 Dawson, D.A. Stochastic evolution equations and related measure processes. J.Multivariate Analysis 5 (1975) 1-52.

2 Dawson, D.A. Measure-valued processes, construction, qualitative behavior and stochastic geometry. Carleton Univ., Lab. Research Stat. Probab. Techn. Report No. 53, 1985.

3 Dawson, D.A., Fleischmann, K., Foley, R.D., and Peletier, L.A. A critical measure-valued branching process with infinite mean. Stoch. Analysis Appl. 4 (1986) 117-129.

4 Durrett, R., Liggett, T.M. Fixed points of the smoothing transformations. Z. Wahrsch. verw. Gebiete 64 (1983)

5 Fleischmann, K. Continuity properties of clustered stochastic point processes and applications to spatially homogeneous branching processes. Colloq. Math. Soc. János Bolyai 24, 49-67, Point Processes and Queueing Problems, Keszthely 1978.

6 Holley, R., Liggett, T.M. Generalized potlatch and smoothing processes. Z. Wahrsch. verw. Gebiete 55 (1981)

7 Iscoe, I. A weighted occupation time for a class of measure-valued branching processes. Probab. Th. Rel. Fields 71 (1986) 85-116.

8 Liggett, T.M. Random invariant measures for Markov chains, and independent particle systems. Z. Wahrsch. verw. Gebiete 45 (1978) 297-313.

9 Shiga, T. Stochastic differential equations for some measure - valued diffusions. Abstracts 5th Working Conf. Stoch. Differential Systems, Eisenach (G.D.R.), Friedrich-Schiller-Univ. Jena 1986, p. 82-85.

ON LARGE DEVIATIONS AND RELATIVE ENTROPY OF MARKOV RANDOM FIELDS

H. Föllmer
Mathematikdepartement
ETH-Zentrum
CH 8092 Zürich

1. Introduction

Let μ be a stationary probability measure on $\Omega = S^I$ where S is a finite state space and I denotes the d-dimensional lattice Z^d. To any $\omega \in \Omega$ and to any finite $V \subset I$ we associate the empirical field

(1.1) $$\rho_V(\omega) = |V|^{-1} \sum_{i \in V} \delta_{\theta_i(\omega)}$$

where θ_i is the shift map on Ω defined by $(\theta_i\omega)(k) = \omega(i+k)$. We are interested in the large deviations of the empirical fields $\rho_{V(n)}$ along the sequence $V(n)$ of d-dimensional cubes.

Assume now that μ is a Markov random field or, more generally, a Gibbs measure with respect to some nice interaction potential. In joint work with Steven Orey [5] it is shown that large deviations of the empirical fields are of the form

(1.2) $$\mu[\rho_{V(n)} \in A] \sim \exp[-|V(n)| \inf_{\nu \in A \cap M(s)} h(\nu;\mu)]$$

where $M(s)$ is the class of stationary probability measures on Ω, and where $h(\nu;\mu)$ is the specific relative entropy of ν with respect to μ. In this note, our purpose is to state the result and to derive an alternative description of the rate function $h(\,.\,;\mu)$ in the case of an attractive interaction. As a corollary, we obtain the absence of phase transition for certain subsystems.

2. Large deviations for Gibbs measures

For $V \subset I$ we denote by ω_V the restriction of $\omega \in \Omega$ to V, by F_V the σ-field generated by $\omega \to \omega_V$, and by μ_V the distribution of ω_V under μ. Let U_V (V finite) be a stationary interaction potential as in [3]. For $\xi, \eta \in \Omega$ we define the conditional energy of ξ on V given η on V^c as

$$E_V(\xi \mid \eta) = \sum_{W \cap V \neq \emptyset} U_W(\zeta)$$

where $\zeta_V = \omega_V$ and $\zeta_{V^c} = \eta_{V^c}$. A stationary probability measure μ is called a *Gibbs measure* with respect to (U_V) if, for any finite V, the conditional distribution of ω_V under μ with respect to the σ-field F_{V^c} is given by

$$\mu_V(\xi_V \mid \eta) = Z_V(\eta)^{-1} \exp[-E_V(\xi \mid \eta)].$$

Among the measures $\nu \in M(s)$, a Gibbs measure ν is characterized by the following variational principle: the specific *relative entropy*

$$h(\nu;\mu) = \lim_n |V(n)|^{-1} H(\nu_{V(n)};\mu_{V(n)}), \tag{2.1}$$

where

$$H(\nu_V;\mu_V) = \sum_{\xi \in S^V} \nu_V(\xi) \log (\nu_V(\xi)/\mu_V(\xi)),$$

assumes in ν its minimal value 0 ; cf. [3], [7]. Behind the existence of the specific relative entropy (2.1) is a spatial version of the Shannon-McMillan-Breiman theorem, i.e., the existence of

$$\lim |V(n)|^{-1} \log(\nu_{V(n)}(\omega_{V(n)})/\mu_{V(n)}(\omega_{V(n)})) \tag{2.2}$$

in $L^1(\nu)$ and even ν-almost surely; cf. [3], [6]. This is one of the crucial steps in obtaining the following spatial analogue to the Donsker-Varadhan theory of large deviations of a Markov process:

(2.3) <u>Theorem (F.-Orey [5])</u> For a subset A in the compact space of probability measures on Ω, the large deviations of the empirical fields satisfy the lower bound

$$\liminf_n |V(n)|^{-1} \log \mu[\rho_{V(n)} \in A] \geq - \inf_{\nu \in A \cap M(s)} h(\nu;\mu) \tag{2.4}$$

if A is open, and the corresponding upper bound if A is closed.

We refer to [5] for the proof, and also for some results on the identification of the minimizing measures on the right hand side of (2.4).

3. Different descriptions of the rate function

The proof in [5] of the large deviations result (2.3) is based on the description of the specific relative entropy in thermodynamical terms:

$$h(\nu;\mu) = - h(\nu) + e_U(\nu) + p_U \ , \tag{3.1}$$

where $h(\nu)$ is the specific entropy of ν,

$$e_U(\nu) = \lim_n |V(n)|^{-1} \int E_{V(n)}(\,.\,|\eta(n))\, d\nu \tag{3.2}$$

is the specific energy of ν, and

$$p_U = \lim_n |V(n)|^{-1} \log Z_{V(n)}(\eta(n)) \ ; \tag{3.3}$$

the last two quantities do not depend on the specific choice of the boundary conditions $\eta(n) \in \Omega$.

Let us now recall the formula

$$(3.4)\qquad h(\nu) = \int H(\nu_0(\,.\,|F^-)(\eta))\,\nu(d\eta) = -\int \log \nu_0(\omega_{\{0\}}|F^-)(\omega)\,\nu(d\omega)$$

for the specific entropy (3.1), where $\nu_0(\,.\,|F^-)(\eta)$ denotes the conditional distribu-distribution of $\omega_{\{0\}}$ under ν with respect to the *spatial past* $F^- = F_{\{i|\,i<0\}}$, and $<$ is the lexicographical order on I ; cf. [3]. (3.4) suggests the formula

$$(3.5)\qquad h(\nu;\mu) = \int H(\nu_0(\,.\,|F^-)(\eta);\mu_0(\,.\,|F^-)(\eta))\,\nu(d\eta)$$

for the specific relative entropy, in analogy to the rate function for Markov processes in [1]. But here the difficulty is that the choice of $\mu_0(\,.\,|F^-)(\eta)$ involves the null sets of μ, while the integration in (3.5) is with respect to ν.

From now on we assume that the interaction is *attractive* in the sense of [7] and [4]. In order to obtain a canonical choice of $\mu_0(\,.\,|F^-)(\eta)$ for any fixed boundary condition $\eta \in \Omega$, let us define $\eta(n)$ as the boundary condition for $W(n) = V(n) \cap \{i|\,i<0\}^c$ which coincides with η on $\{i\,|\,i<0\}$ and which makes $\mu_{W(n)}(\,.\,|\eta(n))$ maximal with respect to the partial order induced by monotone functions on $S^{W(n)}$; cf. [4]. Define $\mu^+(\,.\,|F^-)(\eta)$ as the corresponding limit measure; it may be viewed as the maximal Gibbs measure for the conditional specification induced by η on $\{i|\,i<0\}^c$. Since

$$\lim_n |V(n)|^{-1} \log \mu_{V(n)}(\omega_{V(n)}) = \lim_n |V(n)|^{-1} \log \mu_{V(n)}(\omega_{V(n)}|\eta(n))$$

by a standard estimate of the boundary terms, and since

$$\log \mu_{V(n)}(\omega_{V(n)}|\eta(n)) = |V(n)|^{-1} \sum_{i\in V(n)} [\log \mu_0(\omega_{\{0\}}|F_{W(n,i)})(\eta(n,i))]\circ\theta_i$$

for $W(n,i) = \{j|\,j-i \in W(n)\}$ and the associated shifted boundary condition $\eta(n,i)$, the ergodic theorem for ν and the construction of $\mu^+(\,.\,|F^-)(\eta)$ by monotonicity imply

$$\lim_n \; |V(n)|^{-1} \int \log \mu_{V(n)}(\omega_{V(n)})\, \nu\,(d\omega) \;=\; \int \log \mu^{+}{}_{0}(\omega_{\{0\}} | F^{-})(\omega)\, \nu(d\omega) \;;$$

the details are similar to the proof of the global Markov property for the maximal Gibbs state in [4]. This together with (3.4) leads to the formula

(3.6) <u>Theorem</u> $\quad h(\nu;\mu) \;=\; \int H(\nu_0(\,.\,|F^{-})(\eta);\mu^{+}{}_0(\,.\,|F^{-})(\eta))\, \nu(d\eta)$

In the same way, replacing the maximal measures $\mu^{+}(\,.\,|F^{-})(\eta)$ by the minimal measures $\mu^{-}(\,.\,|F^{-})(\eta)$, we obtain

(3.7) $\quad h(\nu;\mu) \;=\; \int H(\nu_0(.\,|F^{-})(\eta);\mu^{-}{}_0(\,.\,|F^{-})(\eta))\, \nu(d\eta)\,.$

But if ν is a stationary Gibbs measure with the respect to the given interaction, then we have $h(\nu;\mu) = 0$, and the integrands in (3.5) and (3.6) are equal to 0 for ν-almost all η . This implies the absence of phase transition for the conditional system on $\{i| \, i<0\}^{c}$:

(3.8) <u>Corollary</u> For any stationary Gibbs measure ν and for ν-almost all boundary conditions η ,

(3.9) $\quad \mu^{+}(\,.\,|F^{-})(\eta) \;=\; \mu^{-}(\,.\,|F^{-})(\eta)$

References

[1] M.D. Donsker and S.R.S. Varadhan, (1983): Asymptotic evaluation of certain Markov expectations for large time IV, Comm.Pure Appl.Math. 36,183-212

[2] R.S. Ellis, (1985): Entropy, Large Deviations and Statistical Mechanics. Springer, New York-Heidelberg.

[3] H. Föllmer, (1973): On entropy and information gain in random fields. Z. Wahrsch. verw. Geb. 26, 207-217

[4] H. Föllmer, (1979): On the global Markov property. In: Quantum Fields- Algebras, Processes (ed. L.Streit). Springer, Wien-New York

[5] H. Föllmer and S. Orey, (1987): Large deviations for the empirical field of a Gibbs measure. Annals of Probability (to appear).

[6] D. Ornstein and B. Weiss, (1983): The Shannon-McMillan-Breiman theorem for a class of amenable groups. Isr. JM 44, 53-60.

[7] C. Preston, (1976): Random Fields. Springer Lecture Notes in Math. 534

ERROR ESTIMATES FOR FINITE-ELEMENT APPROXIMATION OF THE ZAKAI EQUATION

A. Germani* and M. Piccioni**

* Istituto di Analisi dei Sistemi ed Informatica del CNR, Viale Manzoni 30, 00185 Rome, Italy

** Dipartimento di Matematica, II Università di Roma "Tor Vergata", Via O. Raimondo 00173, Rome, Italy

1. INTRODUCTION

In this paper finite-element approximations of the stochastic partial differential equation

$$\begin{cases} du(t,x)=\left\{ \sum_{i,j=1}^{d} a_{ij}(x)\frac{\partial^2 u}{\partial x_i \partial x_j}(t,x)+\sum_{i=1}^{d} a_i(x)\frac{\partial u}{\partial x_i}(t,x)+a_o(x)u(t,x)\right\}dx \\ \qquad + \sum_{j=1}^{p} b_j(x)u(t,x)dW_t^j , \quad t \in [0,T] \\ u(0,x) = u_o(x) \end{cases} \tag{1.1}$$

will be considered. The motivation comes from nonlinear filtering, where (1.1) describes the evolution of the unnormalized conditional density of a partially observed diffusion [1,2,3] .

For such a problem only finite-difference schemes have been exploited up to now [4,5,6] . The convergence of general Galerkin schemes has been already proved in [7,8] ; here we will show that a special type of finite-element scheme allows to get an explicity computable error estimates, which is usually a very difficult task in nonlinear filtering [9] .

The main difficulty in computing such an error estimate is due to the fact that the unnormalized density must be approximated only with a finite sum of functions of small support. This requires that the equation (1.1) evolves in a space of functions which go to zero at infinity sufficiently fast, namely weighted Sobolev spaces. In addition a substantial modification of the standard finite-element triangulations [10] is required near the boundary.

2. THE ZAKAI EQUATION IN WEIGHTED SOBOLEV SPACES

In this section we recall some basic result about well-posedness of equation (1.1) in weighted Sobolev spaces, obtained by Krylov-Rozovski [11,12,13]. We require the coefficients a's and b's to be bounded and smooth, with all derivatives bounded, and the usual strong ellipticity condition

$$\sum_{i,j=1}^{d} a_{ij}(x) z_i z_j \geq \gamma |z|^2, \quad x \in R^d, \quad z \in R^d \tag{2.1}$$

Moreover $\{W_t^j\}$, $j = 1,\dots,m$ are independent standard Wiener processes, and u_0 is non random.

Let us associate to (2.1) the following sequence of bilinear forms

$$\begin{cases} A_m(u,v) = \sum_{i,j=1}^{d} (a_{ij}\partial_i u, \partial_j v)_m + \sum_{i=1}^{d} (\bar{a}_i \partial_i u, v)_m - (a_0 u, v)_m \\ u,v \in H^{m+1}(R^d), \quad m = 0,1,2,\dots \end{cases} \tag{2.2}$$

where $\bar{a}_i = \sum_{j=1}^{d} \partial_j a_{ij} - a_i$ and

$$(f,g)_m = \sum_{|\alpha| \leq m} (D^\alpha f, D^\alpha g)_0 = \sum_{|\alpha| \leq m} \int_{R^d} (D^\alpha f)(x)(D^\alpha g)(x)\,dx,$$

α being a multiindex of order $|\alpha|$.

The form A_m is bounded on $H^{m+1}(R^d)$ and λ-coercive w.r.t. $H^{m+1}(R^d)$ and $H^m(R^d)$, so that the equation in $H^m(R^d)$

$$\begin{cases} \frac{d}{dt}(u(t),v)_m + A_m(u(t),v) = 0, \quad v \in H^{m+1}(R^d) \\ u(0) = u \in H^m(R^d) \end{cases} \tag{2.3$_m$}$$

has a unique continuous solution for $m = 0,1,2,\dots$, which is independent of m as it results by replacing v by $(I-\Delta)^{-m} v$ in $(2.4)_m$ with $v \in H^1(R^d)$ and using the fact that $(f,g)_m = (f,(I-\Delta^m g)$, $f \in H^m(R^d)$, $g \in H^{2m}(R^d)$. Given that $(I-\Delta)^{-m}$ commutes with every partial derivative, equation $(2.3)_m$ is changed into $(2.3)_0$.

If we consider weights on the space R^d we are able not only to control the degree of smoothness of the solution of (2.3) (and therefore (1.1)), but the way it approaches zero at infinity, too. This can be done by introducing the weighted Sobolev space $H^{m,r}(R^d)$, $m=0,1,\dots,r \geq 0$, of $H^m(R^d)$-functions f such that

$$\|f\|^2_{m,r} = \sum_{|\alpha|\leq m} \int_{R^d} (1+|x|^2)^r |D^\alpha f(x)|^2 dx < +\infty \tag{2.4}$$

which is a Hilbert space with the scalar product

$$(f,g)_{m,r} = \sum_{|\alpha|\leq m} \int_{R^d} (1+|x|^2)^r D^\alpha f(x) D^\alpha g(x)\, dx. \tag{2.5}$$

For some constant $C_{4,m,r}$ it is

$$C^{-1}_{4,m,r}\|v\|_{m,r} \leq \|Jv\|_m \leq C_{4,m,r}\|v\|_{m,r} \tag{2.6}$$

where $J(v)(x) = (1+|x|^2)^{r/2}v(x)$. By using this invertible transformation it is possible to show that if $u_0 \in H^{m,r}(R^d)$, the solution $u(t) = S(t)u_0$ to equation $(2.3)_m$ belongs to $H^{m,r}(R^d)$, so that $\{S(t)\}$ is a Co-semigroup of operators on $H^{m,r}(R^d)$, with the bound

$$\|S(t)\|_{m,r} \leq C^2_{4,m,r}\, e^{\omega_m t} \tag{2.7}$$

for some constant ω_m, where $C_{4,m,0} = 1$, too. This semigroup has as infinitesimal generator the closure, in the corresponding norm, of the partial differential operayor

$$Au = \sum_{i,j=1}^{d} a_{ij}(x)\frac{\partial^2 u}{\partial x_i \partial x_j} + \sum_{i=1}^{d} a_i(x)\frac{\partial u}{\partial x_i} + a_0(x)u. \tag{2.8}$$

3. INTERPOLATION AND APPROXIMATION IN WEIGHTED SOBOLEV SPACES

In this section the basic error estimate for the finite-element approximation in the deterministic case is obtained. Those difficulties due to the unboundedness of the domain are overcome by the use of weighted Sobolev spaces.

Let us first introduce the family of finite-dimensional subspaces of $H^1(R^d)$ in which the approximate solutions lie. For any $R > 1$, let Ω_R be the interior of a polyhedron in R^d which contains the sphere $B(0,R)$. For each $h > 0$ let $\tilde{T}_{h,R}$ be a family of proper polyhedra (elements) with the following properties:

a) they have disjoint interiors;

b) $\bigcup_{K\in \tilde{T}_{h,R}} K = \tilde{\Omega}_{h,R}$ is contained in a sphere $B(0,\delta R)$, for some $\delta > 1$;

c) each face of $K_1 \in \tilde{T}_{h,R}$ is either a face of some $K_2 \in \tilde{T}_{h,R}$ or belongs to the boundary of $\tilde{\Omega}_{h,R}$;

d) if $T_{h,R}$ is the family of those $K \in \tilde{T}_{h,R}$ such that $K \cap \partial\tilde{\Omega}_{h,R} = \emptyset$, then $\bigcup_{K \in T_{h,R}} K = \bar{\Omega}_R$;

e) $\max_{K \in T_{h,R}} \operatorname{diam} K = h$;

f) if round $K = \sup\{r: \exists x_o \in K, B(x_o,r) \subset K\}$, then diam $K \leqslant \sigma$ round K, $\forall K \in T_{h,R}$, for some $\sigma > 0$;

g) any element $K \in \tilde{T}_{h,R}$ is the image of a reference element $\hat{K} \subset R^d$ affine invertible map F_K of R^d into itself.

Now we introduce the type of reference element considered below and state conditions for the maps F_K, $K \in \tilde{T}_{h,R} \setminus T_{h,R}$. The reference element $\hat{K}$ is equipped with a set of nodes $\hat{\Sigma} = \{\hat{a}_1,\ldots,\hat{a}_N\}$, $\hat{a}_i \in \hat{K}$, $i = 1,\ldots,N$, and a subspace $\hat{P}$ of $C^1(K)$ (shape functions), such that $\hat{\Sigma}$ is $\hat{P}$-unisolvent, that is the map $\psi: \hat{P} \to R^N$, with $\psi(\hat{p}) = \hat{p}|_{\hat{\Sigma}}$ is bijective. In the following $\hat{p}_i$ will be the unique function in $\hat{P}$ such that $\hat{p}_i(\hat{a}_j) = \delta_{ij}$, $i,j = 1,\ldots,n$. Moreover the triple $(\hat{K},\hat{\Sigma},\hat{P})$ has to satisfy the following conditions:

i) $\hat{K}$ is contained in the unit cube;

ii) for each $1 \leqslant k \leqslant d$, $1 \leqslant j_1 \leqslant \ldots \leqslant j_k \leqslant d$, the subspace $S_{j_1,\ldots,j_k} = \{x: x_{j_1} = \ldots = x_{j_k} = 0\}$ is a face of $\hat{K}$ and the orthogonal projection $\Gamma_{j_1,\ldots,j_k}(\hat{a}_i)$ on such a face of any $\hat{a}_i \in \hat{\Sigma}$ belongs to $\hat{\Sigma}$;

iii) for each face $\hat{K}'$ of $\hat{K}$ the set $\hat{\Sigma} \cap \hat{K}'$ is unisolvent w.r.t. $\hat{P}|_{\hat{K}'} = \{\hat{p}|_{\hat{K}'}: \hat{p} \in \hat{P}\}$;

iv) given any pair of faces $\hat{K}_1$ and $\hat{K}_2$ of $\hat{K}$ and any affine invertible map $\hat{F}$ such that $\hat{K}_2 = \hat{F}(\hat{K}_1)$, it is $\hat{\Sigma} \cap \hat{K}_2 = \hat{F}(\hat{\Sigma} \cap \hat{K}_1)$, and $\forall p_1 \in \hat{P}\ \exists p_2 \in \hat{P}$ such that $p_1(x) = p_2(\hat{F}(x))$, $\forall x \in \hat{K}_1$;

v) for some integer $k \geqslant k^* = [\frac{d}{2}+1]$, $\hat{P}$ contains the restrictions to $\hat{K}$ of all polynomials of degree not larger than k;

vi) for $1 \leqslant k \leqslant d$, $1 \leqslant j_1 < \ldots < j_k \leqslant d$, for each $\hat{p} \in \hat{P}$ the function $\hat{\hat{p}}$ belongs to $\hat{P}$, where

$$\hat{\hat{p}}(x_1,\ldots,x_n) = \hat{p}(x_1,\ldots,x_{j_1-1},0,x_{j_1+1},\ldots,x_{j_k-1},0,x_{j_k+1},\ldots,x_n).$$

The last conditions on the family $\{\tilde{T}_{h,R}\}$ is summarized by the following:

h) for each $K \in \tilde{T}_{h,R} \setminus T_{h,R}$ there exists $1 \leqslant k \leqslant d$, $1 \leqslant j_1 < \ldots < j_\ell \leqslant d$ such that $F_K(S_{j_1\ldots j_\ell} \cap \hat{K}) \cap \bar{\Omega}_R$ is a singleton, $F_K = F''_K \circ F'_K$, with F''_K affine with the linear part Φ_K such that $0 < \ell_1 \leqslant \|\Phi_K\| \leqslant \ell_2$ and F'_K is a diagonal operator which is the identity on $S_{j_1,\ldots,j_\ell}$ and h times the identity on its orthocomplement.

It is easy to modify families made by simplexes and parallelothopes [10] in order to satisfy the above conditions.

The maps F_K induce on each element a subset of nodes $F_K(\hat{\Sigma}) = \Sigma_K$ and a subspace of $C^1(K)$, namely $P_K = \{p: p = \hat{p} \circ F_K^{-1},\ \hat{p} \in \hat{P}\}$ for $K \in T_{h,R}$ and $P_K = \{p: p = \hat{p} \circ F_K^{-1},\ p = 0 \text{ on } \partial\tilde{\Omega}_{h,R},\ \hat{p} \in \hat{P}\}$ for $K \in \tilde{T}_{h,R} \setminus T_{h,R}$. These allow to define the space of interpolating functions

$$V_{h,R} = \{v : v \in C^1(R^d) \text{ and } \forall K \in \tilde{T}_{h,R},\ v|_K \in P_K,\ v|_{R^d \setminus \tilde{\Omega}_{h,R}} \equiv 0\}.$$

By the above assumptions, it can be shown that $V_{h,R} \subset H^1(R^d)$ [10]. Moreover it is a finite-dimensional subspace of $H^1(R^d)$: if $(\cup\{\Sigma_k : K \in \tilde{T}_{h,R}\}) \cap \text{int}\, \tilde{\Omega}_{h,R} = \{a_1, \ldots, a_I\}$, a basis is given by those functions $\{p_i,\ i = 1, \ldots, I\}$ in $V_{h,R}$ such that $p_i(a_j) = \delta_{ij}$, $j = 1, \ldots, I$.

Finally it is needed to introduce the interpolation operator $\pi_{h,R}$ of $C(R^d)$ onto $V_{h,R}$

$$(\pi_{h,R} v)(x) = \sum_{i=1}^{I} v(a_i) p_i(x)$$

By Sobolev embedding [14] $\pi_{h,R} v$ is defined for any function v in the space $H^{k^*,r}(R^d)$, $r \geq 0$.

THEOREM 3.1. *For any* $v \in H^{k+1}(R^d) \cap H^{k^*+1,r}(R^d)$, $r \geq 0$

$$\|v - \pi_{h,R} v\|_1 \leq C_{6,k} h^k \|v\|_{k+1} + C_{7,r} \frac{R^{d/2}}{(1+R^2)^{r/2}} \|v\|_{k^*+1,r} \tag{3.1}$$

PROOF. First of all

$$\|v - \pi_{h,R} v\|_1 \leq \|1_{\Omega_k}(v - \pi_{h,R} v)\|_1 + \|1_{R^d \setminus \Omega_R} v\|_1 + \|1_{\tilde{\Omega}_{h,R} \setminus \Omega_R} (\pi_{h,R} v)\|_1 \tag{3.2}$$

In view of the standard results on interpolation theory the first term is bounded by $C_{6,k} h^k \|v\|_{k+1}$, whose $C_{6,k}$ could be computed once the reference element is specified [10]. The second one is easily established by

$$\|1_{R^d \setminus \Omega_R} v\|_1^2 \leq \int_{R^d \setminus S(0,R)} |v(x)|^2 dx + \sum_{i=1}^{d} \int_{R^d \setminus S(0,R)} |\partial^i v(x)|^2 dx \leq (1+R^2)^{-r} \|v\|_{1,r}^2 \tag{3.3}$$

The third term can be evaluated by summing up on each element $K \in \tilde{T}_{h,R} \setminus T_{h,R}$

$$\|1_{\tilde{\Omega}_{h,R} \setminus \Omega_R} (\pi_{h,R} v)\|_1^2 = \sum_K \|1_K (\pi_{h,R})\|_1^2 = \sum_K \|\sum_{i=1}^{d} v(F_K(\hat{a}_i)) \hat{p}_i (F_K^{-1}(x))\|_1^2 \tag{3.4}$$

Now there exists $C_8 > 0$ such that

$$\sup_{\hat{x} \in \hat{K}} \max_{i,j} \{|\hat{p}_i(\hat{x})|, |\partial_j \hat{p}_i(\hat{x})|\} \leq C_8$$

so that by assumption b)

$$\sum_K \|\sum_{i=1}^{d} v(F_K(\hat{a}_i)) \hat{p}_i (F_K^{-1}(x))\|_0^2 \leq d C_8^2 \text{vol}(\tilde{\Omega}_{h,R} \setminus \Omega_R) \sup_{x \notin \Omega_R} |v(x)|^2 \leq$$

$$\leq \frac{\pi^{d/2}(\delta^d-1)R^d}{\Gamma(\frac{d}{2}+1)(1+R^2)^r} \sup_{x\in R^d} |Jv(x)|^2 \leq \frac{C_9^2\pi^{d/2}(\delta^d-1)R^d}{\Gamma(\frac{d}{2}+1)(1+R^2)^r} \|Jv\|^2_{k*} \leq$$

$$\leq \frac{C_9^2C_{4,k,r}^2(\delta^d-1)\pi^{d/2}}{\Gamma(\frac{d}{2}+1)} \frac{R^d}{(1+R^2)^r}\|v\|^2_{k*,r} \tag{3.5}$$

where C_9 is given by Sobolev embedding.

The remaining term in (3.4) is

$$\sum_K \int_K |\sum_{i=1}^d v(F_K(\hat{a}_i))\nabla_x\hat{p}_i(F_K^{-1}(x))|^2 dx =$$

$$= \sum_K \int_{\hat{K}} |\sum_{i=1}^d v(F_K(\hat{a}_i))\nabla_{\hat{x}}\hat{p}_i(\hat{x})F_K'^{-1}\Phi_K^{-1}|^2 \det(\Phi_K F_K')d\hat{x} \leq$$

$$\leq b_1^{-1} \sum_K \int_{\hat{K}} |\sum_{i=1}^d v(F_K(\hat{a}_i))\nabla_{\hat{x}}\hat{p}_i(\hat{x})F_K'^{-1}|^2 \det(\Phi_K F_K')d\hat{x} \tag{3.6}$$

Let us consider a single term $K \in \tilde{T}_{h,R} \setminus T_{h,R}$ in the above sum. Let us assume that, without losing generality, assumption h) holds for such a K with $j_1 = 1,\dots,j_\ell = \ell$, so that (3.6) is equal to the sum over K of

$$b_1^{-1}\sum_{j=1}^{\ell} \int_{\hat{K}} (\sum_{i=1}^d v(F_K(\hat{a}_i))\partial_j\hat{p}_i(\hat{x}))^2 \det(\Phi_K F_K)d\hat{x} +$$

$$+h^{-2}b_1^{-1}\sum_{j=\ell+1}^{d} \int_{\hat{K}} (\sum_{i=1}^d v(F_K(\hat{a}_i))\partial_j\hat{p}_i(\hat{x}))^2 \det(\Phi_K F_K)d\hat{x} \tag{3.7}$$

It is easily seen that the first term in (3.7) is bounded by

$$d^3C_8^2 \sup_{x\notin\Omega_R} |v(x)|^2 \mathrm{vol}(K) \leq \frac{d^3C_8^2}{(1+R^2)^r} C_9^2C_{4,k*,r}^2\|v\|^2_{k*,r} \mathrm{vol}(K). \tag{3.8}$$

The last one requires a finer analysis. Let $\hat{\Sigma} \cap S_{1,\dots,\ell} = \{\hat{a}_0^0,\dots,\hat{a}_L^0\}$ and for each $s = 0,1,\dots,L$ $\{\hat{a}_s^0,\hat{a}_s^1,\dots,\hat{a}_s^{m_s}\} = \hat{\Sigma} \cap \Gamma_{1,\dots,\ell}^{-1}(\{\hat{a}_s^0\})$. The same notation is used for the corresponding basis functions $\hat{p}_i$. Therefore, for $j = \ell+1,\dots,d$, by assumption ii)

$$\sum_{i=1}^d v(F_k(\hat{a}_i))\partial_j\hat{p}_i(\hat{x}) = \sum_{s=0}^L \sum_{m=0}^{m_s} v(F_k(\hat{a}_s^m))\partial_j\hat{p}_s^m(\hat{x}) =$$

$$= \sum_{s=0}^L \{v(F_k(\hat{a}_s^0)) \sum_{m=0}^{m_s} \partial_j\hat{p}_s^m(\hat{x}) + \sum_{m=0}^{m_s}(v(F_k(\hat{a}_s^m))-v(F_k(\hat{a}_s^0))\partial_j\hat{p}_s^m(\hat{x})\} \tag{3.9}$$

Now, by unisolvence, $\sum_{m=0}^{m_s} \hat{p}_s^m(\hat{x})$ is the unique basis function which is 1 on $\{\hat{a}_s^m\ ,\ s = 0,\dots,m_s\}$, and zero on the remaining nodes. By assumptions iii) and vi) this is a function only of $(x_1,\dots,x_\ell)$. Moreover v is continuously differentiable and for $j = 1,\dots,d$

$$\sup_{x\notin\Omega_R}|\partial_j v(x)| \leq (1+R^2)^{-r/2} \sup_{x\in R^d} |J\partial_j v(x)| \leq \frac{C_9 C_{4,k^*,r}}{(1+R^2)^{r/2}}\|v\|_{k^*+1,r} \tag{3.10}$$

from which (3.9) is bounded by

$$C_8 \sum_{m=0}^{m_s} |v(F_k(\hat{a}_s^m))-v(F_k(\hat{a}_s^0))| \leq hb_2 dC_{4,k^*,r}C_8C_9(1+R^2)^{-r/2}\|v\|_{k^*+1,r}\ ,$$

so that the second term in (3.7) is bounded by

$$b_1^{-1}d^3(b_2C_{4,k^*,r}C_8C_9)^2\|v\|^2_{k^*+1,r}\ \mathrm{vol}(K)/(1+R^2)^r \tag{3.11}$$

Therefore by (3.8) and (3.11), by summing up on each $K\in\tilde{T}_{h,R}\setminus T_{h,R}$, (3.6) is not larger than

$$\frac{\pi^d b_1^{-1}(1+b_2^2)d^3(C_{4,k^*,r}C_8C_9)^2(\delta^d-1)R^d}{\Gamma(\frac{d}{2}+1)(1+R^2)^r}\|v\|^2_{k^*+1,r}\ . \tag{3.12}$$

By collecting (3.2),(3.3),(3.4),(3.5) and (3.12) the constant $C_{7,r}$ in the required estimate (3.1) is obtained. □

The estimate (3.1) yields the convergence to zero of the interpolation error provided $r > d$ in Theorem 3.1.

By using the above estimate one can obtain an error estimate for the parabolic equation $(2.3)_0$. We consider the following approximation

$$\begin{aligned}&\frac{d}{dt}(u_{h,R}(t),v_{h,R})+A_0(u_{h,R}(t),v_{h,R}) = 0, \qquad v_{h,R}\in V_{h,R}\\ &u_{h,R}(0) = u_{0,h,R}\in V_{h,R}\end{aligned} \tag{3.14}$$

whose unique solution can be written as $u_{h,R}(t) = S_{h,R}(t)u_{0,h,R}$ where $\{S_{h,R}(t)\}$ is a C_0-semigroup of operators on $V_{h,R}$, which is extended to $L^2(R^d)$ by the identity on $V^\perp_{h,R}$, whenever it is needed. Such semigroups have the same bound on the norm than $\{S(t)\}$, because the form A_0 is merely restricted to a subspace of $H^1(R^d)$. The following theorem is thus obtained.

THEOREM 3.2. *Let* $u(t)$ *and* $u_{h,R}(t)$ *be the unique solutions to* $(2.4)_0$ *and* (3.14), *respectively. Then, for any* $T>0$ *the following estimate holds, provided that* $u_0 \in H^{k+3}(R^d) \cap H^{k^*+3,r}(R^d)$, $r \geqslant 0$

$$\|u_{h,R}(t)-u(t)\|_0 \leqslant e^{\omega_0 T}\|u_{0,h,R}-u_0\|_0 + C_{10,k,T}h^k\|u_0\|_{k+3} \tag{3.15}$$

$$+ C_{11,r,T}\frac{R^{d/2}}{(1+R^2)^{r/2}}\|u_0\|_{k^*+3,r}$$

PROOF. Let us choose λ such that A_0 is λ-coercive and define the elliptic projection operator $Q_{h,R}:H^1(R^d) \to V_{h,R}$ by

$$\lambda(Q_{h,R}u-u,v_{h,R})+A_0(Q_{h,R}u-u,v_{h,R}) = 0, \quad v_{h,R} \in V_{h,R} \tag{3.16}$$

A well known result on variational approximation [10] gives a constant K_λ such that

$$\|(I-Q_{h,R})u\|_1 \leqslant K_\lambda\ \delta_1(u,V_{h,R}) \tag{3.17}$$

where δ_1 denotes distance in the H_1-norm. By subtracting $(2.3)_0$ from (3.14) and by using (3.16) it is easily obtained for $v_{h,R} \in V_{h,R}$

$$\frac{d}{dt}(u_{h,R}(t)-Q_{h,R}u(t),v_{h,R})+A_0(u_{h,R}(t)-Q_{h,R}u(t),v_{h,R}) =$$

$$= -\lambda((I-Q_{h,R})u(t),v_{h,R})+((I-Q_{h,R})\frac{du}{dt}(t),v_{h,R})$$

from which, by integrating and applying (3.17)

$$\|u_{h,R}(t)-u(t)\| \leqslant \|(I-Q_{h,R})u(t)\|+\|u_{h,R}(t)-Q_{h,R}u(t)\|_0 \leqslant$$

$$\leqslant e^{\omega_0 t}\|u_{0,h,R}-u_0\|_0 + K_\lambda[e^{\omega_0 t}\delta_1(u_0,V_{h,R}) +$$

$$+ \delta_1(u(t),V_{h,R}) + \frac{e^{\omega_0 t}}{\omega_0}(\lambda \sup_{0\leqslant s\leqslant t}\delta_1(u(s),V_{h,R}) +$$

$$+ \sup_{0\leqslant s\leqslant t}\delta_1(\frac{du}{ds}(s),V_{h,R})] \tag{3.18}$$

Now, by Theorem 3.1 and inequality (3.15)

$$\delta_1(u(t),V_{h,R}) \leqslant \|(I-\pi_{h,R})u(t)\|_1 \leqslant$$

$$\leqslant C_{12,k,t}h^k\|u_0\|_{k+1} + C_{13,r,t}\frac{R^{d/2}}{(1+R^2)^{r/2}}\|u_0\|_{k^*+1,r} \tag{3.19}$$

where $C_{12,k,t} = C_{6,k}e^{\omega_{k+1}t}$ and $C_{13,r,t} = C_{7,r}C^2_{4,k^*+1,r}e^{\omega_{k^*+1}t}$

In the same way, by differentiating $(2.3)_0$ and taking into account that, if A is defined by (2.8), then it is readily computed that $\|Au_0\|_{m,r} \leq \leq C_{14,m}\|u_0\|_{m+2,r}$, so that

$$\delta_1(\frac{du}{dt}(t),V_{h,R})=\delta_1(Au(t),V_{h,R})\leq C_{12,k,t}h^k\|Au_0\|_{k+1} +$$

$$+C_{13,r,t}\frac{R^{d/2}}{(1+R^2)^{r/2}}\|Au_0\|_{k^*+1,r} \leq C_{12,k,t}C_{14,k+1}h^k\|u_0\|_{k+3} +$$

$$+ C_{13,r,t}C_{14,k^*+1}\frac{R^{d/2}}{(1+R^2)^{r/2}}\|u_0\|_{k^*+3,r}$$

from which the desired constants $C_{10,k,T}$ and $C_{11,r,T}$ of the estimate $(3.15)_0$ are computed. □

4. THE STOCHASTIC ESTIMATE

Now we turn to the study of the stochastic equation (2.1). It is possible to define operators on $H^{m,r}(R^d)$

$$(B_ju)(x) = b_j(x)u(x), \quad u \in H^{m,r}(R^d)$$

where $\|B_j\|_{m,r} \leq C_{16,m,r}$.

We will consider the mild solution to (2.1), namely the unique solution in $C(0,T;L^2(H^{m,r}(R^d)))$ of the following equation [1]

$$u(t) = S(t)u_0 + \sum_{j=1}^{p}\int_0^t S(t-s)B_ju(s)dW^j_s \ , \quad t \in [0,T] \tag{4.1}$$

If $u_0 \in H^{m,r}(R^d)$, the following estimate holds

$$\sup_{t\in[0,T]} E\|u(t)\|^2_{m,r} \leq C^2_{4,m,r}\exp(\omega_mT + \frac{d}{2}C^2_{4,m,r}e^{2\omega_mT}C_{16,m,r})\|u_0\|^2_{m,r} =$$

$$= C^2_{17,m,r,T}\|u_0\|^2_{m,r} \tag{4.2}$$

so that, by Sobolev embedding, if $m > \frac{d}{2}$, then the mild solution will be classical, in the sense that (1.1) is satisfied almost surely.

The final step will be to use the estimate (3.15) in a variation of constants argument for the stochastic equation (4.1). This will be approximated by the system of ordinary differential equations

$$\frac{d}{dt}(u_{h,R}(t),p_i)+A_0(u_{h,R}(t),p_i) = \sum_{j=1}^{p}(B_j u_{h,R}(t),p_i)dW_t^j \;,\; i = 1,\dots,I$$

$$u_{h,R}(0) = \pi_{h,R}u_0 \;,\quad t \in [0,T] \tag{4.3}$$

where $\{p_i \;,\; i = 1,\dots,I\}$ is the basis in $V_{h,R}$ introduced in the last section. By integrating (4.3) its mild form is obtained

$$u_{h,R}(t) = S_{h,R}(t)\pi_{h,R}u_0 + \sum_{j=1}^{p}\int_0^t S_{h,R}(t-s)P_{h,R}B_j u_{h,R}(s)dW_s^j \tag{4.4}$$

where $P_{h,R}$ is the orthogonal projection of $L^2(R^d)$ onto $V_{h,R}$. This equation has a unique solution in $C(0,T;L^2(L^2(R^d)))$, too, with the L^2 estimate (4.2) (with $m = 0$, $r = 0$). Now the final estimate can be obtained.

THEOREM 4.1. *Let* $u(t)$ *and* $u_{h,R}(t)$ *be the unique solutions of* (4.1) *and* (4.4), *respectively. Suppose that* $u \in H^{k+3}(R^d) \cap H^{k^*+3,r}(R^d)$, $r \geq 0$. *Then for any* $T > 0$, $0 \leq t \leq T$

$$\{E\|u(t)-u_{h,R}(t)\|^2\}^{1/2} \leq C_{18,k,T}h^k\|u_0\|_{k+3}+C_{19,r,T}\frac{R^{d/2}}{(1+R^2)^{r/2}}\|u_0\|_{k^*+3,r} \tag{4.5}$$

PROOF. By subtracting (4.4) from (4.1) it is obtained

$$u(t)-u_{h,R}(t) = S(t)u_0-S_{h,R}(t)u_{0,h,R} + \sum_{j=1}^{p}\Big(\int_0^t [S(t-s)B_j u(s) + $$
$$- S_{h,R}(t-s)P_{h,R}B_j u_{h,R}(s)]\,dW_s^j \tag{4.6}$$

and by squaring and computing expectations

$$E\|u(t)-u_{h,R}(t)\|_0^2 \leq 2\|S(t)u_0-S_{h,R}(t)u_{0,h,R}\|_0^2 +$$
$$+ 2\sum_{j=1}^{p} E\int_0^t \|S(t-s)B_j u(s)-S_{h,R}(t-s)P_{h,R}B_j u_{h,R}(s)\|_0^2\, ds \tag{4.7}$$

Let us consider a particular j in the above sum

$$E\|S(t-s)B_j u(s)-S_{h,R}(t-s)P_{h,R}B_j u_{h,R}(s)\|_0^2 \leq$$
$$\leq 2E\|(S(t-s)-S_{h,R}(t-s)P_{h,R})B_j u(s)\|_0^2+2e^{2\omega_0 T}C_{16,0,0}^2\|u(s)-u_{h,R}(s)\|_0^2 \tag{4.8}$$

from which, with the position $C_{20} = 4pe^{2\omega_0 T}C_{16,0,0}^2$, it is

$$E\| u(t)-u_{h,R}(t)\|_0^2 \leq c^2(t)+C_{20}\int_0^t E\| u(s)-u_{h,R}(s)\|_0^2 ds, \tag{4.9}$$

$$c(t)=\{2\| S(t)u_0-S_{h,R}(t)\pi_{h,R}u_0\|_0^2+4\sum_{j=1}^{p}\int_0^t E\|(S(t-s)-S_{h,R}(t-s)P_{h,R})B_j u(s)\|_0^2 ds\}^{1/2}. \tag{4.10}$$

By Gronwall inequality, for $0 \leq t \leq T$

$$\{E\| u(t)-u_{h,R}(t)\|_0^2\}^{1/2} \leq C_{21} \sup_{0\leq s\leq T} c(s)$$

where $C_{21} = (C_{20}-1+e^{C_{20}T})^{1/2}\, C_{20}^{-1/2}$. By Theorem 3.2, (2.7) and (4.3)

$$E\|(S(t-s)-S_{h,R}(t-s)P_{h,R})B_j u(s)\|_0^2 \leq$$

$$\leq 3e^{2\omega_0 T} E\|(I-P_{h,R})B_j u(s)\|_0^2 + 3C_{10,k,T}^2 h^{2k} E\| u(s)\|_{k+3}^2 +$$

$$+ 3C_{11,r,T}^2 \frac{R^d}{(1+R^2)^r} E\| u(s)\|_{k^*+3,r}^2 \leq 3e^{2\omega_0 T} E\|(I-\pi_{h,R})B_j u(s)\|_1^2 +$$

$$+3(C_{17,k+3,0,T}\,C_{10,k,T})^2 h^{2k}\| u_0\|_{k+3}^2+3(C_{17,k^*+3,r,T}C_{11,r,T})^2\frac{R^d}{(1+R^2)^r}\| u_0\|_{k^*+3,r}^2 \tag{4.11}$$

By Theorem 3.1 and (4.2) it is obtained

$$E\|(I-\pi_{h,R})B_j u(s)\|_1^2 \leq 2(C_{6,k}C_{16,k+1,0,T}C_{17,k+1,0,T})^2 h^{2k}\| u_0\|_{k+1}^2 +$$

$$+ 2(C_{7,r}C_{16,k^*+1,r,T}C_{17,k^*+1,t,T})^2 \frac{R^d}{(1+R^2)^r}\| u_0\|_{k^*+1,r}^2 \tag{4.12}$$

Finally, by applying Theorem 3.1 to the first term in (4.10) and by using (4.11) and (4.12) it is easy to compute the constants $C_{18,k,T}$ and $C_{19,r,T}$ appearing in (4.5).

COROLLARY 4.1. *Under the assumptions of Section* 3 *and Theorem* 4.1, *if* $r > \frac{d}{2}$, *then for any* $\varepsilon > 0$ *it is possible to compute* h *and* R *such that the corresponding finite-element approximation* (4.4) *guarantees that*

$$\sup_{0\leq t\leq T}\{E\| u(t)-u_{h,R}(t)\|_0^2\}^{1/2} < \varepsilon .$$

REFERENCES

[1] A. GERMANI, M. PICCIONI, *Nonlinear filtering for Markov processes: an* L^2 *approach*, in *Analysis and Optimization of Systems, Part 1*, A. Bensoussan, J.L. Lions, eds., Springer-Verlag, Berlin, 1984.

[2] E. PARDOUX, *Stochastic partial differential equations and filtering of diffusion processes*, Stochastics, 3, 1979, pp. 127-167.

[3] M. ZAKAI, *On the optimal filtering of diffusion processes*, Z. Wahrschein verw. Geb., 11, 1969, pp. 230-243.

[4] H.J. KUSHNER, *A robust discrete state approximation to the optimal nonlinear filter for a diffusion*, Stochastics, 3, 1979, pp. 75-83.

[5] G.B. DI MASI, W.J. RUNGGALDIER, *Continuous-time approximations for the nonlinear filtering problem*, Appl. Math. Optim. 7, 1981, pp. 233-245.

[6] H. KUSHNER, *Probability Methods for Approximations in Stochastic Control and for Elliptic Equations*, Academic Press, New York, 1977.

[7] A. GERMANI, M. PICCIONI, *Finite-dimensional approximations for the equation of nonlinear filtering derived in mild form*, to appear on Applied Mathematics and Optimization.

[8] A. GERMANI, M. PICCIONI, *A Galerkin approximation for the Zakai equation*, in *System Modelling and Optimization*, P. Thoft-Christensen, ed., Springer-Verlag, Berlin, 1984.

[9] G.B. DI MASI, M. PRATELLI, W.J. RUNGGALDIER, *An approximation for the nonlinear filtering problem with error bound*, Stochastics, 14, 1985, pp. 247-271.

[10] P.A. RAVIART, J.M. THOMAS, *Introduction à l'Analyse Numérique des Equations aux Dérivées Partielles*, Masson, Paris, 1983.

[11] N.V. KRYLOV, B.L. ROZOVSKII, *On the Cauchy problem for linear stochastics partial differential equations*, Math. USSR Izvestja, 11, 1977, pp. 1267-1284.

[12] N.V. KRYLOV, B.L. ROZOVSKII, *Stochastic evolution equation*, Journal of Soviet Mathematics, 14, 1981, pp. 1233-1255.

[13] N.V. KRYLOV, B.L. ROZOVSKII, *Stochastic partial differential equations and diffusion processes*, Russian Math. Surveys 37, 1982, pp. 81-105.

[14] H. BREZIS, *Analyse Fonctionnelle*, Masson, Paris, 1983.

SEMIGROUP PROPERTIES OF MARKOV PROCESSES WITH A SEVERAL DIMENSIONAL PARAMETER

L. Hoy
Sektion Mathematik
Technische Hochschule Merseburg
Merseburg 4200, G.D.R.

The theory of Markov processes is closely connected with the concept of Itô integrals for Wiener processes. In the two parameter case the fundamental definitions of Itô integrals and an Itô formula for two parameter stochastic processes were given by Ponomarenko (1) and Gichman (2). The Itô theory in the N parameter case was considered e.g. by Surgailis (3). An Itô calculus for N parameter, d dimensional stochastic processes was given by Imkeller (4). Using the basic concepts and notations of the two parameter Itô calculus, we derive Kolmogorov equations for several types of multiparameter Markov processes. We consider these equations for two parameter birth and death processes and two parameter diffusion processes. Two parameter semigroup properties are obtained. Applications are possible e.g. in the theory of reliability of systems with several components.

1. Basic definitions and notations.

Let $(\Omega, \mathcal{F}, P)$ be a complete probability space, $(R_2^+, \mathcal{B}_2^+)$ the positive orthant of the two parameter Euclidean space R_2 with the Borel σ-algebra $\mathcal{B}_2$. Let the parameters $r = (s,t)$, $r_i = (s_i, t_i)$ be vectors of R_2^+, and $X(r) = X(s,t)$, $X(r_i) = X(s_i, t_i)$ form real valued two parameter stochastic processes on $(\Omega, \mathcal{F}, P)$. We use the order relation $r_1 \leq r_2 \Leftrightarrow \{s_1 \leq s_2 \text{ and } t_1 \leq t_2\}$ and define the interval $[r_1, r_2] = [s_1, s_2] \times [t_1, t_2]$ as the Cartesian product of the components s and t. With these notations the space $(R_2^+, \mathcal{B}_2^+)$ forms a vector lattice, and we have $\max(r_1, r_2) = (\max(s_1, s_2), \max(t_1, t_2))$, $\min(r_1, r_2) = (\min(s_1, s_2), \min(t_1, t_2))$. In the rectangular array $[0,R] = [0,S] \times [0,T]$ we denote the closed interval $[r,R]$ as the future, the parameter set $\{r' : 0 \leq s' \leq s \text{ or } 0 \leq t' \leq t\}$ as the past and the intersection between future and past as the two parameter present time according to r. Furthermore, the difference

$\square_{r'} X(r) = X(s',t') - X(s',t) - X(s,t') + X(s,t)$ is denoted as rectangular or mixed difference. Practically, the concept of a multi-parameter time can be used in stochastic dynamic models with several components.

2. Two parameter birth and death processes

Let the process $X(s,t)$ take values in the state space Γ of integers. With the initial value condition $X(s,0) = X(0,t) = 0$ a.s. we define the Markovian state probabilities $p_i(s,t) = P(X(s,t) = i)$, $i \in \Gamma$. Let $p(s,t) = (p_i(s,t),\ i \in \Gamma)$ form the vector of state probabilities. (I.e., the two parameter process $X(s,t)$ counts the number of random "points" or "events" in the interval $[0,s) \times [0,t)$.) In the following way we define transition probabilities for the process $X(s,t)$ (with s or t fixed):

$$p^1_{j|i}(h,t) = P(\square_{(s+h,t)} X(s,0) = j \mid X(s,t) = i),$$

$$p^2_{j|i}(s,k) = P(\square_{(s,t+k)} X(0,t) = j \mid X(s,t) = i).$$

Furthermore, we assume, that these probabilities fulfil the following assumptions of homogenity and ordinarity:

$$p^1_{+1|i}(h,t) = \lambda_i\ ht + o(ht),$$

$$p^1_{-1|i}(h,t) = \mu_i\ ht + o(ht),$$

$$p^1_{0|i}(h,t) = 1 - (\lambda_i + \mu_i)\ ht + o(ht),$$

$$p^2_{+1|i}(s,k) = \lambda_i\ sk + o(sk),$$

$$p^2_{-1|i}(s,k) = \mu_i\ sk + o(sk),$$

$$p^2_{0|i}(s,k) = 1 - (\lambda_i + \mu_i)\ sk + o(sk).$$

(Here $o(\ )$ denotes the Landau symbol.)
If the integer valued process $X(s,t)$ fulfils the above assumptions, we say, $X(s,t)$ is a two parameter birth and death process.

Theorem 1: Let $X(s,t)$ be a two parameter birth and death process with the transition intensities λ_i and μ_i. Then the vector of state probabilities $p(s,t)$ fulfils the two parameter matrix Kolmogorov equation

$$\frac{\partial^2}{\partial s \partial t} p(s,t) = p(s,t)\ (\Lambda^2\ st + \Lambda),$$

where the matrix

$$\Lambda = \begin{pmatrix} \ddots & & & & \\ & & \ddots & \lambda_{i-1} & \\ \ddots & \mu_i & & -(\lambda_i + \mu_i) & \lambda_i & \ddots \\ & & & \mu_{i+1} & & \ddots \\ & & & & \ddots & \end{pmatrix}$$

is the matrix of Markovian intensities.

Proof: The two parameter matrix Kolmogorov equation follows from the definition of Markovian intensities, if we apply the Markov property twice.

For $\lambda_i = \lambda$, $\mu_i = 0$, $i = 0,1,2,\ldots$ we have the two parameter Poisson process. In the general case, conditions of existence and unity for the solutions of this Kolmogorov equation are observed. If the state space is bounded and the matrix Λ has a finite norm, the solution of the Kolmogorov equation is of the form

$$p(s,t) = p(0,0)\exp(\Lambda st),$$

where $\exp(\Lambda st)$ is the matrix exponential function. This yields the semigroup properties

$$p(s,t) = p(s_0,t)\exp(\Lambda t(s-s_0)),$$

$$p(s,t) = p(s,t_0)\exp(\Lambda s(t-t_0))$$

in s- or t- direction respectively.

We remark, that in the inhomogeneous case the transition probabilities fulfil the conditions of homogenity and ordinarity

$$p^1_{+1|i}(s,s+h,0,t) = \int_0^t \lambda_i(s,v)\, h\, dv + o(ht),$$

$$p^1_{-1|i}(s,s+h,0,t) = \int_0^t \mu_i(s,v)\, h\, dv + o(ht),$$

$$p^2_{+1|i}(0,s,t,t+k) = \int_0^s \lambda_i(u,t)\, k\, du + o(sk),$$

$$p^2_{-1|i}(0,s,t,t+k) = \int_0^s \mu_i(u,t)\, k\, du + o(sk),$$

where the intensities $\lambda_i(u,v)$, $\mu_i(u,v)$ are depending not only on the state i, but also on the parameters $0 \le u \le s$, $0 \le v \le t$. In this case the two parameter Kolmogorov equation contains the terms

${}_0\int^s \lambda_i(u,t)\,du$, ${}_0\int^t \lambda_i(s,v)\,dv$, λ_i instead of $\lambda_i s$, $\lambda_i t$, λ_i

and ${}_0\int^s \mu_i(u,t)\,du$, ${}_0\int^t \mu_i(s,v)\,dv$, μ_i instead of $\mu_i s$, $\mu_i t$, μ_i, respectively.

3. Homogeneous two parameter diffusion processes

In the case of the continuous state space R^1 we proceed in an analogous manner. Let

$$f(x,s_o,t_o,s,t,y) = \frac{\partial}{\partial y} P(X(s,t) - X(s_o,t_o) < y \mid X(s_o,t_o) = x)$$

be the density function of the transition probability from the state $X(s_o,t_o)$ to $X(s,t)$. As usual in the one parameter case, we define the infinitesimal moments $a(s_o,t_o,x) \geq 0$, $b(s_o,t_o,x)$ in the following way:

$$\int_{\bar{D}} f(x,s_o,t_o,s,t,y)\,dy = o(st - s_o t_o), \quad D = \{|y-x| \leq \delta\},$$

$$\int_D (y-x)\, f(x,s_o,t_o,s,t,y)\,dy = b(s_o,t_o,x)\,(st - s_o t_o) + o(st - s_o t_o),$$

$$\int_D (y-x)^2 f(x,s_o,t_o,s,t,y)\,dy = a(s_o,t_o,x)\,(st - s_o t_o) + o(st - s_o t_o).$$

Under these assumptions the process $X(s,t)$ is called two parameter diffusion process. For $a(s_o,t_o,x) = a(x)$, $b(s_o,t_o,x) = b(x)$ the process $X(s,t)$ is a two parameter homogeneous one.

Theorem 2: Let the process $X(s,t)$ be a two parameter diffusion process with the infinitesimal moments $a(s_o,t_o,x)$, $b(s_o,t_o,x)$. Then the transition density of $X(s,t)$ fulfils the following backward Kolmogorov equation:

$$\frac{\partial^2}{\partial s_o \partial t_o} f(x,s_o,t_o,s,t,y) =$$

$$= \frac{1}{4} a^2(s_o,t_o,x)\, s_o t_o \frac{\partial^4}{\partial x^4} f + a(s_o,t_o,x)\, b(s_o,t_o,x)\, s_o t_o \frac{\partial^3}{\partial x^3} f +$$

$$+ \left(b^2(s_o,t_o,x)\, s_o t_o - \frac{1}{2} a(s_o,t_o,x)\right) \frac{\partial^2}{\partial x^2} f - b(s_o,t_o,x) \frac{\partial}{\partial x} f.$$

Proof: If we develop the Taylor series expansion for the function $f(x,s_o,t_o,s,t,y)$ in s- and t- direction and substitute the obtained one parameter Kolmogorov equations one into the other, we obtain the above two parameter Kolmogorov equation.

With the elliptic differential operator

$$L(s_o,t_o,x) = \frac{1}{2}\, a(s_o,t_o,x) \frac{\partial^2}{\partial x^2} + b(s_o,t_o,x) \frac{\partial}{\partial x}$$

this equation is of the form

$$\frac{\partial^2}{\partial s_o \partial t_o} f(x,s_o,t_o,s,t,y) = (L^2(s_o,t_o,x)\, s_o t_o - L(s_o,t_o,x))\, f.$$

For homogeneous two parameter diffusion processes with L_2-bounded operator $L(x)$ the two parameter Kolmogorov equation has the solution

$$f(x,s_o,t_o,s,t,y) = \exp\,(L(x)(st - s_o t_o))\, f(x,s_o,t_o,y)$$

with the initial density $f(x,s_o,t_o,y)$ and $\exp\,(L(x)(st - s_o t_o))$ - the operator exponential function, defined by the Taylor series expansion. This yields the semigroup properties in s- or t-direction

$$f(x,s_o,t_o,s,t_o,y) = \exp\,(L(x)\, t_o(s - s_1))\, f(x,s_o,t_o,s_1,t_o,y)$$

for $s_o \leq s_1 \leq s$, t_o = constant,

$$f(x,s_o,t_o,s_o,t,y) = \exp\,(L(x)\, s_o(t - t_1))\, f(x,s_o,t_o,s_o,t_1,y)$$

for $t_o \leq t_1 \leq t$, s_o = constant.

4. Homogeneous three parameter diffusion processes

The results obtained in the three parameter case are similar to the two parameter one. Let

$$f(x,s_o,t_o,u_o,s,t,u,y) =$$

$$= \frac{\partial}{\partial y} P(X(s,t,u) - X(s_o,t_o,u_o) < y \mid X(s_o,t_o,u_o) = x)$$

be the transition density function of the three parameter diffusion process, defining the infinitesimal moments $a(s_o,t_o,u_o,x)$, $b(s_o,t_o,u_o,x)$ as above. Developing the Taylor series expansion for this transition density, we obtain the following Kolmogorov equation for homogeneous three parameter diffusion processes:

$$\frac{\partial^3}{\partial s_o \partial t_o \partial u_o} f(x,s_o,t_o,u_o,s,t,u,y) =$$

$$= (-L^3(x)\, s_o^2 t_o^2 u_o^2 + 3L^2(x)\, s_o t_o u_o - L(x))\, f.$$

For an L_2-bounded operator $L(x)$ this equation has the general solution

$$f(x,s_o,t_o,u_o,s,t,u,y) =$$

$$= \exp\,(L(x)(stu - s_o t_o u_o))\; f(x,s_o,t_o,u_o,y)\ ,$$

where $f(x,s_o,t_o,u_o,y)$ denotes the initial density of the process. The transition density f fulfils semigroup properties analogous to the above properties in the two parameter case.

5. References

(1) Ponomarenko,L.L.: Stochastic integrals of several dimensional Brownian motion and related equations (in Russian); Probability theory and mathematical statistics 7(1972), 100 - 109

(2) Gichman,Il.I.: On the Itô formula for two parameter stochastic integrals (in Russian); Theory of Random Processes 4(1976), 40 - 48

(3) Surgailis,D.: On L^2- and non-L^2 multiple stochastic integration; in: Stochastic Differential Equations, Proc. 3rd IFIP- WG 7/1 Working Conf., Visegrad 1980, Springer-Verlag, Lecture Notes Control and Information 36, pp. 212 - 226

(4) Imkeller,P.: Itô's formula for continuous (N,d)- processes; Z. Wahrscheinlichkeitstheorie verw. Gebiete 65(1984), 535 - 562

LARGE DEVIATIONS OF A DIFFUSION IN A BISTABLE INFINITE-DIMENSIONAL POTENTIAL

G. Jetschke
Friedrich-Schiller-Universität Jena
Sektion Mathematik
DDR-6900 Jena

1. Deterministic model

Spatial transport and local production of a chemical substance (as well as many other phenomena) usually are described by a nonlinear reaction-diffusion equation of the type

$$\frac{\partial u}{\partial t}(t,x) = \frac{\partial^2 u}{\partial x^2}(t,x) + f\left[u(t,x)\right] , \tag{1}$$

$$t \geq 0,\ x \in [0,L];\ u(0,x) = \varphi(x).$$

We only consider a long thin tube of length L. At the boundaries x=0 and x=L we impose (for simplicity homogeneous) DIRICHLET or NEUMANN boundary conditions (b.c.):

$$u(t,0) = u(t,L) = 0 \qquad \forall t \geq 0 \tag{2D}$$

or

$$\frac{\partial u}{\partial x}(t,0) = \frac{\partial u}{\partial x}(t,L) = 0 \qquad \forall t \geq 0 . \tag{2N}$$

The function $f: \mathbb{R} \longrightarrow \mathbb{R}$ is assumed to be of class C^1 such that

$$V(u) := \int^u f(z)dz \qquad \forall u \in \mathbb{R} \tag{3}$$

(i.e. $V'(u)=f(u)\ \forall u$) is bounded above. A typical example is the cubic nonlinearity

$$f(u) = \mu u - u^3,\ V(u) = \frac{\mu}{2}u^2 - \frac{1}{4}u^4 . \tag{4}$$

Qualitative properties of the flow generated by (1,2) have been discussed by the author in /5/, an extensive treatment is given by HENRY /4/. The fact that (1,2) may be written, using the "potential" or "GINZBURG-LANDAU functional"

$$S(u) := \int_0^L \left\{ \frac{1}{2}|u'(x)|^2 - V\left[u(x)\right] \right\} dx , \tag{5}$$

as gradient system in a suitable function space ($\delta/\delta u$ denotes the FRECHÉT derivative)

$$\frac{du}{dt} = -\frac{\delta S}{\delta u} , \qquad u(o) = \varphi \tag{6}$$

is thereby of essential help.
If DIRICHLET b.c. (2D) are chosen (we will discuss this case mainly) we may take the SOBOLEV space $\overset{\circ}{W}{}_2^{1}(0,L)$ (which can be identified with the absolutely continuous functions satisfying the b.c. and with square integrable derivative).
From the gradient dynamics (6) the existence of unique global solutions for all initial conditions $\varphi \in \overset{\circ}{W}{}_2^{1}(0,L)$ follows immediately. As indecomposable ω-limit sets of the flow only fixed points u^0 occur being stationary points of S (i.e. $(\delta S/\delta u)(u^0)=0$).
For DIRICHLET b.c. (and for NEUMANN b.c. if V(u) is strictly decreasing for large $|u|$) there exist only finitely many fixed points. Isolated minima of S are asymptotically stable fixed points of node type, all others being saddles with finite-dimensional unstable manifolds. Apart from the latter set any initial condition belongs to the basin of attraction of a stable node.

2. Stochastic model: Heuristics

From the many-particle nature of the physical system fluctuations arise inevitably which are modelled on a "mesoscopic" level by adding ad hoc a spatio-temporal GAUSSian white noise ξ in (1) (see /14/,/3/, /1/):

$$\frac{\partial u}{\partial t}(t,x) = \frac{\partial^2 u}{\partial x^2}(t,x)+f\left[u(t,x)\right] + \sqrt{2\varepsilon}\cdot \xi(t,x), \qquad 0, \quad (7)$$

$$E\xi(t,x)=0, \; E\xi(t,x)\xi(t',x')= \delta(t-t')\delta(x-x'). \quad (8)$$

This stochastic partial differential equation is treated very formally by physicists as a diffusion process ("diffusion" here in the sense of stochastic processes!) in a certain function space (usually not specified). Heuristically its transition probability and its one-dimensional distribution are characterized by a formal density with respect to the (non-existing) infinite-dimensional LEBESGUE measure $d^{\infty}q$ and satisfy a formal FOKKER-PLANCK equation /3/,/1/. The "density" $p^{st}(q)$ of the invariant distribution P^{st} of the MARKOV process $(u(t,\cdot))_{t\geq 0}$ can be given explicitly,

$$p^{st}(q) = \tilde{\mathcal{N}}\cdot \exp(-\frac{1}{\varepsilon^2}\, S(q)),\; \tilde{\mathcal{N}}^{-1}:= \int \exp(-\frac{1}{\varepsilon^2}S(q))d^{\infty}q \quad (9)$$

and has a simple structure: Its stationary points are the fixed points of the deterministic equation, especially its maxima coincide with the stable fixed points, which are in this sense the most probable states of the stochastic system.
Furthermore it is concluded that any initial distribution, especially

if peaked around one of the stable fixed points, spreads out to the invariant distribution (9) if $t \to \infty$. Hence, if the system starts near one stable fixed point, in the long run fluctuations allow it to leave its basin of attraction and to reach a neighborhood of another stable fixed point. Such an event is called "tunneling". The mean tunneling time and the most probable tunneling path should be determined by the landscape of S, hence the natural conjecture is that tunneling occurs most probably via the lowest saddle point between initial and final minimum of S.

The rest of the paper is devoted to present this idea rigorously.

3. Stochastic model: Rigorous formulation

The precise meaning of (7) (with (2) and (8)) is the integral equation

$$u_t = T_t\varphi + \int_0^t T_{t-s} f(u_s)ds + \sqrt{2\varepsilon}\cdot\int_0^t T_{t-s}dz_s \qquad (10)$$

and was given f.e. in /6/,/12/. In the case of DIRICHLET b.c. (2D) $(T_t)_{t\geq 0}$ denotes the strongly continuous semigroup of conctractions generated by $A_D := d^2/d^2x^2$ (+b.c.) on

$$C_{oo}[0,L] := \{q \in C[0,L] \mid q(0)=q(L)=0\}$$

which is given by the flow of the linear deterministic equation (1,2D) (with $f\equiv 0$). The last term in (10) is a stochastic integral with respect to an infinite-dimensional (cylindrical) WIENER process (z_t). It can also be derived from a two-parameter WIENER field /7/.

Global existence, uniqueness and path continuity of the solution of (10) were proven by MANTHEY /12/,/13/ (see also his contribution in this volume) for a large class of nonlinearities. Actually he treated the solution as a real-valued two-parameter random field $u(t,x)$, but this is equivalent to (10) (see JETSCHKE /7/). To consider a function-valued process $u_t=u(t,\cdot)$ carries the advantage that the usual MARKOV property arises in a natural way, hence the solution becomes a diffusion process with values in $C_{oo}[0,L]$. Its invariant measure P_ε^{st} is ergodic and given by its density /11/

$$\frac{dP_\varepsilon^{st}}{dQ_\varepsilon}(q) = \mathcal{N}\cdot\exp\left(\frac{1}{\varepsilon^2}\int_0^L V[q(x)]\,dx\right), \quad \mathcal{N}^{-1} := \int_{C_{oo}[0,L]} \exp\left(\frac{1}{\varepsilon^2}\int_0^L V[q(x)]\,dx\right) Q_\varepsilon(dq) \qquad (11)$$

with respect to the centered GAUSSian measure Q_ε on $C_{oo}[0,L]$ with covariance functional

$$B_{Q_\varepsilon}(\delta_x, \delta_{x'}) = \varepsilon^2\cdot\left[\min(x,x') - \frac{xx'}{L}\right] \qquad (12)$$

being the invariant distribution of the linear equation (10) with $f\equiv 0$.

Remember that a GAUSSian process $(b(x))_{x\in[0,L]}$ is called (standard) BROWNian bridge on $[0,L]$ if

$$Eb(x) = 0,\ Eb(x)b(x') = \min(x,x') - \frac{xx'}{L} \quad \forall x,x' \in [0,L], \tag{13}$$

hence Q_ε is the distribution of a BROWNian bridge with parameter ε. Therefore we will call a random process $(u_\varepsilon(x))_{x\in[0,L]}$ with distribution P_ε^{st} given by (11) a "nonlinear" BROWNian bridge.

4. Large deviations of states

Theorem 1: The family $(Q_\varepsilon)_{\varepsilon>0}$ satisfies on $C_{00}[0,L]$ a large deviation principle (LDP) with rate function

$$I(q) = \begin{cases} \frac{1}{2}\int_0^L |q'(x)|^2 dx, & \text{if } q \in \overset{\circ}{W}{}_2^1(0,L) \\ \infty & \text{otherwise.} \end{cases} \tag{14}$$

Proof: See /8/. Since b can be constructed pathwise from a WIENER process w via the continuous map $b(x)=w(x)-x\cdot w(L)/L$, the result follows by applying VARADHAN's contraction principle /15/ to the rate function of the WIENER process.

Theorem 2: For every BOREL set $A \subset C_{00}[0,L]$ it holds

$$\begin{aligned} -\inf_{q\in\overset{\circ}{A}}[F(q)+I(q)] &\le \underline{\lim}_{\varepsilon\downarrow 0}\ \varepsilon^2 \cdot \ln \int_A e^{-F(q)/\varepsilon^2}\ Q_\varepsilon(dq), \\ \overline{\lim}_{\varepsilon\downarrow 0}\ \varepsilon^2 \cdot \ln \int_A e^{-F(q)/\varepsilon^2} Q_\varepsilon(dq) &\le -\inf_{q\in\bar{A}} [F(q)+I(q)] \end{aligned} \tag{15}$$

where $F(q):=-\int_0^L V[q(x)]dx$ ($\overset{\circ}{A}$ denotes the interiour of A and $\bar{A}$ its closure).

Proof: See /8/. $C_{00}[0,L]$ can be replaced by any complete separable metric space if F is bounded below.

Theorem 3: The distributions $(P_\varepsilon^{st})_{\varepsilon>0}$ of the nonlinear BROWNian bridges given by (11) satisfy a LDP with rate function

$$S_0(q) = \begin{cases} S(q)-s_0 & , \text{if } q \in \overset{\circ}{W}{}_2^1(0,L) \\ \infty & \text{otherwise} \end{cases} \tag{16}$$

with $s_0:=\inf\{S(q)\,|\,q\in\overset{\circ}{W}{}_2^1(0,L)\}$, i.e. we have (with the abbreviation $S_0(A):=\inf\{S(q)\,|\,q\in A\}$)

$$-S_0(\overset{\circ}{A}) \le \underline{\lim}_{\varepsilon\downarrow 0}\ \varepsilon^2\cdot\ln P_\varepsilon^{st}(A) \le \overline{\lim}_{\varepsilon\downarrow 0}\ \varepsilon^2\cdot\ln P_\varepsilon^{st}(A) \le -S_0(\bar{A}) \tag{17}$$

for any BOREL set $A \subset C_{00}[0,L]$.
Proof: See /8/. This is a consequence of Theorem 2, since $S(q) = I(q)+F(q)$.

Especially for regular BOREL sets A formula (17) implies

$$P_\varepsilon^{st}(A) = \exp(-\frac{1}{\varepsilon^2}[S_0(A)+O(\varepsilon^2)]),$$

which means that for sufficiently small noise intensity ε preferably such states are realized where S_0 takes as small as possible values. But these are exactly the (local or global) minima of the deterministic potential S given in (5)! Hence (as a by-product) we proved that the formal picture of section 3 is true at least for small noise.

4.A. Most probable states if ε is not small

If the noise is not small the most probable states of the stochastic equation (10) are the stable fixed points of (1) too, since we have

Theorem 4: Suppose $h \in \overset{\circ}{W}{}_2^1(0,L)$ to be twice continuously differentiable and let $K_\delta(h)$ and $K_\delta(0)$ be open balls of radius δ around h and 0, respectively. Then

$$\lim_{\delta \downarrow 0} \frac{P_\varepsilon^{st}[K_\delta(h)]}{Q_\varepsilon[K_\delta(0)]} = \mathcal{N} \cdot e^{-S(h)/\varepsilon^2} \tag{18}$$

with S given by (6) and $\mathcal{N}$ given by (11).
Proof: See /9/. To obtain (18) one has to prove a GIRSANOV-like formula for translates of the GAUSSian measure Q_ε and to estimate the probability of small balls.

5. Large deviations of paths

In order to investigate the influence of small noise on the deterministic flow one has to extend the WENTSELL-FREIDLIN theory /16/ to an infinite-dimensional state space. Important steps into this direction have been worked out by FARIS and JONA-LASINIO /2/. However, they only treat the special cubic nonlinearity (4), DIRICHLET b.c. (2D) and a sufficiently large L.
A more detailed study shows that their results can be extended to a large class of nonlinearities provided we know global existence and uniqueness of the solutions of (10). But this is the case, for example (see /13/) for all functions f being locally LIPSCHITZ continuous and bounded from both sides by decreasing functions g_+ and g_-,

$$g_-(u) \le \min(f(u),0), \qquad \max(f(u),0) \le g_+(u) \qquad \forall u \in \mathbb{R}.$$

Of course the general cubic nonlinearity

$$f(u) = -\alpha u^3 + \beta u^2 - \mu u + \gamma \tag{19}$$

satisfies this condition.

We formulate the results for a bistable potential S. Let f and L be chosen such that the deterministic system (1,2) has exactly two asymptotically stable fixed points $u^{(1)}$ and $u^{(2)}$. Let φ be an initial state near $u^{(1)}$, let U be an open neighborhood of $u^{(2)}$ and denote by $A(\varphi,U;T)$ the transition from φ into U during the time intervall $[0,T]$. Such a tunneling event is caused, after a long-time wandering around $u^{(1)}$, by an especially favourable fluctuation. Denote by Σ the set of points separating the two basins of attraction and by Π_ε the distribution induced by the solutions of (10) on its sample path space.

Theorem 5: There is a neighborhood U of $u^{(2)}$ and a $\varepsilon_0 > 0$ such that for all $\delta > 0$ and every compact set $K \subset C_{oo}[0,L]$ a neighborhood N of $u^{(1)}$ and a time $T > 0$ exist with

$$-\Delta S - \delta \leq \varepsilon^2 \cdot \ln \Pi_\varepsilon \{A(\varphi,U;T)\} \leq -\Delta S + \delta \tag{20}$$

for all $\varphi \in K \cap N$ and all $\varepsilon < \varepsilon_0$ with

$$\Delta S := \min_{q \in \Sigma} S(q) - S(u^{(1)}).$$

Proof: See /2/,/10/.

For smooth functions f (as we assumed) the minimum of $S(q)$, $q \in \Sigma$, is always reached in a saddle point of S, but it may happen that several saddle points have the same "height".

Theorem 6: If noise is sufficiently small tunneling most probably occurs via the lowest saddle point v on Σ. More precisely, the most probable tunneling tube is a neighborhood of the two deterministic heteroclinic orbits joining v with $u^{(1)}$ and $u^{(2)}$ (such that the uphill part of the tunneling elapses opposite to the gradient of S).

Proof: See /10/,/2/. A probabilistic argument shows that the set $A(\varphi,U;T)$ in Theorem 5 can be contracted to a small tube in state space without changing the estimate (20). A topological argument proves that the lowest saddle point always has a one-dimensional unstable manifold. If v_1 is a saddle point on Σ with $\dim W_u(v_1)=2$ then the one-dimensional set $W_u(v_1) \cap \Sigma$ is not attracted by $u^{(1)}$ or $u^{(2)}$ but by another saddle point $v_2 \in \Sigma$. This implies $S(v_2) < S(v_1)$.

A straightforward interpretation of Theorem 6 says that the weakly noisy stochastic system most probably chooses the way of least

resistance (a common behaviour in social life).
The results of sections 4 and 5 are also valid if NEUMANN b.c. (2N) are imposed. However, this requires a separate proof of the LDP for the underlying GAUSSian process ((10) with $\varphi=0, f\equiv 0$).

6. Examples of tunneling paths

Let us discuss the cubic nonlinearities (4), (19) in detail.

6.1. Symmetric case: $f(u)=\mu u-u^3$, $\mu>0$.

(i) DIRICHLET b.c. (2D):
For $(n-1)L_c < L < nL_c$, $n=1,2,\ldots$, $L_c := \pi/\sqrt{\mu}$ there exist exactly 2n-1 fixed points of (1,2D) $u_o(x)\equiv 0$, $\pm u_1(x),\ldots,\pm u_{n-1}(x)$, where $u_k(x)$ has k-1 zeros in (0,L). If $L_c < L$ the system is bistable with $\pm u_1$ as asymptotically stable fixed points. An analysis of $S(q)$, $q\in\Sigma$ shows

$$S(\pm u_2) < S(\pm u_3) < \ldots < S(u_o) \qquad \forall n \tag{22}$$

hence we find

a) tunneling via $u_o(x)\equiv 0$, if $L_c < L < 2L_c$,

b) tunneling via $u_2(x)$ or $-u_2(x)$, if $2L_c < L$.

Typical tunneling paths are displayed in Fig. 1.

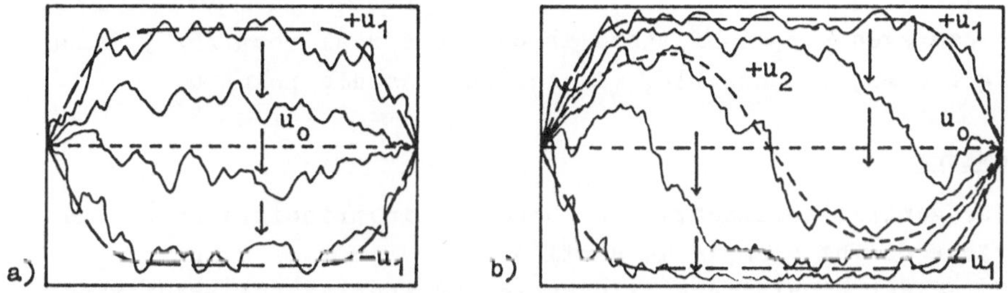

Fig. 1: Typical tunneling paths for $f(u)=\mu u-u^3$ and DIRICHLET b.c. if a) $L_c < L < 2L_c$ or b) $2L_c < L$.

(ii) NEUMANN b.c. (2N):
If $(n-1)L_c < L < nL_c$, $n=1,2,\ldots$ there exist exactly 2n+1 fixed points $u_o(x)\equiv 0$, $\pm u_1(x)\equiv\pm\sqrt{\mu}$, $\pm u_2(x),\ldots,\pm u_n(x)$, where $u_k(x)$ has k-1 zeros in [0,L]. The system is always bistable with $\pm u_1$ as minima of S. The saddle points satisfy (22) too, hence we find

a) tunneling via $u_o(x)\equiv 0$, if $0 < L < L_c$,

b) tunneling via $u_2(x)$ or $-u_2(x)$, if $L_c < L$.

Typical tunneling paths are depicted in Fig. 2.

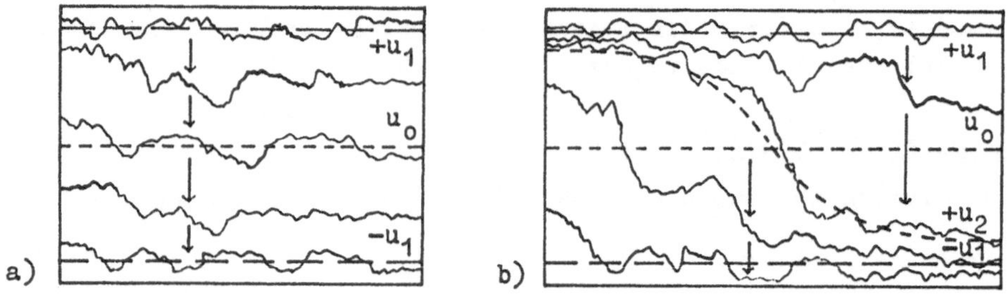

Fig. 2: Typical tunneling paths for $f(u)=\mu u-u^3$ and NEUMANN b.c. if a) $L<L_c$ or b) $L_c<L$.

6.2. General case: $f(u)=-\alpha(u-\alpha_1)(u-\alpha_0)(u-\alpha_2)$.

Assume $\alpha_1<\alpha_0<\alpha_2$. If $\alpha_1<0<\alpha_2$ we get a slight deformation of the situation in 6.1. If $0<\alpha_1$ (or $\alpha_2<0$) the deterministic system (1,2N) always has three fixed points

$$u_1(x)\equiv\alpha_1,\ u_0(x)\equiv\alpha_0,\ u_2(x)\equiv\alpha_2,$$

u_1 and u_2 being stable ones. The system (1,2D) has at most three fixed points u_1,u_0,u_2 with one maximum (or minimum) at $x=L/2$ and

$$u_1(x)<u_0(x)<u_2(x) \qquad \forall x\in(0,L).$$

(This can even happen if there is only one real root /5/.) Hence tunneling occurs, if possible, via the only saddle point u_0.

7. Remarks

(1) For general nonlinearities tunneling in principle shows the same behaviour as the example in section 6.

(2) Numerical simulations illustrate the theoretical predictions (see Fig. 3).

(3) For very large L the values of S do not differ very much for several saddle points, hence tunneling is also possible via other paths.

(4) If non-homogeneous b.c. are given the usual translations create gradient systems with explicitly x-dependent potential S, for which all assertions made above remain valid.

(5) Conjecture: If D is a bounded smooth domain contained in the basin of the attraction of a stable fixed point u^0 of equation (1), then, in the limit of small noise, the most probable exit point from D is the point v, where S is minimal on ∂D.

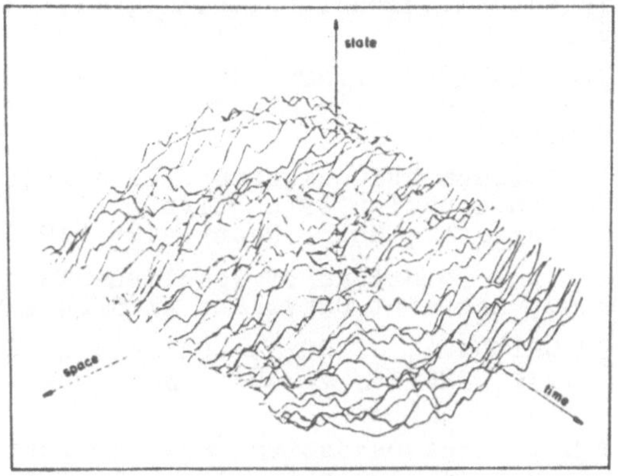

Fig. 3: A detail of a long computer simulation: Tunneling from $+u_1$ to $-u_1$ occurs via $u_0 \equiv 0$ for the cubic nonlinearity $f(u)=u-u^3$, $L=5$, $\varepsilon=0.01$ and DIRICHLET b.c. (compare with Fig. 1a)).

References

/1/ EBELING, W., Y.L. KLIMONTOVICH, Selforganization and Turbulence in Liquids, Teubner-Texte zur Physik, Bd. 2, Leipzig 1984.

/2/ FARIS, W.G., G. JONA-LASINIO, Large fluctuations for a nonlinear heat equation with noise, J. Phys. A: Math. Gen. 15 (1982), 3025-3055.

/3/ HAKEN, H., Synergetics. An Introduction, Springer Series in Synergetics, vol. 1, Berlin-Heidelberg-New York 1978.

/4/ HENRY, D., Geometric theory of semilinear parabolic equations, Lect. Notes Math. 840, Springer 1981.

/5/ JETSCHKE, G., Dissertation, Jena 1978; General stability analysis of dissipative structures in reaction-diffusion equations with one degree of freedom, Phys. Lett. 72 A (1979), 265-268.

/6/ JETSCHKE, G., Different approaches to stochastic parabolic differential equations, Suppl. Rend. Circ. Math. Palermo, II-2 (1982), 161-169.

/7/ JETSCHKE, G., On the equivalence of different approaches to stochastic partial differential equations, Math. Nachr. (1986 to appear).

/8/ JETSCHKE, G., Invariant distribution of a nonlinear stochastic partial differential equation and free energy of statistical physics: I. Large deviations of a nonlinear Brownian bridge, preprint Univ.Jena N/86/11 (1986).

/9/ JETSCHKE, G., Invariant distribution ...: II. Most probable states of a nonlinear Brownian bridge, preprint Univ.Jena N/86/20 (1986).

/10/ JETSCHKE, G., Invariant distribution ...: III. Tunneling in a bistable infinite-dimensional potential, preprint Univ.Jena N/86/42 (1986).

/11/ JETSCHKE, G., R. MANTHEY, Finite-dimensional approximation and invariant measure of a nonlinear stochastic parabolic equation, unpublished manuscript.

/12/ MANTHEY, R., Existence and uniqueness of a solution of a reaction-diffusion equation with polynomial nonlinearity and white noise disturbance, Math. Nachr. 125 (1986), 121-133.

/13/ MANTHEY, R., On the solutions of reaction-diffusion equations with white noise, preprint Univ.Jena N/85/24 (1985).

/14/ NITZAN, A., P. ORTOLEVA, J. ROSS, Nucleation in systems with multiple stationary states, Faraday Symp. Chem. Soc. 9 (1974), 241-253.

/15/ VARADHAN, S.R.S., Large Deviations and Applications, CBMS-NSF Regional Conference Series in Applied Mathematics, SIAM, 1984.

/16/ WENTSELL, A.D., M.I. FREIDLIN, Fluctuations in dynamical systems under the action of random perturbations (in Russian), Moskva 1979.

WHITE NOISE CALCULUS FOR TWO-PARAMETER FILTERING

G. KALLIANPUR[1] and H. KOREZLIOGLU[2]

[1] Department of Statistics, University of North Carolina at Chapel Hill, Chapel Hill, NC 27514, USA.

[2] Ecole Nationale Supérieure des Télécommunications, 46 rue Barrault 75634 Paris Cedex 13, France.

1. INTRODUCTION

This work is an introduction to the white noise calculus for the filtering of two-parameter processes having a Markov property with respect to left and/or lower half-planes.

The filtering problem for two-parameter processes was modelled and solved in the linear Gaussian case by Wong and Tsui [11] and Wong [10]. The same model was used by Korezlioglu, Mazziotto and Szpirglas in [7] for the derivation of nonlinear filtering equations for two-parameter semimartingales ; but no method of solution was proposed in the nonlinear case. The main difficulty in this case is due to the fact that the filtering problem can only be solved by means of filtering techniques for infinite dimensional processes. In the Gaussian linear case, the problem reduces to the resolution of Riccati equations for nuclear operator valued covariances. In the nonlinear case, the passage from two-parameters to infinite dimension was discussed by Korezlioglu in [4] where the horizontal filtering equation was derived for bidirectional diffusion processes, considered as infinite dimensional ones. Although this kind of equation cannot be solved explicitely, approximate solutions can be constructed as for finite dimensional processes.

The observation process considered in the above mentioned works is of the following type :

$$(1.1)\qquad Y_{s,t} = \int_0^s \left(\int_0^t h(u,v,X_{u,v})dv\right) du + W_{s,t}$$

where W is a Brownian sheet on $\mathbb{R}_+^2$, independent of the state process X, and h is a bounded real function, for the nonlinear model, and of the form CX with a matrix-valued function C, in the linear case.

In the formulation of the filtering problem, the rectangle $R_{s,t} = ((u,v)\in \mathbb{R}_+^2\ ;\ u\leqslant s,\ v\leqslant t)$ describes a certain past and according to the displacements $(s,t)\to(s+ds,t)$,

$(s,t) \to (s,t+dt)$ and $(s,t) \to (s+ds,t+dt)$, with $ds > o$ and $dt > o$, three types of presents are defined for $R_{s,t}$, namely its east, north and north-east boundaries. The latter is denoted here by $\partial R_{s,t}$. The filtering equations concern the evolution of the conditional probability distribution of the present values of X, given the past values of Y, and the above displacements yield the horizontal filtering of height t, the vertical filtering of width s and the diagonal filtering. We deal here only with the nonlinear filtering problem, where we suppose h bounded.

In order to simplify the notations, we only consider the case of a real valued Y ; the obtained results can easily be extended to the case of a finite dimensional Y. Let us suppose that X takes its values in an abstract space E endowed with a σ-algebra $\underline{\underline{E}}$. The horizontal filtering problem can be formulated as follows : Let $(\Omega,\underline{\underline{A}},\mathbb{P})$ be a complete probability space on which all random variables and processes considered here are defined, let $\underline{\underline{G}}_{s,t}$ denote the σ-algebra generated by $(Y_{u,v} ; (u,v) \in R_{s,t})$ and X_s the set $(X_{s,v} ; v \leq t)$. It may happen that X_s can be considered as a random variable with values in a function space $S \subset E^{[o,t]}$, for which it is possible to define a conditional probability distribution. In this case the horizontal filtering of X with height t concerns the computation of the conditional distribution of X_s given $\underline{\underline{G}}_{s,t}$. For the observation model (1.1) there is a probability measure $\mathbb{P}_o$ under which Y is a Brownian sheet, independent of X whose probability law is unchanged. $\mathbb{E}^X$ will denote the expectation under the law of X ; $\mathbb{E}$ and $\mathbb{E}_o$, the expectations under $\mathbb{P}$ and $\mathbb{P}_o$, respectively. The likelihood ratio on $R_{s,t}$ is given by

$$(1.2) \qquad \hat{Z}_{s,t} = \exp\left[\int_{R_{s,t}} h(u,v,X_{u,v})dY_{u,v} - \frac{1}{2}\int_{R_{s,t}} h(u,v,X_{u,v})dudv\right]$$

The unnormalized horizontal filter $\hat{\sigma}^t$ of height t for X is defined by $\hat{\sigma}_s^t(F) = \mathbb{E}^X[F(X_s)\hat{Z}_{s,t}]$ for any bounded measurable function F on S. The normalized horizontal filter $\hat{\Pi}^t$ of height t is then defined by the Bayes formula $\hat{\Pi}_s^t(F) = \hat{\sigma}_s^t(F)/\hat{\sigma}_s^t(1) = \mathbb{E}[F(X_s)|\underline{\underline{G}}_{s,t}]$ a.s.. The unnormalized (resp. normalized) filtering equation of height t concerns the evolution of $\hat{\sigma}_s^t$ (resp. $\hat{\Pi}_s^t$) in terms of s. The vertical filtering equations are obtained in a similar way. The diagonal filtering equation, concerning the evolution of the finite dimensional marginal conditional probability distributions of X on $\partial R_{s,t}$ given $\underline{\underline{G}}_{s,t}$, can then be derived by the combination of the horizontal and the vertical filterings.

In this paper we consider the application of the white noise calculus for the derivation of the filtering equations. The approach is entirely based on Kallianpur and Karandikar's work on the white noise calculus in the nonlinear filtering theory (cf [2] and references in there). The horizontal filtering problem is brought to that considered in [3] with an observation noise taking its values in a Hilbert space. All the conclusions of [3] are then valid for the constructed model.

In the next section, we describe the class of Markov processes for which recursive filtering equations can be written and, in the third section, we derive and discuss the filtering equations.

2. A FEW MARKOV PROPERTIES ON THE PLANE

The generic points of $\mathbb{R}_+^2$ are denoted by (s,t) and (u,v), s and u representing the horizontal coordinates and,t and v the vertical coordinates. We write $(s,t)<(u,v)$ if $s\leq u$ and $t\leq v$. An increasing family $\underline{\underline{F}} = (\underline{\underline{F}}_{s,t}\ ;\ (s,t)\in\mathbb{R}_+^2)$, $((s,t)<(u,v)\Rightarrow \underline{\underline{F}}_{s,t}\subseteq \underline{\underline{F}}_{u,v})$, of sub-σ-fields of $\underline{\underline{A}}$ is called a filtration. $\underline{\underline{F}}^1 = (\underline{\underline{F}}^1_s\ ;\ s\in\mathbb{R}_+)$ and $\underline{\underline{F}}^2 = (\underline{\underline{F}}^2_t\ ;\ t\in\mathbb{R}_+)$ are the one-parameter filtrations defined by $\underline{\underline{F}}^1_s = \bigvee_t \underline{\underline{F}}_{s,t}$ and $\underline{\underline{F}}^2_t = \bigvee_s \underline{\underline{F}}_{s,t}$, respectively. We put $\underline{\underline{F}}^X_{s,t} = \sigma(X_{u,v}\ ;\ (u,v)<(s,t))$ for a process X.

$(E,\underline{\underline{E}})$ is a measurable space. If E is a topological space, then $\underline{\underline{E}}$, also denoted $\underline{\underline{B}}(E)$, is always its Borel σ-algebra. b(E) is the space of all $\underline{\underline{E}}$-measurable bounded real functions on E.

DEFINITION 2.1. If for all $n\in\mathbb{N}^*$, all $t_1,\ldots,t_n\in\mathbb{R}_+$, all $f_1,\ldots,f_n\in b(E)$ and all $s'>s$, $E(\prod_{i=1}^n f_i(X_{s',t_i})|\underline{\underline{F}}^1_s)$ is measurable with respect to

i) $\sigma(X_{s,t}\ ;\ t\in\mathbb{R}_+)$, then X is said to be a VA-horizontal Markov process (VA-HM process in short).

ii) $\sigma(X_{s,t}\ ;\ t\leq \max_i t_i)$, then X is said to be a VC-HM process.

iii) $\sigma(X_{s,t_1},\ldots,X_{s,t_n})$, then X is said to be a HM process. □

VA stands for vertically anticipative and VC for vertically causal.

It is obvious that the VA-HM property is larger than that of VC-HM which, in turn, is larger than the HM property. Horizontally anticipative vertical Markov (HA-VM), horizontally causal vertical Markov (HC-VM) and vertical Markov (VM) properties are defined similarly.

DEFINITION 2.1.(continued). iV) A process is said to be a bidirectional Markov (BDM) process if it is both a HM and a VM process. It is said to be a causally bidirectional Markov (CBDM) process if it is both a VC-HM and a HC-VM process. □

BDM processes were considered and examples constructed in [5]. An example of a CBDM process is the solution of the following stochastic differential equation studied by Cairoli in [1].

$$(2.1)\qquad X_{s,t} - X_{s,o} - X_{o,t} + X_{o,o} = \int_{R_{s,t}} a(X_{u,v})\,du\,dv + \int_{R_{s,t}} b(X_{u,v})\,dB_{u,v}$$

where B is a Brownian sheet on $\mathbb{R}_+^2$, independent of the processes $(X_{s,o})$ and $(X_{o,t})$, satisfying the conditions :

$$\int_0^s X_{u,o}^2 du < \infty \text{ and } \int_0^t X_{o,v}^2 dv < \infty \qquad \text{a.s. all } s,t \in \mathbb{R}_+ ,$$

and real functions a and b satisfy the uniform Lipschitz condition :

$$|a(x) - a(y)| + |b(x) - b(y)| \leq c|x-y| , \text{ all } x,y \in \mathbb{R}$$

with a constant c.

DEFINITION 2.2. A real valued process $(X_{s,t})$ is called a bidirectional diffusion (BDD in short) if it is the unique continuous BDM process satisfying the following system S.

$$(S_1) \qquad X_{s,t} = X_{o,t}+\int_0^s f(u,t,X_{u,t})\, du + \int_{u=o}^s g(u,t,X_{u,t})[G_u' dB_u' + \int_{v=o}^t G_{u,v}\, dB_{u,v}]$$

$$(S_2) \qquad X_{s,t} = X_{s,o}+\int_0^t \tilde{f}(s,v,X_{s,v})\, dv + \int_{v=o}^t \tilde{g}(s,v,X_{s,v})[G_v'' dB_v'' + \int_{u=o}^s G_{u,v}\, dB_{u,v}]$$

where, B' and B" are Brownian motions, B is a Brownian sheet such that $X_{o,o}$, B' , B" and B are independent, G', G", G are locally square integrable nonrandom functions and f, $\tilde{f}$, g, $\tilde{g}$ are C^2-functions in x and continuously differentiable in their first two arguments with continuous cross derivatives.□

The multidimensional version of these equations can be written similarly.

In the system (S) studied in [5], G' and G" were taken to be zero ; but the introduction of nonzero G' and G" does not alter the approach. BDM processes satisfying a more general system of horizontal and vertical stochastic differential equations were also obtained by Nualart in [9].

Since X is a solution of S_1 and S_2 at the same time, functions f, $\tilde{f}$, g, $\tilde{g}$ are not independent. In [5] it is proved that $\tilde{g}$ = g, and that f and $\tilde{f}$ can be expressed in terms of g under the hypothesis that g is strictly positive or strictly negative. In this case we have the following result, proved in [4].

PROPOSITION 2.3. There is a real function ϕ on $\mathbb{R}_+^2 \times \mathbb{R}$ such that for each $(s,t) \in \mathbb{R}_+^2$, $\phi(s,t,.) : \mathbb{R} \to \mathbb{R}$ is one-to-one and

$$(2.2) \qquad X_{s,t} = \phi(s,t,M_{s,t}) ,$$

where M is defined by

$$(2.3) \qquad M_{s,t} = M_{o,o} + \int_0^s G_u'\, dB_u' + \int_0^t G_v''\, dB_v'' + \int_{R_{s,t}} G_{u,v}\, dB_{u,v} \quad ;$$

in this expression G', G", G, B', B", B are as in system (S) and $M_{o,o}$ is given by $X_{o,o} = \phi(o,o,M_{o,o})$.□

It is deduced from the above result that all the centered Gaussian BDD's are of the form

(2.4) $\quad X_{s,t} = \Psi_{s,t} M_{s,t}$, with $\Psi_{o,o} = 1$,

where Ψ is a positive real function with continuous derivatives.

Representation (2.2) can be used for the characterization of BDD processes which are homogeneous or stationary in s or/and t. A short discussion of this matter can be found in [4].

3. FILTERING

In order to simplify the notations, we suppose that all the processes of the model are real valued. We also suppose that X is continuous. We put [o,1] = I and denote by C(I) the Banach space of continuous real functions on I. We restrict the parameters to I. For each $s \in I$, $X_s = (X_{s,t} \; ; \; t \in I)$ can be considered as a C(I)-valued random variable. Then $X = (X_s \; ; \; s \in I)$ is a continuous C(I)-valued process. We suppose that X is a VA-HM process on I^2, so that (X_s) considered as an infinite dimensional process is a Markov process with respect to the filtration $(\underline{F}^X_{s,1})_s$. Notice that if X, defined on $\mathbb{R}^2_+$, is a VA-HM process its restriction to $\mathbb{R}_+ \times I$ may not be a VA-HM process. But if it is a VC-HM process, it remains Markovian as C([o,t])-valued process for any height t.

We first consider the horizontal filtering of height 1 for a VA-HM process. $L^2(I)$ (resp. $L^2(I^2)$) denotes the Hilbert space of square integrable functions with respect to the Lebesgue measure on I (resp. on I^2). The scalar product and the norm on $L^2(I)$ are denoted by $(.,.)$ and $||.||$, respectively.

The white noise version of the observation equation (1.1) is

(3.1) $\quad y_{s,t} = h(s,t,X_{s,t}) + w_{s,t}$, with $y,w \in L^2(I)$.

and the corresponding lihelihood ratio at the point s is given by

(3.2) $\quad Z^y_{s,1} = \exp \int_0^s [\int_0^1 (h(u,v,X_{u,v})y_{u,v} - \frac{1}{2} h^2(u,v,X_{u,v}))dv]du$.

Taking into account the isometry between $L^2(I^2)$ and $L^2_{L^2(I)}(I)$, one can consider $h(s,.,X_{s,.})$, $y_{s,.}$ and $w_{s,.}$ as elements of $L^2(I)$ and write the equations (3.1) and (3.2) as follows :

(3.3) $\quad y_s = H_s(X_s) + w_s$, $\qquad y,w \in L^2_{L^2(I)}(I)$

(3.4) $\quad Z^y_{s,1} = \exp \int_0^s [(H_u(X_u),y_u) - \frac{1}{2} ||H_u(X_u)||^2]du$.

This model coincides with the one studied in [3], and, therefore, all the conclusions of [3] can be used for it.

Let Q_u^s be the two-parameter non-homogeneous semigroup on $\underline{\underline{b}}(C(I))$ defined by the relation

$$(Q_s^u F)(x) = \mathbb{E}\,[F(X_s)|X_u = x], \qquad u<s.$$

It is known that $U_s = (s,X_s)$ is a homogeneous Markov process with values in $I \times C(I)$. If (P_s) denotes the transition semigroup associated with (U_s), then we have

$$(P_s F)(u,x) = [Q_{u+s}^u\, F(u+s,.)]\,(x).$$

We denote by L the extended generator of (P_s) with domain $\mathcal{D}$ defined by

$$\mathcal{D}(L) = \{F \in b(IxC(I))\ ;\ w \lim_{s\to 0} \frac{P_s F - F}{s} = LF \text{ exists in } b(IxC(I))\}.$$

The unnormalized horizontal filter $\sigma_s^{1,y}$ of height 1 is given in terms of $y \in L^2(I)$ by

$$\sigma_s^{1,y}(F) = \mathbb{E}^X\,[Z_{s,1}^y\, F(X_s)], \qquad F \in b(C(I))$$

and the normalized horizontal filter $\Pi_s^{1,y}$ of height 1 is given by

$$\Pi_s^{1,y}(F) = \sigma_s^{1,y}(F)/\sigma_s^{1,y}\,(1).$$

For a measure M and a bounded measurable function F, $<F,M>$ denotes the integral of F with respect to M.

We borrow the following results from [3].

THEOREM 3.1. i) $\sigma_s^{1,y}$ satisfies the following equation

$$(3.5) \qquad <F,\sigma_s^{1,y}> = <Q_s^0\, ,\sigma_0^{1,y}> + \int_0^s <[(H_u,y_u) - \frac{1}{2}\,||H_u||^2]\; Q_s^u F\ ,\ \sigma_u^{1,y}> du$$

with $F \in b(C(I))$.

ii) $\sigma_s^{1,y}$ is the unique solution of (3.5) in the class of measures K_s on $\underline{\underline{B}}(C(I))$ satisfying the following condition (3.6).

(3.6) For $B \in \underline{\underline{B}}(C(I))$, $K_0(B) = \mathbb{E}^X 1_B(X_0)$, $K_s(B)$ is a bounded Borel measurable function of s.

iii) Define $\sigma_{s,n}^{1,y}$ inductively as follows : $\sigma_{s,0}^{1,y}\,(B) = \mathbb{E}\; 1_B(X_s)$, and for $n \geq 0$, $<F,\sigma_{s,n+1}^{1,y}>$ is defined by the right hand side of (3.5) with $\sigma_{s,n}^{1,y}$ replacing $\sigma_s^{1,y}$. Then $\sigma_{s,n}^{1,y}$ converges uniformly (in s) to $\sigma_s^{1,y}$ in the total variation norm.

iv) If $y_n \to y$ in $L^2(I)$, then $\sigma_{s,n}^{1,y} \to \sigma_s^{1,y}$ in the total variation norm.□

THEOREM 3.2. i) $\sigma_s^{1,y}$ satisfies the following equation

$$(3.7)\quad <F(s,.),\sigma_s^{1,y}> = <F(s,.),\sigma_0^{1,y}> + \int_0^s <(LF)(u,.),\sigma_u^{1,y}> du + $$
$$+ \int_0^s <[(H_u,y_u) - \frac{1}{2}\,||H_u||^2]\; F(u,.),\ \sigma_u^{1,y} > du$$

where L is the extended generator of (P_s) and $F \in \mathcal{D}(L)$.

ii) $\sigma_s^{1,y}$ is the unique solution of (3.7) in the class of measures satisfying (3.6).

We refer the reader to [3] for the proofs and the derivation of the corresponding equations for $\Pi_s^{1,y}$.

The horizontal filtering equation of height t is obtained by replacing 1 by t in the above setting. The vertical filtering equation of width s can be written similarly.

Now we consider the diagonal filtering equation. The aim of the diagonal filtering is the study of the evolution of the conditional law of $X_{\partial R_{s,t}} = (X_{u,v}$; $(u,v) \in \partial R_{s,t})$ given $(y_{u,v}$; $(u,v) \in R_{s,t})$. Formally speaking, the diagonal filtering equation concerns the variation of this law when (s,t) increases to (s+ds, t+dt) with infinitesimal increments ds and dt. The conditional law of $X_{\partial R_{s,t}}$ is uniquely determined by the conditional laws of $(X_{u_i,v_i}$; $i = 1,\dots,n)$ for all $n = 1,2,\dots$ and all $(u_i,v_i) \in \partial R_{s,t}$. We do not have, for the moment, an equation for the conditional law of $X_{\partial R_{s,t}}$, but we can derive one for the finite dimensional marginal laws. In order to show the method and to simplify notations, we are going to give the diagonal filtering equation for $X_{s,t}$ in terms of the white noise version of the observation on $R_{s,t}$ and explain how one can derive it for the above set $(X_{u_i,v_i}$; $i = 1,\dots,n)$. We shall even consider the filtering of a BDD process which can be represented by (2.2). But in this case, the filtering of X is equivalent to that of the process M of (2.3). Therefore, we shall derive the diagonal filtering equation for X = M.

We put

$$A^y_{s,t} = H(s,t,M_{s,t})y_{s,t} - \frac{1}{2} H^2(s,t,M_{s,t}) ,$$

$$Z^y_{s,t} = \exp \int_{R_{s,t}} A^y_{u,v} \, du \, dv ,$$

$$D_{s,t} = \int_0^s G'^2_u \, du + \int_0^t G''^2_v \, dv + \int_{R_{s,t}} G^2_{u,v} \, du \, dv.$$

For a random variable U we write $\mathbb{E}^M(Z^y_{s,t}U) = (U)_{s,t}$. In the sequel, whenever we use this notation, it is guaranteed that such an expression has a measurable version.

Let F be a C^4-function with bounded derivatives. Then we have, by the Itô formula

$$\begin{aligned} F(M_{s,t})Z^y_{s,t} &= F(M_{o,o}) + \int_0^t F'(M_{o,v}) \, G''_v \, dB''_v + \frac{1}{2} \int_0^t F''(M_{o,v}) \, G''^2_v dv \\ &+ \int_0^s F(M_{u,t})Z^y_{u,t} \left(\int_0^t A^y_{u,v} dv\right) du \\ &+ \int_0^s F'(M_{u,t})Z^y_{u,t} \left(G'_u \, dB'_u + \int_0^t G_{u,v} \, dB_{u,v}\right) \\ &+ \frac{1}{2} \int_0^s F''(M_{u,t})Z^y_{u,t} \frac{\partial D_{u,t}}{\partial u} \, du. \end{aligned}$$

From this we deduce

$$(3.8)\qquad (F(M_{s,t}))_{s,t} = \mathbb{E}^M[F(M_{o,o})] + \frac{1}{2}\int_0^t E^M[F''(M_{o,v})]\, G''^2_v\, dv$$
$$+ \int_{R_{s,t}} (F(M_{u,t})A^y_{u,v})_{u,t}\, du\, dv + \frac{1}{2}\int_{R_{s,t}} (F''(M_{u,t})_{u,t} \frac{\partial D_{u,t}}{\partial u}\, du.$$

This is the unnormalized filtering equation of height t for $M_{s,t}$. The corresponding normalized equation is obtained by developing the ratio $(F(M_{s,t}))_{s,t}/\mathbb{E}^M(Z^y_{s,t})$ in s (cf. [8] for the expression of the normalized filtering equation for $X_{s,t}$ when X is a BDD process). To save space we do not consider the normalized filtering equations.

Now, in order to derive a differential equation for $(F(M_{s,t}))_{s,t}$ in both s and t, we develop, by the Itô formula, $F(M_{u,t})Z_{u,t}\int_0^t A_{u,v}dv$ and $F''(M_{u,t})Z_{u,t}\frac{\partial D_{u,t}}{\partial u}$ as functions of t with constant u and we replace them in (3.8). We then get the following integro-differential equation.

$$(3.9)\qquad \frac{\partial^2 (F(M_{s,t}))_{s,t}}{\partial s\, \partial t} = (F(M_{s,t})A^y_{s,t})_{s,t} + \int_{R_{s,t}} (F(M_{s,t})A^y_{u,t}\, A^y_{s,v})_{s,t}\, du\, dv$$
$$+ \frac{1}{2}(F''(M_{s,t}))_{s,t}\, G^2_{s,t} + \frac{1}{2}\left(\int_0^s (F''(M_{s,t})A^y_{u,t})_{s,t} du\right) \frac{\partial D_{s,t}}{\partial s}$$
$$+ \frac{1}{2}\left(\int_0^s (F''(M_{s,t})A^y_{s,v})_{s,t} dv\right) \frac{\partial D_{s,t}}{\partial t} + \frac{1}{4}(F^{iv}(M_{s,t}))_{s,t} \frac{\partial D_{s,t}}{\partial s} \cdot \frac{\partial D_{s,t}}{\partial t}$$

with $F(M_{s,o})_{s,o} = \mathbb{E}^M[F(M_{o,o})] + \frac{1}{2}\int_0^s \mathbb{E}^M[F''(M_{u,o})]\, G'^2_u\, du$

$F(M_{o,t})_{o,t} = \mathbb{E}^M[F(M_{o,o})] + \frac{1}{2}\int_0^t \mathbb{E}^M[F''(M_{o,v})]\, G''^2_v\, dv.$

The same method can be used for the derivation of the diagonal filtering equation for $(X_{u_i,v_i}\ ;\ i = 1,\dots,n)$ with $(u_i,v_i) \in \partial R_{s,t}$, when X satisfies the equation (2.1) or the system S. The Itô formula in s applied to $F(X_{u_1,v_1},\dots,X_{u_n,v_n})Z_{s,t}$, provides the horizontal filtering equation. The integrands of this equation are then developed in t in order to finally provide the diagonal filtering equation. This would lead to the computation of the conditional distribution of $(X_{u_i,v_i}\ ;\ i = 1,\dots,n)$ based on that of $X_{\partial R_{u,v}}$ for all interior points (u,v) of $R_{s,t}$. The computation of the conditional distribution of $X_{\partial R_{s,t}}$ by means of its finite dimensional marginal distribution is not an easy task. This suggests that numerical approximation methods should be developed. We note that the method used in [6] for the approximation of the nonlinear filter by periodic sampling can be extended to the filtering of BDD processes.

REFERENCES

1. R.Cairoli, Sur une équation différentielle stochastique, C.R.Acad. Paris, t.274, Série, 1739-1742 (1972).

2. G.Kallianpur and R.L.Karandikar, White noise calculus and nonlinear filtering theory, The Ann.Probability, 13(4),1033-1107 (1985).

3. G.Kallianpur and R.L.Karandikar, Measure-valued equations for the optimum filter in finitely additive nonlinear filtering theory, Z.Wahrsch.Verw.Geb. 66, 1-17 (1984).

4. H. Korezlioglu, Passage from two-parameters to infinite dimension, to be published in "Stochastic Partial Differential Equations", Trento Meeting, sept.30-oct.5, 1985, Lect. Notes in Math. Springer-Verlag.

5. H.Korezlioglu, P.Lefort and G.Mazziotto, Une propriété markovienne et diffusions associées, Lect. Notes in Math. 863, 245-274, Springer-Verlag (1981).

6. H.Korezlioglu and G.Mazziotto, Approximation of the nonlinear filter by periodic sampling and quantization, Lect. Notes Control and Inf. Sciences, 62, Springer-Verlag (1984).

7. H.Korezlioglu, G.Mazziotto and J.Szprirglas, Nonlinear filtering equations for two-parameter semimartingales, Stochastic Process. Applications 15, 139-269 (1983).

8. R.Mazumdar and A.Bagchi, A finitely additive white noise approach to filtering for two-parameter diffusion processes, in "Theory and Applications in Nonlinear Control Systems", North Holland (1986).

9. D.Nualart, Two-parameter diffusion processes and martingales, Stochastic Process. Applications 15, 31-54 (1983).

10. E.Wong, Recursive causal linear filtering for two-dimensional random fields, IEEE Trans. IT 24(1), 50-59 (1978).

11. E.Wong and E.T.Tsui, One-sided recursive filters for two-dimensional random fields, IEEE Trans. IT 23(5), 633-637 (1977).

REACTION-DIFFUSION EQUATIONS WITH WHITE NOISE DISTURBANCE

R. Manthey
Friedrich-Schiller-Universität Jena,
Sektion Mathematik
Universitätshochhaus (17. OG)
DDR-6900 Jena

1. Introduction

Reaction-diffusion equations

(RD) $\frac{\partial}{\partial t} u = \Delta u + f(u)$

arise naturally in a variety of models from theoretical physics, chemistry and biology. Such equations describe the behaviour of real systems in which local production of (for example) a chemical substance with the nonlinear rate function f competes with spatial transport by linear diffusion Δ (the LAPLACE operator). The reaction term f is frequently a polynom.

In most cases real systems possess a many-particle structure and the effects one wants to examine may be essentially stochastic, whereas the evolution process generated by equations of type (RD) is deterministic. Quantum effects and other sources of fluctuations are ignored. A more adequate description of such situations is obtained if the fluctuations are modeled by an additional stochastic source term (noise) in the macroscopic equation (RD). From the macroscopic point of view it is impossible to give detailed informations about the properties of the fluctuation process. Therefore, the equation considered by many authors in the physical literature is the following one:

(RDN) $\frac{\partial}{\partial t} u = \Delta u + f(u) + \sigma \xi$,

where ξ is a generalized GAUSSian process which is governed by $E\xi(t,x)=0$ and $E\xi(t,x)\xi(s,y)=\delta(t-s)\delta(x-y)$, i.e. ξ is a space-time white noise on a given probability space $(\Omega, \mathcal{F}, P)$.

2. The initial-boundary value problem to (RDN)

Consider the formal initial-boundary value problem

$$\frac{\partial}{\partial t}u(t,x) = \frac{\partial^2}{\partial x^2}u(t,x) + f(u(t,x)) + \sigma\xi(t,x),$$

$$\text{(S)} \qquad (t,x)\in D=(0,T]\times(0,L), \quad \sigma\geq 0,$$

$$u(0,x) = \varphi(x), \qquad x\in [0,L],$$

$$u(t,0)=r_0(t), \; u(t,L)=r_L(t), \qquad t\in [0,T].$$

Suppose that initial and boundary conditions are compatible and let $\varphi : \Omega\times[0,L] \rightarrow R$ be a sample continuous random process. A sample continuous random field $W=(W(t,x))_{(t,x)\in R_+^2}$ with $EW(t,x)=0$ and $EW(t,x)W(s,y) = (t\ s)\wedge(x\ y)$ for all $t,s,x,y\geq 0$ is called BROWNian sheet. ($a\wedge b$ denotes $\min(a,b)$.) In the sense of generalized random fields it holds $\frac{\partial^2}{\partial x\partial t}W(t,x)= \xi(t,x)$. Let φ be independent of W. By $\mathcal{F}_t$ we will denote the σ-algebra generated by φ and W for $0\leq s\leq t$, $y\in [0,L]$. Let $G: (0,T]\times[0,L]\times[0,L] \rightarrow R_+$ be the GREEN's function to the problem (S) for $f=0$, $r_0=r_L=0$ and $\sigma=0$. It is well known that G has the representation

$$G(t,x,y) = \frac{2}{L}\sum_{n=1}^{\infty} \sin\frac{n\pi x}{L} \sin\frac{n\pi y}{L} \exp(-(\frac{\pi n}{L})^2 t).$$

Assume that r_0 and r_L are non-random and sufficiently smooth.

By a solution to the formal initial-boundary value problem (S) we mean a random field $u=(u(t,x))_{(t,x)\in\bar{D}}$ such that for every fixed $t\in [0,T]$ it is $\mathcal{F}_t$-measurable and satisfies for all $(t,x)\in\bar{D}$ the stochastic integral equation

$$\text{(IS)} \quad u(t,x)=q(t,x)+ \int_0^L G(t,x,y)\varphi(y)dy + \int_0^t\int_0^L G(t-s,x,y)f(u(s,y))dyds$$

$$+ \sigma\int_0^t\int_0^L G(t-s,x,y)dW_{sy} \qquad \text{P-a.s.}$$

(Here q is generated by the boundary conditions.) The last term on the r.h.s. of (IS) is the stochastic WIENER integral of G with respect to the BROWNian sheet W. In order to define this integral we have to verify that

$$\int_0^t\int_0^L G^2(s,x,y)dyds < \infty$$

for all $(t,x)\in\bar{D}$. For $x\in [0,L]$ this is true. Unfortunately, this property is lost in higher space dimensions. At present there is no general theory for this case. Thus we deal only with the one-dimensional situation.

We can motivate the presence of the stochastic integral in (IS) as follows (cf. [1]).

THEOREM 1. Let $(\xi_n)_{n\geq 1}$ be a sequence of mean zero GAUSSian random fields belonging P-a.s. to $L_2(\bar{D})$ and possessing the properties

$$\left(\int_0^{t_1}\int_0^{x_1} \xi_n(s,y)dyds,\dots,\int_0^{t_m}\int_0^{x_m}\xi_n(s,y)dyds\right) \xrightarrow[\text{weakly}]{} (W_{t_1,x_1},\dots,W_{t_m,x_m})$$

and

$$\limsup_n \iint_{\bar{D}} |E\xi_n(s,y)\xi_n(t,x)|\,dyds \leq C < \infty .$$

Then the random field r_n defined by

$$r_n(t,x) = \int_0^t\int_0^L G(t-s,x,y)\,\xi_n(s,y)dyds$$

converges weakly to the field r with

$$r(t,x) = \int_0^t\int_0^L G(t-s,x,y)dW_{sy}.$$

Note that if ξ_n is sufficiently smooth r_n represents the classical solution of the inhomogeneous heat equation. As an example to Theorem 1 consider a mean zero GAUSSian random field ξ_n with the covariance function $E\xi_n(t,x)\xi_n(s,y) = n^2\exp(-2n(|t-s|+|x-y|))$.

We can prove the following theorems concerning existence, uniqueness and sample continuity of a solution to (S) (cf. [2]).

THEOREM 2. Let $f\colon R \longrightarrow R$ be a continuous mapping satisfying the growth condition

$$|f(u)| \leq M(1+|u|).$$

Then there exists at least one solution to (S).

THEOREM 3. Under the conditions of Theorem 2 every solution to (S) possesses a continuous version.

THEOREM 4. Under the conditions of Theorem 2 there is only one solution, if f satisfies

$(\Lambda)\quad |f(u) - f(v)| \leq K_N |u-v|$

for $|u|, |v| \leq N$.

THEOREM 5. Suppose that f: R→R satisfies (Λ). Assume that there exists monotonously nonincreasing functions h and g such that

$$f^+(u) \leq h(u) \text{ and } f^-(u) \geq g(u),$$

where $f^+(u)=\max(f(u),0)$, $f^-(u)=\min(f(u),0)$. Then the problem (S) has a pathwise unique sample continuous solution.

Clearly, $f(u)=\sum_{k=0}^{n} a_k u^k$, $a_n<0$, n odd, has the properties required in the last theorem.
Note that we do not assume any monotonicity of f.

Let u and v be the solutions to (S) for different initial conditions φ and Ψ.

THEOREM 6. The solution u to (S) depends continuously on the initial condition, more precisely, it holds P-a.s.

$$\sup_{(t,x)\in D} |u(t,x)-v(t,x)| \leq e^{KT} \sup_{0\leq x\leq L} |\varphi(x)-\Psi(x)|,$$

where $K=K(\omega)$ is a P-a.s. finite random variable.

It can be proven that the solution to (S) is a MARKOV process in time. This MARKOV property reflects the causality principle in physics and underlines the fact that the equation (RDN) describes an evolution in time direction. For details see [4].

3. The CAUCHY problem to (RDN)

Consider the formal CAUCHY problem

$$\frac{\partial}{\partial t}u(t,x) = \frac{\partial^2}{\partial x^2}u(t,x) + f(u(t,x)) + \sigma\cdot\xi(t,x)$$

$$\text{(C)}\qquad (t,x)\in(0,T]\times R,\quad \sigma\geq 0,$$

$$u(0,x) = \varphi(x),\qquad x\in R.$$

Solutions to (C) are needed for example to describe phenomena studied in theormodynamics (phase transitions) and quantum field theory.
The function H: $(0,T]\times R \to R$ defined by

$$H(t,x) = \frac{1}{2\sqrt{\pi t}} \exp(-x^2/4t)$$

is called the fundamental solution to (C) for $f=0$, $\sigma=0$. Denote by $C(R)$ the space of all continuous functions defined on R and introduce the BANACH space

$$C_e^{(k)} = \{\Psi \in C(R): \sup_{x\in R} |\Psi(x)| \exp(-k|x|) < \infty\}$$

Further, let W be a random function defined on the set $\mathcal{R}$ of all bounded rectangles of the type $(s_1,s_2] \times (y_1,y_2]$ in $R_+ \times R$ possessing the following properties

(i) If $A\in\mathcal{R}$, $W(A)$ is a mean zero GAUSSian random variable with variance $\lambda(A)$, where λ is the LEBESGUE measure on R.

(ii) If $A,B\in\mathcal{R}$, $A\cap B=\emptyset$, $W(A)$ and $W(B)$ are independent.

Note that the BROWNian sheet generates such a set function. Finally, assume that φ and W are independent and introduce the σ-algebra $\mathcal{F}_t = \sigma\{W(A), \varphi(y): A\in\mathcal{B}_0([0,t]\times R), y\in R\}$, where $\mathcal{B}_0$ denotes the set of bounded BOREL sets.

By a solution to (C) we mean a random field $u:\Omega\times[0,T]\times R\to R$ such that it is $\mathcal{F}_t$-measurable for every $t\in[0,T]$ and satisfies the stochastic integral equation

$$\text{(IC)}\quad u(t,x) = \int_R H(t,x-y)\varphi(y)dy + \int_R\int_0^t H(t-s,x-y)f(u(s,y))dsdy + \sigma\int_R\int_0^t H(t-s,x-y)dW_{sy}$$

P-a.s. for every $(t,x)\in[0,T]\times R$.

The stochastic WIENER integral $m(t,x) = \int_0^t\int_R H(t-s,x-y)dW_{sy}$ is a centred GAUSSian random field with

$$Em(t,x)m(s,y) = \int_0^{t\wedge s} H(t+s-2\tau, x-y)d\tau .$$

It can be shown (cf. [3]) that m possesses a continuous version and the property

$$\sup_{x\in R} \frac{|m(t,x)|}{1+|x|} = c_t < \infty, \qquad \text{P-a.s.}$$

for every $t\in[0,T]$.

THEOREM 7. Let φ belong to $C_e^{(k)}$ and let f be as in Theorem 5, where $h(u)$ and $|g(u)|$ are smaller than $C(1+|u|^p)$, where p is a positive real number. Then there exists a solution to (C). Moreover, every solution to (C) belongs to $C_e^{(l)}$ $(l \geq k)$ and possesses a continuous version.

THEOREM 8. Under the conditions of Theorem 7 the solution to (C) is pathwise unique, if f satisfies

$$|f(u)-f(v)| \leq |u-v| B(u,v),$$

where $0 \leq B(u,v) \leq \text{const}(1+\max(|u|^m, |v|^m))$, $\quad m \geq 1$.

For details see [3].

4. Random measures as solutions to (RDN)

In section 2 it was pointed out that there is no hope to get a function-valued solution to (RDN) if the space parameter x has more than one dimension. Indeed, in the linear case it is possible to construct a random measure which can be interpreted as the solution to a given problem related to (RDN). To illustrate this approach consider

$$(C^*)\quad \begin{aligned} &\frac{\partial}{\partial t}u(t,x)=(\Delta u)(t,x)-pu(t,x)+\xi(t,x), \quad (t,x)\in(0,T]\times R^d, \\ &u(0,x)=0, \qquad x\in R^d. \end{aligned}$$

The GAUSSian random function $\pi : \Omega\times[0,T]\times\mathcal{B}_0(R^d)\rightarrow R$ defined by $E\,\pi_t(A)=0$ and

$$E\,\pi_t(A)\pi_s(B) = \int_A \int_B \int_0^{t\wedge s} H_p(t+s-2\tau,x-z)\,d\tau\,dx\,dz$$

is called the solution to (C^*), where H_p denotes the fundamental solution of (C^*) without noise, which corresponds to the operator $\Delta - pI$. If d=1 we observe

$$P\Big(\int_A m(t,x)dx = \pi_t(A)\Big) = 1.$$

Let $A=\bigcup_{i=1}^{\infty} A_i \in\mathcal{B}_0(R^d)$, $A_i\cap A_j=\emptyset$ if $i\neq j$. Then

$$\lim_{N\to\infty} E\Big(\pi_t(A)-\sum_{i=1}^{N}\pi_t(A_i)\Big)^2 = 0.$$

For fixed $A\in\mathcal{B}_0(R^d)$ the random process $(\pi_t(A))_{t\geq 0}$ possesses a continuous version. Further $(\pi)_{t\geq 0}$ is a MARKOV process. For details see [5]. These results can be extended to a more general linear equation.

References

[1] R. Manthey, Weak convergence of solutions of the heat equation with Gaussian noise, Math. Nachr. 123 (1985) 157-168.

[2] R. Manthey, On the solutions of reaction-diffusion equations with white noise, Forschungsergebnisse FSU Jena N/85/24.

[3] R. Manthey, Existence, uniqueness and continuity of solutions to the Cauchy problem for reaction-diffusion equations with white noise, Forschungsergebnisse FSU Jena N/86/4.

[4] Ch. Stiewe, G. Jetschke and R. Manthey, Nonlinear reaction-diffusion equations with white noise disturbance generate strong Markov processes, in preparation.

[5] R. Manthey, Random measures as the solutions to the cable equation with white noise, in preparation.

The Propagation of Chaos for diffusions with bad drift coefficients

by

Masao Nagasawa

1. Let us begin with an equation on $\mathbb{R}^1$

$$(1) \qquad \frac{1}{2}\frac{d}{dx}(\log \phi) = b * \phi$$

where ϕ is a smooth probability density and $b * \phi$ denotes the convolution of b and ϕ. In the equation (1) we assume that ϕ is given and b is unknown. As an example, if ϕ_0 is Gaussian $\phi_0 = ce^{-x^2}$, then the equation (1) has a unique solution $b_0(x) = -x$.

2. Consider a system of interacting Brownian motions

$$(2) \qquad X_i(t) = X_i(0) + B_i(t) + \int_0^t b[X_i(s), U^{(n)}(s)]ds \ , \ i = 1,2, \ldots, n,$$

where $\{X_i(0)\}$ is independent identically distributed according to $u \in P(\mathbb{R}^1)$ (= the space of probability measures on $\mathbb{R}^1$), $\{B_i(t)\}$ is a famiy of independent Brownian motions which are independent of $\{X_i(0)\}$, and $U^{(n)}(t)$ denotes the empirical distribution

$$(3) \qquad U^{(n)}(t) = \frac{1}{n}\sum_{j=1}^{n} \delta_{X_j(t)} \cdot$$

The case of

$$(4) \qquad b[x,u] = b * u \ , \ u \in P(\mathbb{R}^1) \, ,$$

with a Lipschitz continuous b was first treated by McKean (1967). The propagation of chaos, in the sense of McKean, tells us that the law of large number holds

$$(5) \qquad U^{(n)}(t) \to u_t \in P(\mathbb{R}^1), \text{ in law},$$

and as $n \to \infty$ in (2), for $i = 1$, we get McKean-Vlasov's SDE

$$(6) \qquad X(t) = X(0) + B(t) + \int_0^t b[X(s), u_s]ds \ ,$$

where u_t is the probability distribution of $X(t)$.

If u_t has a smooth density ϕ_t with respet to the Lebesgue measure, it satisfies a non-linear *PDE*

$$(7) \qquad \frac{\partial \phi}{\partial t} = \frac{1}{2}\frac{\partial^2 \phi}{\partial x^2} - \frac{\partial}{\partial x}(b[\cdot,\phi]\phi)$$

by Itô's formula. If the process given by (6) attains the stationarity, then it is easy to see from (7) that ϕ and b satisfy the equation (1).

3. The example given in Sec. 1 tells us that the system of Brownian motions with a pair interaction $b_0(x) = -x$ has the Gaussian equilibrium distribution $\phi_0(x) = ce^{-x^2}$ in the McKean-Vlasov limit. As the second example let us consider $\phi_1 = c\,x^2 e^{-x^2}$, which is the distribution of the first excited state of the 1-dim. harmonic oscillator in quantum theory. In order to make the equation (1) meaningful for ϕ_1, we introduce two types of Brownian motions (painted blue and red, respectively) and distribute blue ones on $(-\infty,0]$ according to ϕ_1 and red ones on $[0,\infty)$ according to ϕ_1. Then the equation (1) turns out to be solvable (see Nagasawa-Tanaka (1985)). The propagation of chaos for interacting Brownian motions of two types is discussed in Nagasawa-Tanaka (1986). The case of systems with a moving reflecting boundary (segregating front) is treated in Nagasawa-Tanaka (to appear) under a condition on interactions to be bounded Lipschitz.

4. The last case of moving reflecting boundary can be treated in a different way in terms of Brownian motions without reflection but with a generalized interaction not of the convolution type (4), satisfying the following conditions:

Condition 1. $b[x,u]$ is a bounded measurable function defined on $\mathbb{R}^1 \times P(\mathbb{R}^1)$;

Condition 2. If $u_n \in P(\mathbb{R}^1)$ converges weakly to $u \in P(\mathbb{R}^1)$ which has a strictly positive density (almost everywhere) with respect to the Lebesgue measure on $\mathbb{R}^1$, then $<u_n, b[\cdot, u_n]f>$ converges to $<u, b[\cdot, u]f>$ as $n \to \infty$ for any $f \in C_b(\mathbb{R}^1)$;

Condition 3. The uniqueness holds for $P(\mathbb{R}^1)$–valued solution of the initial value problem

$$(8) \qquad \begin{cases} \frac{d}{dt} <u(t), f> = <u(t), A_{u(t)}f>, & \text{for } f \in C_k^\infty(\mathbb{R}^1), \\ u(0) \in P(\mathbb{R}^1) \end{cases}$$

where $A_u f = \frac{1}{2}\frac{d^2 f}{dx^2} + b[\cdot, u]\frac{df}{dx}$.

Theorem. *Let* $X^{(n)}(t) = (X_1^{(n)}(t), \ldots, X_n^{(n)}(t))$ *be a solution of SDE* (2). *Under the conditions* 1, 2 *and* 3 *the following assertions hold:*

(i) $U^{(n)}(t) = \frac{1}{n} \sum_{i=1}^{n} \delta_{X_i(t)}$ converges in probability to some (non-random) limit $u(t) \in P(\mathbb{R}^1)$ which is a solution of (8);

(ii) For each $m < n$, the process $(X_1^{(n)}(t), \ldots, X_m^{(n)}(t))$ converges as $n \to \infty$ in law to $(X_1(t), \ldots, X_m(t))$, where $\{X_i(t); i = 1, 2, \ldots, m\}$ are mutually independent and each $X_i(t)$ satisfies the McKean-Vlasov's *SDE* (6);

(iii) $u(t)$ in the first assertion coincides with the one in the second assertion.

For detail see Nagasawa-Tanaka (preprint).

References

[1] McKean, H. P., (1967) *Propagation of chaos for a class of non-linear parabolic equations,* Lecture Series in Differential Equations, Catholic Univ., Washington, DC, 41-57.

[2] Nagasawa, M. and Tanaka, H., (1985) *A diffusion process in a singular mean-drift field,* **Z. Wahr. verw. Gebiete 68**, 247-269.

[3] Nagasawa, M. and Tanaka, H., (1986) *Propagation of chaos for diffusing particles of two types with singular mean field interaction,* **Probab. Th. Rel. Fields 71**, 69-83.

[4] Nagasawa, M. and Tanaka, H., (to appear) *Diffusion with interactions and collisions between coloured particles and the propagation of chaos,* **Probab. Th. Rel. Fields**.

[5] Nagasawa, M. and Tanaka, H. (preprint) *On the propagation of chaos for diffusion processes with drift coefficients not of average form.*

APPROXIMATION FOR INFINITE-DIMENSIONAL WIENER PROCESSES IN SEPARABLE HILBERT SPACES

M.Richter
Technical University
DDR-Karl-Marx-Stadt, Psf.964
9010

Let $\mathfrak{X}$ be a separable Hilbert space with the inner product $(.,.)$ and the norm $|x| = (x,x)^{1/2}$, $x \in \mathfrak{X}$. Consider an $\mathfrak{X}$-valued Wiener process $(X_t)_{t\in[0,T]}$ defined on a complete probability space $(\Omega, \mathfrak{F}, P)$ (for the definitions see [1], p. 134). The correlation operator K of the Wiener process is given. In this note we discuss the approximation of the Wiener process in the "state space". Consider an orthogonal projector Q from $\mathfrak{X}$ on a subspace $\underline{\mathfrak{X}}$ ($\underline{\mathfrak{X}} \subset \mathfrak{X}$) and the approximation $(QX_t)_{t\in[0,T]}$ of the Wiener process. With a $(\mathfrak{F}\otimes\mathfrak{F})$ - measurable nonnegative functional L (loss function) defined on $(\mathfrak{X}\times\mathfrak{X})$ we estimate the <u>approximation</u> <u>error</u> by $L(X_t, QX_t)$ in the time t . Then the mean of this error $E\,L(X_t, QX_t)$ describes the <u>approximation risk</u> (in the time t with respect to the loss function L).
Denote by $m_1 \geq m_2 \geq \ldots \geq 0$ - the ordered eigenvalues, $e_1, e_2, \ldots$ - the (orthonormal) eigenfunctions of the correlation operator K and Q^n - the orthogonal projector on the linear subspace spanned by the eigenfunctions $e_1, e_2, \ldots, e_n$.
The next theorem gives the risks $r_i(t,n) = E\,L_i(X_t, Q^nX_t)$, where

$$L_1(X_t,Q^nX_t) = |X_t - Q^nX_t|^2 \qquad (1)$$

$$L_2(X_t,Q^nX_t) = \chi(|X_t - Q^nX_t|^2 > d) \quad , \quad 0 < d < 1 \;,$$
$\chi(A)$ - indicator of A ,

$$L_3(X_t,Q^nX_t) = \exp(|X_t - Q^nX_t|^2)$$

$$L_4(X_t,Q^nX_t) = \chi\left(\frac{|X_t - Q^nX_t|^2}{|X_t|^2} > d\right) \quad , \quad 0 < d < 1 \;.$$

<u>Theorem 1:</u> Let

$$F_n(x;q) = P\left(q\cdot\sum_{j=1}^{n} m_j\cdot Y_j^2 + \sum_{j=n+1}^{\infty} m_j\cdot Y_j^2 < x\right)$$

be the distribution function of the quadratic form of the i.i.d. $N(0;1)$ - random variables $Y_1, Y_2, \ldots$. Then

$$r_1(t,n) = t \cdot \sum_{j=n+1}^{\infty} m_j \tag{2}$$

$$r_2(t,n) = 1 - F_n(\tfrac{d}{t};0) \tag{3}$$

$$r_3(t,n) = \begin{cases} \infty & \text{if } m_{n+1}t \geq 1/2 \\ \prod_{j=n+1}^{\infty} (1 - 2\cdot t\cdot m_j)^{-1/2} & \text{if } m_{n+1}t < 1/2 \end{cases} \tag{4}$$

$$r_4(t,n) = 1 - F_n(0;-\tfrac{d}{1-d}) \tag{5}$$

Proof: Since K is nuclear and positive, the system (e_i) of all eigenfunctions of K forms a complete orthonormal basis for $\mathcal{H}$. Therefore

$$|X_t - Q^n X_t|^2 = \sum_{j=n+1}^{\infty} (X_t,e_j)^2 \quad \text{w.p. } 1 \tag{6}$$

for all $t \in [0,T]$ ([1], p. 134), where $((X_t,e_i),\ t \in [0,T])$ are mutually independent real Wiener processes with

$$E\,(X_t,e_i) = 0 \quad \text{and} \quad E\,(X_t,e_i)(X_s,e_i) = m_i\cdot \min(t,s)\ , \quad 0 \leq t,s \leq T\ .$$

From this (2) and (3) follow directly.
The characteristic function of (1) is

$$\varphi(p) = E\exp(i\cdot p\cdot |X_t - Q^n X_t|^2) = \prod_{j=n+1}^{\infty} (1 - 2\cdot i\cdot p\cdot t\cdot m_j)^{-1/2}\ . \tag{7}$$

$\varphi(p)$ for $p = -i$ yields (4). Since

$$\left(\frac{|X_t - Q^n X_t|^2}{|X_t|^2} > d\right) = (|X_t - Q^n X_t|^2\cdot(1 - d) - d\cdot|Q^n X_t^2| > 0) \tag{8}$$

and using (6) we have shown (5).

Let us now make some remarks about the calculation of the risks given in Theorem 1. Using the trace of the operator K an immediate consequence is $r_1(t,n) = t\cdot(\mathrm{tr}\,K - \sum_{i=1}^{n} m_i)$. With respect to the inequality $x \leq -\ln(1-x) \leq \frac{x}{1-x}$, $0 \leq x < 1$, and

$$\ln r_1(t,n) = -1/2 \cdot \sum_{j=n+1}^{\infty} \ln(1 - 2\cdot t\cdot m_j) ,$$

it yields

$$1 \leq r_3(t,n)\cdot \exp(-r_1(t,n)) \leq \exp(2\cdot t^2 \cdot \sum_{j=n+1}^{\infty} \frac{m_j^2}{1 - 2\cdot t\cdot m_j^2}) \leq$$

$$\exp(r_1(t,n)\cdot \frac{2\cdot t\cdot m_{n+1}}{1 - 2\cdot t\cdot m_{n+1}}) \qquad \text{for } 0 \leq t < \frac{1}{m_{n+1}^2} .$$

Since $F_n(\frac{d}{t};0)$ is the distribution function of a positive-definite form, we can calculate $r_2(t,n)$ with methods given in [2]. Explicite formulas for the risks are given in Theorem 2 and 3 with additional assumptions on the eigenvalues.

Theorem 2: If $m_{n+1} = m_{n+2} > m_{n+3} = m_{n+4} > \dots$,

then

$$r_2(t,n) = \sum_{k=1}^{\infty} \exp(-\frac{d}{t\cdot 2\cdot m_{n+2k}})\cdot \prod_{\substack{j=1\\ j\neq k}}^{\infty} (1 - \frac{m_{n+2j}}{m_{n+2k}})^{-1} , \quad d > 0 \tag{9}$$

and

$$r_4(t,n) = \sum_{k=1}^{\infty} \prod_{\substack{j=1\\ j\neq k}}^{\infty} (1 - \frac{m_{n+2j}}{m_{n+2k}})^{-1} \prod_{l=1}^{n} (1 + t\cdot \frac{d}{1-d}\cdot \frac{m_l}{m_{n+2k}})^{-1/2} .$$

Proof: The equation (9) is an immediate result of [2], Theorem 1. Let f_1 be the density function of $|Q^n X_t|^2$. Since the random values $|Q^n X_t|$ and $|X_t - Q^n X_t|$ are independent, for the distribution function of

$$-\frac{d}{1-d}\cdot |Q^n X_t|^2 + |X_t - Q^n X_t|^2$$

follows

$$F_n(x;-\frac{d}{1-d}) = \frac{d}{1-d}\cdot \int_{-\infty}^{\infty} f_1(-\frac{d}{1-d}\cdot y)\cdot F_n(x-y;0)\, dy .$$

Then from (5) we have

$$r_4(t,n) = 1 - \frac{d}{1-d} \int_{-\infty}^{0} f_1(-\frac{d}{1-d}\cdot y)\cdot F_n(-y;0)\, dy .$$

With respect to (9) this leads to

$$r_4(t,n) = \sum_{k=1}^{\infty} \prod_{\substack{j=1\\ j\neq k}}^{\infty} \left(1 - \frac{m_{n+2j}}{m_{n+2k}}\right)^{-1} \cdot I_{n,k} \quad ,$$

where

$$I_{n,k} = \frac{d}{1-d} \int_{-\infty}^{0} f_1\left(-\frac{d}{1-d}\cdot y\right)\cdot \exp\left(\frac{y}{2\,m_{n+2k}}\right) dy \quad .$$

Analogous to (7) it holds

$$I_{n,k} = E \exp\left(-\frac{d}{1-d}\cdot\frac{1}{2m_{n+2k}}\cdot |Q^n X_t|^2\right) = \prod_{l=1}^{n} \left(1 + t\cdot\frac{m_l}{m_{n+2k}}\cdot\frac{d}{1-d}\right)^{-1/2}$$

The next theorem gives the risk of the relativ approximation error.

<u>Theorem 3</u>: If the eigenvalues of K are equal by pairs, that is $m_1 = m_2 > m_3 = m_4 > \dots$, then for

$$L_5(X_t, Q^{2n}X_t) = \frac{|X_t - Q^{2n}X_t|^2}{|X_t|^2}$$

it holds

$$r_5(t,2n) = \sum_{k=1}^{\infty} \sum_{r=1}^{n} \left[2m_{n+2k} \prod_{\substack{l=1\\ l\neq r}}^{n} \left(1 - \frac{m_{2l}}{m_{2r}}\right) \prod_{\substack{j=1\\ j\neq k}}^{\infty} \left(1 - \frac{m_{n+2k}}{m_{n+2j}}\right)\right]^{-1}$$

$$\left[\frac{\ln\left(\frac{m_r}{m_{n+2k}}\right)}{\left(\frac{m_{2r}}{m_{n+2k}} - 1\right)^2} - \frac{m_{n+2k}}{m_{2r}\left(\frac{m_{2r}}{m_{n+2k}} - 1\right)}\right] \quad .$$

<u>Proof:</u> Denote by $f_2(x)$ the density function of $|X_t - Q^{2n}X_t|^2$. Then

$$r_5(t,2n) = \int_0^{\infty}\int_0^{\infty} \frac{x}{x+y}\cdot f_1(y)\cdot f_2(x)\, dy\, dx \quad . \tag{10}$$

Under the conditions of the theorem it follows

$$f_1(y) = \sum_{r=1}^{n} \left(2\cdot m_{2r} \prod_{\substack{l=1\\ l\neq r}}^{n} \left(1 - \frac{m_{2l}}{m_{2r}}\right)\right)^{-1} \exp\left(-\frac{y}{2m_{2r}}\right) \quad , \; y > 0 \;, \; ([2]) \; .$$

Substituting $f_1(y)$ in (10) and integrating we get

$$r_5(t,2n) = -\sum_{r=1}^{n} \left(2\cdot m_{2r} \prod_{\substack{l=1\\ l\neq r}}^{n} \left(1 - \frac{m_{2l}}{m_{2r}}\right)\right)^{-1} \int_0^{\infty} x\cdot f_2(x)\cdot \exp\left(\frac{x}{2m_r}\right)\cdot Ei\left(-\frac{x}{2m_r}\right) dx \tag{11}$$

([4], p. 325).

From (9) together with (3) we have

$$f_2(x) = \sum_{k=1}^{\infty} (2m_{n+2k} \prod_{\substack{j=1 \\ j \neq k}}^{\infty} (1 - \frac{m_{n+2k}}{m_{n+2j}}))^{-1} \exp(- \frac{x}{2m_{n+2k}}) \quad , \quad x > 0 \quad .$$

Substituting this into (11), from [4] follows the desired result.

If $m_{n+1} \geq m_{n+2} > m_{n+3} \geq m_{n+4} > \ldots$ holds, then Theorem 2 gives lower bounds for the risks r_2 and r_4 owing to the monotonity of the loss functions. In the case that the eigenvalues $m_{n+2}, m_{n+4}, \ldots$ are replaced by the eigenvalues $m_{n+1}, m_{n+2}, \ldots$ in the formulas of Theorem 2, we obtain upper bounds for the risks. In the same way with Theorem 3 we can construct lower and upper bounds for $r_5(t,n)$.

Now let us discuss the approximation of a Wiener process with a given level for the approximation error and risk. For the sequence of projectors $\underline{Q} = (Q^n)_{n=0}^{\infty}$, with $Q^0 x = 0$, $x \in \mathfrak{X}$, and a given level $c > 0$, we consider

$$\underline{n} = \inf\{k \geq 0 : r(t,k) < c\} \quad .$$

Then the approximation $Q^{\underline{n}} X_t$ guarantees a approximation risk less than c with the least "dimension" of approximation-subspace. For the calculation of the "optimal dimension" $\underline{n}$ for the risks r_i , i=1, ...,5 we can use the Theorems 1 to 3. If we take the approximation

$$Q^N X_t = \sum_{n=0}^{\infty} \chi(N = n) \cdot Q^n X_t \qquad \text{with}$$

$$N = \inf\{k \geq 0 : L(X_t, Q^k X_t) < c\} \quad ,$$

the approximation $Q^N X_t$ has for each realization of the Wiener process an approximation error $L(X_t, Q^N X_t)$ less than c with the least approximations-dimension. Because N is a random variable , the projector Q^N and the subspace $Q^N \mathfrak{X}$ are random too. Now let us discuss some properties of N for the approximation error (1). More general results are given in [3].

<u>Theorem 4:</u> If (1) holds, then

1. $\lim_{n \to \infty} P(N > n) = 0$ for $c > 0$
2. $P(N > n) = r_2(t,n)$ with $d = c$.

<u>Proof:</u> The first part follows from

$$(N > n) = \bigcap_{k=0}^{n} (|X_t - Q^k X_t|^2 > c) = (|X_t - Q^n X_t|^2 > c) \qquad (12)$$

and the inequality

$$P(N > n) \leq \frac{1}{c} E\, |X_t - Q^n X_t|^2 = \frac{t}{c} \cdot \sum_{k=n+1}^{\infty} m_k \quad . \qquad (13)$$

The second part of the theorem is a direct consequence of (12) and (3).

With (12) we obtain an upper bound for $E\,N$

$$E\,N = \sum_{n=0}^{\infty} P(N > n) \leq \frac{t}{c} \sum_{n=0}^{\infty} \sum_{j=n+1}^{\infty} m_j = \frac{t}{c} \sum_{n=1}^{\infty} n \cdot m_n \quad .$$

In [3] is proved the following result.

<u>Theorem 5:</u> If there exists a monotone sequence $(m_i^!)$ such that $m_1^! > m_2^! > \ldots$, $m_i^! \geq m_i$, $i=1,2,\ldots$, and that for any $\varepsilon > 0$ there is a n_0 , with

$$- m_n^! \sum_{k=n+1}^{\infty} \ln(1 - \frac{m_k^!}{m_n^!}) < \varepsilon \quad \text{for all} \quad n \geq n_0 \quad , \qquad (14)$$

then $E\,N^s < \infty$ for $s > 0$.

<u>Remark:</u> The condition (13) is true e.g. if $m_n \leq n^{-a}$, $n \geq n_0$, $a > 1$.

<u>References:</u>

[1] Curtain, R.F.; A.J. Pritchard: Infinite-dimensional linear systems theory, Lect. Notes Contr. Inf. Sc., vol. 8, 1978

[2] Richter, M.: Berechnung der Verteilungsfunktion quadratischer Formen von normalverteilten Zufallsgroessen und deren Anwendungen in der Statistik, Math.Oper.Math.Stat.,Ser.Stat.,<u>15</u>,(1984),177-94

[3] Richter, M.: Approximation von Gausschen Zufallselementen, Wiss. Ztschr.Techn.Hochsch.Karl-Marx-Stadt,<u>28</u>,(1986),188-95

[4] Gradstein, I.S.; I.M. Ryshik: Tablicy, integralov, summ, rjadov and proizvedenij, Nauka, Moscow, 1971 (Russ.)

A PREDICTION PROBLEM FOR GAUSSIAN PLANAR PROCESSES WHICH ARE MARKOVIAN WITH RESPECT TO INCREASING AND DECREASING PATHS

RUSSO Francesco
Ecole Polytechnique Fédérale
Département de Mathématiques
1015 Lausanne (Switzerland)

1. INTRODUCTION

This paper is an extension of [DR] which has similar objectives but for a particular case : the Brownian sheet. Let $X = (X_s)_{s \in \mathbb{R}^2_+}$ be a stochastic process as in the title. We first remark that X is not far from a Gaussian planar process with independent increments having a product measure as a variance; an important example is given by the Ornstein-Ulenbeck sheet (cf. section 3).

The principal aim here is to give estimation of the position of X using the information contained in the germ or sharp fields in relation to certain curves termed separation lines by several authors (cf. section 5). Problems of this nature have been considered for Gaussian multiparameter or generalized processes by McKean ([Mc]), Pitt ([P]), Rozanov ([R1]), [R2]), Carraro ([Ca]).

In this context we also remark that X is Markovian with respect to every subset A of $\mathbb{R}^2_+$ (section 4); we give a necessary and sufficient condition for the sharp and germ fields of a separation line to coincide. Finally we point out that our predictors provide solutions of a simple hyperbolic stochastic partial differential equation.

2. SOME SPECIAL CLASSES OF PLANAR PROCESSES

We begin with some notations and conventions. In the whole paper j will always denote an element of $\{1,2\}$, T will be $I_1 \times I_2$, where I_j are closed real intervals with left-end points a_j; sets $I_1 \times \{a_2\}$ and $\{a_1\} \times I_2$ will be called the axes. If $s = (s_1,s_2) \in T$, R_s will stand for the set $[a_1,s_1] \times [a_2,s_2]$. If $\alpha : I \to J$ is a continuous surjective function and I,J are real intervals we note

$$(2.1) \quad \alpha^{-1}(x) = \inf \{ y \mid \alpha(y) > x \}.$$

Let consider a process $X = (X_s)_{s \in T}$ on a complete probability space $(\Omega,\underline{\underline{F}},P)$. Equalities between random variables will generally hold almost surely and σ-fields are always supposed to be completed by P-null sets.

Definition 2.1 We will say that X has property PO, if it is a mean zero Gaussian process with a continuous covariance function not vanishing on the diagonal excepted maybe on the axes. □

For a process X as above, note $\eta : \underline{\underline{B}}_T \to L^2(\Omega,\underline{\underline{F}},P)$ the random (Hilbert) measure of X, where $\underline{\underline{B}}_T$ is the δ-ring of bounded Borel sets of I. Recall that

$$\eta(R_s) = X_s,$$

$$\eta(]s_1,t_1] \times]s_2,t_2]) = X_{t_1,t_2} - X_{s_1,t_2} - X_{t_1,s_2} + X_{s_1,s_2},$$

if $s = (s_1,s_2)$, $t = (t_1,t_2) \in T$. Because of the continuity of X in $L^2(\Omega,\underline{\underline{F}},P)$, the measure of an horizontal or vertical segment not lying on the axes is 0.

Definition 2.2 We say that X satisfies the order Markov property (OMP), if the following property is realised. For every parametrisation $\gamma : [0,1] \to T$ of a curve which is increasing with respect to either of the two orders $\lesssim$ or $\underset{\sim}{\wedge}$, where

$$(s_1,s_2) \lesssim (\text{resp. } \underset{\sim}{\wedge}) \ (t_1,t_2) \iff s_1 \leq t_1,\ s_2 \leq t_2 \ (\text{resp. } s_2 \geq t_2),$$

$(X_{\gamma(u)})_{u \in [0,1]}$ is Markovian (see figure 1). □

Definition 2.3 We say that a process X as above has property P1, if it is with independent increments; this means that $\eta(A)$ and $\eta(B)$ are independent when A and B are disjoint elements of $\underline{\underline{B}}_T$. We say that X has property P2 if the variance ν of X ($\nu(A) = E(\eta(A))^2$, if $A \in \underline{\underline{B}}_T$), is $\nu_1 \otimes \nu_2$ where ν_j are σ-finite measures on the Borel sets of I_j. □

3. EXAMPLES OF PROCESSES HAVING THE OMP

a) If a process $X = (X_s)_{s \in T}$ has properties PO, P1, P2 then it has the OMP for the reasons explained below.

If $z_1 \leq z_2 \leq z_3$ or $z_1 \curlywedge z_2 \curlywedge z_3$, we can easily verify that

(3.1) $\Gamma(z_1,z_3)\Gamma(z_2,z_2) = \Gamma(z_1,z_2)\Gamma(z_2,z_3)$,

where Γ is the covariance of X; moreover (3.1) is equivalent to the conditional independence of X_{z_1} and X_{z_3} given X_{z_2}. In conclusion X has the OMP.

Remark that $\nu_j\,[a_j, a_j + \varepsilon] > 0$ for $\varepsilon > 0$ since the covariance of X does not vanish on the diagonal.

b) The Brownian sheet is an example of process satisfying properties PO,P1,P2 on $T = \mathbb{R}^2_+$; in this case ν is Lebesgue measure.

c) If $(W_s)_{s \in \mathbb{R}^2_+}$ is a Brownian sheet,

$$X_{s_1,s_2} = e^{-(s_1+s_2)} W_{e^{2s_1},e^{2s_2}}$$

gives a process having the OMP (Ornstein-Ulenbeck sheet).□

Remark 3.1 If X has PO and OMP then there exist continuous non decreasing $\alpha_j : I_j \to \mathbb{R}_+$, a continuous function $h : T \to \mathbb{R}$ and a Brownian sheet $(W_s)_{s \in \mathbb{R}^2_+}$ such that X is equivalent in distribution to the process defined by

$$h(s_1,s_2)\, W_{\alpha_1(s_1),\alpha_2(s_2)}.$$

Previous result is due to [C].

Note that $h(s) \neq 0$, if s does not belong to the axes $(E(X_s^2) > 0)$. Consequently there is a process $X' = (X'_s)_{s \in T}$ with properties PO, P1, P2 such that $X_s = h(s)X'_s$, $s \in T$. X' will be called associated process with X. The variance of X is given by

$$\nu = \nu_1 \otimes \nu_2, \quad \nu_j([a_j, s_j]) = \alpha(s_j), \qquad s_j \in I_j. \ \square$$

Remark 3.2 X' (then X) has a continuous version; in fact we can easily see that $(Y_{s_1,s_2}) = (X'_{\alpha_1^{-1}(s_1),\alpha_2^{-1}(s_2)})$ is a Brownian sheet; by the corollary 1.3 of [W2] there is a continuous version Y' of Y; it is then clear that $(Y'_{\alpha_1(s_1),\alpha_2(s_2)})$ gives a continuous version of $(X'_{s_1 s_2})$. □

4. GAUSSIAN SPACES AND SPLITTING σ-FIELDS

Let consider a process $X = (X_s)_{s \in T}$ satisfying P0, and $A \subset T$. Let $\underline{\underline{F}}(A)$ be the σ-field $\sigma(X_s, s \in A)$: this is the sharp field of A; note

$$\underline{\underline{G}}(A) = \bigcap_{\varepsilon > 0} \underline{\underline{F}}(A_\varepsilon),$$

where A_ε is an ε-neighborhood of A; this is the germ field of A.

Note by H(A) the closed linear span in $L^2(\Omega, \underline{\underline{F}}, P)$ of X_s, $s \in A$, and

$$G(A) = \bigcap_{\varepsilon > 0} H(A_\varepsilon).$$

By lemma 3.3 of [M], we know that $\underline{\underline{G}}(A) = \sigma(G(A)$ if A is closed. H(T) is the Gaussian space of X.

Note by K the kernel reproducing space of X : it is composed by functions

$$s \to E(ZX_s),$$

defined on T, where $Z \in H(T)$. More precisely there is an isometry between the Hilbert spaces H(T) and K given by

$$J : Z \to (s \to E(ZX_s))$$

Remark 4.1 If X and X' are as in remark 3.1 then the σ-fields $\underline{\underline{F}}(A)$, $\underline{\underline{G}}(A)$ and the Gaussian subspaces H(A), G(A) are the same for X and X'. □

Remark 4.2 If X has properties P0, P1 then

$$H(T) = \left\{ \int_T g \, d\eta, \ g \in L^2(T, \nu) \right\}$$

$$K = \left\{ t \to \int_{R_t} g d\nu,\ g \in L^2(T,\nu) \right\},$$

where K is an Hilbert space with scalar product

$$< \int_{R.} g_1 d\nu, \int_{R.} g_2 d\nu > = < g_1, g_2 >_{L^2(T,\nu)} \ . \ \square$$

Proposition 4.3 If X has properties P0,P1 then X has the germ Markov property with respect to every $A \subset T$, that is to say $\underline{\underline{F}}(A)$ and $\underline{\underline{F}}(A^c)$ are conditionally independent given $\underline{\underline{G}}(\partial A)$.

Proof. By corollary 2.2 of [Ru], we only have to consider the Markov property with respect to sets A such that $A = \text{Int}\,\bar{A}$. Following [N], th. 1.1, it is enough to verify that K is local that is to say

(i) $<u_1, u_2>_K = 0$ if $u_1, u_2 \in K$ with disjoint supports.

(ii) If $u \in K$, then $u = u_1 + u_2$, where u_1, u_2 are functions with disjoint supports, then $u_1, u_2 \in K$.

The method used is the same as in [N], th. 3.1; we only have to adapt lemma 3.1 in the following way.

Lemma 4.4 Let D be an open set of T, $g \in L^2(T,\nu)$. If $f : T \to \mathbb{R}$, is such that

$$f(t) = \int_{R_t} g d\nu,\ \forall t \in D,$$

and its support is included in D, then

$$f(t) = \int_{R_t \cap D} g d\nu,\ \forall t \in T. \ \square$$

Corollary 4.5 A process X having properties P0 and OMP, is germ Markovian with respect to every $A \subset T$. □

Proof. This result is a consequence of proposition 4.3 and remarks 3.1 and 4.1. □

Remarks 4.6

(i) A process with independent increments (not necessarily Gaussian) vanishing on the axes has the germ Markov property for every bounded set $A \subset T$ ([Ru]).

(ii) Proposition 4.3 is still valid and corollary 4.5 can be generalized for $T = \prod_{j=1}^{n} I_j \subset \mathbb{R}^n$; the new OMP would involve 2^{n-1} partial orders. □

A σ-field $\underline{\underline{A}}$ such that $\underline{\underline{F}}(A)$ and $\underline{\underline{F}}(A^c)$ are conditionally independent given $\underline{\underline{A}}$, is called a <u>splitting field</u> for $\underline{\underline{F}}(A)$ and $\underline{\underline{F}}(A^c)$; it is well-known that there is a minimal splitting field for $\underline{\underline{F}}(A)$ and $\underline{\underline{F}}(A^c)$ ([Mc]) which I note $\underline{\underline{M}}(\partial A)$; by also using the continuity in L^2 of X, it is not difficult to notice the following chain of embeddings

(4.1) $\underline{\underline{G}}(\partial A) \supset \underline{\underline{M}}(\partial A) \supset \underline{\underline{F}}(A) \cap \underline{\underline{F}}(\bar{A}^c) \supset \underline{\underline{F}}(\partial A)$.

5. SEPARATION LINES

The rest of this work will concern domains A whose boundary is a separation line as explained below. Letter C will always denote such a curve.

<u>Definition 5.1</u> A curve C is called a <u>separation line</u>, if it is defined by a continuous parametrisation

$$\varphi = (\varphi_1,\varphi_2) : [0,1] \to T$$

such that φ_1 is non-decreasing, φ_2 is non-increasing, $\varphi_1(0) = a_1$, $\varphi_2(1) = a_2$ (cf. [DR]). □

<u>Remark 5.2</u> In some results we can also consider not necessarily bounded separation lines; in this case C will be the closure of the image of $\varphi = (\varphi_1,\varphi_2) :]0,1[\to T$ such that $\varphi_1(0+) = a_1$, $\varphi_2(1-) = a_2$. The prediction formulas concerning these lines are less elegant than in the bounded case, so we will omit them. □

C splits T into two domains D and D_+ such that

$$\bar{D} = \{(\varphi_1(u),\varphi_2(v)) \mid 0 \leq u \leq v \leq 1\}.$$

C will also be supposed to satisfy

(5.1) $(\varphi_1(u),\varphi_2(u)) \in \text{Int } T, \ \forall u \in]0,1[$

In the rest of the paper we will study in a deep way the relations (4.1) when $A = D$; moreover we will calculate $E(X_s|\underline{\underline{A}})$ for $s \in T$ and $\underline{\underline{A}}$ being a σ-field appearing in (4.1) and analyse some consequences.

6. REPRESENTATION THEOREMS

Let X be a process with properties PO and OMP and X'the associated process with X according to remark 3.1. The minimal splitting field $\underline{\underline{M}}(C)$ is equal to $\underline{\underline{F}}(D) \cap \underline{\underline{F}}(D_+)$. Moreover, if η is the random measure of X, then a system of generators of this σ-field is given by

$$(6.1) \quad \eta_u^j = \eta(D_u^j), \quad 0 \leqslant u \leqslant 1, \qquad j \in \{1,2\},$$

where (see figures 2 and 3)

$$D_u^j = \{(x_1,x_2) \in D \mid (-1)^{j-1} x_j \leqslant (-1)^{j-1} \varphi_j(u)\}.$$

Previous property is proved by [W1] (th. 3.1 and 3.2) for the Brownian sheet; to obtain the case of a process X' having PO,P1,P2, it is enough to replace the white noise W by η in the proof of [W1]; the general case follows from remark 4.1.

The same considerations hold by a little adaptation for a not necessarily bounded separation line.

If ν is the variance of $\eta, \nu = \nu_1 \otimes \nu_2$, we note $\psi_j(u) = \nu_j([0,\varphi_j(u)])$, $0 \leqslant u \leqslant 1$; we have $\psi_j(u) > 0$ if $u \in]0,1[$; ψ_1 is non-decrasing and ψ_2 non-increasing; in order to give a sense to integral formulas that will appear in a short time we extend ψ_j by noting $\psi_1(0-) = \psi_2(1+) = 0$.

Remark 6.1 (η_u^j) are processes with independent increments; by remark 3.2, we can choose a continuous version of these processes in order to have continuous L^2-martingales. However we will not need Ito's formalism because we will only integrate deterministic functions with respect to η^j. The variance of their random (Hilbert) measure is given by

$$\gamma(u) = E(\eta_u^j)^2 = \int_{[0,u]} \psi_{3-j} d\psi_j. \ \square$$

Lemma 6.2 $\underline{\underline{M}}(C) = \underline{\underline{G}}(C)$

Proof. According to remark 4.1 we can consider the associated process X' with X : for X' which satisfies properties PO,P1,P2, the same method as in [DR] concerning the Brownian sheet, can be used. $\square$

Remark 6.3 The result and proof above is valid even if X just has PO and Pl; in a following paper we will see that the result is valid when C is unbounded and it can even be extended when D is a more general domain and X a special generalized process. □

We can easily prove that

$$(6.2)\quad \int_{[0,1]} f_j d\eta^j = \int_D f_j(\varphi_j^{-1}(x_j))d\eta\ (x_1,x_2)$$

for $f_j \in L^2(\psi_{3-j}d\psi_j)$; (6.2) and the identity

$$(6.3)\quad E\left(\int_T g_1 d\eta \int_T g_2 d\eta\right) = \int_T g_1 g_2 d\nu,\ \forall g_1,g_2 \in L^2(T,\nu)$$

will help us to prove the following representation theorem for G(C) and H(C). The method used is the same as in [DR].

Theorem 6.4

(i) $G(C) = \left\{\int_{[0,1]} (f_1 d\eta^1 + f_2 d\eta^2) \mid f_j \in L^2(\psi_{3-j}d\psi_j), j = 1,2\right\}$

(ii) $H(C) = \left\{\int_{[0,1]} f(d\eta^1 - d\eta^2) \mid f \in L^2(\psi_1 d\psi_2) \cap L^2(\psi_2 d\psi_1)\right\}$.

Remarks 6.5

(i) We have uniqueness up to a constant. If (f_1',f_2') is another "representative" of $Z \in G(C)$ in $L^2(\psi_2 d\psi_1) \times L^2(\psi_1 d\psi_2)$, then there is a constant K such that

$$f_1 - f_1' = f_2' - f_2 = K\ d\psi_1 \text{ and } d\psi_2 \text{ a.e.}$$

(ii) If f_1,f_2 (resp. f.) have a bounded variation, $Z \in G(C)$ (resp. $Z \in H(C)$) can be written with the help of deterministic Lebesgue-Stieltjes integrals :

$$Z = \int_{[0,1]} \eta_u^1\ df_1(u) + \int_{[0,1]} \eta_u^2\ df_2(u)$$

$$(\text{resp. } Z = \int_0 X'_{\varphi(u)} df(u)).$$

(iii) In the case of the Brownian sheet, [R1] has given the following representation of G(C).

$$G(C) = \left\{g \in L^2(dx) \mid \frac{\partial^2}{\partial s_1 \partial s_2} g = 0 \text{ on } T \setminus C \text{ the weak sense}\right\}.$$

By using (6.2), we can see that the previous description of G(C) coincides with ours.

Remark 6.6 Given $Y \in H(T)$, we have explicitely calculated f_1,f_2

(resp. f.) so as $Z = E(Y|\underline{\underline{G}}(C))$ (resp. $E(Y|\underline{\underline{F}}(C))$ are described in theorem 6.4. We have to solve the following integral equations.

$$\lambda_j(u) = E(Y\eta_u^j) = E\left[\eta_u^j\left(\int_{[0,1]} f_1 d\eta^1 + \int_{[0,1]} f_2 d\eta^2\right)\right],$$

$$0 \leq u \leq 1, \quad j = 1,2$$

respectively

$$\lambda(u) = E(YX'_{\varphi(u)}) = E\left[X'_{\varphi(u)}\left(\int_{[0,1]} f(d\eta^1 - d\eta^2)\right)\right], \quad 0 \leq u \leq 1.$$

The method used is the same as in [DR].

The explicit formulas will be only given in specific cases.

7. CONSEQUENCES OF THE REPRESENTATION THEOREM

We will consider the same notations as in section 6. An easy consequence of theorem 6.4 and remark 6.5 i) follows (as in [DR]).

Proposition 7.1 The three following properties are equivalent :

(i) $\underline{\underline{F}}(C) = \underline{\underline{G}}(C)$

(ii) $\eta(D)$ is $\underline{\underline{F}}(C)$-measurable

(iii) $d\psi_1$ and $d\psi_2$ give mutually singular measures.

Remark 7.2 If C consists of horizontal and vertical segments, (iii) is trivially verified. □

Concerning the prediction problem, we have to evaluate the expressions of $E(Y|\underline{\underline{G}}(C))$ and $E(Y|\underline{\underline{F}}(C))$ for $Y = X_s, s \in T$. There are two situations :

a) $s \in \bar{D}$, b) $s \in \bar{D}_+$

a) $s \in \bar{D}$ is of the form

$s = (\varphi_1(a), \varphi_2(b))$, $0 \leq a \leq b \leq 1$ (see figure 4)

We obtain

$$E(Y|\underline{\underline{G}}(C)) = h(s)\nu(R_s)\left\{\begin{array}{l}\displaystyle\int_{[a,b]}\left(\frac{\eta^1}{\psi_1}\, d\left(\frac{1}{\psi_2}\right) + \frac{\eta^2}{\psi_2}\, d\left(\frac{1}{\psi_1}\right)\right) \\ \displaystyle + \frac{\eta_a^1}{(\psi_1\psi_2)(a)} - \frac{\eta_b^2}{(\psi_1\psi_2)(b)}\end{array}\right\}$$

$$E(Y|\underline{\underline{F}}(C)) = h(s)\,\nu(R_s) \left\{ \begin{array}{l} - \displaystyle\int_{[a,b]} \frac{d\left(\frac{1}{\psi_1}\right)}{d\left(\frac{\psi_2}{\psi_1}\right)} \left(d\left(\frac{\eta^1}{\psi_1}\right) - d\left(\frac{\eta^2}{\psi_1}\right)\right) \\ \\ + \displaystyle\frac{X_{\varphi(b)}}{\psi_1(b)\psi_2(b)} \end{array} \right\}$$

If $\psi_1(a)\psi_2(b) = 0$ we make the convention that the expression in brackets is zero.

b) $s \in \bar{D}_+$. Without loss of generality we may suppose

$$s = (\varphi_1(a)\varphi_2(b)),\ 0 \leq b \leq a \leq 1 \qquad \text{(see figure 5).}$$

With a similar convention as before we obtain

$$E(Y|\underline{\underline{G}}(C)) = h(s)(\eta_a^1 - \eta_b^2)$$

$$E(Y|\underline{\underline{F}}(C)) = X_{\varphi(a)} - h(s) \int_{[b,a]} \frac{d\psi_2}{d\left(\frac{\psi_2}{\psi_1}\right)} \left(d\left(\frac{\eta^1}{\psi_1}\right) - d\left(\frac{\eta^2}{\psi_1}\right)\right).$$

The structure of these formulas shows us that

$$s = (s_1,s_2) \longrightarrow \left\{ \begin{array}{l} \alpha_1(s_1,s_2) = E(X_s|\underline{\underline{G}}(C)) \\ \alpha_2(s_1,s_2) = E(X_s|\underline{\underline{F}}(C)) \end{array} \right.$$

are both solutions of the stochastic partial differential equation

$$\frac{\partial^2}{\partial s_1 \partial s_2} \left(\frac{\alpha(s)}{h(s)}\right) = 0 \quad \text{on } D_+$$

in the weak sense with boundary conditions $\alpha(s) = X_s$ on C and solutions of the SPDE,

$$\frac{\partial^2}{\partial s_1 \partial s_2} \left(\frac{\alpha(s)}{h(s)\nu(R_s)}\right) = 0 \quad \text{on } D,$$

with the same boundary conditions. In fact, for instance on D_+ $\alpha(s)$ is the sum of two terms the first depending on a and the second on b.

Remark 7.3 If we use the notations of [R2], and X is the Brownian sheet, α_1 and α_2 can be seen as unique solutions of the same SPDE with different boundary conditions.

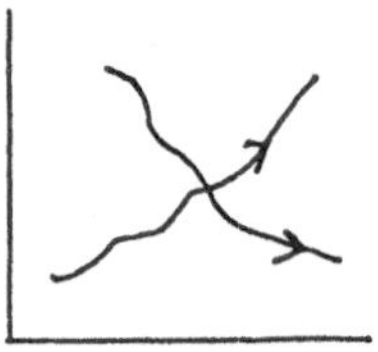

Figure 1

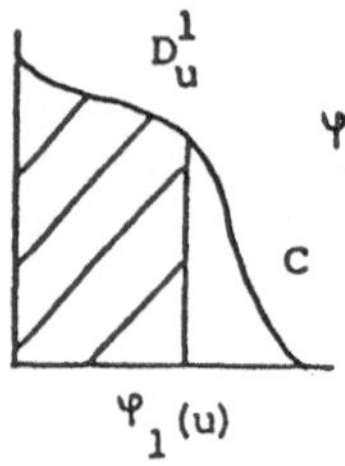

Figure 2

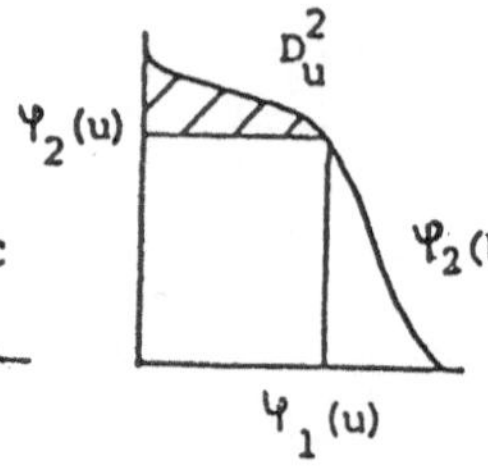

Figure 3

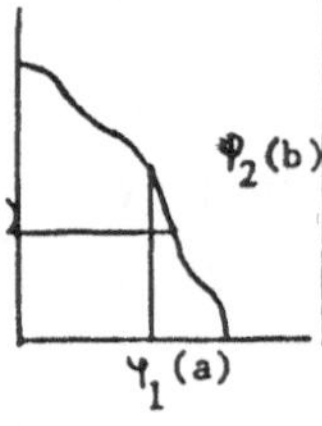

Figure 4

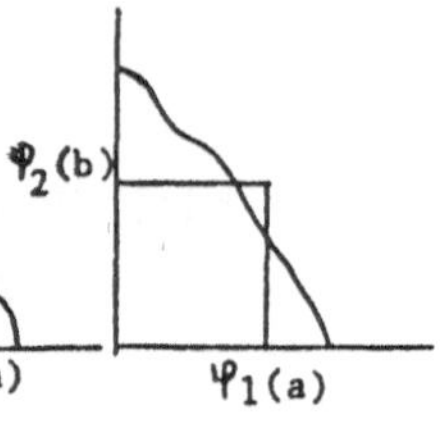

Figure 5

REFERENCES

[C] E. Carnal. Markov Properties for Certain Random Fields (Rapport interne, Ecole Polytechnique Fédérale de Lausanne, 1981)

[Ca] L. Carraro. Problèmes de prédiction pour le processus de Wiener à deux paramètres. Prob. Th. and Rel. Fields, 1986

[DR] R.C. Dalang. F. Russo. A Prediction Problem for the Brownian Sheet (Preprint: J. of Multivariate analysis)

[M] V. Mandrekar. Germ-Field Markov Property for Multiparameter Processes, Sém. de Prob. X, Lect. N in Math. 511, Springer Verlag (1976), p. 78-85

[Mc] H.P. Mc Kean. Brownian Motion with a several-dimensional time, Th. Prob. Appl. 8 (1963), p. 335-354

[N] D. Nualart. Propriedad de Markov para funciones aleatorias Gaussianas, Cuadernos de Estadistica Matematica de la Universidad de Granada, Serie A, Probabilidad, no. 5 (1980), p. 30-43

[P] L.D. Pitt. A Markov Property for Gaussian Processes with Multi-dimensional parameter, Arch. Rat. Mech. Anal. 43 (1971), p. 367-391

[R1] Ju. Rozanov. Markov Random fields, Springer Verlag (1982)

[R2] Ju. Rozanov. Boundary Problems for Stochastic Partial Differential Equations (Preprint, Bibos, n.108)

[Ru] F. Russo. Etude de la propriété de Markov étroite en relation avec les processus planaires à accroissements indépendants, Sém. de Prob. XVIII, Lect. N. in Math. 1059, Springer Verlag (1984), p. 353-387.

[W1] J.B. Walsh. Cours de troisième cycle, Univ. de Paris 6, 1976-77

[W2] J.B. Walsh. Cours de Saint-Flour XIV 1984. Lect. N. in Math. 1180, Springer-Verlag (1986)

ON THE DISTRIBUTION OF FUNCTIONALS OF STOCHASTIC FIELDS

J. vom Scheidt and B. Fellenberg
Ingenieurhochschule Zwickau, PF 35
Zwickau
9541 DDR

A technological problem: temperature propagation

An important problem as to the construction of brakes of vehicles is the determination of the temperature propagation and the stresses in the brakes. An easy but adequate mathematical boundary-initial value problem for the domain G (see Figure 1) can be written in the following form: $u=u(t,x,y,\omega)$ denotes the temperature

$$\frac{\partial u}{\partial t} = a\Delta u \ , \qquad \Delta u = u_{xx} + u_{yy}$$

initial condition: $\quad u(0,x,y) = u_o(x,y)$

boundary conditions: $\quad -\lambda\frac{\partial u}{\partial y}(t,x,y)\Big|_{(\partial G)_1} = P(t,x,\omega) \qquad (1)$

$$\lambda\frac{\partial u}{\partial n}(t,x,y)+\alpha_k(u(t,x,y)-u_a(t,x,y))\Big|_{(\partial G)_k} = 0$$

for k=2,3,4.

u_o, u_a are given temperatures. It is clear from the technological task that P must be a random function. The randomness of P follows from the random surface $(\partial G)_1$ in a brake and also from the random brake power. The temperature conductivity and the thermal conductivity are denoted by a and λ, respectively. The heat transfer coefficient α_k is a further constant.

Figure 1:

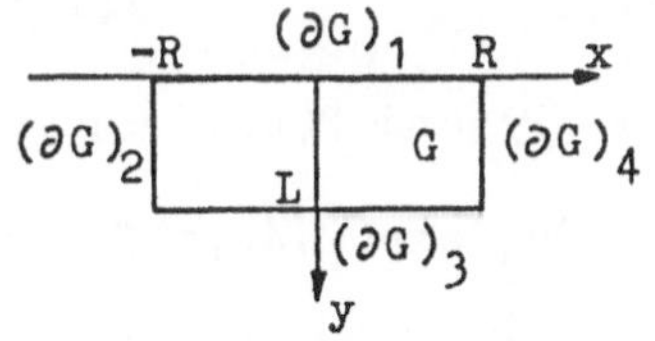

It is possible to find in the technological literature many results of statistical characteristics of the measured temperature u. These temperatures can be measured by direct methods and also by electroanalogy. But the expenditure of time for such experiments is very large and such experiments are very complicated since the temperature is interesting for small values of y (close to the surface $(\partial G)_1$).

In Figure 2 we have plotted some estimates of variances from measured values of u. $\bar{u}$ is defined by $\bar{u}=u-w$ and w denotes the solution of the averaged problem belonging to (1). This averaged problem is obtained if the random function P is replaced by the expectation $\langle P \rangle$ of P. Figure 2 contains already the calculated results from the weakly correlated theory which will be still represented in this paper.

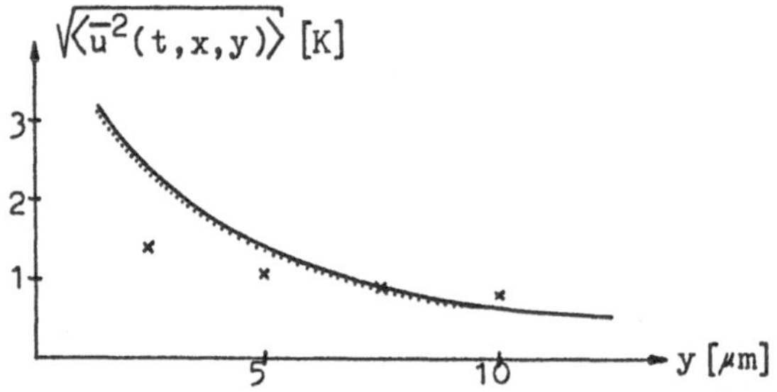

Figure 2: Variances for t=34.1 μs and x=0
× estimates from measured values
calculated values from the weakly correlated theory

It is also possible to consider other technological applications investigated from the same mathematical point of view. Examples would be: vibration problems in connection with automotive engineering (linear and nonlinear vibration problems of n-mass systems, vibration problems of technological construction parts), stochastic problems with respect to fuel-injection of engines, technological eigenvalue problems (e.g. buckling problems, bending vibrations), wave propagation and others.

If we deal with the above temperature propagation then a solution $\bar{u}$ of (1) can be found in the form (cf. also /2/)

$$U(t,\omega) = (\bar{u}(t,\cdot,\cdot),\Psi(\cdot,\cdot)) = \int_0^t \int_{-R}^{R} F(t-\tau,x)\bar{P}(\tau,x,\omega)\, dx\, d\tau \qquad (2)$$

where $\bar{P}$ is defined by $\bar{P}=P-\langle P \rangle$ and F(t,x) is given by

$$F(t,x) = -a \sum_{i,j=1}^{\infty} C(d_j)\exp(-a\Lambda_{ij}t)\varphi_{1i}(x)(f_{ij},\Psi) \quad .$$

φ_{1i}, φ_{2j}, Λ_{ij}, $C(d_j)$ and $f_{ij}(x,y)=\varphi_{1i}(x)\cdot\varphi_{2j}(y)$ follow from the eigenvalue problem:

$$-\Delta f = \Lambda f \ , \quad \frac{\partial f}{\partial y}(x,y)\Big|_{(\partial G)_1} = 0$$

$$\lambda\frac{\partial f}{\partial n}(x,y)+\alpha_k f(x,y)\Big|_{(\partial G)_k} = 0 \ , \quad k=2,3,4.$$

The random function $\bar{P}(t,x,\omega)$ has the property of a weakly correlated function with the correlation length ε. The property of the weak correlation of $f(x,\omega)$, $x\in R^m$, is defined by the fact that $f(x_1,\omega)$ and $f(x_2,\omega)$ are independent if the distance of x_1 and x_2 is greater than ε. The exact definition is given by a seperation of all moments

$$\langle \prod_{i=1}^{n} f(x_i)\rangle \quad \text{for } n=2,3,\ldots,\ x_i\in R^m \quad \text{(cf. /1/ and /4/).}$$

The mathematical theory of weakly correlated functions

We consider now general linear functionals

$$r_{i\varepsilon}(\omega) = \int_{D_i} F_i(x)\, f_\varepsilon(x,\omega)\, dx, \quad i=1,2,\ldots,n$$

where $f_\varepsilon(x,\omega)$ is a weakly correlated random function with $x\in D\supset \bigcup_{i=1}^{n} D_i \subset R^m$. The sample functions of $f_\varepsilon(x,\omega)$ are supposed to be continuous and $\langle |f_\varepsilon(x)|^p\rangle \leq c_p<\infty$, $p=2,3,\ldots$ D_i, $i=1,\ldots,n$ are assumed to be bounded domains from R^m with piecewise smooth boundaries. Let $F_i(x)$ be functions from $C^1(D_i)$.

First we compute the expansion of a k-th moment

$$\langle \prod_{p=1}^{k} r_{i_p\varepsilon}\rangle = \int_{D_{i_1}}\ldots\int_{D_{i_k}} \prod_{p=1}^{k} F_{i_p}(x_p)\langle \prod_{p=1}^{k} f_\varepsilon(x_p)\rangle \; dx_1\ldots dx_k$$

relative to the correlation length ε. For this we deduce expansions of the form

$$\int_{\mathcal{E}(I)} \prod_{i\in I} F_i(x_i)\langle \prod_{i\in I} f_\varepsilon(x_i)\rangle \; dx_1\ldots dx_p$$
$$= {}^pA_{p-1}(F_1,\ldots,F_p)\varepsilon^{m(p-1)} + {}^pA_p(F_1,\ldots,F_p)\varepsilon^{m(p-1)+1} + o(\varepsilon^{m(p-1)+1})$$

and

$$\int_{\mathcal{E}(I_1\cup I_2)} \prod_{i\in I_1\cup I_2} F_i(x_i)\langle \prod_{i\in I_1} f_\varepsilon(x_i)\rangle\langle \prod_{i\in I_2} f_\varepsilon(x_i)\rangle dx_1\ldots dx_{p_1+p_2}$$
$$= A_{p_1p_2}(F_1,\ldots,F_{p_1};F_{p_1+1},\ldots,F_{p_1+p_2})\varepsilon^{m(p_1+p_2-1)} + o(\varepsilon^{m(p_1+p_2-1)})$$

where

$$\mathcal{E}(I) = \{(x_1,\ldots,x_p)\in \mathop{X}_{q=1}^{p} D_q:\ (x_1,\ldots,x_p)\ \varepsilon\text{-neighbouring}\},$$
$$I = \{1,\ldots,p\},\ I_1 = \{1,\ldots,p_1\},\ I_2 = \{p_1+1,\ldots,p_1+p_2\}.$$

After some considerations we obtain

$$\langle \prod_{p=1}^{k} r_{i_p\varepsilon}\rangle = \begin{cases} H_1(m;i_1,\ldots,i_k)\varepsilon^{mk/2} + H_2(m;i_1,\ldots,i_k)\varepsilon^{mk/2+1} \\ \qquad + o(\varepsilon^{mk/2+1}) & \text{for } k \text{ even} \\ \bar{H}_1(m;i_1,\ldots,i_k)\varepsilon^{m(k+1)/2} + \bar{H}_2(m;i_1,\ldots,i_k)\varepsilon^{m(k+1)/2+1} \\ \qquad + o(\varepsilon^{m(k+1)/2+1}) & \text{for } k \text{ odd} \end{cases}$$

and H_p, $\overline{H}_p$ can be written as functions of the statistical characteristics $^2A_1$, $^2A_2$, $^3A_2$, $^3A_3$, $^4\underline{A}_3$ with

$$^4\underline{A}_3(F_1,F_2,F_3,F_4) \doteq {}^4A_3(F_1,F_2,F_3,F_4) - A_{22}(F_1,F_2;F_3,F_4)$$
$$- A_{22}(F_1,F_3;F_2,F_4) - A_{22}(F_1,F_4;F_2,F_3)$$

given below. For example, assuming k even we obtain

$$\langle r_{1\varepsilon}^k\rangle = \begin{cases} (^2A_1\varepsilon)^{k/2}\dfrac{k!}{2^{k/2}(k/2)!}\Big\{1 + \dfrac{k}{2}\Big[\dfrac{^2A_2}{^2A_1} + \dfrac{1}{6}(\dfrac{k}{2}-1)\dfrac{^4\underline{A}_3}{(^2A_1)^2} \\ \qquad + \dfrac{1}{9}(\dfrac{k}{2}-1)(\dfrac{k}{2}-2)\dfrac{(^3A_2)^2}{(^2A_1)^3}\Big]\varepsilon + o(\varepsilon)\Big\} & \text{for } m=1 \\ (^2A_1\varepsilon^m)^{k/2}\dfrac{k!}{2^{k/2}(k/2)!}\Big\{1 + \dfrac{k}{2}\dfrac{^2A_2}{^2A_1}\varepsilon + o(\varepsilon)\Big\} & \text{for } m\geq 2 \end{cases} \quad (3)$$

and an adequate formula for k odd. We have put $^pA_q(F_1,\ldots,\overset{p}{F_1}) \doteq {}^pA_q$. For the determination of the density function of $(r_{1\varepsilon},\ldots,r_{n\varepsilon})$ the transformation

$$\tilde{r}_{i\varepsilon}(\omega) = \frac{1}{\sqrt{\varepsilon^m}}\sum_{p=1}^{i} b_{ip}r_{p\varepsilon}\ , \quad i=1,2,\ldots,n \qquad (4)$$

is carried out and the coefficients b_{ip} are calculated by the equations

$$\sum_{p=1}^{i}\sum_{q=1}^{j} b_{ip}b_{jq}B_{pq} = \delta_{ij}, \qquad i,j=1,2,\ldots,n$$

where

$$B_{pq} \doteq H_1(m;p,q) = {}^2A_1(F_p,F_q)\ .$$

Now the transformed random vector $(\tilde{r}_{1\varepsilon},\ldots,\tilde{r}_{n\varepsilon})$ is considered. For the second moments we obtain

$$\langle\tilde{r}_{i\varepsilon}\tilde{r}_{j\varepsilon}\rangle = \delta_{ij} + o(\varepsilon)\ ,$$

i.e. these second moments are orthonormal in the lowest order. Let $\tilde{p}(\tilde{u}_1,\ldots,\tilde{u}_n)$ denote the density of the random vector $(\tilde{r}_{1\varepsilon},\ldots,\tilde{r}_{n\varepsilon})$. Then we set

$$\tilde{p}(\tilde{u}_1,\ldots,\tilde{u}_n) = \frac{1}{\sqrt{2\pi}^n}\exp(-\frac{1}{2}\sum_{p=1}^{n}\tilde{u}_p^2)\Big\{1 + \sum_{\substack{k_1,\ldots,k_n=0 \\ k\doteq\sum_{s=1}^{n}k_s \geq 2}}^{\infty} (-1)^k\frac{c_{k_1\ldots k_n}}{k_1!\ldots k_n!}H_{k_1}(\tilde{u}_1)\ldots H_{k_n}(\tilde{u}_n)$$

where $H_p(u)$, p=1,2,... denote the Chebyshev-Hermite polynomials defined by

$$H_p(u) \doteq (-1)^p\exp(u^2/2)\frac{d^p}{du^p}\exp(-u^2/2)\ .$$

Furthermore, we can find by comparing two approaches to the characteristic function $\tilde{\varphi}(t_1,\dots,t_n)$ of $(\tilde{r}_{1\varepsilon},\dots,\tilde{r}_{n\varepsilon})$ that $\tilde{\varphi}$ has the form

$$\tilde{\varphi}(t_1,\dots,t_n) = \exp\Big(-\frac{1}{2}\sum_{s=1}^{n} t_s^2\Big)\Big\{1 + \sum_{\substack{k_1,\dots,k_n=0\\ k\geq 2}}^{\infty} (-1)^k c_{k_1\dots k_n} \prod_{j=1}^{n} \frac{(it_j)^{k_j}}{k_j!}\Big\}$$

$$= 1 + \sum_{\substack{k_1,\dots,k_n=0\\ k\geq 2}}^{\infty} \frac{\tilde{\alpha}_{k_1\dots k_n}}{k_1!\dots k_n!} \prod_{j=1}^{n} \frac{(it_j)^{k_j}}{k_j!}$$

where $$\tilde{\alpha}_{k_1\dots k_n} = \Big\langle \prod_{q=1}^{n} \tilde{r}_{q\varepsilon}^{k_q} \Big\rangle \quad .$$

Using the expansion of the moments we can determine now the coefficients $c_{k_1\dots k_n}$ and then the density function $\tilde{p}(\tilde{u}_1,\dots,\tilde{u}_n)$. We have to neglect many considerations and we restrict us to give the results:

For m=1

$$\tilde{p}(\tilde{u}_1,\dots,\tilde{u}_n) = \frac{1}{\sqrt{2\pi}^n} \exp\Big(-\frac{1}{2}\sum_{q=1}^{n} \tilde{u}_q^2\Big)\Big\{1 + \sum_{\substack{k_1,\dots,k_n=0\\ k=3}} {}^3\tilde{A}_2(\{\overbrace{1,\dots,1}^{k_1},\dots,\overbrace{n,\dots,n}^{k_n}\})\, H_{\underline{k}}(\underline{\tilde{u}})\sqrt{\varepsilon}$$

$$+\Big[\sum_{\substack{k_1,\dots,k_n=0\\ k=2}} {}^2\tilde{A}_2(\{\overbrace{1,\dots,1}^{k_1},\dots,\overbrace{n,\dots,n}^{k_n}\})\, H_{\underline{k}}(\underline{\tilde{u}})$$

$$+ \sum_{\substack{k_1,\dots,k_n=0\\ k=4}} {}^4\underline{\tilde{A}}_3(\{\overbrace{1,\dots,1}^{k_1},\dots,\overbrace{n,\dots,n}^{k_n}\})\, H_{\underline{k}}(\underline{\tilde{u}})$$

$$+ \frac{1}{2}\sum_{\substack{k_1,\dots,k_n=0\\ k=6}} \sum_{\substack{p_1,\dots,p_n=0\\ \sum_{s=1}^{n} p_s=3,\ 0\leq k_s-p_s\leq 3}} {}^3\tilde{A}_2(\{\overbrace{1,\dots,1}^{k_1-p_1},\dots,\overbrace{n,\dots,n}^{k_n-p_n}\})\cdot \tag{5}$$

$$\cdot {}^3\tilde{A}_2(\{\overbrace{1,\dots,1}^{p_1},\dots,\overbrace{n,\dots,n}^{p_n}\})\, H_{\underline{k}}(\underline{\tilde{u}})\Big]\varepsilon + o(\varepsilon)\Big\} \quad ,$$

and for m≥2

$$\tilde{p}(\tilde{u}_1,\dots,\tilde{u}_n) = \frac{1}{\sqrt{2\pi}^n} \exp\Big(-\frac{1}{2}\sum_{q=1}^{n} \tilde{u}_q^2\Big)\Big\{1 +$$

$$+\Big[\sum_{\substack{k_1,\dots,k_n=0\\ k=2}} {}^2\tilde{A}_2(\{\overbrace{1,\dots,1}^{k_1},\dots,\overbrace{n,\dots,n}^{k_n}\})\, H_{\underline{k}}(\underline{\tilde{u}}) \quad +$$

$$+ \sum_{\substack{k_1,\dots,k_n=0 \\ k=3}} {}^3\tilde{A}_2(\{1,\dots,\overset{k_1}{\check{1}},\dots,n,\dots,\overset{k_n}{\check{n}}\})\, H_{\underline{k}}(\underline{\tilde{u}})\, \delta_{2m}\Big]\varepsilon + o(\varepsilon)\Big\}$$

where ${}^p\tilde{A}_q$ is defined by

$${}^p\tilde{A}_q(\{1,\dots,\overset{k_1}{\check{1}},\dots,n,\dots,\overset{k_n}{\check{n}}\}) = {}^pA_q(\tilde{F}_1,\dots,\overset{k_1}{\check{\tilde{F}}}_1,\dots,\tilde{F}_n,\dots,\overset{k_n}{\check{\tilde{F}}}_n) \quad ,$$

$$\tilde{F}_i(x) \doteq \sum_{p=1}^{i} b_{ip}F_p(x)$$

and the abbreviation

$$H_{\underline{k}}(\underline{\tilde{u}}) \doteq \prod_{q=1}^{n} \frac{H_{k_q}(\tilde{u}_q)}{k_q!}$$

is used.

In particular, for n=1 we obtain

$$\tilde{p}(\tilde{u}) = \begin{cases} \frac{1}{\sqrt{2\pi}}\exp(-\frac{1}{2}\tilde{u}^2)\Big\{1 + \frac{1}{6}\,{}^3\tilde{A}_2H_3(\tilde{u})\sqrt{\varepsilon} \\ \qquad + \frac{1}{2}\Big[{}^2\tilde{A}_2H_2(\tilde{u}) + \frac{1}{24}\,{}^4\underline{\tilde{A}}_3H_4(\tilde{u}) + \frac{1}{72}({}^3\tilde{A}_2)^2H_6(\tilde{u})\Big]\varepsilon + o(\varepsilon)\Big\} & \text{for } m=1 \\ \frac{1}{\sqrt{2\pi}}\exp(-\frac{1}{2}\tilde{u}^2)\Big\{1 + \frac{1}{2}\,{}^2\tilde{A}_2H_2(\tilde{u})\varepsilon + \frac{1}{6}\,{}^3\tilde{A}_2H_3(\tilde{u})\varepsilon^{m/2}\Big\} + o(\varepsilon) & \text{for } m\geq 2 \end{cases}$$

with ${}^p\tilde{A}_q \doteq {}^pA_q(\tilde{F}_1,\dots,\overset{p}{\check{\tilde{F}}}_1)$.

Now we deal with some special cases. Let $f_\varepsilon(x,\omega)$ be symmetrically distributed. We obtain

$${}^3\tilde{A}_2(\{1,2,3\}) = 0$$

and the terms multiplied by $\sqrt{\varepsilon}$ are zero. Hence, the first approximation (normal distribution) is for assumed symmetric distribution of $f_\varepsilon(x,\omega)$ a better approximation than for non-symmetric distribution of $f_\varepsilon(x,\omega)$. An example of a symmetrically distributed vector field (weakly correlated) is a Gaussian vector field $f_\varepsilon(x,\omega)$. Furthermore, we have for a Gaussian vector field

$${}^4\underline{\tilde{A}}_3(\{1,2,3,4\}) = 0 \ .$$

In this special case (5) leads to

$$\tilde{p}(\tilde{u}) = \frac{1}{\sqrt{2\pi}}\exp(-\frac{1}{2}\tilde{u}^2)\Big\{1 + \frac{1}{2}\,{}^2\tilde{A}_2(\{1,1\})H_2(\tilde{u})\varepsilon\Big\} + o(\varepsilon) \qquad \text{for } n=1$$

$$\tilde{p}(\tilde{u}_1,\tilde{u}_2) = \frac{1}{2\pi}\exp(-\frac{1}{2}(\tilde{u}_1^2+\tilde{u}_2^2))\Big\{1 + \frac{1}{2}\Big[{}^2\tilde{A}_2(\{1,1\})H_2(\tilde{u}_1) + 2\,{}^2\tilde{A}_2(\{1,2\})H_1(\tilde{u}_1)H_1(\tilde{u}_2) + {}^2\tilde{A}_2(\{2,2\})H_2(\tilde{u}_2)\Big]\varepsilon + o(\varepsilon)\Big\} \qquad \text{for } n=2.$$

Now we determine the density function $p(u_1,\dots,u_n)$ of $(r_{1\varepsilon},\dots,r_{n\varepsilon})$. The transformation relation between $(r_{1\varepsilon},\dots,r_{n\varepsilon})$ and $(\tilde r_{1\varepsilon},\dots,\tilde r_{n\varepsilon})$ is given by (4) and we have

$$p(u) = \tilde p\left(\frac{1}{\sqrt{\varepsilon^m}}\,Tu\right)\frac{|\det T|}{\sqrt{\varepsilon^{nm}}}$$

where $u=(u_1,\dots,u_n)^T$, $T=(b_{ip})_{1\le i,p\le n}$ and $b_{ip}\doteq 0$ for $i<p$ is put. We consider $\tilde u^T\tilde u$ and obtain

$$\tilde u^T\tilde u = \frac{1}{\varepsilon^m}\,u^T\underline{B}^{-1}u \qquad \text{with } \underline{B}^{-1} = T^T T.$$

Hence, the 1st approximation $p_o(u)$ of the density function of $(r_{1\varepsilon},\dots,r_{n\varepsilon})$ is given by

$$p_o(u) = \frac{1}{\sqrt{2\pi}^n}\,\frac{1}{\sqrt{\varepsilon^{nm}\det\underline{B}}}\,\exp\left(-\frac{1}{2}\,\frac{1}{\varepsilon^m}\,u^T\underline{B}^{-1}u\right) \quad .$$

This formula describes the density of a Gaussian vector $(\xi_1,\dots,\xi_n)$ with mean zero and correlation relations

$$\langle \xi_p\xi_q\rangle = \varepsilon^m B_{pq} = \varepsilon^m\, {}^2A_1(F_p,F_q) \quad .$$

Hence, the random vector $(r_{1\varepsilon},\dots,r_{n\varepsilon})$ is a normally distributed vector as to the 1st approximation with $\langle r_{p\varepsilon}r_{q\varepsilon}\rangle = \langle \xi_p\xi_q\rangle$. This represents the same result deduced in /1/. In a straight forward way it is possible to give $p(u_1,\dots,u_n)$ as to the higher orders.

Finally, we give some results with respect to the determination of the statistical characteristics ${}^pA_q$. Firstly, we have

$${}^2A_1(F_1,F_2) = \int_{D_1\cap D_2} F_1(x)F_2(x)\overset{2}{a}(x)\,dx$$

for functions F_i from $L_2(D_i)$ where $\overset{2}{a}(x)$ denotes the intensity of the weakly correlated field $f_\varepsilon(x,\omega)$:

$$\overset{2}{a}(x) \doteq \lim_{\varepsilon\downarrow 0}\frac{1}{\varepsilon^m}\int_{K_\varepsilon(x)}\langle f_\varepsilon(x)f_\varepsilon(y)\rangle\,dy \ , \quad K_\varepsilon(x)=\{y\in R^m: |x-y|\le\varepsilon\}.$$

Furthermore, we can write

$${}^3A_2(F_1,F_2,F_3) = \int_{D_1\cap D_2\cap D_3}\prod_{p=1}^{3}F_p(x)\overset{3}{a}(x)\,dx$$

with
$$\overset{3}{a}(x) = \lim_{\varepsilon\downarrow 0}\frac{1}{\varepsilon^{2m}}\int_{M_3(x)}\langle f_\varepsilon(x)f_\varepsilon(x_1)f_\varepsilon(x_2)\rangle\,dx_1dx_2 \quad ,$$

$${}^4\underline{A}_3(F_1,F_2,F_3,F_4) = \int_{D_1\cap D_2\cap D_3\cap D_4}\prod_{p=1}^{4}F_p(x)\overset{4}{a}(x)\,dx$$

with
$$\overset{4}{a}(x) = \lim_{\varepsilon\downarrow 0}\frac{1}{\varepsilon^{3m}}\int_{M_4(x)}\Big[\langle f_\varepsilon(x)f_\varepsilon(x_1)f_\varepsilon(x_2)f_\varepsilon(x_3)\rangle \ -$$

$$- \langle f_\varepsilon(x)f_\varepsilon(x_1)\rangle\langle f_\varepsilon(x_2)f_\varepsilon(x_3)\rangle - \langle f_\varepsilon(x)f_\varepsilon(x_2)\rangle\langle f_\varepsilon(x_1)f_\varepsilon(x_3)\rangle$$
$$- \langle f_\varepsilon(x)f_\varepsilon(x_3)\rangle\langle f_\varepsilon(x_1)f_\varepsilon(x_2)\rangle \Big] dx_1dx_2dx_3$$

and $M_p(x)$, $p=3,4$, can be characterized as sets for which the integrand is not zero. And we have

$$^2A_2(F_1,F_2) = \int_{D_1\cap D_2} F_1F_2\overset{2}{b}(x)\, dx + \int_{(\partial D_2)_1^i} F_1F_2(x) \int_{-1}^{1} \overset{2}{a}(u;x)\, du dS_x$$
$$+ \int_{(\partial D_2)_1^b} F_1F_2(x) \int_{I_{12}(x)} \overset{2}{a}(u;x)\, du dS_x - \int_{(\partial D_2)_1^o} F_1F_2\overset{2}{a}(x)\, dS_x$$

where we have defined:

$$(\partial D_2)_1^o \doteq \{x\in\partial D_2 : x \text{ accumulation point of } D_1\cap D_2\} \quad ,$$
$$(\partial D_2)_1^i \doteq \{x\in\partial D_2 : x\in D_1\} \quad , \quad (\partial D_2)_1^b \doteq \{x\in\partial D_2 : x\in\partial D_1\}$$

$$I_{12}(x) = \begin{cases} (0,1) & \text{if } n(x,D_2) = n(x,D_1) \\ (-1,0) & \text{if } n(x,D_2) = -n(x,D_1) \\ \emptyset & \text{otherwise} \end{cases} \quad \text{for } x\in(\partial D_2)_1^b$$

$$\int_{K_\varepsilon(x)} \langle f_\varepsilon(x)f_\varepsilon(y)\rangle\, dy = \overset{2}{a}(x)\varepsilon^m + \overset{2}{b}(x)\varepsilon^{m+1} + o(\varepsilon^{m+1}),$$

$$\lim_{\varepsilon\downarrow 0}\Big[\frac{1}{\varepsilon^m}\int_{D_i\cap K_\varepsilon(z)} \langle f_\varepsilon(z)f_\varepsilon(y)\rangle\, dy\Big]_{z=x+\varepsilon un} = \overset{i}{a}(u;x) \quad \text{for } x\in\partial D_i \ .$$

The normal vector n as to $x\in\partial D_i$ shows in the interior of D_i and is denoted by $n(x,D_i)$.
Now we have summarized all facts in order to apply this theory to the different technological applications.

Application to temperature propagation

We assume that $\overline{P}(t,x,\omega)$ is weakly correlated as to (t,x). It is also possible to suppose that $\overline{P}(t,x,\omega)$ is partially weakly correlated as to t or x. This assumption is desided by the technological situation. In order to compare with measured values we choose $\alpha_k \to \infty$ and $u_a(t,x,y)=u_a=\text{const}$ for k=2,3,4 in (1). Then the boundary conditions

$$-\lambda\frac{\partial u}{\partial y}(t,x,y)\Big|_{(\partial G)_1} = P(t,x,\omega), \quad u(t,x,y)\Big|_{\partial G\setminus(\partial G)_1} = u_a$$

are obtained and these boundary conditions lead to the compatibility conditions

$$\overline{P}(0,x,\omega) = 0, \quad \overline{P}(t,\pm R,\omega) = 0$$

if we put $u_o(x,y)=u_o=\text{const}$ as the initial temperature. The cause for such compatibility conditions consists in the non-random character of the boundary conditions on ∂G. $a(t,x)$ denotes the intensity of $\overline{P}(t,x)$

$$a(t,x) = \lim_{\varepsilon \downarrow 0} \frac{1}{\varepsilon^2} \int_{K_\varepsilon(t,x)} \langle \overline{P}(t,x)\overline{P}(s,y)\rangle \, ds\, dy$$

which can be calculated using some given physical characteristics. Then it can be deduced for the solution $U(t,\omega)$ that in case of $d_k \to \infty$ it yields (cf. (2))

$$C(d_j) = \sqrt{\frac{2}{L}} \;, \quad \Lambda_{ij} = \left(\frac{i\pi}{2R}\right)^2 + \left(\frac{(2j-1)\pi}{2L}\right)^2 \qquad i,j=1,2,\ldots$$

$$\varphi_{1i}(x) = \sqrt{\frac{1}{R}} \begin{cases} \sin\left(\frac{i\pi x}{2R}\right) & \text{for } i \text{ even} \\ \cos\left(\frac{i\pi x}{2R}\right) & \text{for } i \text{ odd} \end{cases}$$

and

$$(f_{ij},\psi) = \int_G \varphi_{1i}(x)\varphi_{2j}(y)\psi(x,y)\, dxdy$$

where

$$\varphi_{2j}(y) = \sqrt{\frac{2}{L}} \cos\left(\frac{(2j-1)\pi y}{2L}\right) \quad .$$

We want to give some information in order to compute the variance of $\overline{u}(t,x,y,\omega)$. Then it follows $D_1=D_2=D=(0,t)\mathrm{x}(-R,R)$ and

$$(\partial D_2)_1^o = \partial D, \quad (\partial D_2)_1^i = \emptyset, \quad (\partial D_2)_1^b = \partial D, \quad I_{12}(x) = (0,1) \text{ for } x\in\partial D.$$

We take $\overline{P}(t,x,\omega)=\psi_1(t)\psi_2(x)\tilde{P}(t,x,\omega)$ with

$$\psi_1(0)=0, \quad 0\le\psi_1(t)\le 1 \text{ for } 0\le t\le\eta, \quad \psi_1(t)=1 \text{ for } t\ge\eta,$$
$$\psi_2(\pm R)=0, \quad 0\le\psi_2(x)\le 1 \text{ for } -R\le x\le -R+\eta,\ R-\eta\le x\le R,$$
$$\psi_2(x)=1 \text{ for } -R+\eta\le x\le R-\eta,$$

where η is small but arbitrary. Then the compatibility conditions are fulfilled. Furthermore, we consider the correlation function

$$\langle\tilde{P}(t,x)\tilde{P}(s,y)\rangle = \sigma^2 \begin{cases} \left(1-\frac{|t-s|}{\tilde{\varepsilon}}\right)\left(1-\frac{|x-y|}{\tilde{\varepsilon}}\right) & \text{for } \max(|t-s|,|x-y|) \le \tilde{\varepsilon} \\ 0 & \text{otherwise} \end{cases}$$

and we have the real correlation length $\varepsilon=\sqrt{2}\,\tilde{\varepsilon}$ according to the Euclidean metric. Using this special correlation function the further computations are easy. We obtain

$$a(t,x) = \tfrac{1}{2}\sigma^2 \text{ for } t>\eta \;, \; -R+\eta<x<R-\eta \;,$$

$$\lim_{x\to\pm R} a(t,x) = \lim_{t\downarrow 0} a(t,x) = 0, \quad b(t,x) = 0,$$

$$a(u;t,x) = 0 \text{ for } t=0 \text{ and for } x=\pm R,$$

$$a(u;t,x) = \tfrac{1}{4}\sigma^2(1+2\sqrt{2}u-2u^2) \text{ and } \int_0^1 a(u;t,x)du = \tfrac{1}{4}\sigma^2\left(\tfrac{1}{3}+\sqrt{2}\right)$$
$$\text{for } t>\eta \;, \; -R+\eta<x<R-\eta$$

and hence

$$^2A_2(F,F) \approx \frac{1}{4}\sigma^2(\sqrt{2} - \frac{5}{3}) \int_{-R}^{R} F^2(t,x)\,dx .$$

The variance follows from (3)

$$\langle U^2(t)\rangle = {}^2A_1\varepsilon^2 + {}^2A_2\varepsilon^3 + o(\varepsilon^3)$$

where

$$^2A_1 \approx \frac{1}{2}\sigma^2 \int_0^t \int_{-R}^{R} F^2(t-s,x)\,dxds$$

$$= \frac{a}{L}\sigma^2 \sum_{i,j,p=1}^{\infty} \frac{(f_{ij},\Upsilon)(f_{ip},\Upsilon)}{\Lambda_{ij}+\Lambda_{ip}}(1-\exp(-at(\Lambda_{ij}+\Lambda_{ip})))$$

$$^2A_2 \approx \frac{1}{4}\sigma^2(\sqrt{2} - \frac{5}{3}) \int_{-R}^{R} F^2(0,x)\,dx = \frac{\sigma^2 a^2}{2L}(\sqrt{2} - \frac{5}{3})\sum_{i,j,p=1}^{\infty}(f_{ij},\Upsilon)(f_{ip},\Upsilon)$$

A numerical analysis shows the behaviour of the variances contained in Figure 2. Thereby it can be noted that for the considered example the correction (........) with the 2nd order term $^2A_2\varepsilon^3$ is neglectable small in comparison with the 1st term $^2A_1\varepsilon^2$ (——). That means practically the next correction terms are of order ε^4.

In our considerations we have only considered the variance. From the theory we can obtain more information, e.g. all finite-dimensional distribution functions of $\overline{u}(t,x,y,\omega)$. On the other hand it is very difficult to obtain more information from measured results. A large investigation of probabilities can be found in /5/ and of further statistical characteristics in /2/.

References

/1/ vom Scheidt, J. and Purkert, W.: Random Eigenvalue Problems, Akademie-Verlag Berlin, North-Holland New York, 1983.

/2/ vom Scheidt, J.: Random Temperature Fields, Stochastic Analysis (in print).

/3/ Boyce, W.E. and Ning-Mao Xia: The approach to normality of the solutions of random boundary and eigenvalue problems with weakly correlated coefficients, Q. Appl. Math., Vol.Xl,No.4, 1983.

/4/ vom Scheidt, J.: Random Equations of the Mathematical Physics, Akademie-Verlag Berlin (in preparation).

/5/ Fellenberg, B. and vom Scheidt, J.: Probabilistic Analysis of Random Temperature Fields, FMC-Series, No. 19, Karl-Marx-Stadt 1986, 15-24.

FINITE-DIMENSIONAL APPROXIMATION OF STOCHASTIC NAVIER-STOKES-EQUATION

Bjoern Schmalfuss
Ingenieurhochschule Koethen
Abt. Mathematik/Rechentechnik
Koethen
4370 GDR

1. Introduction

The solution of the stochastic NAVIER-STOKES equation (sNSe) is a probability measure (see [1]). This measure is defined on a BOREL-σ-algebra of a function space, which contains the trajectories satisfying the sNSe. Moreover, the measure P is coupled with the distribution of the white noise. We consider finite-dimensional approximations. If we generate finite-dimensional GAUSSian random variables, we can compute the trajectories of these approximations. So we get informations of the distributions of this solutions by statistical methods. Because of the convergence of this distributions to the measure P, we can interpret these values as approximations of the corresponding values of the measure P. This method can be extended to other approximation schemes.

2. Assumptions

Let G be an open bounded connected set in $\mathbb{R}^n$ $(n\geq 2)$. We assume the boundary ∂G is a C^∞-submanifold of $\mathbb{R}^n$. Let $L_2(G)$ be the HILBERT-space of quadratic integrable n-dimensional functions on G. We introduce the set V of all divergence-free C^∞-functions having a compact support on G. The closure of V in $L_2(G)$ is denoted by $\mathcal{H}^0$. The space $\mathcal{H}^0$ is a HILBERT-space and π is the orthonormal projection from $L_2(G)$ into $\mathcal{H}^0$. Let A be the extension of $-\pi\Delta$ (Δ denotes the LAPLACE-operator). The mapping A is a self-adjoint positive-definite operator with eigenvalues $0<\lambda_1 \leq \lambda_2 \leq \ldots$ and corresponding eigenfunctions $e_1, e_2, \ldots$. The finite-dimensional space V_m is spanned by the first m eigenfunctions $e_1, e_2, \ldots, e_m$ of A. We denote by π_m the orthonormal projection of $L_2(G)$ on V_m. Let $\mathcal{H}^s (s\in\mathbb{R})$ be the BANACH-space of (generalized) functions $u = \sum_{i=1}^{\infty} \hat{u}_i e_i$ ($\hat{u}_i \in \mathbb{R}$ $i \in \mathbb{N}$) having a finite norm $\|u\|_s = (\sum_{i=1}^{\infty} \lambda_i^s |\hat{u}_i|^2)^{1/2}$. The space $\mathcal{H}^{-s}$ is the dual space of $\mathcal{H}^s (\forall s \in \mathbb{R})$. We denote the duality by < , >. Let X be a normed space, then the spaces $L_p(0,T;X)$ $(1 \leq p \leq \infty)$, $C(0,T;X)$ are defined in the usual form. The space CL_s contains the functions $u \in C(0,T;\mathcal{H}^{-s})$ with a halfnorm

$$\|u\|_{CL_s} = \sup_{t_1 \neq t_2 \in [0,T]} (\|u(t_1) - u(t_2)\|_{-s} / h(t_1 - t_2)) < \infty$$
$$h(t) = (|t| \ln(2T/|t|))^{1/2} \qquad s > n/2 + 1 \quad .$$

Then we define the spaces $Z = L_2(0,T;\mathcal{H}^0) \cap C(0,T;\mathcal{H}^{-s})$, $UL_s = L_2(0,T;\mathcal{H}^1) \cap CL_s$ $(s > n/2+1)$. This spaces have the norms:

$$\|u\|_Z = \|u\|_{L_2(0,T;\mathcal{H}^0)} + \|u\|_{C(0,T;\mathcal{H}^{-s})} ,$$
$$\|u\|_{UL_s} = \|u\|_{L_2(0,T;\mathcal{H}^1)} + \|u\|_{CL_s} .$$

The injection of UL_s in Z is compact.
By B a bilinear continuous operator is given:

$$B: \mathcal{H}^0 \times \mathcal{H}^0 \longmapsto \mathcal{H}^{-s} \quad (s > n/2+1)$$

satisfying the condition $\langle B(u,v),v \rangle = 0 \ \forall u,v \in \mathcal{H}^s$. We regard the sNSe:

$$\mathcal{A}(u) = dw/dt \ , \ u(0) = u_0 \ ; \ \mathcal{A}(u) := du/dt + \nu Au + B(u,u) - g$$

$$t \in [0,T] \ , \ \nu > 0, \ g \in L_2(0,T;\mathcal{H}^{-1}).$$

Let $\overset{\circ}{V}{}^s$ be the closure of the infinite differentiable (in t) functions of [0,T] into $\mathcal{H}^s$ with compact support in]0,T[in the norm

$$\|u\|_{V^s} := \|du/dt\|_{L_2(0,T;\mathcal{H}^s)} + \|u\|_{L_2(0,T;\mathcal{H}^s)} \quad ,$$

and let V^{-s} be the dual space of $\overset{\circ}{V}{}^s$. Then $\mathcal{A}$ is continuous from Z (or $L_2(0,T;\mathcal{H}^0)$) into V^{-s}. The process w is on the probability space $(\Omega, \mathcal{F}, \mathcal{F}_t, \mathbb{P})$ (with expectation $\mathbb{E}$ ·) a WIENER-process having trajectories in $C(0,T;\mathcal{H}^0)$. The distribution Λ of this process is defined on the BOREL-σ-algebra of continuous functions: $\mathcal{B}(C(0,T;\mathcal{H}^0))$. The covariance of w is the nonnegative nuclear operator Q. Let μ be the distribution of u_0 . The probability measure μ fulfils the conditions

$$\int_{\mathcal{H}^0} \|u\|_0^4 \mu(du) < \infty \quad , \ u_0 \text{ and } w \text{ are independent.} \tag{1}$$

A solution of the sNSe is a probability measure P, which is defined on $\mathcal{B}(Z)$, concentrated on UL_s. It fulfils the following equations:

$$P(u \in UL_s \ \mathcal{A}(u) \in B) = N(B) \quad \forall B \in \mathcal{B}(V^{-s}) \tag{2}$$

$$P(u \in UL_s \; u(0) \in B') = \mu(B') \quad \forall B' \in \mathcal{B}(\mathcal{H}^{-s}) \tag{2'}$$

(see [1]). The measure N is the distribution of dw/dt with the domain $\mathcal{B}(V^{-s})$. Let $\tilde{\mu}$ and $\tilde{\Lambda}$ be the characteristic functionals of u_0 and w, respectively. Then the following condition is equivalent to (2),(2'):

$$\int_Z \exp(i(\langle u(0),v\rangle + \{\mathcal{A}(u),\vartheta\}))P(du) = \tilde{\mu}(v)\tilde{\Lambda}(-d\vartheta/dt) \tag{3}$$

$$\forall v \in \bigcup_{m=1}^{\infty} V_m, \; \vartheta \in \bigcup_{m=1}^{\infty} C_0^{\infty}(0,T;V_m)$$

(see [1]); { , } denotes the V^{-s}, $\mathcal{V}^s$ - duality.

3. Finite-dimensional approximations

Let $k:=T/N$ be the length ot the timesteps of discretization and $t_l := l\cdot k$ $(l=0,\dots,N)$ the points of discretizations in $[0,T]$. Then we consider the finite-dimensional equations:

$$u_{km}^l - u_{km}^{l-1} = -k\,\nu\, A u_{km}^l - k\pi_m B(u_{km}^{l-1}, u_{km}^l) + k\cdot g_{km}^l + w_{km}^l - w_{km}^{l-1}; \tag{4}$$

$$g_{km}^l := 1/k \int_{t_{l-1}}^{t_l} \pi_m g(t)\,dt; \quad w_{km}^l := \pi_m w(t_l) \quad l=1,\dots,N \quad \forall k,m$$

$$w_{km}^0 = 0 \text{ a.s. } \forall k,m; \quad u_{km}^0 := \pi_m u_0; \quad u_{km}^l := \sum_{i=1}^m \hat{u}_{kmi}^l e_i \in V_m \quad \hat{u}_{kmi}^l \in \mathbb{R}.$$

The process $w_m := \pi_m w$ is also a WIENER-process with trajectories in $C(0,T;V_m)$. The covariance of w is the nonnegative nuclear operator $Q_m := \pi_m Q \pi_m$, which fulfils the condition : trace $Q_m <$ trace Q $\forall$ m. Each of these equations (4) is linear in the variables u_{km}^l, because we know u_{km}^{l-1} from the last step of computation. We must solve a linear system of equations (the corresponding matrix is nonsingular). In many cases we can use the special form of this matrix. The solution of these equations is a set of random variables $(u_{km}^0, u_{km}^1, \dots, u_{km}^N)$.
Let u_{km} be a random variable with values in $(L_2(0,T;\mathcal{H}^0), \mathcal{B}(L_2(0,T;\mathcal{H}^0)))$: $u_{km}(t) = u_{km}^l \; \forall t \in]t_{l-1}, t_l]$. The distributions of these random variables are probability measures P_{km}, which are defined on $\mathcal{B}(L_2(0,T;\mathcal{H}^0))$. We must show several a priori estimations:

Theorem 3.1: The solutions of (4) are bounded in the following sense:

$$\mathbb{E}\|u_{km}^N\|_0^2 + \mathbb{E}\|u_{km}\|^2_{L_2(0,T;\mathcal{H}^1)} \le \mathbb{E}\|u_0\|_0^2 + T\cdot \text{trace } Q + 1/\nu \cdot \|g\|^2_{L_2(0,T;\mathcal{H}^{-1})} \tag{5.1}$$

$$\forall k, m$$

$$\mathbb{E}\|u_{km}^{\hat{l}}\|_0^2 \le \mathbb{E}\|u_0\|_0^2 + T\cdot \text{trace } Q + 1/\nu \cdot \|g\|^2_{L_2(0,T;\mathcal{H}^{-1})} \; \forall k,m,\hat{l}=1,\dots,N \tag{5.2}$$

$$\mathbb{E}\sum_{i=1}^N \|u_{km}^{i-1} - u_{km}^i\|_0^2 \le c_1 \quad \forall k,m \tag{5.3}$$

$$\mathbb{E}\|u^l_{km}\|^4_0 \le c_2 \qquad \forall k,m,\ l=1,\dots,N \ . \tag{5.4}$$

Sketch of the proof: We use the properties

$$2\langle a-b,a\rangle = \|a\|^2_0 + \|b\|^2_0 + \|a-b\|^2_0 \quad \forall a,b \in \mathcal{H}^0 ,$$

$$\langle Au,u\rangle = \|u\|^2_1 \quad \forall u \in \mathcal{H}^1$$

and taking the inner product of (4) by u^l_{km} ($l=1,\dots,N$):

$$\|u^l_{km}\|^2_0 - \|u^{l-1}_{km}\|^2_0 + 2k\nu\|u^l_{km}\|^2_1 = 2k\langle g^l_{km},u^l_{km}\rangle - \|u^l_{km} - u^{l-1}_{km}\|^2_0 + \tag{6}$$

$$+ 2\langle w^l_{km} - w^{l-1}_{km}, u^l_{km}\rangle \le k\cdot 1/\nu\cdot\|g^l_{km}\|^2_{-1} + k\nu\|u^l_{km}\|^2_1 - \|u^l_{km} - u^{l-1}_{km}\|^2_0 +$$

$$2\langle w^l_{km} - w^{l-1}_{km}, u^l_{km} - u^{l-1}_{km}\rangle + 2\langle w^l_{km} - w^{l-1}_{km}, u^{l-1}_{km}\rangle$$

$$\le k\cdot 1/\nu\cdot\|g^l_{km}\|^2_{-1} + k\cdot\nu\|u^l_{km}\|^2_1 + \|w^l_{km} - w^{l-1}_{km}\|^2_0 + 2\langle w^l_{km} - w^{l-1}_{km}, u^{l-1}_{km}\rangle \ .$$

Then we compute the mathematical expectation. Because of the independence of the increments $w^l_{km} - w^{l-1}_{km}$ from $\mathcal{F}_{t_{l-1}}$ and the $\mathcal{F}_{t_{l-1}}$ measurability of u^{l-1}_{km}, it is $\mathbb{E}\langle w^l_{km} - w^{l-1}_{km}, u^{l-1}_{km}\rangle = 0 \ \forall k,m$. Moreover, it is $k\cdot\text{trace } Q \ge \mathbb{E}\|w^l_{km} - w^{l-1}_{km}\|^2_0$ and $\|g^l_{km}\|_{L_2(0,T;\mathcal{H}^{-1})} \le \|g\|_{L_2(0,T;\mathcal{H}^{-1})}$ (see [2]). Adding these inequalities for 1 to N or for 1 to $\hat{l}$, we get (5.1), (5.2), respectively. We get property (5.3) in a similar way. Proving (5.4) we use the property

$$(\|u^l_{km}\|^4_0 - \|u^{l-1}_{km}\|^4_0) = (\|u^l_{km}\|^2_0 - \|u^{l-1}_{km}\|^2_0)(\|u^l_{km}\|^2_0 + \|u^{l-1}_{km}\|^2_0) .$$

Then we estimate the first term of the right side by the properties (6), also we use (1), the independence of the increments $w^l_{km} - w^{l-1}_{km}$, and $\mathbb{E}\|w^l_{km} - w^{l-1}_{km}\|^4_0 \le (3k\cdot\text{trace } Q)^2$.

It is to show an estimation of the mathematical expectation of the norm $\|\cdot\|_{CL_S}$. But if we use the random variables u_{km}, it is impossible, because the trajectories are stepfunctions. So we introduce a new class of random variables $\tilde{u}_{km}$, which are defined by

$$\tilde{u}_{km}(t_l) = u^l_{km}$$

$\tilde{u}_{km}(t)$ continuous and linear on each interval $[t_{l-1}, t_l]$.

Let Q_{km} be the distribution of $\tilde{u}_{km}$, which are defined on $\mathcal{B}(Z)$. These measures Q_{km} and random variables $\tilde{u}_{km}$ fulfils the following a priori estimations:

<u>Theorem 3.2:</u> Q_{km}, $\tilde{u}_{km}$ are bounded in the following sense:

$$\mathbb{E}\|\tilde{u}_{km}(t)\|^2_0 \le \mathbb{E}\|u_0\|^2_0 + T\cdot\text{trace } Q + 1/\nu\,\|g\|^2_{L_2(0,T;\mathcal{H}^{-1})} \quad t \in [0,T] \tag{7.1}$$

$$\mathbb{E}\|\tilde{u}_{km}\|^2_{L_2(0,T;\mathcal{H}^1)} = \int_Z \|u\|^2_{L_2(0,T;\mathcal{H}^1)} Q_{km}(du) \le c_3 \tag{7.2}$$

$$\mathbb{E}\|\tilde{u}_{km}\|_{CL_S} = \int_Z \|u\|_{CL_S} Q_{km}(du) \le c_4 \quad \forall k,m\ . \tag{7.3}$$

If we use (5.1),(5.2), we can show (7.1),(7.2), respectively. Property (7.3) follows from (5.4). The relations between u_{km} and $\tilde{u}_{km}$ can be formulated in the following theorem:

Theorem 3.3: The mathematical expectation of the differences $\|u_{km} - \tilde{u}_{km}\|^2_{L_2(0,T;\mathcal{H}^0)}$ converges to zero for $k \longrightarrow 0$, $m \longrightarrow \infty$:

$$\mathbb{E}\,\|\tilde{u}_{km} - u_{km}\|^2_{L_2(0,T;\mathcal{H}^0)} \longrightarrow 0.$$

It can be proofed by

$$\|\tilde{u}_{km} - u_{km}\|^2_{L_2(0,T;\mathcal{H}^0)} = k/3 \cdot \sum_{i=1}^{N} \|u^{i-1}_{km} - u^{i}_{km}\|^2_0 \qquad \text{a.s.}$$

(see [2]) and property (5.3).

4. Compactness and convergence

Our aim is to show the convergence of subsequences $(P_{k'm'})$, $(Q_{k'm'})$ to the solution P of the sNSe. At first we show the relativly compactness of the set $\{Q_{km}, \forall k,m\}$. We will use the theorem of PROCHOROV in the following formulation (see [1]).

Theorem 4.1: Let X, Y be BANACH-spaces. Let Y be separable and $X \subset Y$ with compact injection. A set of probability measures $\{m_i \ ,i \in I\}$, which are defined on $\mathcal{B}(Y)$, is relativly compact in the weak measure topology, if holds $\sup_I(\int_Y \|u\|_X\, m_i(du)) < \infty$. In this case it can be selected a subsequence (i'), which converges to a probability measure m in weak measure convergence:

$$\int_Y f(u) m_{i'}(du) \longrightarrow \int_Y f(u) m(du), \quad f \text{ bounded and continuous on } Y.$$

Because of the separability of Z and the compact injection of UL_S in Z, the set $\{Q_{km} \forall k,m\}$ is relativly compact in weak measure topology (see (7.2), (7.3)). It can be founded a subsequence $(Q_{k'm'})$, which converges weak to a probability measure P for $k' \longrightarrow 0$, $m' \longrightarrow \infty$. Before we show that the measure P is a solution of the sNSe, we should show that also the subsequence $(P_{k'm'})$ converges to P. In this case we extend the measures Q_{km}, P on $\mathcal{B}(L_2(0,T;\mathcal{H}^0))$. We can do it, because of the continuous injection of Z in $L_2(0,T;\mathcal{H}^0)$. Since the assertion of theorem 3.3, there is

$$|\mathbb{E}f(\tilde{u}_{k'm'}) - \mathbb{E}f(u_{k'm'})| = |\int_{L_2} f(u) Q_{k'm'}(du) - \int_{L_2} f(u) P_{k'm'}(du)| \longrightarrow 0 \tag{8}$$

for each uniformly continuous function f on $L_2(0,T;\mathcal{H}^0)$. Using the triangular inequality , we get

$$|\int_{L_2} f(u)P(du) - \int_{L_2} f(u)P_{k'm'}(du)| \leq \tag{9}$$
$$|\int_{L_2} f(u)Q_{k'm'}(du) - \int_{L_2} f(u)P(du)| + |\int_{L_2} f(u)Q_{k'm'}(du) - \int_{L_2} f(u)P_{k'm'}(du)|$$

for each uniformly continuous bounded function f on $L_2(0,T;\mathcal{H}^0)$. The convergence to zero of the right hand side follows by weak measure convergence of $(Q_{k'm'})$ (also the extensions of $Q_{k'm'}$ are converging). The convergence to zero of the second summand of the right hand side follows from (8). The convergence of the left hand side is sufficient for weak measure convergence of $(P_{k'm'})$ to P (see [3] theorem 2.1(ii)).

<u>Theorem 4.2:</u> The measure P fulfils the property (3).

Let γ_m^k $(\forall k,m)$ be operators. γ_m^k maps a piecewise linear function in a stepfunction $(\gamma_m^k \tilde{u}_{km} = u_{km})$. This operator can be extended to a measurable map from Z into $L_2(0,T;\mathcal{H}^0)$. Let $\tilde{w}_{km}$ be the piecewise linearization of w_m . The processes $\tilde{w}_{km}$ have the characteristic functionals $\tilde{\Lambda}_{km}$. We define the mappings $\mathcal{A}_{km} : Z \longrightarrow V^{-s}$

$$\mathcal{A}_{km}(u) := du/dt + \nu A\gamma_m^k u + \Pi_m B(\gamma_m^k u(\cdot - k), \gamma_m^k u) - g_{km}$$

$$g_{km}(t) := g_{km}^l \;\forall t \in]t_{l-1}, t_l] \quad l = 1,\ldots,N.$$

These operators are continuous from Z into V^{-s}. Now we can write our finite-dimensional approximations in the form

$$\mathcal{A}_{km}(\tilde{u}_{km}) = d\tilde{w}_{km}/dt \qquad \tilde{u}_{km}(0) = \Pi_m u_0.$$

Because of the independence of u_0 and w, there is:

$$\int_Z \exp(i(\langle u(0),v\rangle + \{\mathcal{A}_{km}(u),\vartheta\})Q_{km}(du) = \widehat{\Pi_m\mu}(v)\cdot\tilde{\Lambda}_{km}(-d\vartheta/dt) \tag{10}$$
$$\forall v \in \bigcup_{m=1}^{\infty} V_m, \; \vartheta \in \bigcup_{m=1}^{\infty} C_0^{\infty}(0,T;V_m).$$

The functions $\exp(i(\langle u(0),v\rangle + \{\mathcal{A}_{km}(u),\vartheta\}))$ are bounded continuous on Z. Hence the limit of the left hand side of (10) is the left hand side of (3). On the other hand the sequence $\tilde{\Lambda}_{km}(-d\vartheta/dt)\,\widehat{\Pi_m\mu}(v)$ converges to $\tilde{\Lambda}(-d\vartheta/dt)\hat{\mu}(v)$ .It is be shown that the limit P of $(Q_{k'm'})$ fulfils (3). We get also by weak measure convergence that P is concentrated on UL_S. Hence the measure P is a solution of the sNSe.

5. Literature references

[1] Vishik,M.I.,Komech,A.I.,Fursikov,A.V.,UMN (34) 1979,no 5(209)

[2] Temam,R., Navier-Stokes Equation-Theory and numerical analysis, North Holland Publishing Company, Amsterdam New York(1979).

[3] Billingsley,P., Convergence of Probability Measures, John Wiley ...

Stochastic Equations and Diffusions. Approximation of Diffusions

LARGE DEVIATIONS

OF LINEAR STOCHASTIC DIFFERENTIAL EQUATIONS

by

LUDWIG ARNOLD

Institut für Dynamische Systeme, Universität Bremen

D-2800 Bremen 33, FR of Germany

and

WOLFGANG KLIEMANN

Department of Mathematics, Iowa State University,

Ames, Iowa 50011, USA

Contents

Summary

We study parametric perturbations of a linear system by white noise $\dot{W}_1(t),\dots,\dot{W}_m(t)$ and by real noise $\xi(t)$ (a nice Markov process), modelled by the linear stochastic differential equation

$$(1)\quad dx(t) = A_0(\xi(t))x(t)dt + \sum_{i=1}^{m} A_i(\xi(t))x(t) \circ dW_i(t),\ x(0) = x_0 \in \mathbb{R}^d .$$

Under a mild condition every solution $x(t;x_0)$ of (1) has the same exponential growth rate (Lyapunov exponent) λ,

$$\lim_{t\to\infty} \frac{1}{t} \log|x(t;x_0)| = \lambda \quad \text{(law of large numbers)} .$$

(1) is stable if $\lambda < 0$. The key tool for studying the finer properties of (1) is the Lyapunov exponent of the pth moment

$$g(p) = \lim_{t\to\infty} \frac{1}{t} \log E|x(t;x_0)|^p , \quad p \in \mathbb{R} .$$

We have $\lambda = g'(0)$ and

$$\frac{1}{\sqrt{t}}(\log|x(t;x_o)| - \lambda t) \Rightarrow N(0,g''(o)) \quad \text{(central limit theorem)} .$$

It turns out that $g(p)$ also controls the asymptotics of $P(\sup_{t\in\mathbb{R}^+}|x(t;x_o)| \geq C)$ for $C\to\infty$ and of $P(|\frac{1}{t}\log|x(t;x_o)| - \lambda| \geq \varepsilon)$ for $t\to\infty$ (large deviations). (1) is universally stable (i.e. for any noise $\xi(t)$) if $\gamma = \lim_{p\to\infty}\frac{g(p)}{p} < 0$. The random oscillator is treated in detail, with a universal stability diagram. The generalization to the smaller Lyapunov exponents and to nonlinear systems is sketched.

1. Introduction

Numerous applications in science and engineering lead to models in which the entries of A in $\dot{x} = Ax$ are perturbed by noise (parametric or multiplicative noise).

Pathwise and p^{th} mean stability theory was systematically studied in the context of Lyapunov exponents by Arnold, Kliemann and Oeljeklaus [2] (real noise case) and Arnold, Oeljeklaus and Pardoux [3] (white noise case). To make the paper more self-contained we will present the basic results of [2] and [3] in sections 1 and 2.

The aim of this paper is to study the more subtle, but nevertheless practically very relevant asymptotic problems connected with large deviations. For example, even a stable system can get destroyed by a large excursion before it gets the chance of settling down.

As it will become clear later, the main tool for studying those problems will be p^{th} mean stability. This is implicitly contained in Chapter 7 of Freidlin and Wentzell [21] and was used by Le Page [23] and Bougerol and Lacroix [13] for products of random matrices and also implicitly by Stroock [30] for our case (Stroock actually treats a formally more general case, but see Remark 1.1). The difference between Stroock's paper and ours is: (i) We will treat the case where a combination of real and white noise is present, while Stroock only considers the white noise case. (ii) We work under a Lie algebra condition which is more realistic and less restrictive then Stroock's. (iii) By feeding in our previous results on p^{th} mean stability (see [2], [3]) we can apply existing large deviations theorems (in particular Theorem II. 6.1 of Ellis [19]) which quickly yield the results. (iv) We have expressions for large deviation quantities that lend themselves to computation (see e.g. sections 4 and 6).

We will deal with the following linear stochastic differential equation in $\mathbb{R}^d$:

$$(1.1) \quad dx = A_0(\xi(t))x\,dt + \sum_{i=1}^{m} A_i(\xi(t))x \circ dW_i(t) , \quad x_0 \in \mathbb{R}^d .$$

Here

(i) $\xi(t)$ is a stationary diffusion process on a finite-dimensional compact connected (Riemannian) C^∞ manifold M given by

$$(1.2)\quad d\xi(t) = X_0(\xi(t))dt + \sum_{j=1}^{r} X_j(\xi(t)) \circ dV_j(t) .$$

(ii) The vector fields $X_0,\dots,X_r$ are C^∞ and satisfy

$$\dim LA(X_1,\dots,X_r)(\xi) = \dim M \quad \text{for all} \quad \xi \in M,$$

$LA(Z)$ denoting the Lie algebra generated by the set Z of vector fields.

(iii) $A_i: M \to gl(d,\mathbb{R})$ = linear space of real dxd matrices, are C^∞, $i = 0,\dots,m$.

(iv) The Wiener processes W_i, V_j are independent, and "o" denotes the Stratonovich stochastic integral.

Assumption (ii) assures that $\xi(t)$ is ergodic, and the invariant probability ρ has a positive C^∞ density.

The model (1.1) is referred to as "unified treatment" of real and white noise. If M consists of one point, we are in the white noise case $dx = A_0x\,dt + \Sigma A_ix \circ dW_i$, while $A_i \equiv 0$, $i = 1,\dots,m$, describes the real noise case $\dot{x} = A_0(\xi(t))x$.

Introducing polar coordinates $s = x/|x| \in S^{d-1}$, $|x| \in \mathbb{R}^+$ in $\mathbb{R}^d \setminus \{0\}$ and identifying s and $-s$ converts (1.1) into an equation on real projective space $\mathbb{P}^{d-1}$

$$(1.3)\quad ds = h_0(\xi(t),s)dt + \sum_{i=1}^{m} h_i(\xi(t),s) \circ dW_i, \quad s_0 = \frac{x_0}{|x_0|},$$

and

$$(1.4)\quad |x(t;x_0)| = |x_0| \exp\{\int_0^t Q(\xi(\tau),s(\tau;s_0))d\tau + \sum_{i=1}^{m} \int_0^t q_i(\xi(\tau),s(\tau;s_0))dW_i\} .$$

Here

$$q_i(\xi,s) = \langle s,A_i(\xi,s)s\rangle ,$$

$$h_i(\xi,s) = A_i(\xi)s - q_i(\xi,s)s$$

is the projection of the linear vector field $A_i(\xi)x$ onto $\mathbb{P}^{d-1}$, $i = 0,\dots,m$, and

$$(1.5) \quad Q(\xi,s) = q_o(\xi,s) + \sum_{i=1}^{m} \{\tfrac{1}{2}<s,(A_i(\xi) + A_i'(\xi))A_i(\xi)s> - q_i(\xi,s)^2\} .$$
$$= q_o + \tfrac{1}{2} \sum_{i=1}^{m} h_i(q_i).$$

The pair $(\xi(t),s(t))$ is a diffusion process on $M \times \mathbb{P}^{d-1}$ with generator

$$L = G + h_o + \tfrac{1}{2} \sum_{i=1}^{m} h_i^2 ,$$

where $G = X_o + \frac{1}{2} \sum_{j=1}^{r} X_j^2$ is the generator of $\xi(t)$.

Remark 1.1. Note that this procedure works if the vector fields in (1.1) are just homogeneous in $\mathbb{R}^d$.

To evaluate (1.4) for $t \to \infty$ we need to know the ergodic behaviour of $(\xi(t),s(t))$.

The following hypoellipticity assumption will be used throughout the paper:

(H) $\dim LA(X_o+h_o,X_1,\dots,X_r,h_1,\dots,h_m)(\xi,s) = \dim M + d-1$ for all $(\xi,s) \in M \times \mathbb{P}^{d-1}$,

which under (ii) and in case the Lie algebra in (H) is integrable is equivalent to

$$\dim LA(h_o(\xi,\cdot),\dots,h_m(\xi,\cdot),\ \xi \in M)(s) = d-1$$

for all $s \in \mathbb{P}^{d-1}$, cf. San Martin [26].

For many results we will need the ellipticity assumption

(E) $\dim LA(X_o+h_o+\frac{\partial}{\partial t}, X_1,\dots,X_r,h_1,\dots,h_m)(\xi,s,t) = \dim M + d$ for all $(\xi,s,t) \in M \times \mathbb{P}^{d-1} \times \mathbb{R}$.

(E) implies (H), and is equivalent to (H) in case $M \times \mathbb{P}^{d-1}$ has compact universal covering space (which is true if M is simply connected and $d \neq 2$).

An assumption even stronger then (E) is the nondeneracy condition

(N) $\dim LA(X_1,\dots,X_r,h_1,\dots,h_m)(\xi,s) = \dim M + d-1$ for all $(\xi,s) \in M \times \mathbb{P}^{d-1}$, which under (ii) and in case the Lie algebra in (N) is integrable is equivalent to

$$\dim LA(h_1(\xi,\cdot),\ldots,h_m(\xi,\cdot),\xi \in M)(s) = d-1$$

for all $s \in \mathbb{P}^{d-1}$. Of course, (N) implies (E).

Theorem 1.2. *Assume* (H). *Then the diffusion process* $(\xi(t),s(t))$ *given by* (1.2), (1.3) *admits a unique invariant probability* μ *on* $M \times \mathbb{P}^{d-1}$ *with marginal* ρ *on* M. *Moreover,* μ *has a* C^∞ *density.*

For a proof see [2], [3].

Note, however, that in general $\operatorname{supp} \mu = M \times C$, $C \subsetneqq \mathbb{P}^{d-1}$. But $C = \mathbb{P}^{d-1}$ and the density of μ is positive in case (N) is true.

2. Basic results on Lyapunov exponents

If not otherwise stated, the following results are simultaneous generalizations of the real noise case in [2] and the white noise case in [3], with essentially the same proofs.

The pathwise Lyapunov exponent of a solution of (1.1) is defined as

$$\lambda(x_0) = \limsup_{t\to\infty} \frac{1}{t} \log |x(t;x_0)|, \qquad x_0 \neq 0 .$$

Theorem 2.1. *Assume* (H). *Then for each* $x_0 \neq 0$.

$$\lambda(x_0) = \lim_{t\to\infty} \frac{1}{t} \log |x(t;x_0)| = \lambda \quad \text{a.s.}$$

where

$$\lambda = \int_{M\times\mathbb{P}^{d-1}} Q(\xi,s)d\mu .$$

Q *is defined by* (1.5) *and* μ *is the unique invariant probability of* $(\xi(t),s(t))$. *Moreover,* λ *is the top Lyapunov exponent from Oseledec's theorem.*

□

The Lyapunov exponent of the pth moment of a solution of (1.1) is defined by

$$g(p;x_0) = \limsup_{t\to\infty} \frac{1}{t} \log E|x(t;x_0)|^p, \quad p \in \mathbb{R},\ x_0 \neq 0 .$$

Theorem 2.2. *Assume* (E). *Then*

(i) $g(p;x_0) = \lim_{t\to\infty} \frac{1}{t} \log E|x(t;x_0)|^p = g(p) \in \mathbb{R}$

for all $p \in \mathbb{R}$ *and* $x_0 \neq 0$.

(ii) $g(p)$ *is the principal eigenvalue of*

$$L(p) = L + pX + pQ + \frac{p^2}{2} R,$$

where $X = \sum_{i=1}^{m} q_i h_i$ *and* $R = \sum_{i=1}^{m} q_i^2$, *and* $g(p)$ *strictly dominates the spectrum.*

(iii) $g: \mathbb{R} \to \mathbb{R}$ *is convex and analytic with* $g(0) = 0$ *and* $g'(0) = \lambda$.

(iv) $\gamma: \mathbb{R} \to \mathbb{R}$ *defined by* $\gamma(0) = \lambda$, $\gamma(p) = g(p)/p$ *for* $p \neq 0$, *is increasing and analytic.*

□

Define

$$\gamma_- = \lim_{p\to-\infty} \gamma(p) , \qquad \gamma_+ = \lim_{p\to\infty} \gamma(p) .$$

Then either $\gamma(p) \equiv \lambda$ (iff $g(p) = \lambda p$) or $\gamma(p)$ is strictly increasing (iff $g(p)$ is strictly convex). In the latter case

$$\gamma_- < \lambda = \gamma(0) < \gamma_+ .$$

The importance of γ_+ lies in the fact that under $\gamma_+ < 0$ not only the decay rate λ of the sample paths of a solution is negative, but the moments of any order $p > 0$ are exponentially decaying with rate $g(p) < 0$. Similarly for γ_- and $|x(t;x_0)|^{-1}$. Let

$$N := \{A_0(u_0) + \sum_{i=1}^{m} u_i A_i(u_0) : u_0 \in M, u_1,\dots,u_m \in \mathbb{R}\} \subset gl(d, \mathbb{R}).$$

The group G and semigroup S generated by N, i.e.

$$G := \{e^{t_n B_n} \cdots e^{t_1 B_1} : \ t_i \in \mathbb{R},\ B_i \in N,\ i = 1,\dots,n \in \mathbb{N}\}$$

and

$$S := \{e^{t_n B_n} \dots e^{t_1 B_1} : t_i > 0,\ B_i \in N,\ i = 1,\dots,n \in \mathbb{N}\}$$

are called system group and system semigroup , resp. Let

$$H := \text{group generated by } \{\sum_{i=1}^{m} u_i A_i(\xi) : \xi \in M,\ u_1,\dots,u_m \in \mathbb{R}\} .$$

We have

$$S \subset G \subset Gl(d, \mathbb{R})$$

and

$$\bar{H} \subset \bar{S} \subset \bar{G} \subset Gl(d, \mathbb{R}) \quad \text{(closure in } GL(d, \mathbb{R})).$$

Let G^o and H^o be the groups generated by the corresponding sets, but with $A_i(\xi)$ replaced by

$$A_i^o(\xi) := A_i(\xi) - (\tfrac{1}{d} \operatorname{trace} A_i(\xi))\mathrm{id},\ i = 0,\dots,m.$$

Clearly, $\overline{H^o} \subset \overline{G^o} \subset Sl(d, \mathbb{R})$.

The relation between λ, $\gamma_\pm$ and $\pm\infty$ can be completely characterized by the compactness or noncompactness of the above groups as follows.

Theorem 2.3. Assume (E).

1. *If* $\bar{H}$ *is not compact then*

$$-\infty = \gamma_- < \lambda < \gamma_+ = \infty .$$

2. *If* $\bar{H}$ *is compact (thus* $H = H^o$*) then*

$$-\infty < \gamma_- \le \lambda \le \gamma_+ < \infty .$$

In particular:

2.1.1 *If* $\bar{G}$ *is not compact and* $\overline{G^o}$ *is not compact then*

$$\gamma_- < \lambda < \gamma_+ .$$

2.1.2 *If* $\bar{G}$ *is not compact, but* $\overline{G^o}$ *is compact then*

a) *in case* $\frac{1}{d}\text{trace } A_0(\xi) \equiv c$ *then*

$$\gamma_- = \lambda = \gamma_+ = c \ ,$$

b) *in case* $\frac{1}{d}\text{trace } A_0(\xi) \not\equiv c$ *then*

$$\gamma_- = \frac{1}{d}\min_{\xi\in M} \text{ trace } A_0(\xi) < \lambda = \frac{1}{d}\int_M \text{trace } A_0(\xi)d\rho < \gamma_+ = \frac{1}{d}\max_{\xi\in M} \text{ trace } A_0(\xi).$$

2.2 *If* $\bar{G}$ *is compact (thus* $G = G^0$*) then*

$$\gamma_- = \lambda = \gamma_+ = 0 \ .$$

Remark 2.4. (i) $\bar{G}$ is compact if and only if there exists a $T \in Gl(d, \mathbb{R})$ such that

$$T\{A_i(\xi): \xi \in M, \ i = 0,\dots,m\}T^{-1} \subset SO(d, \mathbb{R}) \ ,$$

$so(d, \mathbb{R})$ = skewsymmetric real $d{\times}d$ matrices, similarly for the other groups.

(ii) If $G = G^0$ then $g(-d) = 0$. This gives the additional information that $\lambda = 0$ if $\bar{G}$ is compact and $\lambda > 0$ otherwise, see [2], [3].

(iii) If $\bar{H}$ is compact and (N) holds then the support theorem (see section 4) implies that

$$\gamma_+ = \max_{M\times \mathbb{P}^{d-1}} q_0(\xi,s) = \max_{\xi\in M} \text{ spec } \frac{A_0(\xi)+A_0'(\xi)}{2} \ ,$$

similarly for γ_- (here we have assumed w.l.o.g. that $\bar{H} \subset SO(d, \mathbb{R})$).

(iv) If $\bar{H}$ is not compact then $g(p)$ typically grows like p^2, cf. [3], Remark 3.2, and Remark 3.8 of this paper. See also Simon [29] and Baxendale and Stroock [9]. In any case

$$|g(p)| \le c_1|p| + c_2p^2 \ .$$

Proof of Theorem 2.3: 1. If $\bar{H}$ is not compact then the proof of Proposition 3.1 (iii) in [3] and the equation $\beta = \gamma_+$ from section 4 yield $\gamma_+ = \infty$.

2. If $\bar{H}$ is compact, choose a basis in which $\bar{H} \subset SO(d, \mathbb{R})$. Then $\gamma_+ \le \max q_0(\xi,s)$ can be read off from (1.4).

2.1.1 If $\lambda = \gamma_+$ then, because of analyticity, $g(p) \equiv \lambda p$. This implies $g''(0) = 0$

which in turn implies $\overline{G^o}$ compact (see Theorem 2.4), contrary to the assumption.

2.1.2 Choose a coordinate system in which $\overline{G^o} \subset SO(d,\mathbb{R})$. Then

$$g(p) = \lim_{t\to\infty} \frac{1}{t} \log E \exp p \int_0^t (\frac{1}{d} \operatorname{trace} A_0(\xi(\tau)))d\tau$$

and

$$\lambda = \frac{1}{d} E \operatorname{trace} A_0(\xi(0)),$$

and the statements follow from the support theorem for the nondegenerate diffusion process $\xi(t)$ on M.

2.2 is clear after the choice of a coordinate system for which $\bar{G} \subset SO(d,\mathbb{R})$.

□

To the law of large numbers (Theorem 2.1) there corresponds the following central limit theorem.

Theorem 2.5. *Assume* (E). *Then for* $t \to \infty$ *and each* $x_0 \neq 0$

$$\frac{1}{\sqrt{t}}(\log|x(t;x_0)| - \lambda t) \Rightarrow N(0,g''(0)),$$

where $g''(0)$ *can be represented by*

$$g''(0) = \int_{M\times\mathbb{P}^{d-1}} (\sum_{j=1}^{r} (X_j\varphi)^2 + \sum_{i=1}^{m} (q_i - h_i\varphi)^2)d\mu$$

and φ *is a solution of* $L\varphi = Q-\lambda$. *In particular,* $g''(0) = 0$ *if and only if* $\gamma_- = \lambda = \gamma_+$.

Proof: The theorem (except for the last statement) was proved for the real noise case in [2] and for the white noise case under (N) by Baxendale [8]. Baxendale's proof works under (E) for the general case. It remains to be proved that under (E) $g''(0) = 0$ entails $\gamma_- = \gamma_+$.

Assume $g''(0) = 0$. Since L is hypoelliptic, φ is C^∞, thus for all $(\xi,s) \in M \times C = \operatorname{supp} \mu$

$$X_j\varphi = 0,\ j = 1,\dots,r,\quad h_i\varphi = q_i,\quad i = 1,\dots,m .$$

Since the Lie algebra generated by $X_1,\dots,X_r$ is full by assumption, the first set of equations yields $\varphi = \varphi(s)$ in $M \times C$. Hence, by using the second set of equations,

$$L\varphi = h_o\varphi + \frac{1}{2}\sum_{i=1}^{m} h_i^2\varphi = h_o\varphi + \frac{1}{2}\sum_{i=1}^{m} h_i(q_i) = h_o\varphi + Q - q_o ,$$

and, on the other hand, $L\varphi = Q - \lambda$, which results in the equation

$$h_o\varphi = q_o - \lambda \quad \text{in} \quad M \times C .$$

Now fix an arbitrary $s_o \in \operatorname{int} C \neq \emptyset$. Then with

$$v := \operatorname{grad} \varphi(s_o) \in T_{s_o} \mathbb{P}^{d-1}$$

$h_i\varphi = q_i$ reads, using $\langle s_o, v\rangle = 0$,

$$\begin{aligned} h_i\varphi(s_o) &= \langle h_i, v\rangle = \langle A_i(\xi)s_o - q_i s_o, v\rangle \\ &= \langle A_i(\xi)s_o, v\rangle = \langle A_i(\xi)s_o, s_o\rangle . \end{aligned}$$

This means that for all $\xi \in M$

$$\langle (A_o(\xi)-\lambda)s_o, v-s_o\rangle = 0 \quad \text{and} \quad \langle A_i(\xi)s_o, v-s_o\rangle = 0, \quad i = 1,\dots,m .$$

In other words $(A_o(\xi)-\lambda\, id)s_o$, $A_1(\xi)s_o,\dots,A_m(\xi)s_o \perp v-s_o$ for all $\xi \in M$, thus

$$LA(A_o(\xi)-\lambda\, id, A_1(\xi),\dots,A_m(\xi), \xi \in M)(s_o) \perp v-s_o.$$

Since $\langle v-s_o, s_o\rangle \neq 0$ and the group generated by the above matrices acts transitively on $\mathbb{P}^{d-1}$ (which is already implied by (H)), we can use Remark 3.2 in [3]. The above group is thus contained in a compact group in $Gl(d,\mathbb{R})$. In particular, this group is equal to G^o, and $\frac{1}{d}$ trace $A_o(\xi) \equiv \lambda$. One checks directly by choosing a coordinate system in which $G^o \subset SO(d,\mathbb{R})$ that $\gamma_- = \lambda = \gamma_+$.

□

A quantity of practical importance is given by the probability with which a solution exceeds a safety threshold $C > 0$ in the course of time.

Theorem 2.6. *Assume* (E).

(i) *If* $\lambda < 0 < \gamma_+$, *then there is a unique* $a > 0$ *with* $g(a) = 0$ *and a* $K \geq 1$ *such that for all* $0 < |x_o| < C$

$$\frac{1}{K}\left(\frac{|x_o|}{C}\right)^a \leq P\left(\sup_{t\in\mathbb{R}^+} |x(t;x_o)| \geq C\right) \leq K\left(\frac{|x_o|}{C}\right)^a .$$

(ii) *If* $\lambda \le \gamma_+ \le 0$ *then there exists a* $K \ge 1$ *such that with probability one*

$$\sup_{t\in\mathbb{R}^+} |x(t;x_0)| \le K|x_0|.$$

Proof: The proof is given by Baxendale [8] under (N) for the white noise case. It relies on the fact that if $L(p)\varphi_p = g(p)\varphi_p$ (see Theorem 2.2) then $\varphi_a(\xi(t),s(t;s_0))\ |x(t;x_0)|^a$ is a martingale and under (N) $\varphi_a > 0$ on $M \times \mathbb{P}^{d-1}$. Under (E), $\varphi_a > 0$ on $\operatorname{supp}\mu = M \times C$ which suffices for the proof.

□

3. Level-1 large deviations

We will now establish a large deviation principle for $\log|x(t;x_0)|/t$. It turns out that the main work consists in proving the existence of $g(p)$ as a limit and its smoothness as a function of p. But this is Theorem 2.2. Then we can apply well-known theorems.

A natural candidate for the level-1 entropy function (rate function) is the Legendre-Fenchel transform of $g(p)$:

$$I(r) = \sup_{p\in\mathbb{R}} (rp-g(p)), \quad r \in \mathbb{R}.$$

Proposition 3.1. *Assume* (E).

(A) *If* $\gamma_- = \gamma_+$ *then*

$$I(r) = \begin{cases} 0, & r = \lambda, \\ \infty, & r \ne \lambda. \end{cases}$$

(B) *If* $\gamma_- < \gamma_+$ *then the following holds:*

(i)
$$I(r) = \begin{cases} \text{finite}, & r \in (\gamma_-,\gamma_+), \\ \infty, & r \notin [\gamma_-,\gamma_+]. \end{cases}$$

(ii) I *is strictly convex and analytic on* (γ_-,γ_+).

(iii) $I(r) \ge 0$, $I(r) = 0$ *iff* $r = \lambda$, *and* $I'(\lambda) = 0$.

(iv) $I''(\lambda) = 1/2g''(0)$, i.e.

$$I(r) = \frac{(r-\lambda)^2}{2g''(0)} + O((r-\lambda)^3) \quad \textit{for } |r-\lambda| \textit{ small.}$$

(v) *I is strictly decreasing on* (γ_-,λ) *and strictly increasing on* (λ,γ_+).

Proof. See section 2 and convex analysis (cf. Ellis [19], Theorems VI. 5.6 and VII. 2.1).

□

This yields the following large deviation principle.

Theorem 3.2. *Assume* (E). *Then for every* $x_o \in \mathbb{R}^d \setminus \{0\}$

(i) $$\limsup_{t\to\infty} \frac{1}{t} \log P\left(\frac{1}{t} \log \frac{|x(t;x_o)|}{|x_o|} \in F\right) \le -\inf_{r\in F} I(r)$$

for each closed set $F \subset \mathbb{R}$ *and*

(ii) $$\liminf_{t\to\infty} \frac{1}{t} \log P\left(\frac{1}{t} \log \frac{|x(t;x_o)|}{|x_o|} \in G\right) \ge -\inf_{r\in G} I(r)$$

for each open set $G \subset \mathbb{R}$.

Proof. Choose any sequence $a_n = t_n \uparrow \infty$ and put

$$W_n = \log \frac{|x(t_n;x_o)|}{|x_o|} = \int_0^{t_n} (Qd\tau + \sum_{i=1}^m q_i dW_i).$$

Then Theorem II. 6.1 of Ellis [19] applies, thus the statements hold for the subsequence (t_n). Since

$$\lim_{t_n\to\infty} \frac{1}{t_n} \log E|x(t_n;x_o)|^p = g(p),$$

$g(p)$ and $I(r)$ are independent of (t_n). The statements thus hold for continuous time.

□

Remark 3.3.

(i) Theorem 3.2 holds uniformly in the sense that in (i) and (ii) $P(\cdot)$ can be replaced by

$$\sup_{x_0/|x_0|\in C} P(\cdot) \quad \text{and} \quad \inf_{x_0/|x_0|\in C} P(\cdot) ,$$

resp. Here C is defined through $\operatorname{supp} \mu = M \times C \subset M \times \mathbb{P}^{d-1}$. This follows from the fact that $g(p)$ is a uniform limit in C (cf. [3], proof of Lemma 2.5) and from the proof of Theorem II. 6.1 in Ellis [19].

(ii) In particular, we have a uniform large deviation principle with respect to $x_0 \in \mathbb{R}^d \setminus \{0\}$ if $C = \mathbb{P}^{d-1}$. This is certainly true under (N). See Remark 3.8.

(iii) We expect that in the case $-\infty < \gamma_- < \gamma_+ < \infty$ typically $I(\gamma_\pm) = \infty$. In fact,

$$I(\gamma^+) = \sup_{p\in\mathbb{R}} (\gamma^+ p - g(p)) = \lim_{p\uparrow\infty} (\gamma^+ p - g(p)),$$

and in nice situations

$$g(p) = \gamma^+ p + c_1\sqrt{p} + c_2 + \dots \qquad \text{for } p \to \infty , \quad c_1 < 0$$

(cf. Simon [29]).

Corollary 3.4. *For each* $x_0 \neq 0$ *and* $\varepsilon > 0$

$$\lim_{t\to\infty} \frac{1}{t} \log P(|\frac{1}{t} \log \frac{|x(t;x_0)|}{|x_0|} - \lambda| \geq \varepsilon) = c(\varepsilon),$$

where

$$c(\varepsilon) = -\min(I(\lambda+\varepsilon), I(\lambda-\varepsilon)) < 0 .$$

Corollary 3.5. *Let* $f: (0,\infty) \to (0,\infty)$ *with* $\lim_{t\to\infty} \frac{1}{t} \log f(t) = 0$ (e.g. $f(t) \equiv c > 0$).

(A) *If* $\gamma_- = \gamma_+ = \lambda$, *then for each* $x_0 \neq 0$

$$\lim_{t\to\infty} \frac{1}{t} \log P(\frac{|x(t;x_0)|}{|x_0|} \geq f(t)) = \begin{cases} 0, & \lambda > 0, \\ -\infty, & \lambda < 0, \end{cases}$$

$$\lim_{t\to\infty} \frac{1}{t} \log P(\frac{|x(t;x_0)|}{|x_0|} \leq f(t)) = \begin{cases} 0 & \lambda < 0, \\ -\infty, & \lambda > 0. \end{cases}$$

(B) *If* $\gamma_- < \gamma_+$, *then for each* $x_0 \neq 0$

$$\lim_{t\to\infty} \frac{1}{t} \log P(\frac{|x(t;x_0)|}{|x_0|} \geq f(t)) = \begin{cases} 0 & , \lambda \geq 0, \\ -I(0), & \lambda < 0, \gamma_+ \neq 0, \end{cases}$$

$$\lim_{t\to\infty} \frac{1}{t} \log P\left(\frac{|x(t;x_0)|}{|x_0|} \le f(t)\right) = \begin{cases} -I(0), & \lambda > 0, \gamma_- \neq 0, \\ 0 & , \lambda \le 0. \end{cases}$$

Proof. We prove the first statement of part (B). For any $\delta > 0$ there exists $t_\delta > 0$ such that

$$-\delta < \frac{1}{t} \log f(t) < \delta$$

for all $t \ge t_\delta$. Hence for $t \ge t_\delta$

$$P\left(\frac{1}{t}\log\frac{|x(t;x_0)|}{|x_0|} > \delta\right) \le P\left(\frac{|x(t;x_0)|}{|x_0|} \ge f(t)\right) \le P\left(\frac{1}{t}\log\frac{|x(t;x_0)|}{|x_0|} \ge -\delta\right).$$

By Theorem 3.1

$$-\inf_{r>\delta} I(r) \le \liminf_{t\to\infty} \frac{1}{t} \log P\left(\frac{|x(t;x_0)|}{|x_0|} \ge f(t)\right) \le$$

$$\le \limsup_{t\to\infty} \frac{1}{t} \log P\left(\frac{|x(t;x_0)|}{|x_0|} \ge f(t)\right) \le -\inf_{r\ge-\delta} I(r) .$$

If $\lambda \ge 0$ then $\inf_{r\ge-\delta} I(r) = 0$ and $\lim_{\delta\downarrow 0} \inf_{r>\delta} I(r) = 0$.

If $\lambda < 0$ and $\gamma_+ \neq 0$ then I is continuous at $r = 0$ and both $\inf_{r>\delta} I(r)$ and $\inf_{r\ge-\delta} I(r)$ converge to $I(o)$.

□

Remark 3.6. (i) Note that $-I(o) = \min_{p\in\mathbb{R}} g(p)$.

(ii) In the case $\lambda = 0$ more information is available for the white noise case, cf. Bougerol ([10],[11]).

Remark 3.7. Le Page [23] (see also Bougerol and Lacroix ([13], Theorem 6.1)) proves the analogue of Corollary 3.4 for products of iid random matrices. However, under his assumption one can only be sure that $g(p)$ is smooth in a neighborhood of $p = 0$, so that Le Page gets the result only for small $\varepsilon > 0$.

Remark 3.8. Stroock [30] proves the uniform version of Theorem 3.2 (see Remark 3.3 (ii)) under condition (N) for the white noise case $A_i(\xi) = A_i$, $i=1,\dots,m$ (for the formally slightly more general case of homogeneous rather then linear vector fields). However, (N) is often too restrictive for applications as typically just one parameter is perturbed by noise (thus $m = 1$). Stroock bases his discussion on a quantity which in the unified treatment takes on the form

$$\alpha = \inf_{\varphi \in C^\infty(M \times \mathbb{P}^{d-1})} \int \left(\sum_{j=1}^{r} (X_j \varphi)^2 + \sum_{i=1}^{m} (h_i \varphi - q_i)^2 \right) d\mu .$$

By Theorem 2.5

$$0 \le \alpha \le g''(0) .$$

Stroock proves that under (N) $\alpha = 0$ iff $\gamma_+ < \infty$ and $\alpha > 0$ iff $\gamma_+ = \infty$. Also, there is a constant $c > 0$ such that for $\alpha = 0$

$$c(r-\lambda)^2 \le I(r) \text{ or, equivalently, } g(p) \le \lambda p + \frac{1}{2c} p^2$$

and for $\alpha > 0$

$$c(r-\lambda)^2 \le I(r) \le \frac{1}{2\alpha} (r-\lambda)^2 \quad \text{for all} \quad r \in \mathbb{R}$$

or, equivalently,

$$\alpha p^2 + \lambda p \le g(p) \le \lambda p + \frac{1}{2c} p^2 \quad \text{for all} \quad p \in \mathbb{R} .$$

We have characterized the appearance of the cases $\alpha = 0$ and $\alpha > 0$ directly in terms of the matrices $A_i(\xi)$, $i=1,\dots,m$, $\xi \in M$, in Theorem 2.5: $\alpha = 0$ iff $\bar{H}$ is compact, $\alpha > 0$ otherwise. Theorem 2.5 and 2.6 also tell us that $0 = \alpha = g''(0)$ iff $\gamma_- = \gamma_+$, while $0 = \alpha < g''(0)$ iff $\gamma_- < \gamma_+$.

<u>Level-1 vector versions</u>

Assume (E). We can derive the large deviation principle for

$$\log \frac{|x(t;x_0)|}{|x_0|} = \int_0^t Q d\tau + \sum_{i=1}^{m} \int_0^t q_i dW_i$$
$$= \int_0^t q_0 d\tau + \sum_{i=1}^{m} \int_0^t q_i \circ dW_i$$

(see equs. (1.4) and (1.5)) from level-1 <u>vector</u> versions via a contraction principle. We just sketch things as proofs remain unchanged.

(i) Define for $\vec{p} = (p_0, p_1, \dots, p_m) \in \mathbb{R}^{m+1}$

$$g_1(\vec{p}) := \lim_{t\to\infty} \frac{1}{t} \log E \exp(p_o \int_0^t Q d\tau + \sum_{i=1}^m p_i \int_0^t q_i dW_i) .$$

This function has properties completely analogous to those of g: It is convex, and as the principal eigenvalue of

$$L_1(\vec{p}) = L + p_o Q + \sum_{i=1}^m (p_i q_i h_i + \frac{1}{2} p_i^2 q_i^2)$$

it is analytic with

$$\frac{\partial g_1}{\partial p_i}\Big|_{\vec{p}=0} = \begin{cases} \lambda, & i=0, \\ 0, & i=1,\dots,m. \end{cases}$$

The Legendre transform

$$I_1(\vec{r}) = \sup_{\vec{p}\in \mathbb{R}^{m+1}} (\langle\vec{r},\vec{p}\rangle - g_1(\vec{p})), \quad \vec{r} \in \mathbb{R}^{m+1},$$

is the entropy function of a large deviation principle for the vector

$$X(t) = (\int_0^t Q d\tau, \int_0^t q_1 dW_1, \dots, \int_0^t q_m dW_m)$$

(just use the (m+1)-dimensional version of Theorem II. 6.1 of Ellis [19]).

Theorem 3.2 can be recovered from this because

$$\log \frac{|x(t;x_o)|}{|x_o|} = \langle X(t),\vec{1}\rangle , \quad \vec{1} = (1,\dots,1) \in \mathbb{R}^{m+1},$$

via the contraction principle

$$g(p) = g_1(p\vec{1}), \quad p \in \mathbb{R},$$

or

$$I(r) = \inf_{\vec{r}\in \mathbb{R}^{m+1}:\langle\vec{r},\vec{1}\rangle=r} I_1(\vec{r}) , \quad r \in \mathbb{R}.$$

(ii) Similarly for

$$g_2(\vec{p}) := \lim_{t\to\infty} \frac{1}{t} \log E \exp(p_o \int_0^t q_o d\tau + \sum_{i=1}^m p_i \int_0^t q_i \circ dW_i).$$

This is the principal eigenvalue of

$$L_2(\vec{p}) = L + p_o q_o + \sum_{i=1}^m (\frac{1}{2} p_i h_i(q_i) + p_i q_i h_i + \frac{1}{2} p_i^2 q_i^2)$$

and as such is analytic and convex with

$$\frac{\partial g_2}{\partial p_i}\Big|_{\vec{p}=0} = \begin{cases} Eq_o & , i=0, \\ \frac{1}{2} Eh_i(q_i), & i=1,\dots,m. \end{cases}$$

The Legendre transform $I_2(\vec{r})$ of $g_2(\vec{p})$ is the entropy function of a large deviation principle for the vector

$$Y(t) = \left(\int_0^t q_o d\tau, \int_0^t q_1 \circ dW_1, \dots, \int_0^t q_m \circ dW_m\right).$$

Since $\langle Y(t),\vec{1}\rangle = \log \frac{|x(t;x_o)|}{|x_o|}$ we can again recover Theorem 3.2 via the contraction principle $g(p) = g_2(p\vec{1})$ or

$$I(r) = \inf_{\vec{r}\in \mathbb{R}^{m+1}:\langle\vec{r},\vec{1}\rangle=r} I_2(\vec{r}), \quad r \in \mathbb{R}.$$

In section 5 we will give level-2 principles which will contract to the one's presented here.

4. Periodic noise and random noise

It turns out that, in contrast to λ and $g(p)$, γ_- and γ_+ do not depend on the statistics of the noise, but just on its range. We formulate all results for γ_+ only. Completely analogous results can be obtained for γ_- by reversing time (see [3], Corollary 4.1).

We associate with (1.1) a deterministic control system

$$(4.1) \quad \dot{x} = (A_o(u_o(t)) + \sum_{i=1}^m u_i(t)A_i(u_o(t))x, \quad x_o \in \mathbb{R}^d,$$

where the set of admissible controls is

$$\mathcal{U} = \{u = (u_o,u_1,\dots,u_m): \mathbb{R}^+ \to M \times \mathbb{R}^m \text{ piecewise constant}\} .$$

The set of possible matrices on the r.h.s. of (4.1) is exactly N, and G and S are the system group and semigroup of (4.1), resp. (see section 2 for definitions). S describes the action of (4.1) forward in time. Let

$$S_t = \{g \in S: \sum_{i=1}^n t_i \le t\}, \quad t > 0 .$$

We have for $0 < t_1 < t_2$

$$S_{t_1} \subset S_{t_2} \subset S \subset G \subset Gl(d, \mathbb{R}).$$

The exponential growth rate of $(S_t, t > 0)$ can be defined by

$$\beta := \lim_{t\to\infty} \frac{1}{t} \log r(S_t), \quad r(S_t) = \sup_{g\in S_t} r(g),$$

where $r(g)$ is the spectral radius of $g \in Gl(d, \mathbb{R})$, or by

$$\delta := \lim_{t\to\infty} \frac{1}{t} \log\| S_t\| , \quad \| S_t\| = \sup_{g\in S_t}\| g \|,$$

where $\|g\|$ is the operator norm of $g \in Gl(d, \mathbb{R})$. Of course $\beta \le \delta$, and it was proved in [2] (Theorem 5.1) and [3] (Theorem 3.1) that, under (E), $\beta \le \gamma_+ \le \delta$.

Theorem 4.1. *Assume* (E). *Then*

$$\beta = \gamma_+ = \delta .$$

Proof. The real noise case was proved by Colonius and Kliemann ([17], Theorem 4.1). By Proposition 3.1 in [3], $\beta = \infty$ iff $\bar{H}$ is not compact iff $\gamma_+ = \infty$. The statement thus remains to be proved in the case $\bar{H}$ compact. Assume w.l.o.g. that a coordinate system has been introduced in which all $A_i(\xi)$, $\xi \in M$, $i=1,\dots,m$, are skewsymmetric. Then for a given admissible control u the Lyapunov exponent of the solution of (4.1) is

$$\lambda(x_o,u) := \limsup_{t\to\infty} \frac{1}{t} \log|x(t;x_o,u)| = \limsup_{t\to\infty} \frac{1}{t} \int_0^t q_o(u_o(\tau),s(\tau;s_o,u))d\tau.$$

Define

$$\kappa := \sup_{x_o\neq 0} \sup_{u\in\mathcal{U}} \lambda(x_o,u), \quad \kappa_p := \sup_{x_o\neq 0} \sup_{u\in\mathcal{U}_p} \lambda(x_o,u)$$

where

$$\mathcal{U}_p = \{u \in \mathcal{U}: u \text{ periodic}\}.$$

Now the periodicity principle of Colonius and Kliemann ([17], Corollary 3.3) applies and gives $\kappa_p = \kappa$. The proof is completed by the following lemma.

Lemma 4.2. *Assume* (H) *and that* $\bar{H}$ *is compact. Then*

(i) $\beta = \kappa_p$,

(ii) $\delta = \kappa$.

Proof: The proof of (i) is the same as in the real noise case (Lemma 4.3 in [2]). In (ii), $\kappa \leq \delta$ follows from the definitions. It remains thus to show that $\delta \leq \kappa$.

Fix an open set $U \subset \text{int } C$ with $\bar{U} \subset \text{int } C$. Then $T < \infty$, where

$$T := \sup_{x \in C, y \in \bar{U}} \inf_{u \in \mathcal{U}} \{t > 0: s(t;x,u) = y\}$$

(see [17], Proposition 2.11). By the definition of δ there exist for given $\varepsilon > 0$ $s_0 \in U$, $u \in \mathcal{U}$, $t_0 > 0$ such that

$$\frac{1}{t_0+T} \left(\int_0^{t_0} q_0(u_0(\tau), s(\tau;s_0,u))d\tau + T \min_{M \times C} q_0 \right) > \delta - \varepsilon .$$

Hence there exists a periodic control u^p with period $t_1 \leq t_0 + T$ such that

$$\frac{1}{t_1} \int_0^{t_1} q_0(u_0^p(\tau), s(\tau;s_0,u^p))d\tau > \delta - \varepsilon ,$$

i.e. $\delta \leq \kappa_p \leq \kappa$. □

Remark 4.3. Theorem 4.1 tells us that γ_+ actually only depends on the semigroup S and thus only on the set N of matrices. $\gamma_+ = \delta$ means that γ_+ is the top growth rate of (4.1) for any possible right hand side, while $\beta = \gamma^+$ tells us that this growth rate can be achieved by a Floquet exponent for a periodic right hand side. Theorem 4.1 also implies that always $\lambda \leq \beta$, i.e. as far as exponential growth rate and destabilization is concerned a stochastic system is always less effective then good periodic switching in the same set N of matrices. However, the definition of γ_+ tells us that very high moments of a solution of (1.1) are able to "see" the optimal growth rate corresponding to the set $N \subset gl(d, \mathbb{R})$.

Corollary 4.4. *Assume* (E). *Then*

$$\gamma_+ = \sup_{A(\cdot)\text{stationary}} \lambda(A(\cdot)) ,$$

where $\lambda(A(\cdot))$ *is the top Lyapunov exponent of* $\dot{x} = A(t)x$, $A(\cdot)$ *a stationary process with values in* N.

Proof: For $\varepsilon > 0$ pick a $u \in \mathcal{U}_p$ with period T and an $x_0 \neq 0$ with $\lambda(x_0,u) > \gamma_+ - \varepsilon$. Now choose

$$A(t) = A_0(\eta_0(t)) + \sum_{i=1}^{m} \eta_i(t)A_i(\eta_0(t)) , \quad t \in \mathbb{R}^+,$$

where $\eta(t) = u(t+\tau)$ (τ uniformly distributed in $[0,T]$) is a stationary process with values in $M \times \mathbb{R}^m$, and $\lambda(A(\cdot)) > \gamma_+ - \varepsilon$.

□

So $\gamma_+ < 0$ assures stability if we know nothing about the noise except its range (universal stability).

There is still another interesting consequence of Theorem 4.1 which sheds new light on γ_+.

Let X be a random variable with $|X|^p \in L_1(P)$ for all $p \in \mathbb{R}$. Define

$$\|X\|_p = \begin{cases} \text{ess inf}|X| , & p = -\infty , \\ \exp E \log|X| \text{ (geometric mean)}, & p = 0 \\ (E|X|^p)^{1/p}, & 0 < |p| < \infty, \\ \text{ess sup}|X| , & p = \infty. \end{cases}$$

Then $p \to \|X\|_p$ is a continuous and increasing function on $[-\infty,\infty]$ into $[0,\infty]$ (Loève [24], p. 160). Recall that $\gamma(p) = g(p)/p$ is also continuous and increasing on $[-\infty,\infty]$ with $\gamma(0) = \lambda$ and $\gamma(\pm\infty) = \gamma_\pm = \lim_{p\to\pm\infty} \gamma(p)$.

The above definition of $\|X\|_p$ and formula (1.4) immediately yield for any $x_0 \neq 0$ and $p \in \mathbb{R}$

$$\gamma(p) = \lim_{t\to\infty} \frac{1}{t} \log \|x(t;x_0)\|_p .$$

Corollary 4.5. *Assume* (E). *Then for any* $x_0 \neq 0$

$$\gamma_\pm = \lim_{t\to\infty} \frac{1}{t} \log \|x(t;x_0)\|_{\pm\infty} .$$

Proof: We restrict ourselves to γ_+. We need to show that we can interchange the limits for $t \to \infty$ and for $p \to \infty$ if applied to $\frac{1}{t} \log \|x(t;x_0)\|_p$. In fact,

$$\lim_{p\to\infty} \lim_{t\to\infty} \frac{1}{t} \log \| x(t;x_0)\|_p = \gamma_+$$

by definition, and

$$\overline{\lim_{t\to\infty}} \lim_{p\to\infty} \frac{1}{t} \log \|x(t;x_0)\|_p = \overline{\lim_{t\to\infty}} \frac{1}{t} \log \|x(t;x_0)\|_\infty$$

$$= \overline{\lim_{t\to\infty}} \frac{1}{t} \text{ ess sup}\left(\int_0^t q_0 d\tau + \sum_{i=1}^m \int_0^t q_i \circ dW_i\right)$$

$$(4.2) \quad = \overline{\lim_{t\to\infty}} \sup_{u\in U} \frac{1}{t} \int_0^t (q_0(u_0(\tau),s(\tau;s_0,u)) + \sum_{i=1}^m u_i(\tau) q_i(u_0(\tau),s(\tau;s_0,u)))d\tau$$

(by the support theorem of Stroock and Varadhan [31])

$$= \delta$$

by the definition of δ and since the r.h.s. of (4.2) is independent of s_0 (first steer s_0 into int C and than maximize). We also see that the $\overline{\lim}$ is actually a lim, and Theorem 4.1 gives $\delta = \gamma_+$.

□

5. Level-2 descriptions

We now look at the deviation of an appropriate random occupation measure of the process $(\xi(t),s(t))$ from its unique invariant probability μ. We try to make the choice such that we have a contraction principle to the level-1 quantities of section 3.

Assume (E) for the whole section.

Let $\mathcal{C} = C(M \times \mathbb{P}^{d-1})$. Define for fixed $x_0 \neq 0$ and for $\vec{V} = (V_0,V_1,\ldots,V_m) \in \mathcal{C}^{1+m}$

$$g(\vec{V}) = \lim_{t\to\infty} \frac{1}{t} \log E \exp \sum_{i=0}^m \int_0^t V_i(\xi(\tau),s(\tau;s_0))dW_i(\tau),$$

where $dW_0(\tau) = d\tau$. The function $g\colon \mathcal{C}^{1+m} \to \mathbb{R}$ is convex and equal to the principal eigenvalue of

$$L(\vec{V}) = L + \sum_{i=1}^m V_i h_i + V_0 + \frac{1}{2} \sum_{i=1}^m V_i^2 .$$

The convex conjugate of g is defined on $M^{1+m} = (\mathcal{C}^{1+m})^*$, $M = \mathcal{C}^*$= signed measures on $M \times \mathbb{P}^{d-1}$, by

$$I(\vec{\nu}) = \sup_{\vec{V}\in\mathcal{C}^{1+m}} (\langle\vec{\nu},\vec{V}\rangle - g(\vec{V})),$$

$\vec{\nu} = (\nu_o, \nu_1, \ldots, \nu_m) \in M^{1+m}$, and, conversely,

$$(5.1)\quad g(\vec{V}) = \sup_{\vec{\nu} \in M^{1+m}} (\langle \vec{\nu}, \vec{V} \rangle - I(\vec{\nu})).$$

The entropy functionals in section 3 can be obtained from $I(\vec{\nu})$ by contraction. In fact, e.g.

$$g_1(\vec{p}) = g(p_o Q, p_1 q_1, \ldots, p_m q_m) ,$$
$$g(p) = g_1(p\,\vec{1}),$$

and

$$I_1(\vec{r}) = \inf_{\vec{\nu} \in M^{1+m}} \{I(\vec{\nu}) : (\int Q d\nu_o, \int q_1 d\nu_1, \ldots, \int q_m d\nu_m) = \vec{r}\} ,$$
$$I(r) = \inf_{\vec{\nu} \in M^{1+m}} \{I(\vec{\nu}) : \int Q d\nu_o + \sum_{i=1}^{m} \int q_i d\nu_i = r\} .$$

(5.1) gives a level-2 characterization of γ_+ via

$$\frac{g(p)}{p} = \sup_{\vec{\nu} \in M^{1+m}} (\int Q d\nu_o + \sum_{i=1}^{m} q_i d\nu_i - \frac{I(\vec{\nu})}{p}) .$$

For $p \to \infty$ we obtain

$$\gamma_+ = \sup_{\vec{\nu} : I(\vec{\nu}) < \infty} \int Q d\nu_o + \sum_{i=1}^{m} q_i d\nu_i .$$

Thus $\gamma_+ = \infty$ indicates that $I(\vec{\nu}) < \infty$ for measures with arbitrarily big mass.

We do not write down explicitly a level-2 principle for $I(\vec{\nu})$. Let us just sketch things since they are beyond the scope of this paper.

A natural candidate for a quantity with entropy functional $I(\vec{\nu})$ is

$$\vec{\nu}_T(\vec{B}) = (\int_0^T 1_{B_o}(\xi(t), s(t; s_o))dt, \ldots, \int_0^T 1_{B_m}(\xi(t), s(t; s_o))dW_m) ,$$

$\vec{B} = (B_o, B_1, \ldots, B_m)$ with $B_i \in \mathcal{B}$ = Borel sets in $M \times \mathbb{P}^{d-1}$. By the ergodic theorem, for $T \to \infty$, $\frac{1}{T} \nu_{T,o}(\omega, \cdot) \to \mu$ a.s. It is also well-known that $\frac{1}{T} \nu_{T,i}(\omega, \cdot) \to 0$ a.s., $i = 1, \ldots, m$, i.e. $\frac{1}{T} \vec{\nu}_T(\omega, \cdot) \to (\mu, 0, \ldots, 0)$ a.s. Thus, under appropriate conditions, the above functional will describe a large deviation principle of the form

$$\frac{1}{T} \log P(\omega: \frac{1}{T} \vec{\nu}_T(\omega,\cdot) \in A) \sim -\inf_{\vec{\nu}\in A} I(\vec{\nu}) .$$

Note that the restriction of the above to $\{\vec{V} = (V_0,0,\ldots,0): V_0 \in C\} \simeq C$ leads to a "classical" level-2 principle for the occupation measure $\nu_{T,0}$ (Donsker and Varadhan [18], Gärtner [22]). However, this principle does not contract to our level-1 principles because it does not contain enough information about the terms $\int q_i dW_i$ (except in the case where $\bar{H}$ is compact and w.l.o.g. $Q = q_0, q_1 = \ldots = q_m = 0$, e.g. in the real noise case).

In particular (cf. also Theorem 2.2), $g(p)$ is the principal eigenvalue of $L(p) = L + pX + pQ + \frac{p^2}{2} R$. Now $A_p = L + pX$ is for each $p \in \mathbb{R}$ the generator of a Markov process on $M \times \mathbf{P}^{d-1}$. Its entropy functional is defined by

$$I_p(\nu) = \sup_{u\in\mathcal{D}(A_p):u>0} - \int \frac{A_p u}{u} d\nu ,$$

$\nu \in M_1$ = probability measures on $M \times \mathbf{P}^{d-1}$ (Donsker and Varadhan [18]), and

$$g(p) = \sup_{\nu\in M_1} (\int (pQ + \frac{p^2}{2} R)d\nu - I_p(\nu)).$$

Let us look again at the case $\gamma_+ < \infty$, assume $q_1 = \ldots = q_m = 0$. Then $R = 0$, $X = 0$ and $A_p = L$ for all $p \in \mathbb{R}$. Thus

$$(5.2) \quad \gamma_+ = \sup_{\nu\in M_1 : I_0(\nu)<\infty} \int q_0 d\nu .$$

It does not seem easy to describe the set of ν's for which $I_0(\nu) < \infty$ if L is only hypoelliptic. For example, in the real noise case $L = G + h_0$ we find by testing against functions $u = u_1(\xi)u_2(s)$ that

$$I_0(\nu) \ge I_G(\nu_1) + I_{\bar{h}}(\nu_2) ,$$

where we have factorized

$$\nu(d\xi,ds) = \rho_2(ds|\xi)\nu_1(d\xi) = \rho_1(d\xi|s)\nu_2(ds),$$

and

$$\bar{h}(s) = \int_M h_0(\xi,s)\rho_1(d\xi|s)$$

is the averaged vector field on $\mathbf{P}^{d-1}$. Of course,

$$I_{\bar{h}}(\nu_2) = \begin{cases} 0 , & \text{if } \nu_2 \text{ is } \bar{h}\text{-invariant} , \\ \infty , & \text{otherwise} , \end{cases}$$

while $I_G(\nu_1) < \infty$ can be described more easily since G is nondegenerate (cf. Pinsky [25]).

The representation (5.2) does not tell us for which ν's we are close to the sup. More insight is obtained by looking again at periodic noise. We have (Colonius and Kliemann [17])

$$(5.3)\quad \gamma_+ = \sup_{u \in \mathcal{U}_{pp}} \frac{1}{T}\int_0^T q_o(u_o(t),s(t;u(\cdot)))dt ,$$

where

$$\mathcal{U}_{pp} = \{u \in \mathcal{U}: (u,s(\cdot;u)) \text{ is T-periodic for some } T > 0\}.$$

Let $\mu_{u,s}$ be the occupation measure sitting on the periodic orbit $\{(u(t),s(t;u(\cdot))): 0 \le t \le T\} \subset M \times \mathbf{P}^{d-1}$, then (5.3) can be written as

$$(5.4)\quad \gamma_+ = \sup_{u \in \mathcal{U}_{pp}} \int q_o d\mu_{u,s} .$$

This formula gives us a clue as to how the measures with $I_o(\nu) < \infty$ that yield values of $\int q_o d\nu$ close to γ_+ look like: Take the occupation measure of a suboptimal periodic solution and mollify it. Those measures are far from being product measures, see the example in section 6.

Formula (5.4) also tells us that $\gamma_+ = \max q_o$ iff there is a periodic solution that can stay arbitrarily close to $\{(\xi,s): q_o(\xi,s) = \max q_o\}$.

6. Example: The random oscillator

Consider the damped linear oscillator with random restoring force

$$(6.1)\quad \ddot{y} + 2\beta\dot{y} + (1 + \sigma F(\xi(t)))y = 0 ,$$

where $\beta \in \mathbb{R}$ is a damping constant and $\sigma \in \mathbb{R}$ is a strength parameter for the noise. We can assume $\sigma \in \mathbb{R}^+$, as $\sigma < 0$ reduces to that case if we replace F by $-F$.

Assume that $\xi(t)$ is an ergodic diffusion process with invariant probability ρ on a compact manifold M, let

$$F: M \to \mathbb{R} \text{ with } EF(\xi(t)) = \int_M F(\xi)d\rho(\xi) = 0 .$$

Assume an analytic situation. Then (E) is satisfied iff $F \not\equiv$ const. An example would be $F(\xi) = \cos \xi$ and $\xi(t) = W(t) + \tau \pmod{2\pi}$, where $W(t)$ is a Wiener process on $\mathbb{R}$ and τ is uniformly distributed in $[0,2\pi)$ and independent of $W(\cdot)$ (Brownian motion on S^1).

The canonical form of (6.1) is, with $x = (y,\dot{y})'$,

$$(6.2) \quad \dot{x} = A(\xi(t))x, \; A(\xi) = \begin{pmatrix} 0 & 1 \\ -1-\sigma F(\xi) & -2\beta \end{pmatrix} .$$

The level curves of the Lyapunov exponent $\lambda = \lambda(\beta,\sigma)$ and, in particular, the stability diagram in the β,σ plane can be found in Arnold and Kliemann [1], p. 65, and, together with asymptotic results for σ small and large, in Arnold, Papanicolaou and Wihstutz [4]. Of course, the numerical calculation had to be based on simulation.

The characterizations given in section 4 make it possible to explicitly calculate $\gamma_+ = \gamma = \gamma(\beta,\sigma)$ for the above system. It will turn out that γ only depends on A and B, where

$$A := \min_{\xi\in M} F(\xi) < 0 < \max_{\xi\in M} F(\xi) =: B ,$$

and on no other information about the noise. We have chosen the real noise case because whenever white noise is present things become trivial since $\gamma(\beta,\sigma) = \infty$ if $\sigma \neq 0$ (cf. Theorem 2.3).

Instead of investigating (6.2) we first transform (6.1) via $\bar{y} = y \exp \beta t$ into

$$\ddot{\bar{y}} + (1-\beta^2 + \sigma F(\xi(t)))\bar{y} = 0$$

with canonical form

$$(6.3) \quad \dot{\bar{x}} = \begin{pmatrix} 0 & 1 \\ -1+\beta^2-\sigma F(\xi(t)) & 0 \end{pmatrix} \bar{x} , \quad \bar{x} = \begin{pmatrix} \bar{y} \\ \dot{\bar{y}} \end{pmatrix} .$$

An elementary calculation yields

$$g(p) = -\beta p + \bar{g}(p) ,$$

thus $\lambda = -\beta + \bar{\lambda}$ and

$$\gamma = -\beta + \bar{\gamma} .$$

It thus suffices to analyze (6.3). Projection onto $\mathbf{P}^1$ yields, with $s = (\cos \varphi, \sin \varphi)$,

$\varphi \in [0,\pi)$ and by means of formula (5.3)

$$(6.4)\quad \bar{\gamma} = \sup \frac{1}{T}\int_0^T q(\varphi(t),F(t))\,dt,$$

where

$$q(\varphi,F) = \frac{1}{2}(\beta^2-\sigma F)\sin 2\varphi$$

and

$$(6.5)\quad \dot{\varphi}(t) = h(\varphi(t),F(t)),\ h(\varphi,F) = -\sin^2\varphi + (-1-\sigma F+\beta^2)\cos^2\varphi .$$

The sup in (6.4) is taken over all $F: \mathbb{R}^+ \to [A,B]$, $\varphi: \mathbb{R}^+ \to \mathbb{P}^1$ (solution of (6.5)) both with period T, $T > 0$ arbitrary.

We have to distinguish between 3 cases:

Case (i) $-\infty < \frac{\beta^2-1}{\sigma} < A$: By looking at the geometry of the problem one can convince oneself that the optimal strategy has to have the following feedback form:

$$F = \begin{cases} F_0, & 0 \le \varphi < \pi/2, \\ B, & \pi/2 \le \varphi < \pi, \end{cases}$$

with $F_0 \in [A,B)$ to be determined. Then

$$\gamma = -\beta + \frac{1}{T_{F_0}+T_B}\left(\int_0^{\pi/2} \frac{q(\varphi,F_0)}{|h(\varphi,F_0)|}\,d\varphi + \int_{\pi/2}^{\pi} \frac{q(\varphi,B)}{|h(\varphi,B)|}\,d\varphi\right),$$

where

$$T_F = \int_0^{\pi/2} \frac{d\varphi}{|h(\varphi,F)|} = \frac{\pi}{2}\,\frac{1}{\sqrt{1+\sigma F-\beta^2}}$$

and

$$T_B = \int_{\pi/2}^{\pi} \frac{d\varphi}{|h(\varphi,B)|} = \frac{\pi}{2}\,\frac{1}{\sqrt{1+\sigma B-\beta^2}} .$$

Note that the invariant probability corresponding to this solution is highly non-product. Maximizing γ by choosing F_0 yields

$$\gamma = \gamma_1 = -\beta + \frac{1}{\pi}\,\frac{\sqrt{1+\sigma A-\beta^2}\sqrt{1+\sigma B-\beta^2}}{\sqrt{1+\sigma A-\beta^2}+\sqrt{1+\sigma B-\beta^2}}\log\frac{1+\sigma B-\beta^2}{1+\sigma A-\beta^2}$$

for $0 \le \sigma < \frac{1-\delta^2}{-A+\delta^2 B}(1-\beta^2)$ (choose $F_0 = A$), where

$\delta = 0.2785$ = unique solution of $-\log \delta = 1+\delta$.

If

$$\frac{1-\delta^2}{-A+\delta^2 B}(1-\beta^2) \le \sigma < \frac{1}{-A}(1-\beta^2)$$

we have to choose $F_0 = \delta^2 B - \frac{1-\delta^2}{\sigma}(1-\beta^2)$ and obtain $T_{F_0} = T_B/\delta$ and finally

$$\gamma = \gamma_2 = -\beta + \varepsilon\sqrt{1+\sigma B-\beta^2}$$

with $\varepsilon = \frac{2\delta}{\pi} = 0.1773$.

Case (ii) $A \le \frac{\beta^2-1}{\sigma} < B$: Now the optimal regime is either the one of case (i) for γ_2 or just sitting at $F = A$, $\varphi = \arctan\sqrt{-1-\sigma A+\beta^2}$, thus

$$\gamma = \max(\gamma_2,\gamma_3) ,$$

where

$$\gamma_3 = -\beta + \sqrt{-1-\sigma A+\beta^2} .$$

This yields

$$\gamma = \begin{cases} \gamma_2 = -\beta + \varepsilon\sqrt{1+\sigma B-\beta^2} & \text{if condition } (*) \text{ holds,} \\ \gamma_3 = -\beta + \sqrt{-1- A+\beta^2} & \text{if condition } (\bar{*}) \text{ holds,} \end{cases}$$

where

$$(*) = \begin{cases} \frac{1}{-A}(1-\beta^2) \le \sigma < \frac{1+\varepsilon^2}{-A-\varepsilon^2 B}(1-\beta^2) \text{ and } \beta^2 < 1 & \text{if } -A-\varepsilon^2 B > 0, \\ \frac{1}{-A}(1-\beta^2) \le \sigma < \infty \text{ and } \beta^2 < 1 \text{ if } -A-\varepsilon^2 B = 0, \\ \max(\frac{1}{-A}(1-\beta^2), \frac{1+\varepsilon^2}{-A-\varepsilon^2 B}(1-\beta^2)) \le \sigma < \infty \text{ if } -A-\varepsilon^2 B < 0, \end{cases}$$

and

$$(\bar{*}) = \begin{cases} \max(\frac{1+\varepsilon^2}{-A-\varepsilon^2 B}(1-\beta^2), \frac{1}{-B}(1-\beta^2)) \le \sigma & \text{if } -A-\varepsilon^2 B > 0, \\ \frac{1}{-B}(1-\beta^2) \le \sigma < \infty \text{ and } \beta^2 \ge 1 & \text{if } -A-\varepsilon^2 B = 0, \\ \frac{1}{-B}(1-\beta^2) \le \sigma \le \frac{1+\varepsilon^2}{-A-\varepsilon^2 B}(1-\beta^2) \text{ and } \beta^2 \ge 1 & \text{if } -A-\varepsilon^2 B < 0 . \end{cases}$$

<u>Case (iii)</u> $B \le \frac{\beta^2-1}{\sigma} < \infty$: Now the optimal regime is "switching on the worst possible frozen system", i.e. $F = A$ and $\varphi = \arc\tan \sqrt{-1-\sigma A+\beta^2}$, thus

$$\gamma = \gamma_3 = -\beta + \sqrt{-1-\sigma A+\beta^2}$$

which holds for $0 \le \sigma \le \frac{1}{-B}(1-\beta^2)$ and $\beta^2 \ge 1$.

Summing up, we have in the typical case $-A-\varepsilon^2 B > 0$ (e.g. if $A = -B$)

$$\gamma(\beta,\sigma) = \begin{cases} \gamma_1 & \text{for } 0 \le \sigma < \frac{1-\delta^2}{-A+\delta^2 B}(1-\beta^2), \\ \gamma_2 & \text{for } \frac{1-\delta^2}{-A+\delta^2 B}(1-\beta^2) \le \sigma < \frac{1+\varepsilon^2}{-A-\varepsilon^2 B}(1-\beta^2), \\ \gamma_3 & \text{for } \frac{1-\varepsilon^2}{-A-\varepsilon^2 B}(1-\beta^2) \le \sigma < \infty, \end{cases}$$

see Figure 1. Similarly for the other cases.

For $\sigma \downarrow 0$ we obtain

$$\gamma(\beta,\sigma) = \begin{cases} -\beta + \frac{B-A}{2\pi\sqrt{1-\beta^2}}\sigma + o(\sigma) & \text{for } |\beta| < 1, \\ -\beta + \sqrt{\beta^2-1} + \frac{-A}{2\sqrt{\beta^2-1}}\sigma + o(\sigma) & \text{for } |\beta| > 1, \end{cases}$$

this way recovering the deterministic values on the β axis.

Compare this with

$$\lambda(\beta,\sigma) = \begin{cases} -\beta + c_1^2\sigma^2 + O(\sigma^3), & |\beta| < 1, \\ -\beta + \sqrt{\beta^2-1} - c_2^2\sigma^2 + O(\sigma^3), & |\beta| > 1, \end{cases}$$

(see [4]).

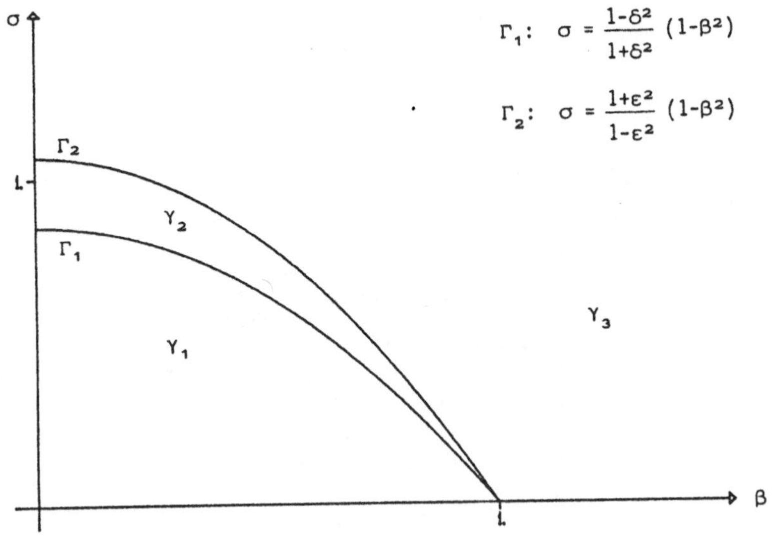

Figure 1: The different regimes for $\gamma(\beta,\sigma)$ for the case $B = -A = 1$

The ultimate aim of this analysis are the level curves of $\gamma(\beta,\sigma)$ in the β,σ plane. They can now be easily calculated, see Figure 2. These curves are straight lines in the area where γ_3 applies. In particular, the curve $\gamma(\beta,\sigma) = 0$ is

$$\sigma = \begin{cases} \frac{1}{-A} & \text{in the } \gamma_3 \text{ region,} \\ \frac{1+\varepsilon^2}{\varepsilon^2 B}\beta^2 - \frac{1}{B} & \text{in the } \gamma_2 \text{ region ,} \end{cases}$$

and more complicated in the γ_1 region. Note that for small β and σ the $\gamma = 0$ curve has the form

$$\sigma = \frac{2\pi}{B-A}\beta + \ldots$$

while the $\lambda = 0$ curve starts like $\sigma = c\sqrt{\beta} + \ldots$

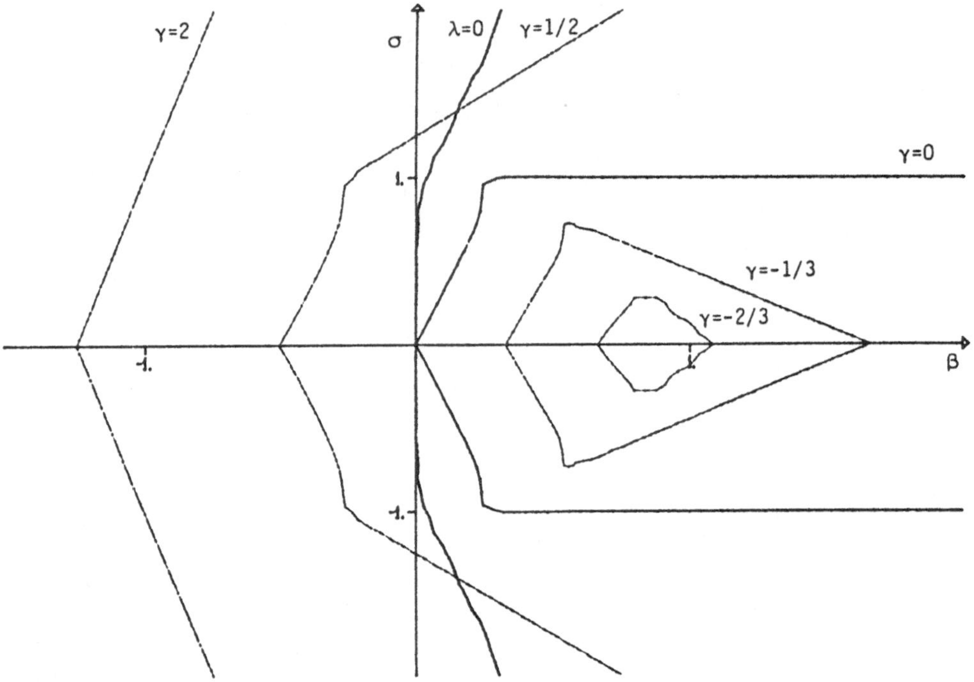

Figure 2: Level curves of γ for the random oscillator in the case $B = -A = 1$. Also shown is the curve $\lambda = 0$ for the case $F(\xi) = \cos \xi$, $\xi(t)$ = standard Brownian motion on S^1.

We stress again that the γ level curves are the same for any noise process with the same A and B (while $\lambda(\beta,\sigma)$ depends on the particular noise). Consequently, inside the $\gamma = 0$ curve of the β,σ plane any system with range $(F) = [A,B]$ is stable with probability 1, and all its positive moments are also stable. We thus have a criterion for stability which is true even if the noise is statistically worst possible (i.e. close to the optimal periodic regime). So, in the $\gamma < 0$ region we have universal stability, and this region is biggest possible!

7. Generalization to flag manifolds and nonlinear stochastic flows

We only briefly sketch things here since on the basis of the preceding they should be fairly obvious.

Full Lyapunov spectrum, flag manifolds

The full spectrum $\lambda_1 \geq \lambda_2 \geq \dots \geq \lambda_d$ of (1.1) from Oseledec's theorem (see Arnold and Wihstutz [6]) can be obtained by measuring the growth rate of the volume of a k-frame moved by the linear system, $k = 1,\dots,d$. The case $k = 1$ is the one considered above. The general case is most elegantly dealt with by 'projecting' the linear system onto the compact manifold of flags of subspaces V_k in $\mathbb{R}^d$,

$$\mathbb{F} = \{\mathbb{f} = (V_1 \subset \dots \subset V_d) : \dim V_k = k,\ k = 1,\dots,d\}.$$

The linear system (1.1) acts on $\mathbb{F}$ via $\mathbb{f} \to \phi(t)\,\mathbf{f} = \mathbf{f}(t) = (\phi(t)V_1 \subset \dots \subset \phi(t)V_d)$, $\phi(t)$ = fundamental matrix of (1.1).

Introduce 'polar coordinates' with 'angular part' $\mathbf{f}(t) \in \mathbb{F}$ and 'radial parts'

$$D^{(k)}(t) = |\mathrm{vol}(\phi(t)V_k)| = |\det \phi(t)|_{V_k}| \in \mathbb{R}^+,$$

$k = 1,\dots,d.$

As in the case $k = 1$, $\mathbb{f}(t)$ satisfies a stochastic differential equation

$$(7.1)\quad d\,\mathbb{f}(t) = \mathbb{F}Y_0(\xi(t),\mathbb{f}(t))dt + \sum_{i=1}^{m} \mathbb{F}Y_i(\xi(t),\mathbb{f}(t)) \circ dW_i,$$

$$\mathbb{f}(0) = \mathbb{f}_0 \in \mathbb{F},$$

which is decoupled from the radial parts. Here $\mathbb{F}Y_i(\xi,\mathbf{f})$ = 'projection' of the linear vector field $Y_i(\xi,x) = A_i(\xi)x$ onto $\mathbb{F}$. Again, $(\xi(t),\mathbb{f}(t))$ is a diffusion process with generator $\mathbb{F}L = G + \mathbb{F}Y_0 + \frac{1}{2}\sum_{i=1}^{m}(\mathbb{F}Y_i)^2$. The radial parts are given by

$$D^{(k)}(t) = |\mathrm{vol}(V_k)|\exp\Big(\int_0^t Q^{(k)}(\xi(\tau);\mathbf{f}(\tau;\mathbb{f}_0))d\tau + \sum_{i=1}^{m}\int_0^t P_i^{(k)}dW_i\Big)$$

(for $Q^{(k)}$ and $P_i^{(k)}$ see Baxendale [7]).

Now assume the analogue of condition (H) for (7.1),(1.2). Then there exists a unique invariant probability μ on $M \times \mathbb{F}$ with marginal ρ on M, and for each fixed $\mathbb{f}_0 \in \mathbb{F}$ and for $k = 1,\dots,d$

$$\lim_{t\to\infty} \frac{1}{t} \log D^{(k)}(t) = \int_{M\times\mathbb{F}} Q^{(k)}d\mu = \lambda_1 + \dots + \lambda_k \quad P\text{-a.s.}$$

In particular, for $k = d$ we obtain the average exponent

$$\lim_{t\to\infty} \frac{1}{t} \log \det \phi(t) = E \operatorname{trace} A_0(\xi(0)) = \lambda_1 + \ldots + \lambda_d \quad P\text{-a.s.}$$

Collecting all equations into one gives, with $\vec{Q} = (Q^{(1)},\ldots,Q^{(d)})$,

$$\Lambda = (\lambda_1, \lambda_1+\lambda_2, \ldots, \lambda_1+\ldots+\lambda_d) = \int_{M\times \mathbb{F}} \vec{Q}(\xi,\mathbb{f})d\mu,$$

which is a formula for the whole Lyapunov spectrum. The function

$$g(\vec{p}) = \lim_{t\to\infty} \frac{1}{t} \log E(D^{(1)}(t))^{p1} \ \ldots \ (D^{(d)}(t))^{pd}, \quad \vec{p} \in \mathbb{R}^d,$$

plays the same rôle for large deviations from $\frac{1}{t} \log D^{(k)}(t) \to \Lambda$ as the $g(p)$ from above.

An equivalent way of getting the whole spectrum consists of lifting (1.1) to the exterior products $(\mathbb{R}^d)^{\wedge k}$ and watching the growth of one solution, see Bougerol and Lacroix [13] and Bougerol [12]. San Martin [27] introduces a polar decomposition on the system group level which is adapted to the system.

Nonlinear stochastic flows

For basic notions and results see Carverhill ([14],[15],[16]), Baxendale [7] and San Martin and Arnold [28]. Let $\xi(t)$ on M be as above. Let (for simplicity) N be a compact Riemannian C^∞ manifold, and let a stochastic flow of diffeomorphisms $x(t;x_0)$ be defined on N by a stochastic differential equation

$$(7.2) \quad dx(t) = Y_0(\xi(t),x(t))dt + \sum_{i=1}^{m} Y_i(\xi(t),x(t)) \circ dW_i,$$

$$x(0;x_0) = x_0 \in N,$$

such that (1.2), (7.2) has a uniquely ergodic solution with invariant probability π on $M \times N$.

We linearize the flow $x(t;\cdot)$ to get a flow $(Tx)(t)$ on $M \times TN$, $TN = \bigcup_{x\in N} \{x\} \times T_xN$ = tangent bundle over N, given by (1.2) and

$$(7.3) \quad d(Tx)(t) = TY_0(\xi(t),(Tx)(t))dt + \sum_{i=1}^{m} TY_i(\xi(t),(Tx)(t)) \circ dW_i,$$

$$(Tx)(0) = v \in TN .$$

This flow is being projected onto $M \times \mathbb{P}N$, $\mathbb{P}N = \bigcup_{x\in N} \{x\} \times \mathbb{P}(T_xN)$ = projective bundle over N, which yields

$$d(\mathbb{P}x)(t) = \mathbb{P}Y_0(\xi(t),(\mathbb{P}x)(t))dt + \sum_{i=1}^{m} \mathbb{P}Y_i(\xi(t),(\mathbb{P}x)(t)) \circ dW_i ,$$

$$(\mathbb{P}x)(0) = \mathbb{P}v \in \mathbb{P}N.$$

Under a very natural generalization of condition (H) to this nonlinear case (see San Martin and Arnold [28]) the diffusion process $(\xi(t), \mathbb{P}x(t))$ has a unique invariant probability μ on $M \times \mathbb{P}N$ with marginal π on $M \times N$, and for each $v \in TN$, $v \neq 0$,

$$(7.4) \quad \lim_{t\to\infty} \frac{1}{t} \log \|Tx(t)(v)\| = \lambda = \int_{M\times \mathbb{P}N} Q d\mu \text{ P-a.s.},$$

where λ is the top Lyapunov exponent of the flow (7.3) (for Q see Carverhill [15] or Baxendale [7]).

The groups G and H from section 2 also have natural generalizations, and large deviations around (7.4) are controlled by

$$g(p) = \lim_{t\to\infty} \frac{1}{t} \log E\|Tx(t)(v)\|^p, \quad p \in \mathbb{R} .$$

For the whole spectrum $\lambda_1 \geq \ldots \geq \lambda_d$ of (7.3) we can work on $M \times \mathbb{F}N$, $\mathbb{F}N = \bigcup_{x\in N} \{x\} \times \mathbb{F}(T_xN)$ = flag bundle over N, like in the first part of this section. A particular case is again the average exponent $\lambda_1 + \ldots + \lambda_d$, to which Elworthy and Stroock [20] have applied Stroock's results [30].

Acknowledgement. The authors are grateful to D. Stroock and J. Gärtner for valuable discussions. The figures were produced by K.-U. Schaumlöffel.

References

[1] Arnold, L., and Kliemann, W.: Qualitative theory of stochastic systems, in: Probabilistic Analysis and Related Topics, Vol. 3, ed. Bharucha-Reid, Academic Press, New York, 1-79 (1983).

[2] Arnold, L., Kliemann, W., and Oeljeklaus, E.: Lyapunov exponents of linear stochastic systems, in [5], 85-125.

[3] Arnold, L., Oeljeklaus, E., and Pardoux, E.: Almost sure and moment stability for linear Ito equations, in [5], 129-159.

[4] Arnold, L., Papanicolaou, G., and Wihstutz, V.: Asymptotic analysis of the Lyapunov exponent and rotation number of the random oscillator and applications, SIAM J. Appl. Math. 46, 427-450 (1986).

[5] Arnold, L., and Wihstutz, V. (editors): Lyapunov Exponents, Proceedings of a Workshop, Lecture Notes in Mathematics Vol. 1186, Springer-Verlag, Berlin, Heidelberg, New York, Tokyo (1986).

[6] Arnold, L., and Wihstutz, V.: Lyapunov exponents - a survey, in [5], 1-26.

[7] Baxendale, P.: The Lyapunov spectrum of a stochastic flow of diffeomorphisms, in [5], 322-337.

[8] Baxendale, P.: Moment stability and large deviations for linear stochastic differential equations, Preprint, Mathematics Department, University of Aberdeen (1985).

[9] Baxendale, P., and Stroock, D.: forthcoming

[10] Bougerol, P.: Oscillation des produits de matrices alêatoires dont l'exposant de Lyapunov est nul, in [5], 27-36.

[11] Bougerol, P.: Tightness of products of random matrices and stability of linear stochastic systems, Annals of Probability (to appear).

[12] Bougerol, P.: Systèmes linêaires a coefficients Markoviens,1. Comparaison des exposants de Lyapunov, Preprint, Paris (1986).

[13] Bougerol, P., and Lacroix, J.: Products of random matrices with applications to Schrödinger operators, Birkhäuser, Boston, Basel, Stuttgart (1985).

[14] Carverhill, A.: Flows of stochastic dynamical systems: ergodic theory, Stochastics 14, 273-318 (1985).

[15] Carverhill, A.: A formula for the Lyapunov numbers of a stochastic flow. Application to a perturbation theorem, Stochastics 14, 209-226 (1985).

[16] Carverhill, A.: Survey: Lyapunov exponents for stochastic flows on manifolds, in [5], 292-307.

[17] Colonius, F., and Kliemann, W.: Infinite time optimal control and periodicity, Preprint # 240, Institute for Mathematics and its applications, University of Minnesota (April 1986).

[18] Donsker, M.D., and Varadhan, S.R.S.: Asymptotic evaluation of certain Markov process expectations for large time, I, Commun. Pure Appl. Math. 28, 1-47 (1975).

[19] Ellis, R.S.: Entropy, large deviations, and statistical mechanics, Springer, New York, Berlin, Heidelberg, Tokyo (1985).

[20] Elworthy, K.D., and Stroock, D.W.: Large deviation theory for mean exponents of stochastic flows (appendix to Carverhill, A., Chappell, M., and Elworthy,K.D.: Characteristic exponents for stochastic flows), in: Albeverio, S., Blanchard, Ph., and Streit, L. (eds.): Stochastic Processes - Mathematics and Physics, Lecture Notes in Mathematics 1158, Springer, Berlin, Heidelberg, New York, Tokyo, 72-80 (1986).

[21] Freidlin, M.I., and Wentzell, A.D.: Random perturbations of dynamical systems, Springer, New York, Berlin, Heidelberg, Tokyo (1984).

[22] Gärtner, J.: On large deviations from the invariant measure, Theory of Probability and its Applications 22, 24-39 (1977).

[23] Le Page, E.: Théorèmes limites pour les produits de matrices aléatoires, in: Heyer, H. (ed.): Probability measures on groups, Lecture Notes in Mathematics 928, Springer, Berlin, Heidelberg, New York, 258-303 (1982).

[24] Loève, M.: Probability theory, Van Nostrand, Princeton, N.J. (1963).

[25] Pinsky, R.: On evaluating the Donsker-Varadhan I-functional, Annals of Prob. 13, 342-362 (1985).

[26] San Martin, L.: A note on transitivity of coupled control systems, in [5], 126-128.

[27] San Martin, L.: PhD Thesis, Control Theory Centre, University of Warwick (1986).

[28] San Martin, L., and Arnold, L.: A control problem related to the Lyapunov spectrum of stochastic flows, Matemática Aplicada e computacional 5 no. 1 (1986).

[29] Simon, B.: Semiclassical analysis of low lying eigenvalues, I. Non-degenerate minima: asymptotic expansions, Ann. Inst. Henri Poincaré (section A: Physique théorique), 38, 295-307 (1983).

[30] Stroock, D.W.: On the rate at which a homogeneous diffusion approaches a limit, an application of the large deviation theory of certain stochastic integrals, to appear in: Annals of Probability 14, No. 3 (1986).

[31] Stroock, D.W., and Varadhan, S.R.S.: On the support of diffusion processes with applications to the strong maximum principle, Proc. 6th Berkeley Symp. Math. Statist. Probab. 3, 333-359 (1972).

ON THE SEMIMARTINGALE DECOMPOSITION OF QUASIDIFFUSIONS WITH NONNATURALE SCALE

G. Burkhardt
Technische Universität
Sektion Mathematik
DDR-8027 Dresden

U. Küchler
Humboldt-Universität
Sektion Mathematik
DDR - 1086 Berlin

1. Introduction

Assume $W = (W_t, \underline{\underline{F}}_t, P_x)$ $(t \geq 0,\ x \in R)$ is a standard Wiener process on the real line R and $l^W(t,x)$ its local time normalized by

$$\int_0^t f(W_s)d_s = 2\int_R l^W(t,x)f(x)dx \qquad (t \geq 0,\ f \text{ bounded}).$$

Let m be a nondecreasing function from R to $\bar{R}$ and define E_m to be the (closed) set where m increases. Put $a_o = -\infty$, $b_o = \inf E_m$, $a_1 = \sup E_m$, $b_1 = \infty$ and assume $b_o < 0 < a_1$, $0 \in E_m$.
Suppose p to be a strictly increasing continuous function from (b_o, a_1) to R with $p(0) = 0$. Define for $t \geq 0$ $\tilde{S}_t := \int l^W(t, p(x))m(dx)$,
$\tilde{T}_t := \inf \{u > 0 / \tilde{S}_u > t\}$. Then $(\tilde{T}_t)^R_{t \geq o}$ is an $\underline{\underline{F}}$-time change ($\underline{\underline{F}} := (\underline{\underline{F}}_t)_{t \geq o}$).
The general theory of time changes and space transformations of Markov processes yields, that

$$X = (X_t, \underline{\underline{G}}_t, P_x) := (p^{-1}(W_{\tilde{T}_t}), \underline{\underline{F}}_{\tilde{T}_t}, P_x) \quad (x \in E_m) \tag{1.1}$$

is a standard Markov process with state space E_m. It is called a quasidiffusion (or gap diffusion) with speed measure m and scale p. The scale is called natural if $p(x) = x$. In general X is not continuous but it is skip free in the sense

$$(\min(X_{t-}, X_t),\ \max(X_{t-}, X_t)) \cap E_m = \emptyset \quad (t > 0) \quad \text{a.s.}$$

Remark that $\tilde{X} := (W_{\tilde{T}_t}, \underline{\underline{G}}_t, P_x)$ $(x \in p(E_m))$ is a quasidiffusion on $p(E_m)$ with speed measure m and natural scale.
In the following we restrict ourselves to $P = P_o$ and assume that the following hypothesis holds:

"p^{-1} is the difference of two convex functions" (1.2)

Then X is a $\underline{\underline{G}}$-semimartingale with respect to P. We shall study the decomposition $X = M+A$ into a local martingale M and a locally bounded variation process A. For natural scale this has been done in [1].

2. Results

2.1. Let us start with some notations. The closed set E_m has the representation $E_m = R \setminus (\bigcup_{k \in K} (a_k, b_k))$ where K is an index set of nonnegative integers with a_i, $b_i (i = 0, 1)$ as above and $0 \in K (1 \in K)$ if and only if $b_o > -\infty$ ($a_1 < \infty$ respectively). Put $K' := K \setminus \{0, 1\}$.

2.2. Introduce for $K \in K'$ and $t > 0$

$$A_k(t) := \sum_{s \leq t} \mathbb{1}\{X_{s-} = a_k, X_s = b_k\}$$

$$B_k(t) := \sum_{s \leq t} \mathbb{1}\{X_{s-} = b_k, X_s = a_k\} .$$

Then the skip-freeness of X implies

$$\{s \leq t \,/\, \Delta X_s \neq 0\} = \bigcup_{k \in K'} \{s = t \,/\, \Delta A_k(s) \neq 0 \ \text{ or } \Delta B_k(s) \neq 0\} .$$

2.3. Define the local time $l(t,x)$ of X by

$$l(t,x) := l^W(\tilde{T}_t, p(x)) \quad (x \in E_m, \ t \geq 0) .$$

Then l is continuous in $(t,x) \in (0,\infty) \times E_m$ and we have

$$\int_0^t f(X_s) ds = \int_{E_m} f(x) l(t,x) m(dx) \quad (t \geq 0, \ f \text{ bounded}).$$

It can be shown, that $(p(b_k) - p(a_k)) A_k(t) - l^W(\tilde{T}_t, p(a_k))$ $(t \geq 0)$ for every $k \in K'$ is a local martingal with respect to $\underline{\underline{G}}$ (see [1], an analogous result holds for $B_k(\cdot)$, $k \in K'$).

2.4. Introduce U_t by

$$U_t := \int_0^t \mathbb{1}_{p(E_m)}(W_s) \, dW_s \quad (t \geq 0) .$$

Then $\tilde{M}^c := (U_{\tilde{T}_t})_{t \geq 0}$ is a continuous $\underline{\underline{G}}$-local martingale and (because $\tilde{T}$ is U-continuous in the sense of [3]) it holds (see [1])

$$\langle U_{\tilde{T}} \rangle_t = \langle U \rangle_{\tilde{T}_t} = \tilde{T}_t^c := \tilde{T}_t - \sum_{s \leq t} \Delta \tilde{T}_s \quad (t \geq 0) .$$

Now we are ready to formulate the

2.5. PROPOSITION: Assume p is twice continuous differentiable. Then the quasidiffusion X defined by (1.1) is a $\underline{\underline{G}}$-semimartingale with respect to P and has the decomposition $X = M^c + M^d + A$ where

a) M^c is a continuous $\underline{\underline{G}}$-local martingale with

$$M_t^c = \int_0^t \frac{d\tilde{M}_s^c}{p'(X_{s-})} \quad \text{and} \quad \langle M^c \rangle_t = \int_0^t \frac{d\tilde{T}_s}{(p'(X_{s-}))^2} \quad (t \geq 0) \tag{2.1}$$

b) M^d is a purely discontinuous $\underline{\underline{G}}$-local martingale given by

$$M_t^d = \sum_{k \in K'} (b_k - a_k) \cdot \left[(A_k(t) - q_k l(t,a_k)) - (B_k(t) - q_k l(t,b_k)) \right] \quad (t \geq 0)$$

where $q_k := (p(b_k) - p(a_k))^{-1}$ $(k \in K')$.

c) A is a continuous $\underline{\underline{G}}$-adapted process with locally integrable variation defined by

$$A_t = -\frac{1}{2} \int_0^t \frac{p''(X_{s-})}{p'(X_{s-})} \, d\langle M^c \rangle_s + \sum_{k \in K'} (b_k - a_k) q_k - \frac{1}{p'(a_k)} \quad l(t,a_k)$$

$$- \sum_{k \in K'} \left((b_k - a_k) q_k - \frac{1}{p'(b_k)} \right) l(t,b_k) + \frac{1}{p'(b_o)} \, l(t,b_o) - \frac{1}{p'(a_1)} \, l(t,a_1)$$

with $p'(b_o) = \lim_{x \downarrow b_o} p'(x)$, $p'(a_1) = \lim_{x \uparrow a_1} p'(x)$.

2.6. The proof follows by an application of Ito's formula for semimartingales ([2], p. 47) to $X = p^{-1}(\tilde{X})$ and by using the decomposition of the quasidiffusion with natural scale X given in [1] .
One can derive an analogy of Proposition 2.5. for all p satisfying (1.2) by using the generalisation of Ito's formula [2], p. 186). But in this formula Meyers local time $L_t^x(X)$ appears, which in general does not coincide with $l(t,x)$. If m is strictly increasing, we have $L_t^x(X) = (p'(x))^{-1} l(t,x)$ with $p' = \frac{p'_r + p'_l}{2}$ where p'_r and p'_l denote the right respectively left hand side derivative of p. The general relation between $L_t^x(X)$ and $l(t,x)$ and its applications to the semimartingale decomposition of X for general scale p will be studied in a further note.

References:

[1] Burkhardt, G., Küchler, U.; The semimartingale decomposition of one-dimensional quasidiffusions with natural scale, Preprint Humboldt-Universität Berlin, (1984), Nr.90, submitted for publication

[2] Jacod, J.; Calcul Stochastique et Problèmes de Martingales. Lecture Notes in Math. Vol. 714. Springer Berlin Heidelberg New York, (1979)

[3] Kazamaki, N.; Changes of time, stochastic integrals, and weak martingales. Z.Wahrscheinlichkeitstheor. Verw. Geb: 22, 25-32 (1972)

TIME REVERSAL OF GAP DIFFUSIONS

H. Langer, W. Schenk
Sektion Mathematik der TU Dresden
Mommsenstr. 13
Dresden, DDR-8027

1. Let s be a real continuous increasing function on $(0,1)$ and m,k two nonnegative finite or σ-finite measures on $(0,1)$ such that $\operatorname{supp} k \subset \operatorname{supp} m$ and $0,1 \in \operatorname{supp} m$. We use Feller's classification of the boundaries 0 and 1. E.g., 0 is $(s,m+k)$-regular if $|s(0)| < \infty$ and $(m+k)((0, 1/2]) < \infty$, $(s,m+k)$-exit if $|s(0)| < \infty$, $(m+k)((0, 1/2]) = \infty$ and $\int_0^{1/2}(m+k)ds < \infty$ (here we identify a measure with one of its generating functions) etc., see [M]. If the boundary $j = 0$ or 1 is regular we impose a boundary condition

$$(-1)^{j+1} \varkappa_j f(j) + \pi_j(D_s^+f)(j) = 0 \tag{1}$$

($\varkappa_j, \pi_j \geqq 0$; $\varkappa_j + \pi_j = 1$, $(D_s^+f)(x)$ is the right derivative of f with respect to s at x and $(D_s^+f)(j) := \lim\limits_{\substack{x \to j \\ x \in (0,1)}} (D_s^+f)(x)$). By C^0 we denote the set of all continuous functions f on $[0,1]$ with the following properties:

(a) On each component of the open set $(0,1)\setminus \operatorname{supp} m$ the function f is linear in s;

(b) If the boundary $j = 0$ or 1 is $(s,m+k)$-exit, -natural or -regular with $\pi_j = 0$ then $f(j) = 0$.

In C^0 we consider the operator A given formally by

$$Af := \frac{dD_s^+f - f\,dk}{dm}$$

and a boundary condition (1) at each $(s,m+k)$-regular boundary j. E.g., the equation $\lambda f - Af = g$ ($\lambda > 0$, $g \in C^0$) means that

$$\lambda f dm - dD_s^+f + f\,dk = g\,dm \text{ on } (0,1), \tag{2}$$

that f satisfies at $j = 0$ the boundary condition (1) if 0 is $(s,m+k)$-regular, $f(0) = 0$ if 0 is $(s,m+k)$-exit or -natural, $(D_s^+f)(0) = 0$ if 0 is $(s,m+k)$-entrance or -natural, and a corresponding condition at $j=1$.

Theorem 1. The operator A is the infinitesimal generator of a strongly continuous nonnegative contraction semigroup $(T_t)_{t\geq 0}$ in C^0.

In the proof of Theorem 1 and in the sequel we make use of two solutions $\chi_0(.;\lambda)$, $\chi_1(.;\lambda)$ $(\lambda > 0)$ of the homogeneous equation corresponding to (2) where, e.g., χ_0 is nondecreasing and satisfies the same boundary conditions at 0 as the solution f of (2).

According to Theorem 1, with the operator A there is associated a Hunt process $X = (\Omega, \mathfrak{F}, \mathfrak{F}_t, X_t, \theta_t, \zeta, P_x)$, see [C], [D], [BG]. In particular, $\zeta(\omega)$ denotes the life time of the path $\omega \in \Omega$. The state space E of X is the set supp $m \cap (0,1)$, augmented by those boundaries $j = 0$ or 1 which are (s,m+k)-regular with $\pi_j \neq 0$ or -entrance. The process X is called the gap diffusion associated with s,m,k and the boundary condition (1).

2. We suppose that the following condition is satisfied:

(K) $k \neq 0$ or, if $k = 0$, for at least one boundary $j = 0$ or 1 we have $|s(j)| < \infty$ and, if j is (s,m+k)-regular, $\varkappa_j > 0$.

Then the potential kernel K(x,y) (with respect to m) of the operator A exists:

$$K(x,y) := \begin{cases} u_0(x)\ u_1(y) & 0 < x \leq y < 1, \\ u_0(y)\ u_1(x) & 0 < y \leq x < 1, \end{cases}$$

where u_0, u_1 are certain nonnegative solutions of the equation $dD_s^+u - u dk = 0$. It can also be shown that for $\lambda \downarrow 0$ we have

$$G(x,y;\lambda) \uparrow K(x,y) \quad (0 < x,y < 1)$$

where $G(x,y;\lambda)$ denotes the Green's kernel of $\lambda I - A$.

In order to assure that the life time ζ of X is finite we need an additional condition:

(K_1) No (s,m+k)-natural boundary $j = 0$ or 1 is (s,k)-regular or -exit.

Theorem 2. For any $\lambda > 0$, $x \in E$ and any continuous function g on [0,1] we have

$$E_x(e^{-\lambda\zeta}\, g(X_{\zeta-});\ \zeta < \infty) = \int_0^1 G(x,y;\lambda) g(y)\, dk(y) +$$

$$+ |W(\chi_0,\chi_1)|^{-1}(\chi_1(x;\lambda)(D_s^+\chi_0)(0;\lambda)g(0) - \chi_0(x;\lambda)(D_s^+\chi_1)(1;\lambda)g(1)).$$

Moreover, $P_x(\zeta = \infty) = 1$ for all $x \in E$ if and only if $k = 0$ and neither boundary $j = 0$ or 1 is (s,m+k)-regular with $\varkappa_j > 0$ or -exit; $P_x(\zeta < \infty) = 1$ for all $x \in E$ if and only if (K) and (K_1) hold.

Corollary. If the process X is started with the initial distribution ν on E we have $E_\nu g(X_{\zeta-}) = \int_E g d\hat{\nu}$ with

$$d\hat{\nu} = h_\nu\, dk + \int_{0-}^{1+} u_1 d\nu (D_s^+ u_0)(0) d\varepsilon_0 - \int_{0-}^{1-} u_0 d\nu (D_s^+ u_1)(1) d\varepsilon_1 \,, \tag{3}$$

where ε_j denotes the unit mass in j. Here, e.g. the last term in (3) is only $\neq 0$ if 1 is (s,m+k)-regular with $\varkappa_1 > 0$ or -exit.

Recall that a measure μ on E such that

$$E_x g(X_{\zeta-}) = \int_E K(x,y) g(y) d\mu(y) \tag{4}$$

holds for all bounded continuous functions g on E is called the equilibrium measure of E (with respect to X). It is a consequence of Theorem 2 that such an equilibrium measure for E exists if and only if there is no (s,m+k)-exit or -regular boundary j with $\pi_j = 0$, and that in this case it is given by

$$d\mu = dk + \frac{\varkappa_0}{\pi_0} d\varepsilon_0 + \frac{\varkappa_1}{\pi_1} d\varepsilon_1 \,,$$

where we put $\frac{\varkappa_j}{\pi_j} = 0$ if j is (s,m+k)-entrance or -natural.

In case of an (s,m+k)-exit or -regular boundary j with $\pi_j = 0$ a relation of the form (4) holds only if μ is generalized in the sense that it is not only a measure (or mass) but can contain the derivative of a measure (or a dipole) at j.

3. Now let ν be any probability measure on E and h_ν the corresponding potential: $h_\nu(x) = \int_E K(y,x) d\nu(y)$ $(x \in (0,1))$. It satisfies the equation $dD_s^+ h_\nu - h_\nu dk = -d\nu$ on (0,1).

Theorem 3. Suppose that (K), (K_1) hold, let ν be a probability measure on E and denote by R_λ, $\lambda > 0$, the resolvent of X (or of A). Then the superharmonic transformation $\hat{R}_\lambda$:

$$\hat{R}_\lambda f := h_\nu^{-1} R_\lambda (h_\nu f), \quad \lambda > 0,$$

is the resolvent of the gap diffusion associated with $\hat{s}$, $\hat{m}$, $\hat{k}$ given by

$$\hat{s}(x) := \int_{1/2}^{x} h_\nu^{-2} ds, \quad \hat{m}(x) := \int_{1/2}^{x} h_\nu^{2}\, dm, \quad \hat{k}(x) := \int_{1/2}^{x} h_\nu d\nu .$$

If the boundary j is (s,m+k)-regular with $\pi_j > 0$ then j is $(\hat{s},\hat{m}+\hat{k})$-regular with boundary condition

$$(D_{\hat{s}}^+ \hat{f})(j) + (-1)^{j+1} h_\nu(j) \nu(\{j\}) \hat{f}(j) = 0, \tag{5}$$

if j is (s,m+k)-regular with $\pi_j = 0$ or -exit then it is $(\hat{s},\hat{m}+\hat{k})$-

entrance, if j is $(s,m+k)$-natural then it is $(\hat{s},\hat{m}+\hat{k})$-natural or -entrance, if j is $(s,m+k)$-entrance then it is $(\hat{s},\hat{m}+\hat{k})$-natural, -entrance, -exit or -regular with $\hat{\pi}_j = 0$.

__Corollary.__ The resolvents R_λ and $\hat{R}_\lambda$ are in duality with respect to the (σ-finite) measure $h_\nu dm$:

$$\int_0^1 \hat{R}_\lambda f\cdot g\, h_\nu dm = \int_0^1 f(R_\lambda g)h_\nu dm \qquad (6)$$

for all nonnegative bounded continuous functions f,g on $\overline{E}$.

The proofs of Theorems 1 and 2 use standard arguments (comp. [I], [IK], [D]), the proof of Theorem 3 uses explicit calculations, the difficulty arising from the fact that h_ν may be 0 or ∞ at the boundaries. Because of lack of space they will be published elsewhere.

4. In the rest of this note we always suppose that the conditions (K) and (K_1) are satisfied, hence

$$P_x(\zeta < \infty) = 1 \quad \text{for all } x \in E.$$

Also, $P_x(0 < \zeta) = 1$ for all $x \in E$ (see [C]), and we can restrict our considerations to the set $\Omega_0 := [0 < \zeta < \infty]$, as $P_x(\Omega_0) = 1$ for all $x \in E$. If $\Gamma \subset E$ we define

$$\gamma_\Gamma(\omega) := \sup\{t > 0\colon X_t(\omega) \in \Gamma\} \quad (\sup \emptyset := 0).$$

The process X is transient (see [C]), that is $P_x(\gamma_\Gamma < \infty) = 1$ for each compact set $\Gamma \subset E$ and all $x \in E$. Further, the paths have left limits and $X_{\gamma_\Gamma(\omega)-}(\omega) \in \overline{\Gamma}$ (the closure of Γ) for all $\omega \in \Omega_0$, $\Gamma \in \mathcal{B}_E$. If the boundary j is $(s,m+k)$-regular with $\pi_j = 0$ or -exit then for some paths ω we have $X_{\gamma_E(\omega)-}(\omega) = X_{\zeta(\omega)-}(\omega) \in \overline{E}\setminus E$.

Now we reverse the paths of X (with respect to ζ):

$$\hat{X}_t(\omega) := X_{(\zeta(\omega)-t)-}(\omega) \qquad (0 \leq t < \zeta(\omega),\ \omega \in \Omega_0).$$

These paths are right continuous on $[0, \zeta)$ and have left limits on $(0, \zeta]$. The values $\hat{X}_\cdot(\omega)$ are in $\overline{E}$, in particular, if j is an $(s,m+k)$-exit or -regular boundary with $\pi_j = 0$ (hence not in the state space E of X), it can be in the range of $\hat{X}_0(.)$. Evidently, $[\hat{X}_t \in \Gamma,\ 0 \leq t < \zeta] \in \mathfrak{F}$ for all $\Gamma \in \mathcal{B}_{\overline{E}}$, $t \geq 0$, and it makes sense to speak of the finite dimensional distributions of $\hat{X}$ with respect to the measure P_ν on Ω_0, where ν is an arbitrary initial distribution.

Theorem 4. Let s, m, k, boundary conditions for each regular boundary $j = 0$ or 1 and an initial distribution ν be given as above and suppose that the conditions (K), (K_1) are satisfied. Then the finite dimensional distributions of $\hat{X}$ with respect to the measure P_ν on Ω_0 coincide with the finite dimensional distributions of the gap diffusion associated with $\hat{s}$, $\hat{m}$, $\hat{k}$ of Theorem 3, boundary conditions (5) for each $(\hat{s},\hat{m}+\hat{k})$-regular boundary and with the initial distribution $\hat{\nu}$ from (3).

In the proof of this theorem, which will be given in the rest of this n°, we make essential use of some ideas of [C]. If $\varepsilon > 0$ define $\psi_\varepsilon(x) := \frac{1}{\varepsilon} P_x\,(0 < \zeta \leqq \varepsilon)$. Then ψ_ε is a bounded universally measurable function on E. If f is a continuous nonnegative function on $\overline{E}$ we consider the integral

$$\int_0^\infty \frac{1}{\varepsilon} E_x(f(X_t);\ t < \zeta \leqq t + \varepsilon)dt.$$

The relations $\zeta \circ \theta_t = (\zeta - t)^+$, $[t < \zeta \leqq t + \varepsilon] = [0 < \zeta \circ \theta_t \leqq \varepsilon]$ and the Markov property of X imply

$$\frac{1}{\varepsilon} E_x(f(X_t);\ t < \zeta \leqq t + \varepsilon) = \frac{1}{\varepsilon} E_x(f(X_t)\chi_{[t<\zeta\leqq t+\varepsilon]}) = \tag{7}$$
$$= \frac{1}{\varepsilon} E_x(f(X_t)E_{X_t}\,\chi_{[0<\zeta\leqq\varepsilon]}) = E_x(f(X_t)\psi_\varepsilon(X_t)),$$

hence

$$\int_0^\infty \frac{1}{\varepsilon} E_x(f(X_t);\ t < \zeta \leqq t + \varepsilon)dt = \int_E K(x,y)f(y)\psi_\varepsilon(y)dm(y). \tag{8}$$

On the other hand

$$\int_0^\infty \frac{1}{\varepsilon} E_x(f(X_t);\ t < \zeta \leqq t + \varepsilon)dt = \int_{[\zeta>0]} \frac{1}{\varepsilon} \int_{(\zeta-\varepsilon)^+}^{\zeta} f(X_t)dtdP_x. \tag{9}$$

Choosing $f = 1$ it follows that

$$\int_0^\infty \frac{1}{\varepsilon} E_x\,\chi_{[t<\zeta\leqq t+\varepsilon]}dt = \int_{[\zeta>0]} \frac{1}{\varepsilon} \int_{(\zeta-\varepsilon)^+}^{\zeta} dt\ dP_x \leqq P_x(\zeta > 0) = 1,$$

which, together with (7) and (8) yields

$$\int_0^\infty E_x(\psi_\varepsilon(X_t))dt = \int_E K(x,y)\psi_\varepsilon(y)dm(y) \leqq 1. \tag{10}$$

Therefore also the integrals in (8) exist and are finite for each continuous function f on $\overline{E}$. If $\varepsilon \downarrow 0$ the right hand side of (9) tends to $\int_{[\zeta>0]} f(X_{\zeta-})dP_x$, and with (8) we find

$$\int_{[\zeta>0]} f(X_{\zeta-})dP_x = \lim_{\varepsilon\downarrow 0} \int_E K(x,y)f(y)\psi_\varepsilon(y)dm(y).$$

Integrating this relation with respect to an arbitrary initial distribution ν it follows

$$\int_{[\zeta>0]} f(X_{\zeta-})dP_\nu = \lim_{\varepsilon\downarrow 0} \int_E f(y)h_\nu(y)\psi_\varepsilon(y)dm(y).$$

On the other hand, the expression on the left hand side equals $\int_{\overline{E}} fd\hat{\nu}$ with $\hat{\nu}$ given by (3). Hence, if $\varepsilon\downarrow 0$ the measures $h_\nu\psi_\varepsilon dm$ converge *-weakly to the measure $\hat{\nu}$. Therefore the theorem will be proved if we show that for arbitrary continuous functions $f_0,\dots,f_n$ on $\overline{E}$ and $0 < s_1 < \dots < s_n$ the relation

$$E_\nu(f_0(X_{\zeta-})f_1(X_{(\zeta-s_1)-}) \dots f_n(X_{(\zeta-s_n)-});\ 0 < s_n < \zeta) =$$

$$= \lim_{\varepsilon\downarrow 0} \int_{\overline{E}} \dots \int_{\overline{E}} \hat{P}(s_1;\ x_0,dx_1)\hat{P}(s_2-s_1;\ x_1,dx_2)\dots \hat{P}(s_n-s_{n-1};x_{n-1},dx_n)$$

$$\cdot\ f_0(x_0)f_1(x_1)\dots f_n(x_n)h_\nu(x_0)\psi_\varepsilon(x_0)dm(x_0) \qquad (11)$$

holds; here $\hat{P}$ denotes the transition function of the gap diffusion associated with $\hat{s}$, $\hat{m}$ etc. We consider $n = 1$, the general case follows by induction.

Thus, let f_0, f_1 be continuous functions on $\overline{E}$, $s > 0$. We first assume additionally that f_0, f_1 vanish near 0 and 1, and consider

$$\int_s^\infty \frac{1}{\varepsilon} E_x(f_0(X_t)f_1(X_{t-s});\ t < \zeta \leq t + \varepsilon)dt. \qquad (12)$$

As $f_0(X_t)f_1(X_{t-s})$ is $\mathfrak{F}_t$-measurable, the Markov property implies that this integral equals

$$\int_s^\infty E_x(f_0(X_t)f_1(X_{t-s})\frac{1}{\varepsilon} E_{X_t}\chi_{[0<\zeta\leq\varepsilon]})dt = \int_0^\infty E_x(f_0(X_{t+s})f_1(X_t)\psi_\varepsilon(X_{t+s}))dt=$$

$$= \int_0^\infty E_x(f_1(X_t)E_{X_t}(f_0(X_s)\psi_\varepsilon(X_s)))dt = \int_0^\infty \int_E P(t;x,dy)f_1(y)(T_s(f_0\psi_\varepsilon))(y)dt=$$

$$= \int_{\overline{E}} f_1(y)(T_s(f_0\ \psi_\varepsilon))(y)K(x,y)dm(y).$$

On the other hand, the integral (12) can be transformed as follows:

$$\int_s^\infty \frac{1}{\varepsilon} E_x(f_0(X_t)f_1(X_{t-s});\ t < \zeta \leq t + \varepsilon)dt =$$

$$= \int_s^\infty \frac{1}{\varepsilon} \int_{\Omega_0} f_0(X_t)f_1(X_{t-s})\chi_{[t<\zeta\leq t+\varepsilon]}dP_x\ dt =$$

$$= \int_{[\zeta>s]} \frac{1}{\varepsilon} \int_{(\zeta-\varepsilon)^+}^{\zeta} f_0(X_t) f_1(X_{t-s}) dt \, dP_x,$$

that is we find

$$\int_{[\zeta>s]} \frac{1}{\varepsilon} \int_{(\zeta-\varepsilon)^+}^{\zeta} f_0(X_t) f_1(X_{t-s}) dt \, dP_x = \int_{\overline{E}} f_1 \, T_s(f_0 \, \psi_\varepsilon) K(x,\cdot) dm.$$

Integrating this relation with respect to ν and observing (6) it follows

$$\int_{[\zeta>s]} \frac{1}{\varepsilon} \int_{(\zeta-\varepsilon)^+}^{\zeta} f_0(X_t) f_1(X_{t-s}) dt \, dP_\nu =$$

$$= \int_{\overline{E}} f_1 \, T_s(f_0 \, \psi_\varepsilon) h_\nu \, dm = \int_{\overline{E}} (\hat{T}_s f_1) f_0 \, \psi_\varepsilon \, h_\nu \, dm.$$

As the measure $\psi_\varepsilon \, h_\nu \, dm$ is finite (see (10)) this relation extends to all continuous functions f_0, f_1 on $\overline{E}$ and, letting $\varepsilon \downarrow 0$, we get

$$E_\nu(f_0(X_{\zeta-}) f_1(X_{(\zeta-s)-}); \; 0 < s < \zeta) =$$

$$= \lim_{\varepsilon \downarrow 0} \int_{\overline{E}} \int_{\overline{E}} f_0(x_0) \hat{P}(s, x_0; dx) f_1(x) \psi_\varepsilon(x_0) h_\nu(x_0) dm(x_0).$$

Thus the relation (11) is proved for $n = 1$. The induction is left to the reader.

Remark 1. Theorem 4 contains some results of [Sh], [Sa]. It is also possible to prove Theorem 4 using the general results on time reversal (see [N], [CW], [Me]). However, then some boundary points need special consideration as h_ν can be 0 or ∞ there and they belong to the state space of X without belonging to the state space of $\hat{X}$.

Remark 2. The considerations in the proof of Theorem 4 remain valid if ζ is replaced by an arbitrary cooptional time γ (see [C]) of X with $P_x(0 < \gamma < \infty) = 1$ for all $x \in E$. In particular, this cooptional time can be the quitting time γ_Γ of a set $\Gamma \in \mathcal{B}_E$. However, in general, in order to find the quitting kernel $L_\Gamma(x,B) := P_x(\gamma_\Gamma > 0, X_{\gamma_\Gamma -} \in B)$ or the quitting time distribution it seems to be simpler first to reverse the given process X with respect to its life time ζ and to observe that

$$\gamma_\Gamma = \hat{\zeta} - \hat{\tau}_\Gamma \, , \quad X_{\gamma_\Gamma -} = \hat{X}_{\hat{\tau}_\Gamma} \, ,$$

where $\hat{\tau}_\Gamma$ denotes the first hitting time of Γ for the reversed process $\hat{X}$ and $\hat{\zeta}(=\zeta)$ denotes its life time.

References.

[M] P. Mandl, Analytical Treatment of One-dimensional Markov Processes, Prag and Berlin - Heidelberg - New York (1968)

[C] K.L. Chung, Lectures from Markov Processes to Brownian Motion, Berlin (1981)

[D] E.B. Dynkin, Markov Processes, Vol. I and II. Berlin - Göttingen - Heidelberg (1965)

[BG] R.M. Blumenthal, R.K. Getoor, Markov Processes and Potential Theory, New York and London (1968)

[I] K. Ito, Verojatnostnyje processy II, Moskva (1963) (russ.)

[IK] K. Ito, H.P. McKean, Diffusion Processes and their Sample Paths, New York and London (1965)

[Sh] M.I. Sharpe, Some Transformations of Diffusions by Time Reversal, Ann. Prob., 8 (1980), 1157-1162

[Sa] P. Salminen, One Dimensional Diffusions and their Duals, Acta Acad. Åboensis, Ser. B, 41, 5 (1981)

[N] M. Nagasawa, Time Reversions of Markov Processes, Nagoya Math. J. 24, (1964), 177-204

[CW] K.L. Chung, J.B. Walsh, To Reverse a Markov Process, Acta Math. 123 (1969) 225-251

[Me] P.A. Meyer, Processus de Markov: la Frontiere de Martin, LN in Math. 77, Berlin (1968)

GENERALIZED SECOND ORDER DIFFERENTIAL OPERATORS AND NONCONSERVATIVE ONE-DIMENSIONAL QUASIDIFFUSIONS WITH NATURAL BOUNDARIES

J.-U. Löbus
Friedrich-Schiller-Universität Jena
Sektion Mathematik
Universitätshochhaus (17. OG)
Jena, 6900, GDR

1. Introduction

One-dimensional quasidiffusions (also called gap diffusions) are generalisations of one-dimensional diffusions as well as birth- and death-processes. They can be introduced by their infinitesimal generators $D_m D_p^+ \frac{\cdot}{\pi}$ or $(dD_s^+ . - .dk)/dn$, cf. /1,4,6,7,8,10/, which are generalisations of ordinary second order differential operators $a\, d^2.\,+ ./dx^2 + b\, d./dx - c$, where a,b,c are measurable functions with $c \geq 0$, cf. /1,3/. By investigation of the equation $D_m D_p^+ \frac{f}{\pi} = g$ a classification of natural boundaries is given. From a probabilistic point of view the distribution of the life time of a particle starting near a boundary is characterized by this. On the other hand operators $D_m D_p^+ \frac{\cdot}{\pi}$ are classified with respect to the behaviour of solutions f of the equation $\lambda f - D_m D_p^+ \frac{f}{\pi} = g$, $\lambda \in (0,\infty)$, at the boundaries.

2. Preliminaries

<u>Definitions and Notations.</u> Let $E \neq \emptyset$ be a compact subset of the extended real axis $\bar{R} := [-\infty,\infty]$. Put $r_0 := \inf(x : x \in E)$ and $r_1 := \sup(x : x \in E)$. Suppose r_0 and r_1 not to be isolated in E and assume $r_0 < 0 < r_1$.
Let $\dot{C}$ be the space of all real continuous functions on $\dot{E} := E \setminus \{r_0, r_1\}$. For $f \in \dot{C}$ denote by $\bar{f} : (r_0, r_1) \to R$ that function which is linear in the components of $(r_0, r_1) \setminus \dot{E}$ and whose restriction to $\dot{E}$ is f.

Remark 2.1. If a function $f : S \to R$, $\dot{E} \subseteq S \subseteq E$, or $f : (r_0, r_1) \to R$ has a (not necessarily finite) limit in r_j then put $f(r_j) := \lim_{x \to r_j} f(x)$, j=0,1.
For $\dot{E} \subseteq S \subseteq E$ concentrate all continuous functions $f : S \to R$ having finite limits in r_0, r_1, and satisfying $f(r_j) = 0$ if $r_j \notin S$ to the space C_S (j=0,1). The norm in C_S is $\|f\| := \sup_{x \in S} |f(x)|$.

If for $\varphi, f:(r_0,r_1)\longrightarrow R$, where f is strictly increasing, for $x\in(r_0,r_1)$, there exists the limit $\lim\limits_{y\downarrow x}\frac{\varphi(y)-\varphi(x)}{f(y)-f(x)}$ then it will be denoted by $(D_f^+\varphi)(x)$, see /1/, appendix §7.

Generalized Second Order Differential Operators. Suppose we are given strictly increasing functions $q,t\in\dot{C}$ with $\bar{q}(0)=\bar{t}(0)=0$ and nondecreasing, right-continuous functions $m,n,k: (r_0,r_1)\longrightarrow R$ with

$$m(0)=n(0)=k(0)=0 \text{ and } \operatorname{supp}(k)\subseteq\operatorname{supp}(n)=\operatorname{supp}(m)=\dot{E}$$

($\operatorname{supp}(\varphi)$ denotes the closure of all points belonging to (r_0,r_1) in which a nondecreasing function $\varphi:(r_0,r_1)\rightarrow R$ is strictly increasing). Moreover set $p:=\bar{q}$, $s:=\bar{t}$ and choose a positive function $\pi\in\dot{C}$ in such a manner that $(\frac{1}{\pi})$ is concave with respect to p (cf. /1/, appendix §8).

We say that an element $u\in\dot{C}$ belongs to the domain of definition $\mathcal{D}(D_mD_p^+\frac{\cdot}{\pi})$ ($\mathcal{D}((dD_s^+.-.dk)/dn)$) if there exist a function $v\in\dot{C}$ and real number a, so that for $x\in(r_0,r_1)$ the relation

$$(D_p^+(\overline{\tfrac{u}{\pi}}))(x) = \int_0^x \bar{v}\, dm + a \quad ^{1)} \tag{2.1}$$

$$((D_s^+\bar{u})(x)-\int_0^x \bar{u}\, dk = \int_0^x \bar{v}\, dn + a)$$

holds. Then we set $D_mD_p^+\frac{u}{\pi}=:v$ ($(dD_s^+u-udk)/dn=:v$).
Moreover we define $H(D_mD_p^+\frac{\cdot}{\pi}):=\{h\in\dot{C}:\ a,b\in R:h(x)=\pi(x)(a+bp(x)),x\in\dot{E}\}$.

Assertions Concerning $D_mD_p^+\frac{\cdot}{\pi}$ and $(dD_s^+.-.dk)/dn$. In the following we summarize some facts which are extensively proved in /9/, cf. also /1,2,4,5,10,11/.

Lemma 2.2. Let $y\in R$, $y^+:=\inf(x\in\dot{E}:x>y)$. For any $f_0,f_1\in\mathcal{D}(D_mD_p^+\frac{\cdot}{\pi})$ and $\varepsilon>0$ there exists a function $f\in\mathcal{D}(D_mD_p^+\frac{\cdot}{\pi})$ satisfying

$$f(x)=f_0(x),\ (D_mD_p^+\tfrac{f}{\pi})(x)=(D_mD_p^+\tfrac{f_0}{\pi})(x),\qquad x\in(r_0,y-\varepsilon)\cap\dot{E},$$

$$f(x)=f_1(x),\ (D_mD_p^+\tfrac{f}{\pi})(x)=(D_mD_p^+\tfrac{f_1}{\pi})(x),\qquad x\in(y^++\varepsilon,r_1)\cap\dot{E}.$$

Theorem 2.3. Let $a\in R$ and m,p,π and n,s,k be functions defining operators $D_mD_p^+\frac{\cdot}{\pi}$ and $(dD_s^+.-.dk)/dn$, respectively. The equation

$$(D_s^+\bar{u})(x) = \int_0^x \bar{u}\, dk + a \tag{2.2}$$

has a solution $u\in\dot{C}$. If (2.2) with $u=\pi$ and

---

1) for a,b R put $\int_a^b := \int_{(a,b]}$ $(a<b)$ and $\int_a^b := -\int_{(b,a]}$ $(a>b)$

$\beta := (I(r_0)-I(r_1))/(p_\lambda(r_1)-p_\lambda(r_0))$,

$I(x) := \int_0^x \int_0^y \bar{g}\, dm_\lambda(t)dp_\lambda(y)$

(the relation $\int_0^{r_j} m_\lambda dp_\lambda < \infty$ holds), $j=0,1$.

Lemma 2.4. <u>Let</u> $g \in C_{\bar{E}}^{\bullet}$ <u>and</u> $\lambda \in (0,\infty)$. <u>Then</u> $f_0(g)$ <u>is the only solution of the problem</u>

$$\lambda f - D_m D_p^{+} \frac{f}{\pi} = g, \qquad f \in C_{\bar{E}}^{\bullet} .$$

Lemma 2.5. <u>The domain of definition</u> $\mathcal{D}(D_m D_p^{+} \frac{\cdot}{\pi} | C_{\bar{E}}^{\bullet})$ <u>of the operator</u> $D_m D_p^{+} \frac{\cdot}{\pi} | C_{\bar{E}}^{\bullet}$ <u>is dense in</u> $C_{\bar{E}}^{\bullet}$.

Lemma 2.6. <u>If</u> $f \in \mathcal{D}(D_m D_p^{+} \frac{\cdot}{\pi})$ <u>then the validity of</u> $f(x) \geq \overline{f}(x)$ <u>in a neighbourhood of</u> $x_0 \in \bar{E}$ <u>and</u> $\dot{f}(x_0) \geq 0$ implies $(D_m D_p^{+} \frac{f}{\pi})(x_0) \leq 0$.

Using the preceding three assertions by the HILLE-YOSIDA-theorem in the form of /1/, Theorem 2.10., it follows:

Theorem 2.7. <u>The operator</u> $D_m D_p^{+} \frac{\cdot}{\pi} | C_{\bar{E}}^{\bullet}$ <u>is the infinitesimal generator of a Standard process with state space</u> $\dot{\bar{E}}$.

A stochastic process mentioned in Theorem 2.7. we will call a (non-conservative) one-dimensional quasidiffusion.

3. Classification of Natural Boundaries

From (2.1) it is seen that the equation

$$D_m D_p^{+} \frac{F}{\pi} = -e, \qquad e \in \dot{C} \tag{3.1}$$

has a solution belonging to $\dot{C}$. First we consider operators $D_m D_p^{+} \frac{\cdot}{\pi}$ which fulfill one of the following conditions:

3.1.) There exist functions $e \in \dot{C}$ and $F \in \mathcal{D}(D_m D_p^{+} \frac{\cdot}{\pi})$ which have a finite limit $e(r_1) := \lim_{x \to r_1} e(x) \neq 0$ and $F(r_1) := \lim_{x \to r_1} F(x)$ and satisfy the equation (3.1).

3.2.) There exist functions $e \in \dot{C}$ and $F \in \mathcal{D}(D_m D_p^{+} \frac{\cdot}{\pi})$ with finite limits $e(r_1) := \lim_{x \to r_1} e(x) = 0$, $F(r_1) := \lim_{x \to r_1} F(x) \neq 0$ fulfilling (3.1).

Lemma 3.1. <u>Let one (and only one) of the conditions 3.1.). 3.2.) be valid and let</u> e, F <u>be the functions mentioned in these conditions. Then there exist elements</u> $\varphi_1 \in \mathcal{D}(D_m D_p^{+} \frac{\cdot}{\pi})$ and $\eta_1 \in \dot{C}$ <u>with</u>

$$p(x) = \int_0^x (\bar{\pi})^{-2} ds, \quad m(x) = \int_0^x \bar{\pi}\, dn, \qquad x\in\dot{E} \tag{2.3}$$

<u>are valid then the operators $D_m D_p^{+}\frac{\cdot}{\pi}$ and $(dD_s^{+}.-.dk)/dn$ coincide.</u>

From this theorem it follows that for $\lambda\in(0,\infty)$ there exist functions $m_\lambda, p_\lambda, \pi_\lambda$ which enable the definition of an operator $D_{m_\lambda} D_{p_\lambda} \frac{\cdot}{\pi_\lambda}$ of such a kind that

$$D_{m_\lambda} D_{p_\lambda} \frac{\cdot}{\pi_\lambda} = (dD_s^{+}.-.d(k+\lambda n))/dn = D_m D_p^{+}\frac{\cdot}{\pi} - \lambda I \tag{2.4}$$

(I denotes the identity operator in $\dot{C}$) is fulfilled.
Without loss of generality we can assume π, π_λ to be decreasing in $(r_0,x')\cap\dot{E}$ and increasing in $(x',r_1)\cap\dot{E}$ for some $x'\in\dot{E}$. Then it follows that $\pi(r_0), \pi(r_1) > \pi(x')$ if

$$\mathbb{1} \notin H(D_m D_p^{+}\frac{\cdot}{\pi}) \tag{2.5}$$

($\mathbb{1}(x)=1, x\in\dot{E}$) is valid. In this case the p-concavity of $\overline{(\frac{1}{\pi})}$ implies $|p(r_j)| < \infty$. Hence, we always have $|p_\lambda(r_j)| < \infty$, $j=0,1$.
Define functions $h_\lambda^d, h_\lambda^i \in \dot{C}$ and in the case (2.5) $h^d, h^i \in \overset{o}{C}$ by

$$h^d(x) := \pi(x)(p(r_1)-p(x)), \quad h_\lambda^d(x) := \pi_\lambda(x)(p_\lambda(r_1)-p_\lambda(x)), \tag{2.6}$$

$$h^i(x) := \pi(x)(p(x)-p(r_0)), \quad h_\lambda^i(x) := \pi_\lambda(x)(p_\lambda(x)-p_\lambda(r_0)), \tag{2.7}$$

$x\in\dot{E}$, $\lambda\in(0,\infty)$. The relation $\lim_{x\to r_1} h^d(x)/h^i(x)=0$ induces the following cases:

$$h^d(r_1) = 0, \; h^i(r_1) < \infty \tag{2.8}$$

$$0 < h^d(r_1) < \infty, \; h^i(r_1) = \infty \tag{2.9}$$

$$h^d(r_1) = 0, \; h^i(r_1) = \infty. \tag{2.10}$$

Henceforth, <u>let</u> r_0, r_1 <u>always be natural boundaries</u>, i.e.

$$h_\lambda^i(r_0) = h_\lambda^d(r_1) = 0, \; h_\lambda^i(r_1) = h_\lambda^d(r_0) = \infty \tag{2.11}$$

or equivalently (if for $u=\pi$ and a certain $a\in R$ (2.2), (2.3) are valid)

$$\int_0^{r_j} (n+k) ds = \int_0^{r_j} s\, d(n+k) = \infty, \quad j=0,1.$$

Denote by $D_m D_p^{+}\frac{\cdot}{\pi}|C_S$, $\dot{E}\subseteq S\subseteq E$, the restriction of the operator $D_m D_p^{+}\frac{\cdot}{\pi}$ to C_S in the sense of Remark 2.1. Define for $g\in C$ having finite limits in r_0 and r_1 the function $f_0 = f_0(g)\in\dot{C}$ by

$$\begin{aligned} f_0(x) &= -\pi_\lambda(x)(I(x)+\alpha+\beta p_\lambda(x)) = \\ &= -\pi_\lambda(x)(I(x)-I(r_j)+\beta(p_\lambda(x)-p_\lambda(r_j))) \end{aligned} \tag{2.12}$$

$x\in\dot{E}$, where

$$\alpha := (p_\lambda(r_0)I(r_1)-p_\lambda(r_1)I(r_0))/(p_\lambda(r_1)-p_\lambda(r_0)),$$

$\varphi_1(r_o)=\eta_1(r_o)=0$, $\eta_1(r_1)=1$ and $\varphi_1(r_1)=1/(\lambda+e(r_1)/F(r_1))$ (put $e(r_1)/0=e(r_1)\cdot\infty$ and $1/\pm\infty=0$) satisfying

$$\lambda\varphi_1 - D_m D_p^+ \frac{\varphi_1}{\pi} = \eta_1, \quad \lambda\in(0,\infty). \tag{3.2}$$

Proof. 1° Let 3.1.) be valid. Because of $\mathcal{O}\in\mathcal{D}(D_m D_p^+\frac{\cdot}{\pi})$ ($\mathcal{O}(x)=0, x\in\dot{E}$) Lemma 2.2. implies the existence of functions $e_1\in\dot{C}$ and $F_1\in\mathcal{D}(D_m D_p^+\frac{\cdot}{\pi})$ with $e_1(r_o)=F_1(r_o)=0$, $e_1(r_1)=1$ and $F_1(r_1)=F(r_1)/e(r_1)$ so that

$$D_m D_p^+\frac{F_1}{\pi} = -e_1. \tag{3.3}$$

2° If 3.2.) is valid then by the same argument there exist functions $e_1\in\dot{C}$ and $F_1\in\mathcal{D}(D_m D_p^+\frac{\cdot}{\pi})$ with $e_1(r_o)=F_1(r_o)=e_1(r_1)=0$ and $F_1(r_1)=1$ fulfilling (3.3).

3° For those functions e_1, F_1 always $e_1(r_1)+\lambda F_1(r_1)=0$, $\lambda\in(0,\infty)$ holds. The assertion is proved by the definition of

$$\varphi_1 := F_1/(e_1(r_1)+\lambda F_1(r_1))$$

and

$$\eta_1 := (e_1+\lambda F_1)/(e_1(r_1)+\lambda F_1(r_1)).$$

Lemma 3.2. The conditiones 3.1.) and 3.2.) cannot be valid simultaneously.

Proof. We suppose that both conditions 3.1.) and 3.2.) hold simultaneously. Let e, F be the functions described in 3.1.). By Lemma 3.1. there exist functions φ_1^1, η_1^1 and φ_1^2, η_1^2 satisfying (3.2) with

$$\varphi_1^1(r_1) = 1/(\lambda+e(r_1)/F(r_1)), \tag{3.4}$$
$$\varphi_1^2(r_1) = 1/\lambda \tag{3.5}$$

($\lambda\in(0,\infty)$) and $\eta_1^1(r_1)=\eta_1^2(r_1)=1$ (set $e(r_1)/0=e(r_1)\cdot\infty$ and $1/\pm\infty=0$). We put $g:=\eta_1^1-\eta_1^2$. Now it is seen that (3.4) and (3.5) contradict Lemma 2.4.

Corollary 3.3. The validity of 3.1.) implies (2.5) and (2.10).

Proof. We set $F:=\mathbb{1}$ (if $\mathbb{1}\in H(D_m D_p^+\frac{\cdot}{\pi})$) or $F:=h^w$ (if (2.5) and (2.8) are valid) or $F:=h^f$ (if (2.5) and (2.9) are valid). This means, that condition 3.2.) is fulfilled. The above assertion now follows from Lemma 3.2.

Corollary 3.4. The validity of 3.1.) implies $\int_0^{r_1} m\,dp<\infty$.

Proof. By $|p(r_j)|<\infty$, $j=0,1$, (cf. chapter 2), $\pi(r_1)=\infty$ (Corollary 3.3. and $p(r_1)-p(r_o)<\infty$) and

$$F(x) = \pi(x)(-\int_0^x \int_0^y \bar{e}(t)dm(t)dp(y) + a+bp(x)), \qquad x\in\dot{E}$$

with suitable real numbers a,b this is obtained (e and F denote the functions described in 3.1.)).

Lemma 3.5. <u>Let</u> $e_o \in \dot{C}$ <u>with</u> $e_o(r_1) := \lim_{x\to r_1} e_o(x) = 0$. <u>If 3.1.) is valid then the equation</u>

$$D_m D_p^+ \frac{f}{\pi} = -e_o \qquad (3.6)$$

<u>has a solution</u> F_o <u>with</u> $F_o(r_1) := \lim_{x\to r_1} F_o(x) = 0$.

Proof. Let e and F be the functions mentioned in condition 3.1.). Without loss of generality we assume $e(r_1)=1$. Moreover we choose $\varepsilon > 0$ and put $\delta := \varepsilon/2(|F(r_1)| + \varepsilon)$. There exists a point $x_\varepsilon \in \dot{E}$ in such a manner that

$$|e_o(x)|, |e(x)-e(r_1)| < \delta, \quad x\in [x_\varepsilon, r_1)\cap\dot{E}. \qquad (3.7)$$

We suppose that for every solution f of (3.6) the relation $\limsup_{x\to r_1} f(x) > \varepsilon$ is fulfilled.

From the corollaries 3.3. and 3.4. it is seen that (without loss of generality) the function F has a representation

$$F(x) = \pi(x) \int_x^{r_1} \int_{x_\varepsilon}^y \bar{e}(t)dm(t)dp(y), \qquad x\in\dot{E}.$$

Furthermore we can define a function F_o by

$$F_o(x) = \pi(x) \int_x^{r_1} \int_{x_\varepsilon}^y \bar{e}_o(t)dm(t)dp(y), \qquad x\in\dot{E}.$$

Then from (3.7) we get

$$|F_o(x)| \leq \delta |F(x)| /(1-\delta) = |F(x)|/(|F(r_1)| + \varepsilon) \leq$$
$$\leq \varepsilon |F(x)| / |F(r_1)|, \qquad x\in [x_\varepsilon, r_1)\cap\dot{E}.$$

This is a contradiction to the above supposition.

Corollary 3.6. <u>The conditions 3.1.) and</u>

3.3.) There exists a function $f\in\mathcal{D}(D_m D_p^+ \frac{\cdot}{\pi})$ having a finite limit $F(r_1) := \lim_{x\to r_1} F(x)$ and satisfying

$$D_m D_p^+ \frac{F}{\pi} = -\mathbb{1} \qquad (3.8)$$

<u>are equivalent.</u>

Remark 3.7. Let 3.3.) be valid. From the Corollaries 3.3. and 3.4. we conclude that for any solution F of (3.8) the limit $F(r_1) := \lim_{x\to\infty} F(x)$

exists and either $|F(r_1)| = \infty$ or

$$F(r_1) = \lim_{x\to r_1} \pi(x) \int_x^{r_1} m\, dp \qquad (>0). \tag{3.9}$$

Now Lemma 3.2., Corollary 3.6. and Remark 3.7. induce a subdivision of natural boundaries:

Definition 3.8. (a) If the condition 3.2.) is fulfilled then we call r_1 a conservative boundary.
(b) If 3.3.) is valid then r_1 is said to be a killing boundary.
(c) A killing boundary we will call weakly killing and strongly killing, if for the function F mentioned in 3.3.) $F(r_1)\neq 0$ and $F(r_1)=0$ is valid, respectively.
(d) A natural boundary r_1 which is not conservative and not killing is said to be an indefinite boundary.

These criterions are rather general. For this reason some convenient sufficient conditions are given which can easily be verified. Let $F(r_1)$ be defined by (3.9). Put

$$l_1 := \begin{cases} 0 & r_1 \text{ is conservative} \\ 1/F(r_1) & r_1 \text{ is killing} \end{cases} \tag{3.10}$$

(with $1/0=\infty$). We mention that, if r_1 is weakly (strongly) killing, then it follows $0<l_1<\infty$ $(l_1=\infty)$.

Theorem 3.9. (a) If $\mathbb{1}\in H(D_m D_p^+ \frac{\cdot}{\pi})$ or if $\mathbb{1}\notin H(D_m D_p^+ \frac{\cdot}{\pi})$ and one of the conditions (2.8), (2.9) is fulfilled then r_1 is a conservative boundary.
(b) Suppose that (2.5) and (2.10) are valid and that there exists the finite limit $L_1 := \lim_{x\to r_1} -m(x)/(D_p^+(\frac{1}{\pi})(x)$. If $L_1>0$ $(L_1=0)$ then r_1 is weakly (strongly) killing. Moreover, $L_1=1/l_1$ (set $1/\infty=0$).
(c) Let (2.5), (2.10) and (2.2) (with $u=\pi$ and a certain $a\in R$) be valid and let $k(dx), n(dx)$ be the BOREL-measures generated by the interval functions

$$k(a,b] := k(b)-k(a), n(a,b] := n(b)-n(a), \qquad a,b\in(r_o,r_1).$$

The existence of the RADON-NIKODYM-derivative $\frac{dk}{dn}$ and the limit $L_1' := \lim_{x\to r_1} (\frac{dk}{dn})(x)$ implies that r_1 is conservative or killing and $L_1'=l_1$.
(d) If (2.5) and (2.10) are fulfilled and the equation (3.8) has a solution F with $0<\liminf_{x\to r_1} F(x) < \limsup_{x\to r_1} F(x)$ then r_1 is an indefinite boundary.

By this theorem examples of conservative, weakly and strongly killing boundaries can be constructed. Employing assertion (d) by

$$E = [-1,1], \quad \pi(x)=1/\sqrt{1-x^2}, \quad p(x)=x \text{ and}$$

$$m(x) = \frac{x}{\sqrt{1-x^2}}(4+\sin(\tfrac{1}{2}\ln(1-x^2))+\cos(\tfrac{1}{2}\ln(1-x^2))), \qquad x \in \dot{E}$$

an example of an indefinite boundary r_1 is given.

4. The Equation $\lambda f - D_m D_p^+ \frac{f}{\pi} = g$

Theorem 4.1. <u>Let</u> $\lambda \in (0,\infty)$ <u>and</u> $g \in \dot{C}$ <u>with</u> $g(r_0)=0$, $0<|g(r_1)|<\infty$ <u>be given. Then the problem</u>

$$\lambda f - D_m D_p^+ \frac{f}{\pi} = g, \quad f \in \dot{C}, \quad f(r_0)=0, \quad |f(r_1)| < \infty, \tag{4.1}$$

<u>has a solution</u> f <u>iff the boundary</u> r_1 <u>is conservative or killing. This solution is unique and representable by (2.12)</u>

$$f = f_0 = f_0(g).$$

<u>Moreover (with</u> l_1 <u>defined in (3.10) and</u> $1/\infty=0$),

$$f(r_1) = g(r_1)/(\lambda + l_1).$$

Proof. 1° Let r_1 be conservative or killing and let φ_1, η_1 be the functions mentioned in Lemma 3.1. (recall also Lemma 3.2.). We decompose g into

$$g = g_0 + g(r_1)\,\eta_1.$$

Then $g_0(r_1)=0$ is valid. Moreover, from Lemma 2.4. it is found that the function

$$f := f_0(g_0)+g(r_1)\eta_1 \tag{4.2}$$

solves the equation (4.1).

On the other hand, using (2.4), it follows that any solution f of (4.1) has a representation

$$f = f_0(g) + a h_\lambda^d + b h_\lambda^i,$$

with suitable real numbers a, b, cf. (2.6), (2.7), (2.12). Furthermore,

$$\lim_{x\to r_0} ((f_0(g)(x)+bh_\lambda^i(x))/h_\lambda^d(x)) = \lim_{x\to r_1} ((f_0(g)(x)+ah_\lambda^d)/h_\lambda^i(x)) = 0.$$

By this from (2.11) the infinity of $|f(r_0)|$ and $|f(r_1)|$ in the case of a=0 and b=0, respectively, is obtained. On account of the existence of a solution of (4.1), namely (4.2), it is seen that only $f=f_o(g)$ satisfies (4.1).

2° Assume that the problem (4.1) is solvable. Then from $\lambda f-g=D_mD_p^+\frac{f}{\pi}$ it can be concluded that one of the conditions 3.1.), 3.2.) must be valid. Corollary 3.6. implies that r_1 is conservative or killing.

5. Quasidiffusions with Generator $D_mD_p^+\frac{\cdot}{\pi}|C_S$, $\dot{E}\subseteq S\subseteq E$

We remark that the boundary r_o can be caracterized and classified in a similar way as r_1.

Theorem 5.1. If r_j is a conservative or weakly killing boundary then the operator $D_mD_p^+\frac{\cdot}{\pi}|C_{E\setminus\{r_{1-j}\}}$ is the infinitesimal generator of a Standard process with state space $E\setminus\{r_{1-j}\}$, $j=0,1$. Provided that both boundaries r_o,r_1 are conservative or weakly killing $D_mD_p^+\frac{\cdot}{\pi}|C_E$ is the generator of a Standard process on E.

Proof. 1° From the Lemmata 2.5., 3.1. and from Corollary 3.6. it follows that $\mathcal{D}(D_mD_p^+\frac{\cdot}{\pi}|C_S)$, $\dot{E}\subseteq S\subseteq E$ is dense in C_S.

2° If r_1 is a conservative boundary then by Lemma 3.2. for $f\in\mathcal{D}(D_mD_p^+\frac{\cdot}{\pi}|C_E)$ the relation $(D_mD_p^+\frac{f}{\pi})(r_1)=0$ is got.

Using the Lemmata 3.2., 3.5. and noting $l_1>0$ (see (3.9)) in the case of a weakly killing boundary for $f\in\mathcal{D}(D_mD_p^+\frac{\cdot}{\pi}|C_E)$ with $f(r_1)>0$ $(=0)$ the validity of $(D_mD_p^+\frac{f}{\pi})(r_1)>0$ $(=0)$ is found.

A similar result can be obtained for the boundary r_o.

3° The assertions 2.4., 2.6., 4.1. and /1/ Theorem 2.10. complete the proof.

The processes described in Theorem 5.1. we will also call (nonconservative) one-dimensional quasidiffusions.

6. Probabilistic interpretation

Denote by $\mathcal{L}(\varphi)$ the LAPLACE-transform $\int_0^\infty e^{-\lambda t}\varphi(t)dt$ and by I_S the identity operator in C_S, $\dot{E}\subseteq S\subseteq E$.

Let $\dot{E}\subseteq S\subseteq E$ and let $X=(X_t,\zeta,P_x)$ be a quasidiffusion (with life time ζ) whose generator is $D_mD_p^+\frac{\cdot}{\pi}|C_S$. Choose a suitable sequence $(g_n)_{n=1,2,\dots}$, $g_n\in C_S$ with $g_n(x)\to 1$, $x\in S$. Employing

$$\mathcal{L}\left(\int_S g(y)P_x(X_t\in dy)\right) = (\lambda I_S-D_mD_p^+\frac{\cdot}{\pi}|C_S)^{-1}g(x),\quad g\in C_S, x\in S, \quad (6.1)$$

$\lambda\in(0,\infty)$, the relation

$$\mathcal{L}(P_x(\zeta>t)) = f_o(\lambda)(x), \qquad x\in S, \tag{6.2}$$

can easily be verified.

Theorem 6.1. Let l_1 be the number defined in (3.10).
(a) If r_1 is a conservative or weakly killing boundary and X is a quasidiffusion with generator $D_m D_p^{+}\frac{\bullet}{\pi}|C_S$, $E\setminus\{r_o\}\subseteq S\subseteq E$, then

$$P_{r_1}(\zeta>t) = \exp(-l_1 t), \qquad t\in(0,\infty).$$

(b) Let r_1 be conservative or killing and X be a quasidiffusion with generator $D_m D_p^{+}\frac{\bullet}{\pi}|C_S$, $\mathring{E}\subseteq S\subseteq E$. Then

$$\lim_{x\to r_1} P_x(\zeta>t) = \exp(-l_1 t), \qquad t\in(0,\infty).$$

Proof. (a) Noting that $1/(\lambda+l_1)$ is the LAPLACE-transform of $\exp(-l_1 t)$ this follows from (6.2) and Theorem 4.1.
(b) Let $\varepsilon>0$. We put $c(x,t):=P_x(\zeta>t)-\exp(-l_1 t)$, $x\in\mathring{E}, t\in(0,\infty)$ and assume that there exists $t_o\in(0,\infty)$ with

$$\limsup_{x\to r_1} |c(x,t_o)| > \varepsilon . \tag{6.3}$$

The monotony of $P_x(\zeta>t)$ in t implies for $x\in\mathring{E}$, $0<t<t'<\infty$ the inequality

$$c(x,t')-c(x,t) \leqq a(t'-t) \tag{6.4}$$

with $a=0$ and $a=l_1$ in the case of a conservative and killing boundary r_1, respectively.
Applying the convergence theorem for LAPLACE-transforms (see for instance /3/ chapter XIII, Theorem 2a) and using Theorem 4.1. and (6.1) the relation

$$\left|\int_t^{t'} c(x,s)ds\right| < \varkappa(x_o,t,t') \tag{6.5}$$

for $t,t'\in(0,\infty)$, $x_o,x\in\mathring{E}$, $x_o<x$ is obtained. Thereby the function $\varkappa(x,t,t')$: $\mathring{E}\times(0,\infty)^2\longrightarrow(0,\infty)$ satisfies $\lim_{x\to r_1}\varkappa(x,t,t')=0$ for any pair $(t,t')\in(0,\infty)^2$.
The relations (6.4) and (6.5) induce for $0<t'<t_o<t<\infty$ and $x_o,x\in\mathring{E}$, $x_o<x$ the inequalities

$$c(x,t_o)(t-t_o) = \int_{t_o}^{t} c(x,t_o)du =$$
$$= \int_{t_o}^{t}(c(x,t_o)-c(x,u))du + \int_{t_o}^{t} c(x,u)du \geqq$$

$$\geq -\frac{a}{2}(t-t_0)^2 - \varkappa(x_0,t_0,t)$$

and analogously

$$c(x,t_0)(t_0-t') \leq \frac{a}{2}(t_0-t')^2 + \varkappa(x_0,t',t_0).$$

By this the estimation

$$-\frac{a}{2}(t-t_0) - \varkappa(x_0,t_0,t)/(t-t_0) \leq c(x,t_0) \leq$$

$$\leq \frac{a}{2}(t_0-t') + \varkappa(x_0,t',t_0)/(t_0-t'), \qquad x\in(x_0,r_1)\cap\dot{E}, \tag{6.6}$$

is got. Now, for $a>0$ let $t<t_0+\varepsilon/a$ and $t'>t_0-\varepsilon/a$ and choose $x_0\in\dot{E}$ of such a kind that $\max(\varkappa(x_0,t_0,t)/(t-t_0),\ \varkappa(x_0,t',t_0)/(t_0-t'))<\varepsilon/2$ is valid. Then (6.6) contradicts (6.3), i.e. $\lim_{x\to r_1} c(x,t)=0$, $t\in(0,\infty)$.

Remark 6.2. Let m,p,π of such a kind that for $S=E\setminus\{r_0\}$ and $S=\dot{E}$ the operator $D_mD_p^+\frac{\cdot}{\pi}\Big|C_S$ is the infinitesimal generator of a quasidiffusion $X^{(S)}=(X_t^{(S)}, \zeta^{(S)}, P_x^{(S)})$. Then relation (6.1) and Theorem 6.1. (a) yield

$$P_x^{(E\setminus\{r_0\})}(X_t^{(E\setminus\{r_0\})}\in B) = P_x^{(\dot{E})}(X_t^{(\dot{E})}\in B), \quad x\in\dot{E}, B\in\sigma(\dot{E}),$$

$$P_{r_1}^{(E\setminus\{r_0\})}(X_t^{(E\setminus\{r_0\})}\in B)=\begin{cases}\exp(-l_1 t) & r_1\in B\\ 0 & r_1\notin B\end{cases}, \quad B\in\sigma(E\setminus\{r_0\}).$$

Similar results hold if $D_mD_p^+\frac{\cdot}{\pi}\Big|C_S$, $S=E\setminus\{r_1\}$ or $S=E$ is a generator.

References

/1/ DYNKIN, E.B., Markov processes I,II, Berlin-Göttingen-Heidelberg 1965.

/2/ FELLER, W., On second order differential operators, Ann. Math. 61, 90-105, 1955.

/3/ FELLER, W., An Introduction to probability theory and its application II, New York-London-Sidney-Toronto, 1971.

/4/ GROH, J. Eine Klasse eindimensionaler Markovprozesse, Dissertation, TU Dresden, 1972.

/5/ ITÔ, K., Random processes II (russ.), Moscow, 1963.

/6/ ITÔ, K., H.P. McKEAN, Diffusion processes and their sample path, Berlin-Heidelberg-New York, 1965.

/7/ KÜCHLER, U., Some asymptotic properties of the transition densities of one-dimensional quasidiffusions, Publ. RIMS, Kyoto Univ. 16, 245-268, 1980.

/8/ LANGER, H., L. PARTZSCH, D. SCHÜTZE, Über verallgemeinerte gewöhnliche Differentialoperatoren mit nichtlokalen Randbedingungen und die von ihnen erzeugten Markovprozesse, Publ. RIMS, Kyoto Univ. 7, 659-702, 1971/72.

/9/ LÖBUS, J.-U., On generalized second order differential operators, (in preparation).

/10/ MANDL, P., Analytical treatment of one-dimensional Markov processes, Berlin-Heidelberg-New York, 1968.

/11/ SATO, K., Levy measures for a class of Markov semigroups in one dimesion., Trans. Amer. Math. Soc., 148, 211-231, 1970.

ON THE CONVERGENCE OF DIFFUSIONS

R. Mikulevičius
Institute of Mathematics and Cybernetics, Akademijos str.,4
Vilnius, Lithuania, USSR

Using [1], we consider the convergence of diffusion processes and corresponding parabolic operators in non-variational form. With the help of analytic methods similar results were obtained by L.Alyushina and N. Krylov (their paper will be published in "Teor.veroyatn. primen." or "Matem.zametki"). The case of sufficiently "good" limit operators was considered in [8].

1. Limit Distributions of Certain Sequences of Semimartingales

Let $\mathcal{X}$ be a Polish space and we have on a probability space $(\Omega, \mathcal{F}, P)$ a sequence of continuous processes $\tilde{Z}^n_t = (Z^n_t, L^n_t)$ with values in $\tilde{\mathcal{X}} = \mathcal{X} \times \mathbb{R}$; $L^n_t = \int_0^t \ell^n_s \, ds$ $(t \geq 0)$ where ℓ^n is $\mathcal{B}([0,\infty)) \otimes \mathcal{F}$ measurable and bounded. The canonical process on the space of trajectories $C = C_{[0,\infty)}(\mathcal{X})$ we denote by $\tilde{Z}_t = (Z_t, L_t)$ $(t \geq 0)$. Let $\mathcal{D} = \sigma(\tilde{Z}_u, u \geq 0)$, $\bar{C} = [0,\infty) \times C$, $\bar{\mathcal{D}} = \mathcal{B}([0,\infty)) \times \mathcal{D}$. For each $t > 0$ we define the measures on $(\bar{C}, \bar{\mathcal{D}})$ by setting $\mu^{n,t}(f) = E\int_0^t f(s, \tilde{Z}^n) dL^n_s$, $f \in \bar{\mathcal{D}}$.

Lemma 1. Let $\sup_n |\ell^n| \leq K < \infty$, $\tilde{Z}^n(P) \to \tilde{P}$. Then $\tilde{P}$ -a.s. L is of bounded variation, there is $\bar{\mathcal{D}}$ -measurable function ℓ such that $|\ell| \leq K$, $dL_t = \ell_t \, dt$ and for each $t > 0$ $\mu^{n,t} \to 1_{[0,t]} dL d\tilde{P}$.

Proof. Fix $t > 0$. Since $|\ell^n| \leq K$ and $C, \bar{C}$ are Polish it follows from Prohorov's theorem that $\{\mu^{n,t}, n \in \mathbb{N}\}$ is relatively compact and for every continuous bounded function f on $\bar{C}$ $|\mu^{n,t}(f)| \leq K E\int_0^t |f(s, \tilde{Z}^n)| ds$. Therefore for each limit point μ of $(\mu^{n,t})$ $|\mu(f)| \leq K \tilde{P}\int_0^t |f(s, \tilde{Z})| ds$, that is, there exists $\bar{\mathcal{D}}$ -measurable function $\hat{\ell}$ such that $|\hat{\ell}| \leq K$, $d\mu = d\hat{\ell} \, d\tilde{P}$ and $\hat{\ell}_s = 0$ for $s > t$. Let g be continuous bounded function on C , $s < t$, $g^s = 1_{[0,s]} g$. Then $\mu^{n,t}(g^s) = E L^n_s \, g(\tilde{Z}^n)$, that is, $\mu(g^s) = \tilde{P} L_s g = \tilde{P} g \int_0^s \ell_u \, du$. Hence $L_s = \int_0^s \hat{\ell}_u \, du$ $\tilde{P}$-a.s. and $d\mu = 1_{[0,t]} \, dL \, dP$. Since t is arbitrary, the lemma is proven.

Let $(\Omega, \mathcal{F}, P)$ be a probability space with a sequence of filtrations $\mathbb{F}^n = (\mathcal{F}^n_t)_{t\geq 0}$ and $\mathbb{R}^d$ -valued $\mathbb{F}^n$-adapted continuous semimartingales X^n be such that $X^n_t = X^n_0 + \int_0^t b^n_s\,ds + M^n_t$, where $M^n \in \mathcal{M}_{loc}(\mathbb{F}^n, P)$ ($\mathcal{M}_{loc}(\mathbb{F}^n, P)$ is the set of local $(\mathbb{F}^n, P)$-martingales); $\langle M^{ni}, M^{nj}\rangle_t = \int_0^t a^{nij}_s\,ds$, $1 \leq i,j \leq d$, (b^n, a^n) is $\mathcal{P}(\mathbb{F}^n)$-measurable $\mathbb{R}^d \times S^+_d$ -valued function ($a^n = (a^{nij})$, $\mathcal{P}(\mathbb{F}^n)$ is $\mathbb{F}^n$-predictable σ-algebra, S^+_d is the set of positive definite $d\times d$ -matrices). Suppose that for some constant $K < \infty$ $|a^n| + |b^n| \leq K$.

Let $A(s,x)$ $((s,x) \in [0,\infty)\times\mathbb{R}^d)$ be a family of convex closed subsets of $\tilde{R} = \mathbb{R}^d \times \mathbb{R}^{d^2}$ such that $\sup\{|y| : y \in A(s,x), (s,x) \in [0,\infty)\times\mathbb{R}^d\} < \infty$. For each $z \in \tilde{R}$ set $F^z(s,x) = \sup_{y\in A(s,x)} \langle z, y\rangle$ $((s,x)\in[0,\infty)\times\mathbb{R}^d)$, where $\langle\cdot,\cdot\rangle$ is a scalar product on $\tilde{R}$. Suppose that F^z is measurable for every z.

Let X denote the canonical process on $C = C_{[0,\infty)}(\mathbb{R}^d)$. Let $\mathcal{D} = \sigma(X_u, u\geq 0)$, $\mathcal{D}_t = \sigma(X_u, u \leq t)$, $\mathbb{D} = (\mathcal{D}_{t+})_{t\geq 0}$.

<u>Proposition 1</u>. Suppose that $\{X^n_0(P), n \in \mathbb{N}\}$ is relatively compact and one of the following conditions is satisfied:

a) F^z is continuous for each $z \in \tilde{R}$;

b) there is $c > 0$ such that $(a^n\eta, \eta) \geq c|\eta|^2$ for each $\eta \in \mathbb{R}^d$.

Then $\{X^n(P), n\in\mathbb{N}\}$ is relatively compact, and if μ is a limit point of $(X^n(P))$, there is $\tilde{R}$ -valued $\mathcal{P}(\mathbb{D})$-measurable function (b,a) such that $(b_s, a_s) \in A(s, X_s)$ $ds\,d\mu$-a.s. and

$$X_t = X_0 + \int_0^t b_s\,ds + M_t,\quad M \in \mathcal{M}_{loc}(\mathbb{D},\mu),$$
$$\langle M^i, M^j\rangle_t = \int_0^t a^{ij}_s\,ds \quad (1\leq i,j\leq d). \tag{1}$$

<u>Remark 1</u>. If the condition b) is satisfied, it follows from Krylov's theorem (see [2]) that there is a constant $\mathcal{N}$ depending only on c, K, d, T such that

$$P\int_0^T f(s, X^1_s)\,ds \leq \mathcal{N}\left(\int_0^T\int |f(s,x)|^{d+1}\,ds\,dx\right)^{1/(d+1)} \tag{2}$$

for every measurable positive f on $[0,\infty)\times\mathbb{R}^d$.

Proof of the proposition 1. Set $B^n_t = \int_0^t b^n_s\,ds$, $A^n_t = \int_0^t a^n_s\,ds$, $\tilde{X}^n_t = (X^n_t, B^n_t, A^n_t)$ $(t \geq 0)$. The canonical process on $\tilde{C} = C_{[0,\infty)}(\mathbb{R}^d\times\mathbb{R}^d\times\mathbb{R}^{d^2})$ we denote by $\tilde{X} = (\hat{X}, B, A)$ and let $\tilde{\mathcal{D}} = \sigma(\tilde{X}_s, s \geq 0)$.

For each $\varepsilon>0$, $\delta>0$, $T>0$, $T_n\in\mathcal{T}(\mathbb{F}^n)$ ($\mathcal{T}(\mathbb{F}^n)$ is the set of $\mathbb{F}^n$-stopping times) and some N non-depending on n

$$P(\sup_{T_n\le u\le T_n+\delta}|\tilde{X}^n_{(T_n+u)\wedge T}-\tilde{X}^n_{T_n\wedge T}|>\varepsilon)\le N\delta/\varepsilon^2 .$$

Therefore $\{\tilde{X}^n(P), n\in\mathbb{N}\}$, $\{X^n(P), n\in\mathbb{N}\}$ are relatively compact. If $\bar{P}$ is a limit point of $(X^n(P))$, then there is a subsequence (n_k) and a measure $\tilde{P}$ on $(\tilde{C},\tilde{\mathcal{D}})$ such that $\tilde{X}^{n_k}(P)\to\tilde{P}$, $X^{n_k}(P)\to\bar{P}$, $\hat{X}(\tilde{P})=\hat{P}$. According to the Lemma 1, there is $\tilde{R}$-valued $\tilde{\mathcal{D}}\otimes\mathcal{B}([0,\infty))$-measurable function $(\tilde{b},\tilde{a})$ such that $dB_t=\tilde{b}_t\,dt$, $dA_t=\tilde{a}_t\,dt$ and for each $t>0$

$$Z^n(dB^n_{\cdot\wedge t}\,dP)\to dB_{\cdot\wedge t}\,d\tilde{P},\qquad Z^n(dA^n_{\cdot\wedge t}\,dP)\to dA_{\cdot\wedge t}\,d\tilde{P}, \tag{3}$$

where $Z^n:(s,\omega)\to(s,\tilde{X}^n_{\cdot}(\omega))$ is a map from $[0,\infty)\times\Omega$ into $[0,\infty)\times\tilde{C}$.

Let $s<t$, and let g be $\mathcal{D}_s$-measurable bounded continuous function, $f\in C_0^\infty(\mathbb{R}^d)$. Then $E[f(X^n_t)-f(X^n_s)-\int_s^t\sum_i D_i f(X^n_u)\,dB^{ni}_u-\int_s^t\frac{1}{2}\sum_{i,j}D^2_{ij}f(X^n_u)\,dA^{nij}_u]\,g(X^n)=0$ and therefore by (3)

$$\tilde{P}[f(\hat{X}_t)-f(\hat{X}_s)-\int_s^t(\sum_i\tilde{b}^i_u D_i f(\hat{X}_u)+\sum_{i,j}\tilde{a}^{ij}_u D^2_{ij}f(\hat{X}_u))\,du]\,g(\hat{X})=0. \tag{4}$$

For every continuous bounded positive function G on $[0,\infty)\times\tilde{C}$ and each $z\in\tilde{R}$, $T>0$

$$E\int_0^T\langle(b^n_s,a^n_s),z\rangle G(s,\tilde{X}^n)\,ds\le E\int_0^T F^z(s,X^n_s)G(s,\tilde{X}^n)\,ds.$$

Therefore

$$\tilde{P}\int_0^T\langle(\tilde{b}_s,\tilde{a}_s),z\rangle G(s,\tilde{X})\,ds\le\tilde{P}\int_0^T F^z(s,\hat{X}_s)G(s,\tilde{X})\,ds \tag{5}$$

(the right-hand side of (4) converges because of the continuity of F^z or of the estimate (2)).

Let $\hat{\mathcal{D}}_t=\sigma(\hat{X}_s, s\le t)$, $\hat{\mathbb{D}}=(\hat{\mathcal{D}}_{t+})_{t\ge0}$. Given $\mathcal{P}(\hat{\mathbb{D}})$, there is $\mathcal{P}(\mathbb{D})$-measurable $\tilde{R}$-valued function (b,a) such that $(b(s,\hat{X}),a(s,\hat{X}))$ is $ds\,d\tilde{P}$-conditional expactation of $(\tilde{b}_s,\tilde{a}_s)$. Therefore (4) is true if we replace $\tilde{a}_s$, $\tilde{b}_s$ by $a(s,\hat{X})$, $b(s,\hat{X})$. It follows from (5) that $(b(s,\hat{X}),a(s,\hat{X}))\in A(s,\hat{X}_s)$ $ds\,d\tilde{P}$-a.s. Thus $\hat{X}(\tilde{P})=\bar{P}$ has desired properties.

Let $\Phi\in C_b^2(\mathbb{R}^d)$, $G=\{\Phi>0\}$, $D\Phi(x)\ne0$ for each

$x \in \{\Phi = 0\}$ and $\Phi(x) = -1$ for sufficiently large $|x|$.

Lemma 2. Let $P(X_0^n \in G) = 1$ $(n \in \mathbb{N})$ and the condition b) of the Proposition 1 is satisfied.

Then for each limit point μ of $(X^n(P))$ the function $\tau = \inf(t : X_t \notin G)$ is continuous μ -a.s.

Proof. Let μ be a limit point of $(X^n(P))$ and (1) is satisfied. For $R>0$ set $\tau_R = \inf(t : |X_t| > R)$. Using Itô formula for $\exp(\lambda |x|^2)$ with $\lambda > 0$ large enough, we obtain $\mu\tau_R < \infty$ i.e. $\mu(\tau_R < \infty) = 1$ for each $R > 0$.

There is $\varepsilon_0 > 0$ such that $\inf\{|D\Phi(x)| : |\Phi(x)| \le \varepsilon_0\} > 0$.

Using Itô formula, we obtain for some $\lambda > 0$, $k > 0$

$$\mu[\exp\{-\lambda\Phi(X_{\tau_\varepsilon \wedge \tau_\delta})\} - \exp\{-\lambda\Phi(X_\tau)\}] \geqslant$$

$$\geqslant k\mu\int_\tau^{\tau_\varepsilon \wedge \tau_\delta} \exp\{-\lambda\Phi(X_u)\}\,du \geqslant k\exp(-\lambda\delta)\mu[\tau_\varepsilon \wedge \tau_\delta - \tau] \tag{6}$$

for each $\varepsilon \in (-\varepsilon_0, 0)$, $\delta \in (0, \varepsilon_0)$, where $\tau = \inf(t : \Phi(X_t) = 0)$,

$\tau_\varepsilon = \inf(t : \Phi(X_t) \le \varepsilon)$, $\tau_\delta = \inf(t \geqslant \tau : \Phi(X_t) \geqslant \delta)$.

Let $\tau_{0-} = \lim\limits_{\varepsilon \uparrow 0} \tau_\varepsilon$, $\tau_{0+} = \lim\limits_{\delta \downarrow 0} \tau_\delta$. Then by (6) $\mu(\tau_{0-} \wedge \tau_{0+} = \tau) = 1$. On the other hand, we have from (6)

$$(e^{-\lambda\varepsilon} - 1)\mu(\tau_\varepsilon < \tau_\delta) + (e^{-\lambda\delta} - 1)(\mu(\tau_\delta < \tau_\varepsilon) \geqslant 0 .$$

Letting $\varepsilon \uparrow 0$, we obtain $\mu(\tau_\delta \le \tau_{0-}) = 0$ for each $\delta \in (0, \varepsilon_0)$, that is, $\tau_{0+} \geqslant \tau_{0-}$ μ -a.s. Therefore $\tau_{0-} = \tau$ μ -a.s., and the Lemma 2 is proved.

Lemma 3. Let the conditions of the Lemma 2 are satisfied, and let μ be a limit point of $(X^n(P))$, $(s, x_0) \in [0,\infty) \times \mathbb{R}^d$, $r > 0$ $B_r = \{x : |x - x_0| \le r\}$, $\mu(X_s \in B_r) = 0$, $\tau = \inf(t \geqslant s, X_t \in B_r)$.

Then $\mu(\tau \le s + \delta) > 0$ for each $\delta > 0$.

Proof. Let $f : [0,\infty) \to \mathbb{R}$ be a smooth function such that $f|_{[0,r)} > 0$, $f(a) = r - a$, if $a > r/2$. Set $H(x) = f(|x - x_0|)$ $(x \in \mathbb{R}^d)$. On $\mathbb{R}^d \setminus B_r$ $|DH| = 1$ and $D^2 H$ is bounded.

Let $Y_t = H(X_t)$ $(t \geqslant 0)$. Suppose $P(\tau \le s+\delta) = 0$ for some $\delta > 0$. There is a measure $\tilde{\mu}$ on $\mathcal{D}_{s+\delta}$ equivalent to the measure $\mu|_{\mathcal{D}_{s+\delta}}$ such that $Y_t = Y_s + M_t$ $(s \le t \le s+\delta)$; M is a $((\mathcal{D}_{u+})_{s \le u \le s+\delta}, \tilde{\mu})$ -martingale and $\langle M \rangle_\cdot =$ $= \int_s^\cdot a_u^{ij} D_i H(X_u) D_j H(X_u)\,du$ (for μ (1) is satisfied).

It is clear that $\langle M \rangle_{s+\delta} \geqslant c\delta$. Let $K_u = \inf(t \geqslant s : \langle M \rangle_t > u)$ $(u \in [s, s+\delta c])$. Set $\tilde{Y}_u = Y_{K_u} = Y_s + M_{K_u}$ $(u \in [s, s+\delta c])$. Then $\mu(\tilde{Y}_u < 0) = 1$,

$M_{K.} \in \mathcal{M}_{loc}((\mathcal{D}_{K_u+}), \tilde{\mu})$, $\langle M_{K.}\rangle_u = u-s$ $(u \in [s, s+c\delta])$.
Thus, we have a contradiction. The lemma is proven.

2. Convergence of Operators

Fix $0<\lambda<K<\infty$. Let $L=\{A=A_s=A(a,b)=$
$=\frac{1}{2}\sum_{i,j=1}^{d} a^{ij}(s,x)D^2_{ij} + \sum_{i=1}^{d} b^i(s,x)D_i$: $(s,x)\in[0,\infty)\times\mathbb{R}^d$; $b=(b^i)$, $a=(a^{ij})$ are measurable, $|a|+|b|\leq K$, $\lambda|\eta|^2 \leq (a\eta,\eta)$ for each $\eta\in\mathbb{R}^d\}$, $SL=\{A(a,b)\in L$: a, b are smooth$\}$.

We denote by $\mathcal{H}$ the set of all subsets $G\subset\mathbb{R}^d$ such that $G=\{\Phi>0\}$, $\Phi\in C^2_0(\mathbb{R}^d)$, $D\Phi(x)\neq 0$ for each $x\in\partial G=\bar{G}\setminus G$.

If $A\in SL$, $f\in L_{d+1}(G_T)$ $(G\in\mathcal{H}$, $G_T=(0,T)\times G)$ then there is a unique function $R(A)f$ belonging to the Sobolev space $W^{2,1}_{d+1}(G_T)$ and satisfying the following equation:

$$(\partial_t + A)R(A)f = f, \quad R(A)f\big|_{\partial_T G} = 0,$$

where $\partial_T G = \{T\}\times\bar{G}\cup[0,T]\times\partial G$ (see [4]).

Definition. Let $A^n\in SL$, $A\in L$. We say $A^n\to A$, if for each $T>0$, $G\in\mathcal{H}$ there is a bounded operator $R: L_{d+1}(G_T)\to$ $\to C(\bar{G}_T)$ ($C(\bar{G}_T)$ is the space of all continuous functions on $\bar{G}_T$ with the supremum norm) such that for each $f\in L_{d+1}(G_T)$, $u\in W^{2,1}_{d+1}(G_T)$ $R(A^n)f\to Rf$ uniformly on $\bar{G}_T$ and $R(\partial_t+A)u=u$ provided $u|_{\partial_T G}=0$ (cf.[3]).

3. Relative Compactness of Families of Distributions of Diffusion Processes

We denote by M_1 the space of probability measures on $(C,\mathcal{D})$ with the topology of weak convergence, that is, with the topology generated by Lévy-Prohorov metric ρ. If $A\in SL$, then for each $(s,x)\in[0,\infty)\times\mathbb{R}^d$ there is a unique $P_{s,x}\in M_1$ such that $P_{s,x}(X_u=x, \forall u\leq s)=1$ and for each $f\in C^\infty_0(\mathbb{R}^d)$

$$f(X_\cdot) - \int_s^{\cdot} A_u f(X_u)\,du \in \mathcal{M}_{loc}((\mathcal{D}_{t+})_{t\geq s}, P_{s,x}).$$

The function $F^A: (s,x)\to P_{s,x}$ from $[0,\infty)\times\mathbb{R}^d$ into M_1 is continuous. We denote by $\mathfrak{A}$ the set of all continuous M_1-valued functions on $[0,\infty)\times\mathbb{R}^d$ with the topology of compact convergence.

<u>Theorem 1</u>. The set $\{F^A : A\in SL\}$ is relatively com-

pact in $\mathfrak{A}$.

Proof. Let $T>0$, $n\in\mathbb{N}$, $t_i = iT/n$, $i=0,\dots,n$. For $w\in C$ we define $w^n\in C$: $w^n_t = [w_{t_k}(t_{k+1}-t) + w_{t_{k+1}}(t-t_k)]/(t_{k+1}-t_k)$, if $t_k\le t\le t_{k+1}$, $w^n_t = w_T$, if $t\geq T$. Thus, $k_n : w\to w^n$, $\lambda_T : w\longrightarrow w_{\cdot\wedge T}$ are maps from C into itself, $I_n : w\longrightarrow (w_{t_i})_{0\le i\le n}$ is a map from C into $\mathbb{R}^{n+1}$. Put $P^1 = F^A_{s,x}$, $P^2 = F^{A}_{s',x'}$ for $A\in SL$, (s,x), $(s',x')\in[0,\infty)\times\mathbb{R}^d$. Similarly as in the case of Lemma 2 in [7] we find that for each $\varepsilon>0, T>0, n\in\mathbb{N}$

$$\rho(P^1,P^2)\le 2\varepsilon + \sum_{i=1}^{2} P^i(d(X,\lambda_T(X))>\varepsilon) + \rho(\lambda_T(P^1),\lambda_T(P^2)) \le$$

$$\le 4\varepsilon + \sum_{i=1}^{2} P^i(d(X,\lambda_T(X))>\varepsilon) + \sum_{i=1}^{2}\lambda_T(P^i)(d(X,k_n(X))>\varepsilon) +$$

$+\rho(k_n\circ\lambda_T(P^1), k_n\circ\lambda_T(P^2))$, where $d(x,y) = \sum_k 2^{-k}(\sup_{u\le k}|x_u - y_u|\wedge 1)$. It is easy to see that $\rho(k_n\circ\lambda_T(P^1), k_n\circ\lambda_T(P^2))\le \rho(I_n(P^1), I_n(P^2))$ (we use same letter ρ to denote Lévy-Prohorov distance between $I_n(P^1)$, $I_n(P^2)$ as well),

$P^i(d(\lambda_T(X),k_n(X))>\varepsilon)\le\lambda_T(P^1)(\sup_k \sup_{t_k\le u\le t_{k+1}}|X_u - X_{t_k}|>\varepsilon/3)\le$ $\le N\varepsilon^{-4}n^{-1}$ for some N non-depending on A, s, x, s', x'. We denote by $\mathcal{H}$ the set of all differentiable functions f on $\mathbb{R}^{d(n+1)}$ such that $|f|_\infty\le 1$, $|Df|_\infty\le 1$. By Lemma 3 in [7] there is a constant $\mathfrak{N}$ non-depending on A, s, x, s', x' such that

$$\rho(I_n(F^A_{s,x}), I_n(F^A_{s',x'}))\le$$

$$\le \mathfrak{N}\sup_{f\in\mathcal{H}}|F^A_{s,x}f(X_{t_0},\dots,X_{t_n}) - F^A_{s',x'}f(X_{t_0},\dots,X_{t_n})|^{1/2}. \qquad (7)$$

Since $F^A_{s,x}f(X_{t_0},\dots,X_{t_n}) = F^A_{s,x}F^A_{t_0,X_{t_0}}\dots F^A_{t_{n-1},X_{t_{n-1}}}f(X_{t_0},\dots,X_{t_n})$ and T, n are arbitrary, the theorem follows easily from Ascoli's theorem, estimation (2) and the following statement:

Lemma 4. Let $\mathcal{X}$ be a metrisable space and let Θ be a family of functions from $\mathcal{X}\times\mathbb{R}^d$ into $\mathbb{R}$ equicontinuous on $\mathcal{X}\times\{x : |x|\le n\}$ for each $n\in\mathbb{N}$ such that $\sup\{|f|_\infty : f\in\Theta\}\le 1$.

Then for each $t>0$, $n\in\mathbb{N}$ $\{(s,x,y)\to F^A_{s,x}f(y,X_t) = \int f(y,X_t(w))F^A_{s,x}(dw) : f\in\Theta, A\in SL\}$ is equicontinuous on $[0,t]\times\{x : |x|\le n\}\times\mathcal{X}$.

Proof. Let $\varepsilon>0$, $g\in C_0^\infty(\mathbb{R}^d)$, $g\geqslant 0$, $\operatorname{supp} g\subset\{x:|x|\leqslant 1\}$,

$\int g(x)dx=1$, $g_\varepsilon(x)=\varepsilon^{-d}g(x/\varepsilon)$,

$f_\varepsilon(x,y)=\int f(y,z)g_\varepsilon(x-z)dz \quad (x\in\mathbb{R}^d, f\in\Theta)$. For $N>1$ we choose a smooth function $g_N:\mathbb{R}^d\longrightarrow[0,1]$ such that $g_N\big|_{\{|x|\leqslant N\}}=0$, $g_N\big|_{\{|x|>N+1\}}=1$. Then for each $\varepsilon>0$, $N>1$

$$|F^A_{s,x}(f(y,X_t)-f_\varepsilon(y,X_t))|\leqslant 2F^A_{s,x}(|X_t|>N)+$$

$$+\sup_{\substack{y,|x-x'|\leqslant\varepsilon\\|x|\leqslant N}}|f(y,x)-f(y,x')|,\quad F^A_{s,x}f_\varepsilon(y,X_t)=$$

$$=f_\varepsilon(y,x)+v(s,x)+F^A_{s,x}\int_s^t(1-g_N(X_u))A_uf_\varepsilon(y,X_u)du, \tag{8}$$

where $v(s,x)=F^A_{s,x}\int_s^t g_N(X_u)A_uf_\varepsilon(y,X_u)du$. There is $C(\varepsilon)<\infty$ non-depending on A,s,x,f such that

$$\left|F^A_{s,x}\int_s^t(1-g_N(X_u))A_uf_\varepsilon(y,X_u)du\right|\leqslant(|x|+C(\varepsilon))C(\varepsilon)|t-s|/N \tag{9}$$

It follows from [4], (2) and Itô formula that

$$(\partial_t+A)v=g_NAf_\varepsilon,\quad v(t,\cdot)=0.$$

Now the statement follows obviously from (8), (9) and the Theorem 4.3 in [1].

4. Convergence of Diffusions

We investigate now the connection between the convergence of diffusion processes and corresponding operators.

Theorem 2. Let $A^n\in SL$ and F^{A^n} converges to F in $\mathfrak{A}$. Then

a) $(X_t,(F_{s,x}))$ is strongly Markovian with respect to $\mathbb{D}$;

b) there is $A=A(a,b)\in L$ such that for each $f\in C_0^\infty(\mathbb{R}^d)$, $(s,x)\in[0,\infty)\times\mathbb{R}^d$

$$f(X_\cdot)-f(X_s)-\int_s^\cdot A_uf(X_u)du\in\mathcal{M}_{loc}((\mathcal{D}_{t+})_{t\geqslant s},F_{s,x}); \tag{10}$$

for each (s,x) $(b(s,x),a(s,x))$ belongs to the convex closure of $\{(b^n(s,x),a^n(s,x)):n\in\mathbb{N}\}$;

c) if $A\in L$ satisfies (10), then $A^n\to A$.

Proof. Let $f:\mathbb{R}^d\to\mathbb{R}$ be a continuous bounded function, $0\leqslant\tau<u$. Let g be a $\mathcal{D}_\tau$-measurable continuous bounded function. Then

$F^{A^n}_{s,x}(f(X_u)|\mathcal{D}_\tau) = F^{A^n}_{\tau,X_\tau} f(X_u)$, $F^{A^n}_{s,x} f(X_u) g =$
$= F^{A^n}_{s,x}\, g\, F^{A^n}_{\tau,X_\tau} f(X_u)$. Since $F^{A^n}_{s,x} f(X_u) g \to F_{s,x}\, g F_{\tau,X_\tau} f(X_u)$
uniformly on compact sets, $F_{s,x} f(X_u) g = F_{s,x}\, g\, F_{\tau,X_\tau} f(X_u)$.
Because of the continuity of F , the statement a) is obvious.

Since (t, X_t) is a homogeneous strong Markov process, b) follows immediately from the Proposition 1 and [6].

Let A satisfy (10). Let $G \in \mathcal{H}$, $T>0$, $\tau = \inf(t : X_t \notin G)$.
Let f be a continuous bounded function on $[0,\infty) \times \mathbb{R}^d$,
$h^u = \int_{u\wedge T}^{(u\vee\tau)\wedge T} f(s,X_s)\,ds$, $u \geq 0$, $\mathcal{H} = \{\nu \in M_1 :$ h^u is ν-a.s. continuous$\}$. It is clear that $\nu \to \nu(h^u)$ is continuous on $\mathcal{H}$. By Lemma 2 $\{F^{A^n}_{s,x}$, $F_{s,x} : (s,x) \in \bar{G}_T$, $n \geq 1\}$ is a compact subset of $\mathcal{H}$. That is, $\forall u$ $F^{A^n}_{s,x} h^u \to F_{s,x} h^u$ uniformly on $\bar{G}_T$. Since $|h^u - h^{u'}| \leq 2|f|_\infty |u-u'|$ for each u, u',
$F^{A^n}_{s,x} h^s \to F_{s,x} h^s$ uniformly on $\bar{G}_T$. If $f \in L_{d+1}(G_T)$

there is by (2) a sequence of continuous bounded functions (f_k) such that $\sup_{n,(s,x)\in\bar{G}_T} (F^{A^n}_{s,x} + F_{s,x}) \int_s^{\tau\wedge T} |f_k(u,X_u) - f(u,X_u)|\,du \to 0$.
That is, $R(A^n) f(s,x) \to R f(s,x) = -F_{s,x} \int_s^{\tau\wedge T} f(u,X_u)\,du$
uniformly on $\bar{G}_T$. It follows from (2) that R is bounded. If

$u \in W^{2,1}_{d+1}(G_T)$, $u|_{\partial_T G} = 0$ then by Itô formula -

$-u(s,x) = F_{s,x} \int_s^{\tau\wedge T} (\partial_\tau + A_\tau) u(\tau, X_\tau)\,d\tau$. The Theorem 2 is proven.

Let M^d_1 be the set of probability measures on $\mathbb{R}^d$. For $\nu \in M^d_1$, $A \in L$ we define $\mathcal{S}(\nu, A) = \{P \in M_1 : P(X_0 \in dx) = \nu(dx)$ and for each $f \in C^\infty_0(\mathbb{R}^d)$ $f(X_\cdot) - \int_0^\cdot A_u f(X_u)\,du \in \mathcal{M}_{loc}(\mathbb{D}, P)\}$.

Corollary 1. Let $A^n \in SL$; $\nu^n, \nu \in M^d_1$, $\nu^n \to \nu$, $P^n \in \mathcal{S}(\nu^n, A^n)$.
Then for each limit point P of (P^n) there is $A \in L$ such that $P \in \mathcal{S}(\nu, A)$ and X is strong Markov process with respect to $(\mathbb{D}, P)$.

Proof. Since $P^n = \int F^{A^n}_{0,x}\, \nu^n(dx)$ for each n , the statement follows obviously from Theorems 1, 2.

Remark 2. It follows obviously from (2) and the Theorem 1 that for each $\nu \in M^d_1$, $A = A(a,b) \in L$ there is $P \in \mathcal{S}(\nu, A)$ such that X is a Feller process with respect to $(\mathbb{D}, P)$. Indeed, it is sufficient to choose $A(a^n, b^n) \in SL$ such that $a^n \to a$,
$b^n \to b$ $ds\,dx$ a.s.

Theorem 3. Let $A^n \in SL$, $A \in L$, $\nu^n, \nu \in M_1^d$, $P^n \in \mathcal{S}(\nu^n, A^n)$, $P \in \mathcal{S}(\nu, A)$, $P^n \to P$.
Then $A^n \to A$.

Proof. For each n $P^n = \int F_{0,x}^{A^n} \nu^n(dx)$. Then for some subsequence (n_k) $F^{A^{n_k}} \to F$ in $\mathcal{O}\!\iota$ and

$$P = \int F_{0,x} \nu(dx). \tag{11}$$

By Theorem 2 there is $\tilde{A} \in L$ such that $F, \tilde{A}$ satisfy (10) and $A^{n_k} \to \tilde{A}$. Because of (11), $P \in \mathcal{S}(\nu, \tilde{A})$. That is, for each $f \in C_0^\infty(\mathbb{R}^d)$ $\int_0^{\cdot} A_u f(X_u)du = \int_0^{\cdot} \tilde{A}_u f(X_u)du$ P-a.s. Therefore for each finite $\tau \in \mathcal{T}(\mathbb{D})$ $F_{\tau, X_\tau}(\int_\tau^{\tau+\cdot} (A_u f(X_u) - \tilde{A}_u f(X_u))du = 0) = 1$ P-a.s. Because of the continuity of F, it follows by Lemma 3 $F_{s,x}(\int_s^{\cdot} (A_u f(X_u) - \tilde{A}_u f(X_u))du = 0) = 1$ for each s, x. That is, F, A satisfy (10) and $A^{n_k} \to A$ by Theorem 2. The statement follows now from the Theorem 1.

Theorem 4. Let $A^n \in SL$, $A \in L$, $A^n \to A$. Then there is $F \in \mathcal{O}\!\iota$ such that $F^{A^n} \to F$ and F, A satisfy (10).

Proof. Let F be a limit point of (F^{A^n}). Let $f : [0,\infty) \times \mathbb{R}^d \to \mathbb{R}$ be a continuous bounded function, $G \in \mathcal{H}$, $\tau = \inf(t : X_t \notin G)$, $T > s$. Since $R(A^n) f(s,x) = -F_{s,x}^{A^n} \int_s^{\tau \wedge T} f(u, X_u)du$ ($R(A^n)$ corresponds to G, T), $Rf(s,x) = -F_{s,x} \int_s^{\tau\wedge T} f(u, X_u)\,du$. Since T, G are arbitrary and F is Markovian, it follows that F is a unique limit point. Thus, $F^{A^n} \to F$.

Let $f \in C_0^\infty(\mathbb{R}^d)$, $\operatorname{supp} f \subset G \in \mathcal{H}$, $T > 0$, $\varepsilon > 0$, $T - \varepsilon > 0$.

Let $f_\varepsilon : [0,T] \to [0,1]$ be a smooth function such that $f_\varepsilon(T) = 0$, $f_\varepsilon\big|_{[0,T-\varepsilon]} = 1$, $g_\varepsilon(s,x) = f_\varepsilon(s) f(x)$ ($(s,x) \in \bar{G}_T$).
It is clear that $g_\varepsilon \in W_{d+1}^{2,1}(G_T)$. Set $v^n(s,x) = F_{s,x}^{A^n} \int_s^{\tau\wedge T} (\partial_r + A_r) g_\varepsilon(r, X_r)dr$ ($\tau = \inf(t : X_t \notin G)$). Since $A^n \to A$, $v^n(s,x) \to -R(\partial_t + A) g_\varepsilon(s,x) = -g_\varepsilon(s,x)$. On the other hand, $v^n(s,x) \to F_{s,x} \int_s^{\tau\wedge T} (\partial_r + A_r) g_\varepsilon(r, X_r)\,dr$ because of the convergence F^{A^n} to F.
Intêgrating by parts and letting $\varepsilon \downarrow 0$ we obtain

$$F_{s,x} f(X_{\tau\wedge T}) - f(x) = F_{s,x} \int_s^{\tau\wedge T} A_u f(X_u)du \quad \text{for each}$$

$(s,x) \in [0,\infty) \times \mathbb{R}^d$. Since G, T are arbitrary, the Theorem 4 is proven.

Theorem 5. Let $A^n \in SL$, $A \in L$, $A^n \to A$, $\nu^n, \nu \in M_1^d$, $P^n \in \mathcal{S}(\nu^n, A^n)$, $\nu^n \to \nu$.

Then there exists $P \in \mathcal{S}(\nu, A)$ such that $P^n \to P$.

Proof. Since $P^n = \int F_{0,x}^{A^n} \nu^n(dx)$ the statement is an obvious consequence of Theorem 4.

Theorem 6. Let $A^n = A(a^n, b^n) \in SL$, $A = A(a,b) \in L$, $\nu^n, \nu \in M_1^d$, $P^n \in \mathcal{S}(\nu^n, A^n)$, $P \in \mathcal{S}(\nu, A)$, $P^n \to P$.

Then $\nu^n \to \nu$ and for each $t > 0$

$$\sup_{u \le t} \Big(\Big| \int_0^u (a^n(r, X_r) - a(r, X_r))\, dr \Big| + \\ + \Big| \int_0^u (b^n(r, X_r) - b(r, X_r))\, dr \Big| \Big) \xrightarrow{P^n} 0 \tag{12}$$

Proof. It follows from Theorem 3-5 that there is $F \in \mathfrak{A}$ such that $F^{A^n} \to F$, $A^n \to A$ and F, A satisfy (10). It is clear that $P^n = \int F_{0,x}^{A^n} \nu^n(dx)$, $P = \int F_{0,x}\, \nu(dx)$. Let $t > 0$, $p \in \mathbb{N}$, $t_i = it/p$ $(i = 0, \ldots, p)$. Set $B_u^{n1} = \int_0^u b^n(r, X_r)\, dr$,

$B_u^1 = \int_0^u b(r, X_r)\, dr$, $B_u^{n2} = \int_0^u a^n(r, X_r)\, dr$, $B_u^2 = \int_0^u a(r, X_r)\, dr$,

$\overline{B}^{npi} = \sum_k P^n(\overline{B}_{t_{k+1}}^{ni} - \overline{B}_{t_k}^{ni} \mid \mathcal{D}_{t_k+})$, $\overline{B}^{ni} = B^{ni} - B^i$, $i = 1, 2$.

It is easy to see (Lemma 3 in [5]) that for each $\varepsilon > 0$

$$\lim_p \overline{\lim_n}\, P^n(|\overline{B}^{npi} - \overline{B}_t^{ni}| > \varepsilon) = 0 \, . \tag{13}$$

If $0 \le k < p$, $F_{t_k,x}^{A^n}(\overline{B}_{t_{k+1}}^{n1} - \overline{B}_{t_k}^{n1}) = F_{t_k,x}^{A^n}[X_{t_{k+1}} - X_{t_k} - (B_{t_{k+1}}^1 - B_{t_k}^1)]$ converges by (2) to $F_{t_k,x}[X_{t_{k+1}} - X_{t_k} - (B_{t_{k+1}}^1 - B_{t_k}^1)] = 0$ uniformly on compact sets. It follows by (13) that $\overline{B}_t^{n1} \xrightarrow{P^n} 0$. Since t is arbitrary,

$$\sup_{u \le t} |\overline{B}_u^{n1}| \xrightarrow{P^n} 0 \, . \tag{14}$$

Fix $z \in \mathbb{R}^d$. Set $M_t^n = (z, X_t - B_t^{n1})$, $M_t = (z, X_t - B_t^1)$.

If $0 \le k < p$, we have by (14)

$$P^n[(M^n_{t_{k+1}} - M^n_{t_k})^2 | \mathfrak{D}_{t_k+}] - P^n[(M_{t_{k+1}} - M_{t_k})^2 | \mathfrak{D}_{t_k+}] \xrightarrow{P^n} 0$$

On the other hand, $F^{A^n}_{t_k,x}(M_{t_{k+1}} - M_{t_k})^2 \to F_{t_k,x}(M_{t_{k+1}} - M_{t_k})^2 =$

$= F^A_{t_k,x}((B^2_{t_{k+1}} - B^2_{t_k})z, z)$ uniformly on compact sets. Therefore for each p $\bar{B}^{np2} \xrightarrow{P^n} 0$. Hence we have by (13) $\bar{B}^{n2}_t \xrightarrow{P^n} 0$. Since t, z are arbitrary, the statement of the Theorem 6 is obvious.

We have by Theorem 6 the following trivial

Corollary 2. If the assumptions of Theorem 6 are satisfied and a, b are continuous, then $P^n \to P$ if and only if $\nu^n \to \nu$ and (12) is true.

References

1 Крылов Н.В., Сафонов М.В. Некоторое свойство решений параболических уравнений с измеримыми коэффициентами. - Изв.АН СССР, сер. матем., 1980, т. 44, № I, с. I6I-I75.

2 Анулова С., Прагараускас Г. О слабых марковских решениях стохастических уравнений. - Лит.матем.сб., 1977, т. I7, № 2, с. 5-26.

3 Крылов Н.В. О G -сходимости эллиптических операторов в недивергентной форме. - Матем.заметки, 1985, т. 37, № 4, с. 522-527.

4 Ладыженская О.А., Солонников В.А., Уральцева Н.Н. Линейные и квазилинейные уравнения параболического типа. - М.:Наука, 1967.

5 Кубилюс К., Микулявичюс Р. Необходимые и достаточные условия сходимости семимартингалов и точечных процессов. II. - Лит.матем. сб., 1984, т. 24, № 4, с. 99-II5.

6 Çinlar E., Jacod J., Protter P., Sharpe M.J. Semimartingales and Markov processes. - Z.Wahrscheinlichkeitstheorie verw. Gebite, 1980, B. 54, S. 161-219.

7 Kubilius K., Mikulevičius R. On the rate of convergence of distributions of semimartingales. - In: Proceedings of the 4th Vilnius Conference on Probability Theory and Math. Statistics (to appear).

8 Mahno S.Y. On weak convergence of solutions of stochastic equations. - Preprint 1985.

DERIVATIVE FREE NUMERICAL METHODS FOR STOCHASTIC DIFFERENTIAL EQUATIONS

E. Platen
Academy of Sciences of the GDR
Institute of Mathematics, Mohrenstr. 39
Berlin, DDR 1086

1. Diffusion process

The diffusion process is more and more used as mathematical model in many fields of application. Therefore the numerical investigation of this process became an important question. The paper gives a short survey on some Monte-Carlo simulation methods based on time discrete diffusion approximations. For simplicity the following presentation is restricted to the one-dimensional diffusion process driven by a one-dimensional Wiener process. But most of the mentioned results are generalized to the multidimensional case.

Let be given on a filtered probability space $(\Omega, \underline{\underline{F}}, \underline{F}, P)$ a diffusion process $X=\{X_t\}_{t\in[0,T]}$ which is defined by the Itô equation:

$$(1)\qquad X_t = x + \int_0^t a(X_s)\,ds + \int_0^t b(X_s)\,dW_s,$$

$t \in [0, T]$, where x denotes the initial value at time 0 and $W=\{W_t\}_{t\in[0,T]}$ the standard Wiener process. The drift and diffusion coefficients a and b, resp., are assumed to be Lipschitz continuous.

Simulation studies and theoretical investigations show that one can not simply apply well-known numerical methods for deterministic ordinary differential equations to simulate approximate trajectories of diffusion processes.

The proposed simulation methods are based on a finite time discretization:

(2) $0=:\tau_0<\tau_1<\ldots<\tau_{i_T}:=T$

of the considered time interval $[0,T]$ which may be equidistant, but in the general case also a step size control during the simulation of an approximate trajectory is possible (see [10]). Each approximate trajectory must be recursively computed at the above discretization points. Instead of the diffusion process itself one uses so called time discrete approximations of the diffusion which allow an easy computation of the increments of approximate trajectories.

The simplest time discrete approximation is the so called Euler approximation:

(3) $Y_{i+1}=Y_i + a_i\Delta_i + b_i\Delta W_i,$

with $Y_0=x$, $\Delta_i=\tau_{i+1}-\tau_i$, $\Delta W_i=W_{\tau_{i+1}}-W_{\tau_i}$, $i\in\{0,\ldots,i_T-1\}$, where we set also in the following $h_i=h(Y_i)$ for any function h. The computer has to generate the normal distributed random variable ΔW_i at each time step and one obtains finally the values of an approximate trajectory at the discretization points. A sufficiently large sample of such trajectories can be used to approximate functionals of the diffusion itself.

It turned out that the Euler approximation converges rather slowly with respect to useful convergence criteria. Several authors, for instance Milstein [1], Rao, Borwankar, Ramakrishna [2], Wagner, Platen [3], Clark, Cameron [4], Rümelin [5], Talay [6] and Platen [7] developed better time discrete approximations. We will consider in the following some of these time discrete approximations with respect to the so called mean square criterion and the mean criterion.

2. Mean square criterion

If one needs pathwise approximate trajectories, then one can estimate the order of convergence of a time discrete approximation Y with respect to the mean square criterion:

(4) $$E \max_i | X_{\tau_i} - Y_i |^2 \leq K \delta^{\gamma},$$

where K is a constant, δ denotes the maximum step size of the time discretization and $\gamma \in (0,\infty)$ is the order of mean square convergence.

For instance the Euler approximation shows first order mean square convergence. In Milstein [1] one can find the following second order mean square approximation:

(5) $$Y_{i+1} = Y_i + a_i \Delta_i + b_i \Delta W_i + b_i b_i' \cdot (\Delta W_i^2 - \Delta_i)/2.$$

By the use of the so called stochastic Taylor formula derived in [3], [8] and [9] it is shown in [3] and [10] that for any desired order of mean square convergence one finds corresponding mean square Taylor approximation, where the Euler approximation represents the first order and the above scheme the second order mean square Taylor approximation. The third order mean square Taylor approximation has the form:

(6) $$\begin{aligned} Y_{i+1} = Y_i &+ a_i \Delta_i + b_i \Delta W_i + b_i b_i' (\Delta W_i^2 - \Delta_i)/2 \\ &+ b_i a_i' \Delta Z_i + (a_i b_i' + b_i^2 b_i''/2)(\Delta W_i \Delta_i - \Delta Z_i) \\ &+ (b_i'' b_i + (b_i')^2) b_i \Delta W_i (\Delta W_i^2/3 - \Delta_i)/2 \\ &+ (a_i a_i' + b_i^2 a_i''/2) \Delta_i^2/2, \end{aligned}$$

where

(7) $$\Delta Z_i = \int_{\tau_i}^{\tau_{i+1}} \int_{\tau_i}^{s_2} dW_{s_1} ds_2 .$$

To prove the above mentioned convergence one has to assume smoothness and Lipschitz properties of a and b.

For the implementation of simulation algorithms on a computer it is desireable to avoid derivatives of a and b. Therefore derivative free algorithms are of special interest. In [7] conditions are described under which general time discrete approximations achieve a given order of mean square convergence. For instance one has the following second order derivative-free mean square approximation:

$$(8)\qquad Y_{i+1}=Y_i + a_i\Delta_i + b_i\Delta W_i + \left(b(Y_i+b_i\,\Delta_i^{1/2}) - b_i\right)\Delta_i^{-1/2}(\Delta W_i^2 - \Delta_i)/2.$$

From the approach in [10] which uses the stochastic Taylor expansion it becomes clear that one needs for higher order schemes more multiple stochastic integrals as $\Delta_i, \Delta W_i, \Delta Z_i, \ldots$ which represent random variables. These random variables must be generated by **pseudo** random number generators on the computer. A more detailed consideration of multiple stochastic integrals especially for the multidimensional Wiener process can be found in [11] . The time discrete approximation of the Itô process with Poissonian jump component is treated in [12] .

3. Mean convergence

In many applications one does not need pathwise approximation. Often one wishes only to approximate functionals, that means, the expectation of functions of the diffusion. In this case one can use the following so called mean convergence criterion:

$$(9)\qquad | E\, g(X_T) - E\, g(Y_{i_T}) | \leq K\delta^{\gamma},$$

where g is some smooth function of polynomial growth. $\gamma \in (0,\infty)$ denotes the order of mean convergence, δ is the maximum step size and K is a constant. This type of convergence was studied by Milstein [13] , Talay [6] and also in [7] . Under sufficient smoothness assumptions on a and b it follows from the above mentioned results that the Euler approximation shows first order mean convergence. In [14] the order of mean convergence of the Euler approximation under Hölder conditions on a and b is studied.

Talay [6] proposed a class of second order mean approximations. In [7] for any order of mean convergence corresponding mean Taylor approximations are introduced by the use of the stochastic Taylor formula. For instance, the second order mean Taylor approximation is of the form:

$$(10)\qquad Y_{i+1}=Y_i + a_i\Delta_i + b_i\Delta W_i + b_i b_2'(\Delta W_i^2 - \Delta_i)/2 + b_i a_i'\Delta Z_i + (b_i' a_i + b_i'' b_i^2/2)(\Delta W_i\Delta_i - \Delta Z_i) + (a_i' a_i + a_i'' b_i^2/2)\Delta_i^2/2,$$

where we used the notation as before. To show the desired mean convergence of higher order mean Taylor approximation one needs sufficient smoothness and boundedness properties of a and b. Corresponding results for the Itô process with jump component are proved in [15] .

A numerically more convenient derivative free second order mean approximation is proposed in [7] :

$$(11) \qquad Y_{i+1}=Y_i + (a(Y_i + a_i\Delta_i + b_i\Delta W_i) + a_i)\Delta_i/2$$
$$+ (b(\bar{Y}_i^+) + b(\bar{Y}_i^-) + 2b_i)\Delta W_i/4$$
$$+ (b(\bar{Y}_i^+) - b(\bar{Y}_i^-))(\Delta W_i^2 - \Delta_i)\Delta_i^{-1/2}/4,$$

with

$$\bar{Y}_i^{\pm}=Y_i + a_i\Delta_i \pm b_i\Delta_i^{1/2}.$$

One notes that at each step only the random variable ΔW_i appears and five values of a and b, resp., are necessary. In [7] one can find conditions under which quite general time discrete approximations show a given order of mean convergence.

4. Conclusions

The most important problem for the practical application is the nessecary computer time to approximate for instance a given functional of a diffusion. Simulation studies in [16] with the above mentioned schemes show that higher order schemes are in some cases more efficient than for instance the simple Euler scheme.

Furthermore, the derivative free methods show similar convergence behaviour as the Taylor approximations. But they are much easier to implement on the computer. It seems that it is only possible to construct efficient Monte-Carlo methods for special classes of simulation problems. Theoretical investigations concerning the order of convergence as in [1] - [7] can only give some hints for the choice of a useful simulation algorithm. Open questions are concerning the step size control and numerical stability of algorithms as well as the choice of an adapted approximation scheme.

5. References

[1] Milstein, G.N.: Approximate integration of stochastic differential equations. Theor. Prob. Appl., XIX, 3, (1974), 583-588.

[2] Rao, N.J.; Borwankar, J.D.; Ramakrishna, D.: Numerical solution of Itô integral equations. SIAM J. Control, 12, 1, (1974), 124-139.

[3] Wagner, W.; Platen, E.: Approximation of Itô integral equations. Preprint ZIMM, AdW der DDR, Berlin (1978).

[4] Clark, J.M.C.; Cameron, R.J.: The maximum rate of convergence of discrete approximations for stochastic differential equations. Lect. Notes in Control and Inform. Sc. 25, Springer Verlag (1980), 162-171

[5] Rümelin, W.: Numerical treatment of stochastic differential equations. Report Nr. 12, Univ. Bremen (1980).

[6] Talay, D.: Efficient numerical schemes for the approximation of expectations of functionals of the solution of a S.D.E. and applications. Lect. Notes in Control and Inf. Sc., 61, Springer Verlag (1984).

[7] Platen, E.: Zur zeitdiskreten Approximation von Itôprozessen, Diss. B, AdW der DDR, IMath, (1984).

[8] Platen, E.; Wagner, W.: On a Taylor formula for a class of Itô processes. Probab. and Math. Statistics, Vol. 3, Fasc. 1, (1982), 37-51.

[9] Platen, E.: A generalized Taylor formula for solutions of stochastic equations. SANKHYA, A, 44, 2, (1982), 163-172.

[10] Platen, E.: An approximation method for a class of Itô processes. Lituvos Matem. Rink., XXI, 1, (1981), 121-133.

[11] Liske, H.; Platen, E.; Wagner, W.: About mixed multiple Wiener integrals. Preprint, P-Math-23/82, IMath, AdW der DDR, Berlin (1982).

[12] Platen, E.: An approximation method for a class of Itô processes with jump component. Lietuvos Matem. Rink., XXII, 2, (1982), 124-136.

[13] Milstein, G.N.: A second order method for integration of stochastic differential equations. Theor. Prob. Appl. XXIII, 2, (1976), 414-419.

[14] Mikulevicius, R.; Platen, E.: Rate of convergence of the Euler approximation for diffusion processes. Preprint, P-Math/86, IMath, AdW der DDR, Berlin (1986).

[15] Mikulevicius, R.; Platen, E: Time discrete Taylor approximations for Itô processes with jump component. Preprint, P-Math/86, IMath, AdW der DDR, Berlin (1986).

[16] Liske, H.; Platen E.: Simulation studies on time discrete diffusion approximations. To appear.

ON THE NUMBER OF CROSSINGS OF A PARTLI REFLECTING HYPERPLANE BY A MULTIDIMENSIONAL WIENER PROCESS

N. PORTENKO, S. YEFIMENKO
Mathematical institute of the Ukrainian Academy of Sciences,
Kiev State University
Kiev, USSR

Let $S=\{x\in R^d:(x,\nu)=0\}$ be a hyperplane in a d-dimensional euclidian space R^d, $\nu\in R^d$ a fixed unit normal vector to S. Define a Wiener process with partly reflecting screen on the hyperplane S as a continuous homogeneous stong Markov process $(x(t), \mathcal{M}_t, P_x)$ on R^d, generated by a semigroup of operators $(T_t)_{t>0}$ acting on a bounded measurable function $\varphi(x), x\in R^d$ as follows

$$T_t\varphi(x)=T_t^0\varphi(x)+\int_0^t d\tau\int_S g(t-\tau,x,y)\frac{\partial T_\tau^0\varphi(y)}{\partial\nu}q(y)d\sigma_y \quad (I)$$

here $g(t,x,y)=(2\pi t)^{-d/2}\exp\{-\frac{|y-x|^2}{2t}\}$, $T_t^0\varphi(x)=\int_{R^d}\varphi(y)g(t,x,y)dy$, $\frac{\partial}{\partial\nu}$ stands for a derivative in the direction ν, $q(y)$ is a given continuous function on S, such that $|q(y)|\le 1$, the inner integral in (I) being the surface integral [1]. The function q is called a reflectin coefficient. If $q(x)\equiv 1$, then the part of the process $x(t)$ over the subset $\mathcal{D}_+\cup S$, where $\mathcal{D}_+=\{x\in R^d:(x,\nu)>0\}$ is the Wiener process with the instanteneous reflecting on the surface S. If $q(x)\equiv -1$ then the analogous statement takes place for the part of the process $x(t)$ over the set $\mathcal{D}_-\cup S$, where $\mathcal{D}_-=\{x\in R^d:(x,\nu)<0\}$. If $q(x)\equiv 0$ the semigroup (I) obviously defines a standard Wiener process on R^d.

For $n=1,2,\ldots$ denote $t_k^{(n)}=k/n$, $\xi_k^{(n)}=x(t_k^{(n)})$, $k=0,1,2,\ldots$. Let us say that the sequence $(\xi_k^{(n)})_{k\ge 0}$ crosses the surface S at a moment j if $(\xi_j^{(n)},\nu)(\xi_{j+1}^{(n)},\nu)<0$. Denote by $\eta_k^{(n)}$ a number of crossings of the surface S by the sequence $\xi_0^{(n)},\ldots,\xi_k^{(n)}$. Obviously,

$$\eta_k^{(n)}=\sum_{j=0}^{k-1}h(\xi_j^{(n)},\xi_{j+1}^{(n)}),\quad k\ge 1, \quad (2)$$

where $h(x,y)=1_{\mathcal{D}_-}(x)1_{\mathcal{D}_+}(y)+1_{\mathcal{D}_+}(x)1_{\mathcal{D}_-}(y)$ ($1_\Gamma(x)$ means the indicator of the set Γ). WE shall be interested in the limit behaviour of $\eta_n^{(n)}$ as $n\to\infty$. Corresponding limit theorems for the Wiener process and some other types of processes (diffusional, in particular) can be found in [2], [3]. Let us notice that in the case $q(x)\not\equiv 0$

the corresponding process $X(t)$ is not a diffusional one, so the results of [3] cannot be strictly applied to it. Nevertheless, as we shall showw, the method worked out in [3] with some modifications can be used in the considered situation.

First of all note that according to the A.V.Skorokhod's lemma from [4] the limiting distribution of variables $\eta_n^{(n)}/\sqrt{n}$ as $n\to\infty$ exists if and only if those exists for variables $\zeta_n^{(n)}/\sqrt{n}$, where

$$\zeta_k^{(n)}=\sum_{j=0}^{k} v_n(\xi_j^{(n)}),\ v_n(x)=M_x h(\xi_0^{(n)},\xi_1^{(n)}),$$

$x\in R^d,\ n=1,2,\dots,\ k=0,1,2,\dots$. So the limiting distributions of variables $\eta_n^{(n)}/\sqrt{n}$ and $\zeta_n^{(n)}/\sqrt{n}$ coicide if only they exist.

As in the papers [2] , [3] the sequence $\sqrt{n}\,v_n(x)$ converges weakly to the generalized function $\sqrt{\frac{2}{\pi}}\,\delta_S(x)$ as $n\to\infty$, $\delta_S(x)$ being defined by the relation

$$\int_{R^d}\varphi(x)\sqrt{\frac{2}{\pi}}\,\delta_S(x)\,dx=\int_S\varphi(x)\,d\sigma$$

for every continuous function with compact support $\varphi(x),\ x\in R^d$. However, unlike these papers we have to calculate integrals of discontinuous functions multiplied by $\sqrt{n}\,v_n(x)$. For such cases the following lemma is useful.

LEMMA I. Let $\varphi(x),\ x\in R^d$ be a real-valued measurable function for which

$$\sup_{\rho\in R^1}\int_{S_\rho}|\varphi(x)|\,d\sigma<\infty \tag{3}$$

(hereafter $S_\rho=\{x\in R^d:(x,\nu)=\rho\}$ for $\rho\in R^1$) and let the following limits exist for $x\in S$

$$\varphi(x+)=\lim_{\substack{y\to x\\ y\in\mathcal{D}_+}}\varphi(y),\qquad \varphi(x-)=\lim_{\substack{y\to x\\ y\in\mathcal{D}_-}}\varphi(y)$$

Then

$$\lim_{n\to\infty}\sqrt{n}\int_{R^d}\varphi(x)v_n(x)\,dx=\sqrt{\frac{2}{\pi}}\int_S\Big[\frac{1+q(y)}{2}\varphi(y-)+\frac{1-q(y)}{2}\varphi(y+)\Big]d\sigma \tag{4}$$

and the following inequality is true

$$\sup_n\sqrt{n}\int_{R^d}|\varphi(x)|\,v_n(x)\,dx\le\sqrt{\frac{2}{\pi}}\sup_{\rho\in R^1}\int_{S_\rho}|\varphi(x)|\,d\sigma \tag{5}$$

Proof. Represent $v_n(x)$ as follows $v_n(x)=v_n'(x)+v_n''(x)$, where $v_n'(x)=1_{\mathcal{D}_+}(x)T_{1/n}1_{\mathcal{D}_-}(x)$, $v_n''(x)=1_{\mathcal{D}_-}(x)T_{1/n}1_{\mathcal{D}_+}(x)$.

Let $\varphi(x)$ satisfies the conditions of the lemma. Then

$$\sqrt{n}\int_{R^d}\varphi(x)v_n'(x)\,dx=\sqrt{n}\int_{\mathcal{D}_+}\varphi(x)\,dx\int_{\mathcal{D}_-}g(1/n,x,y)\,dy+$$

$$+\sqrt{n}\int_{\mathcal{D}_+}\varphi(x)dx\int_0^{1/n} g(\tau,x,y)\,\frac{\partial T^0_{1/n-\tau}1_{\mathcal{D}_-}(y)}{\partial\nu}\,q(y)d\sigma_y = I_1 + I_2$$

Supposing $\rho(z)=(2\pi)^{-d/2}\exp\{-|z|^2/2\}$ for $z\in R^d$ one can write

$$I_1=\int_{\mathcal{D}_-}\rho(z)\Big[\sqrt{n}\int_0^{-(z,\nu)/\sqrt{n}} d\rho\int_{S_\rho}\varphi(x)d\sigma\Big]dz$$

and

$$\lim_{n\to\infty} I_1 = -\int_{\mathcal{D}_-}(z,\nu)\,\rho(z)dz\int_S\varphi(x+)d\sigma=\frac{1}{\sqrt{2\pi}}\int_S\varphi(x+)d\sigma \tag{6}$$

Besides

$$|I_1|\le\frac{1}{\sqrt{2\pi}}\sup_{\rho>0}\int_{S_\rho}|\varphi(x)|d\sigma \tag{7}$$

Let's now consider I_2 . A simple calculations leads to the formula

$$\frac{\partial T^0_\tau 1_{\mathcal{D}_-}(y)}{\partial\nu}=-\frac{1}{\sqrt{2\pi\tau}},\quad y\in S$$

So

$$I_2=\sqrt{\frac{n}{2\pi}}\int_0^{1/n}\phi(\tau)\left(\tfrac{1}{n}-\tau\right)^{-\frac{1}{2}}d\tau, \tag{8}$$

where

$$\phi(\tau)=-\int_{\mathcal{D}_+}\varphi(x)dx\int_S g(\tau,x,y)q(y)d\sigma_y=$$

$$=-\int_0^\infty\frac{e^{-\rho^2/2\tau}}{\sqrt{2\pi\tau}}d\rho\int_{S_\rho}\varphi(x)d\sigma_x\int_S(2\pi\tau)^{-\frac{d-1}{2}}\exp\Big\{-\frac{|y-\Pi x|^2}{2\tau}\Big\}q(y)\,d\sigma_y \tag{9}$$

(Πx denotes a projection of x on S , relation $|y-x|^2=|y-\Pi x|^2+\rho^2,\ y\in S,\ x\in S_\rho$ is used). The last equality gives

$$|\phi(\tau)|\le\frac{1}{2}\sup_{\rho>0}\int_{S_\rho}|\varphi(x)|d\sigma,$$

the previous one implies $\lim_{n\to\infty} I_2=\sqrt{\frac{2}{\pi}}\,\phi(0+)$, moreover $\phi(0+)=-\frac{1}{2}\int_S\varphi(x+)q(x)d\sigma.$ Together with (6) and (7) this fact shows that

$$\lim_{n\to\infty}\sqrt{n}\int_{R^d}v_n'(x)\varphi(x)dx=\frac{1}{\sqrt{2\pi}}\int_S\varphi(x+)(1-q(x))d\sigma$$

and

$$\sqrt{n}\int_{R^d}v_n'(x)|\varphi(x)|dx\le\sqrt{\frac{2}{\pi}}\sup_{\rho>0}\int_{S_\rho}|\varphi(x)|d\sigma$$

Analogously, one can obtain

$$\lim_{n\to\infty}\sqrt{n}\int_{R^d}v_n''(x)\varphi(x)dx=\frac{1}{\sqrt{2\pi}}\int_S\varphi(x-)(1+q(x))d\sigma$$

and

$$\sqrt{n}\int_{R^d} v_n''(x)|\varphi(x)|\,dx \le \sqrt{\frac{2}{\pi}}\sup_{\rho<0}\int_{S_\rho}|\varphi(x)|\,d\sigma$$

The proof of the lemma is concluded.

REMARK I. The statement of the lemma, of cours, means that the sequence of functions $\sqrt{n}\,v_n(x)$ converges to the generalized function $\sqrt{\frac{2}{\pi}}\delta_S(x)$ as $n\to\infty$, but it contains more information than last statement.

COROLLARY. Let the function $\varphi(x)$, $x\in R^d$, be bounded and continuous. Then setting $\psi_n(x)=\sqrt{n}\,v_n(x)\varphi(x)$ one can obtain

$$\lim_{n\to\infty} T_t\psi_n(x)=\sqrt{\frac{2}{\pi}}\int_S g(t,x,y)\varphi(y)(1-q^2(y))\,d\sigma_y \tag{I0}$$

and

$$T_t|\psi_n|(x)\le\sqrt{\frac{2}{\pi t}}\,\|\varphi\| \tag{II}$$

($\|\varphi\|$ denotes $\sup_x|\varphi(x)|$).

Proof. We have $T_t\psi_n = A_1+A_2$, where

$$A_1=T_t^\circ\psi_n(x),\quad A_2=\int_0^t d\tau\int_S g(t-\tau,x,y)\,\frac{\partial T_\tau^\circ\psi_n(y)}{\partial\nu}\,d\sigma_y$$

According to lemma I

$$\lim_{n\to\infty} A_1=\sqrt{\frac{2}{\pi}}\int_S g(t,x,y)\varphi(y)\,d\sigma_y, \tag{I2}$$

and

$$\int_{R^d} g(t,x,y)|\psi_n(y)|\,dy\le\|\varphi\|\sup_\rho\int_{S_\rho} g(t,x,y)\,dy\le\frac{\|\varphi\|}{\sqrt{2\pi t}} \tag{I3}$$

For A_2 the following representation takes place

$$A_2=\int_{R^d}\psi_n(z)\,\frac{\partial F(t,x,z)}{\partial\nu_z}\,dz,$$

where

$$F(t,x,z)=-\int_0^t d\tau\int_S g(t-\tau,x,y)g(\tau,y,z)q(y)\,d\sigma_y$$

(we make use of the fact $\nabla_y g(\tau,y,z)=-\nabla_z g(\tau,y,z)$). The function $F(t,x,z)$ represents a simple-layer potential (see e.g. [5]). Its normal derivative has a jamp on the surface S :

$$\lim_{z\to z_0\pm}\frac{\partial F(t,x,z)}{\partial\nu_z}=\mp g(t,x,z_0)q(z_0)$$

despite of $z_0\in S$. Here $z\to z_0+$ (respectively $z\to z_0-$) denotes $z\to z_0$ and $z\in\mathcal{D}_+$ (respectively $z\in\mathcal{D}_-$). Then by lemma I

$$\lim_{n\to\infty}A_2=-\sqrt{\frac{2}{\pi}}\int_S\varphi(y)\left[\frac{1+q(y)}{2}g(t,x,y)q(y)-\frac{1-q(y)}{2}g(t,x,y)q(y)\right]d\sigma_y=$$

$$= -\sqrt{\frac{2}{\pi}} \int_S \varphi(y) q^2(y) g(t,x,y)\, d\sigma_y$$

This relation together with (I2) leads to (IO). Moreover, according to (5)

$$|A_2| \le \|\varphi\| \sup_\rho \int_{S_\rho} \left| \frac{\partial F(t,x,z)}{\partial \nu_z} \right| d\sigma_z \le$$

$$\le \|\varphi\| \sup_\rho \int_{S_\rho} d\sigma_z \int_0^t d\tau \int_S \frac{|(z,\nu)|}{\tau} g(\tau,z,y) g(t-\tau,x,y) |q(y)| d\sigma_y \le$$

$$\le \|\varphi\| \sup_\rho \int_0^t \frac{|\rho| e^{-\rho^2/2\tau}}{\sqrt{2\pi\tau^3}} d\tau \int_S g(t-\tau,x,y)\, d\sigma_y \le$$

$$\le \|\varphi\| \sup_\rho \int_0^t \frac{|\rho| e^{-\rho^2/2\tau}}{\sqrt{2\pi\tau^3}} \frac{d\tau}{\sqrt{2\pi(t-\tau)}} \le \frac{\|\varphi\|}{\sqrt{2\pi t}}$$

This relation together with (I3) implies (II).

For an arbitrary positive constant K denote $H(K)$ a class of all real-valued measurable functions $\varphi(s,x)$ defined on $[0,\infty)\times R^d$ for which $\|\varphi\| = \sup\limits_{s,x} |\varphi(s,x)| \le K$. For $\varphi \in H(K)$, $n=1,2,\dots$, $t \ge 0$, $x \in R^d$ consider

$$f_n(t,x;\varphi) = \sqrt{n} \int_0^t ds \int_{R^d} v_n(y) \varphi(s,y) P(t-s,x,dy),$$

where $P(t,x,\Gamma)$ is the transient probability generating the semigroup (I). Let $f_n(t,x;\varphi)$ be considered as the function of (t,x), and $\varphi \in H(K)$, $n=1,2,\dots$ being parameters.

LEMMA 2. For arbitrary fixed $K>0$, $T>0$ the family of functions $\{ f_n(\cdot,\cdot;\varphi),\ \varphi \in H(K),\ n=1,2,3,\dots \}$ is equicontinuous on $(t,x) \in [0,T]\times R^d$.

Proof. Let $T>0$, $K>0$ be fixed, $t\in[0,T]$, $\varphi \in H(K)$. Then (argument φ is omitted) $f_n(t,x) = f_n'(t,x) + f_n''(t,x)$ where

$$f_n'(t,x) = \sqrt{n} \int_0^t ds \int_{R^d} v_n(y) g(t-s,x,y) \varphi(s,y)\, dy,$$

$$f_n''(t,x) = \sqrt{n} \int_0^t ds \int_0^{t-s} d\tau \int_S g(t-s-\tau,x,y) \frac{\partial T_\tau^0 v_n(\cdot)\varphi(s,\cdot)}{\partial \nu}(y) q(y)\, d\sigma_y$$

Let $0 \le t < t' \le T$. Then using the same arguments as above while evaluating A_2 one obtain

$$|f_n'(t',x)-f_n'(t,x)|\le C_1\sqrt{\delta}+C_2\sqrt{t'-t}+$$
$$+K\cdot 1_{[\delta,T)}(t)\sqrt{n}\int_0^{t-\delta}d\tau\int_{R^d}v_n(y)|g(t'-\tau,x,y)-g(t-\tau,x,y)|dy \tag{I4}$$

Hereafter C_i stands for a constant which can be dependent on T and K , δ is an arbitrary positive number. Since such $\bar{s}\in[t-\tau,t'-\tau]$ exists that

$$g(t'-\tau,x,y)-g(t-\tau,x,y)=\frac{\partial g}{\partial s}(\bar{s},x,y)\,(t'-t)$$

then because of the inequality (5) andthe following one

$$\left|\frac{\partial g}{\partial s}(s,x,y)\right|\le C_3\,s^{-\frac{d+2}{2}}\exp\left\{-\mu\frac{|y-x|^2}{s}\right\}$$

valid for cfrtain constans $C_3>0$, $\mu>0$ and all $s>0$, $x,y\in R^d$ we have

$$1_{[\delta,T)}(t)\sqrt{n}\int_0^{t-\delta}d\tau\int_{R^d}v_n(y)|g(t'-\tau,x,y)-g(t-\tau,x,y)|\,dy\le$$

$$\le C_3(t'-t)1_{[\delta,T)}(t)\sup_{\rho}\int_0^{t-\delta}d\tau\int_{S_\rho}\delta^{-\frac{d+2}{2}}\exp\left\{-\mu\frac{|y-x|^2}{T}\right\}d\sigma_y\le$$

$$\le C_4\,\delta^{-\frac{d+2}{2}}(t'-t)$$

We made use of the fact that inequalities $0\le\tau\le t-\delta<T$ and $t-\tau\le\bar{s}\le t'-\tau$ imply $\delta\le\bar{s}\le T$, so

$$\left|\frac{\partial g(\bar{s},x,y)}{\partial s}\right|\le C_3\,\delta^{-\frac{d+2}{2}}\exp\left\{-\mu\frac{|y-x|^2}{T}\right\}$$

Substituting obtained inequality into (I4) we find that $f_n'(t,x)$ is equicontinuous on $t\in[0,T]$ uniformly on $x\in R^d$, $\varphi\in H(K)$ and $n=1,2,\ldots$.

Further using the definition of function $F(t,x,z)$ as above we can write

$$f_n''(t,x)=\sqrt{n}\int_0^t ds\int_{R^d}v_n(z)\,\varphi(s,z)\,\frac{\partial F(t-s,x,z)}{\partial \nu_z}\,dz$$

Assuming again $0\le t<t'\le T$ we have

$$|f_n''(t',x)-f_n''(t,x)|\le C_5\sqrt{\delta}+C_6\sqrt{t'-t}+ \tag{I5}$$

$$+C_7\, 1_{[\delta,T)}(t)\sqrt{n}\int_0^{t-\delta} ds \int_{R^d} v_n(z) \left| \frac{\partial F(t'-s,x,z)}{\partial \nu_z} - \frac{\partial F(t-s,x,z)}{\partial \nu_z} \right| dz$$

In order to evaluate the integral in the right side of (I5) make use of the following representation

$$\frac{\partial F(t'-s,x,z)}{\partial \nu_z} - \frac{\partial F(t-s,x,z)}{\partial \nu_z} = Q_1 + Q_2 + Q_3 ,$$

where

$$Q_1 = \int_{t-s}^{t'-s} d\tau \int_S g(t'-s-\tau,x,y)\, g(\tau,y,z) \frac{(z,\nu)}{\tau}\, d\sigma_y ,$$

$$Q_2 = \int_{t-s-\delta}^{t-s} d\tau \int_S [g(t'-s-\tau,x,y) - g(t-s-\tau,x,y)]\, g(\tau,y,z) \frac{(z,\nu)}{\tau}\, d\sigma_y ,$$

$$Q_3 = \int_0^{t-s-\delta} d\tau \int_S [g(t'-s-\tau,x,y) - g(t-s-\tau,x,y)]\, g(\tau,y,z) \frac{(z,\nu)}{\tau}\, d\sigma_y$$

While evaluating integrals of Q_1 and Q_2 the following equality is necessary ($y \in S$)

$$\int_{S_\rho} g(\tau,y,z) \frac{|(z,\nu)|}{\tau}\, d\sigma_z = \frac{|\rho|}{\sqrt{2\pi\tau^3}} e^{-\rho^2/2\tau}$$

Then

$$1_{[\delta,T)}(t)\sqrt{n}\int_0^{t-\delta} ds \int_{R^d} v_n(z) |Q_1| dz \le$$

$$\le 1_{[\delta,T)}(t) \sup_\rho \int_0^{t-\delta} ds \int_{t-s}^{t'-s} \frac{|\rho| e^{-\rho^2/2\tau}}{\sqrt{2\pi\tau^3}}\, d\tau \int_S g(t'-s-\tau,x,y)\, d\sigma_y \le$$

$$\le 1_{[\delta,T)}(t) \sup_\rho \int_0^{t-\delta} ds \int_{t-s}^{t'-s} \frac{|\rho| e^{-\rho^2/2\tau}}{\sqrt{2\pi\tau^3}} \frac{d\tau}{\sqrt{2\pi(t'-s-\tau)}} \le C_8\, \delta^{-1}\sqrt{t'-t}$$

Analogously, $1_{[\delta,T)}(t)\sqrt{n}\int_0^{t-\delta} ds \int_{R^d} v_n(z) |Q_2| \le$

$$\leq 1_{[\delta,T)}(t)\sup_{\rho}\int_0^{t-\delta}ds\int_{t-s-\delta}^{t-s}\frac{|\rho|e^{-\rho^2/2\tau}}{\sqrt{2\pi\tau^3}}[(t-s-\tau)^{-\frac{1}{2}}+(t'-s-\tau)^{-\frac{1}{2}}]d\tau \leq$$

$$\leq C_9\sqrt{\delta}(-\ln\delta)$$

Finally, in the same way as while evaluating the integral in the right side of (I4) we obtain

$$1_{[\delta,T)}(t)\sqrt{n}\int_0^{t-\delta}ds\int_{R^d}v_n(z)|Q_3|dz \leq C_{10}\,\delta^{-\frac{d+2}{2}}(t'-t)$$

Substituting all received inequalities into (I5) we find that $f_n''(t,x)$ is also equicontinuous on $t\in[0,T]$ with $n=1,2,\ldots,\ x\in R^d,\ \varphi\in H(K)$.

Quite analogous reasonings can be applied to the difference $f_n(t,x_1)-f_n(t,x_2)$ in order to get evaluates uniform on $t\in[0,T]$, $n=1,2,\ldots,\ \varphi\in H(K)$, which show that $f_n(t,x)$ is equicontinuous on x. This argument completes the proof of lemma 2.

THEOREM. Variables $\eta_n^{(n)}/\sqrt{n}$ have as $n\to\infty$ a limit distribution $(x\in R^d,\ a\in R^1)$

$$\lim_{n\to\infty} P_x\{\eta_n^{(n)}/\sqrt{n} < a\} = F_x(a) \tag{I6}$$

with a characteristic function $\int_{-\infty}^{\infty}e^{i\theta a}F_x(da),\ \theta\in R^1$, coinciding with the function $u(1,x;\theta)$, where $u(t,x;\theta)$ for $t\geq 0,\ x\in R^d$, $\theta\in R^1$ can be found from the integral equation

$$u(t,x;\theta)=1+i\theta\sqrt{\frac{2}{\pi}}\int_0^t d\tau\int_S g(t-\tau,x,y)\,u(\tau,y;\theta)(1-q^2(y))\,d\sigma_y \tag{I7}$$

Proof. For $t\geq 0$, $x\in R^d$, $\theta\in R^1$, $n=1,2,\ldots$ we set ($[\alpha]$ stands for an entier of $\alpha\in R^1$)

$$u_n^*(t,x;\theta)=M_x\exp\{i\theta\,\eta_{[nt]}^{(n)}/\sqrt{n}\}$$

Then (see [3]) the function u_n^* satisfies the equation

$$u_n^*(t,x;\theta)=1+n\int_0^{\frac{[nt]+1}{n}}ds\int_{R^d}\left(1-\exp\left\{-\frac{i\theta v_n(y)}{\sqrt{n}}\right\}\right)u_n^*(s,y;\theta)\times$$

$$\times P\left(\frac{[nt]-[ns]}{n}, x, dy\right)$$

which can be rewritten in the form

$$u_n^*(t,x;\theta) = \varepsilon_n(t,x;\theta) + 1 +$$

$$+ i\theta\sqrt{n}\int_0^t ds \int_{R^d} v_n(y)\, u_n^*(s,y;\theta)\, P\left(\frac{[nt]-[ns]}{n}, x, dy\right)$$

where $\varepsilon_n(t,x;\theta) \to 0$ as $n\to\infty$ uniformly on $x\in R^d$ and locally uniformly on t and θ. Then if u_n^{**} denotes a solution of the next equation

$$u_n^{**}(t,x;\theta) = 1 + i\theta\sqrt{n}\int_0^t ds \int_{R^d} v_n(y)\, u_n^{**}(s,y;\theta)\, P\left(\frac{[nt]-[ns]}{n}, x, dy\right)$$

so

$$\lim_{n\to\infty} [u_n^*(t,x;\theta) - u_n^{**}(t,x;\theta)] = 0$$

Denote by u_n a solution of the equation

$$u_n(t,x;\theta) = 1 + i\theta\sqrt{n}\int_0^t d\tau \int_{R^d} v_n(y)\, u_n(\tau,y;\theta)\, P(t-s, x, dy) \qquad (18)$$

A bounded solution of this equation (similarly as the solution of the previous one) exists and is unique, this fact can be easily verified using the method of successive approximations. By lemma 2

$$\lim_{n\to\infty} [u_n^{**}(t,x;\theta) - u_n(t,x;\theta)] = 0$$

The convergence in this relation being uniform on x and locally uniform on t and θ.

Further, the equation (18) and uniform boundness of function $u_n(t,x;\theta)$ imply by lemma 2 that for a fixed $\theta\in R^1$ the family of functions $\{u_n(\cdot,\cdot;\theta),\ n=1,2,\ldots\}$ is equicontinuous on $x\in R^d$, $t\in[0,T]$ for every $T<\infty$. So a subsequence $n_k\to\infty$ can be chosen such that $u_{n_k}(t,x;\theta)$ converges to a certain function $u(t,x;\theta)$ uniformly on $x\in R^d$ and locally uniformly on $t\geq 0$. Then the passage to the limit in the equation (18) using the corollary of lemma 1 leads to a conclusion that every limit function for the sequence $(u_n(t,x;\theta))_{n\geq 1}$ is a solution of the equation (I7). However, the last one has the unique solution. Consequently, all the sequence $(u_n(t,x;\theta))_{n\geq 1}$ converges to the solution of the equation (I7) as $n\to\infty$. The theorem is proved.

REMARK 2. For $q(x)\equiv 0$ the proved statement coincides with the one obtained in [2], [3]. For $q(x)\equiv C$, $|C|\leq 1$, the solution of (I7) can be easily found and the limit distribution as well. It equals

$$F_x(a)=1_{(0,\infty)}(a)\sqrt{\frac{2}{\pi}}\int_0^{\frac{a+\gamma r(x)}{\gamma}} e^{-\frac{z^2}{2}}dz, \quad a\in R^1, x\in R^d,$$

where $r(x)$ stands for a distance of x to the surface S, $\gamma=\sqrt{\frac{2}{\pi}}(1-C^2)$.

REFERENCES

I. Portenko N.I. Generalized diffusion processes.- Kiev,Naukova dumka, I982.

2. Gihman I.I. The asymptotic distributions for number of level-crossingsby a random function.- Visnyk Kiev. Universit., ser. astr., mathem., mech., v.I, no.I, I958, p.25-6I.

3. Portenko N.I. Nonnegative additive functionales of Markov processes and some limit theorem.- Theory of stochastic processes, Kiev, Naukova dumka, v. I, I973, p. 86-I07.

4. Skorokhod A.V. Some limit theorems for additive functionals of sequence of independend random variables.-Ukrainian mathem. journ., Kiev, v. XIII, no.4, I96I, p. 67-78.

5. Ladyjenskaja O.A., Solonnikov V.A., Uraltseva N.N.-Linear and quasilinear equations of parabolyc type.- Moskow, Nauka, I967.

ON CONVERGENCE RATES OF APPROXIMATE SOLUTIONS OF STOCHASTIC EQUATIONS

W. Römisch
Sektion Mathematik
Humboldt-Universität
DDR-1086 Berlin, PSF 1297

A. Wakolbinger
Institut für Mathematik
Johannes Kepler Universität
A-4040 Linz, Austria

Abstract: Continuity properties of the mappings $\mu \to \mu F^{-1}$ and $\mu \to \int h d\mu$ with respect to bounded Lipschitz distance on probability measures are investigated. The results are applied to the case where $x=F(z)$ is the solution of the differential equation $dx(s) = f(x(s))ds+g(x(s))dz(s)$ and $h(z)$ is some functional of x.

1. Introduction

Consider the mapping S_1 which turns $z \in C^1[0,1]$ into the solution $x=S_1(z)$ of the (scalar) integral equation

$$x(t) = x_0 + {}_0\!\int^t f(x(s))ds + {}_0\!\int^t g(x(s))dz(s) \qquad (1.1.)$$

Under certain conditions on f and g, S_1 extends to a mapping S defined on all bounded, measurable $z:[0,1]\to \mathbf{R}$, which is continuous with respect to several metrics [1, 16, 11, 14]. Under the mapping S, the distribution μ of a random input z is carried into an output distribution μS^{-1}, and the mapping $\mu \to \mu S^{-1}$ inherits certain continuity properties with respect to suitable metrics on the space of probability distributions, which serve to obtain convergence rates, e. g. if Wiener measure is approximated by a sequence of "simpler" measures. Another natural question is the continuous dependence of certain moments like $\int \|x\|_\infty (\mu S^{-1})(dx)$ on the input distribution μ.

In Section 2, problems of this type are treated in the general framework of a mapping F between two separable metric spaces (Z, d_Z) and (X, d_X), following the line of research in [17, 18, 19, 5, 13]. As a metric on the space of probability distributions we will consider the bounded Lipschitz distance $\beta_Z(\mu,\nu)$ (cf. [2, 4]) defined by

$$\beta_Z(\mu,\nu) := \sup\{ \,|\int \psi d(\mu-\nu)|: \psi:Z\to\mathbf{R}, \|\psi\|_{BL} \leq 1\} \qquad (1.2.)$$

where

$$\|\psi\|_{BL} := \|\psi\|_\infty + \sup\{ |\psi(z_1)-\psi(z_2)|/d_Z(z_1,z_2) : z_1, z_2 \in Z,\ z_1 \neq z_2 \} \qquad (1.3.)$$

In Section 3, these results are applied to the mapping S, thus obtaining, in particular, convergence rates of the output distributions resp. certain moments of these, if the input distribution μ is approximated by a sequence of distributions μ_n. Similar results may be obtained in higher dimensions under additional restrictions on the coefficient function g in (1.1.) (see [14]; for related results which do not hinge on these restrictions but consider more or less special approximations of semimartingale inputs,see, e. g., [7, 9, 12, 10].

2. General results

Let (Z, d_Z) be a separable metric space, and 0 be a fixed element of Z. For any locally Lipschitz continuous mapping G from (Z, d_Z) into some other metric space (Y, d_Y) we put

$$L_G(r) := \|G|_{K_r}\|_L := \sup\{ d_Y(G(z_1),G(z_2))/d_Z(z_1,z_2) : z_1,z_2 \in K_r,\ z_1 \neq z_2 \} \tag{2.1.}$$

where $K_r := \{z \in Z : d_Z(z, 0) \le r\}$.

For any real valued and locally Lipschitz continuous mapping h defined on Z we put

$$B_h(r) := \|h|_{K_r}\|_\infty$$
$$BL_h(r) := \|h|_{K_r}\|_{BL} = L_h(r) + B_h(r)$$

Note that L_G as well as BL_h are nondecreasing and left continuous.

For any probability measure μ on the σ-algebra $B(Z)$ of Borel subsets on Z we put

$$\varepsilon_\mu(r) := \mu(Z-K_r) \tag{2.2.}$$

noting that ε_μ is nonincreasing, right continuous and tends to zero for $r \to \infty$.

The following theorem improves Theorem 2 in [5]. There, one can find also a similar result for the Prokhorov metric instead of the bounded Lipschitz metric.

Theorem 1. Let F be a locally Lipschitz continuous mapping from (Z, d_Z) into some other separable metric space (X, d_X).Then there holds for any two probability measures μ, ν on $B(Z)$:

$$\beta_X(\mu F^{-1},\nu F^{-1}) \le \inf\{\beta_Z(\mu,\nu)[4+\max\{1,L_F(r)\}] + 4\varepsilon_\mu(r-1) : r > 1\} \tag{2.3.}$$

and

$$\beta_X(\mu F^{-1},\nu F^{-1}) \le \beta_Z(\mu,\nu)[8+\max\{1,L_F(1+\varepsilon_\mu^{-1}(\beta_Z(\mu,\nu)))\}] \tag{2.4.}$$

where $\varepsilon_\mu^{-1}(t) := \inf\{r>0 : \varepsilon_\mu(r) < t\}$ $(t > 0)$.

The proof of Theorem 1 (and also that of Theorem 2 below) is based on the following key

Lemma 1.a) Let $h: Z \to R$ be locally Lipschitz continuous. Then there holds for any two probability measures μ, ν on $B(Z)$ and and $r>0$:

$$\left|\int_Z h\, d(\mu-\nu)\right| \le \int_{Z-K_r} (|h|+B_h(r))d(\mu+\nu) + BL_h(r)\, \beta_Z(\mu,\nu). \tag{2.5.}$$

b) If, in addition, h is bounded, then there holds for any $r>1$:

$$\left|\int_Z h\, d(\mu-\nu)\right| \le \beta_Z(\mu,\nu)[4\|h\|_\infty + BL_h(r)] + 4\|h\|_\infty \varepsilon_\mu(r-1) \tag{2.6.}$$

Proof: According to [2, Lecture 7] there exists, for any $r>0$, a bounded, Lipschitz continuous extension h_r of $h|_{K_r}$ to the whole of Z, having the properties

$$\|h_r\|_{BL} = BL_h(r),\ \|h_r\|_\infty = B_h(r) \tag{2.7}$$

For any fixed $r>1$ we thus obtain the following chain of inequalities:

$$|\int_Z h\, d(\mu-\nu)|$$
$$\leq |\int_Z (h-h_r)d\mu| + |\int_Z h_r\, d(\mu-\nu)| + |\int_Z (h-h_r)d\nu|$$
$$\leq \int_{Z-K_r} (|h|+|h_r|)d(\mu+\nu) + \|h_r\|_{BL}\, \beta_Z(\mu,\nu)$$
$$\leq \int_{Z-K_r} (|h|+ B_h(r))d(\mu+\nu) + BL_h(r)\, \beta_Z(\mu,\nu),$$

showing the validity of (2.5.). Under the assumption of b), we may proceed in our estimate by

$$\leq 2\, \|h\|_\infty(\mu(Z-K_r) + \nu\,(Z-K_r)) + BL_h(r)\, \beta_Z(\mu,\nu).$$

The mapping $\phi:Z\to R_+$ defined by $\phi(z) := \min\{1, d_Z(z, K_{r-1})\}$ obeys

$$\|\phi\|_{BL} \leq 2 \text{ and } 1_{Z-K_r} \leq \phi \leq 1_{Z-K_{r-1}}.$$

This leads to

$$\nu\,(Z-K_r) \leq \int_Z \phi d\nu \leq \int_Z \phi d\mu + \|\phi\|_{BL}\, \beta_Z(\mu,\nu) \leq \mu(Z-K_{r-1}) + 2\, \beta_Z(\mu,\nu).$$

Thus we get

$$|\int_Z h\, d(\mu-\nu)| \leq 4\, \|h\|_\infty\, \mu(Z-K_{r-1}) + \beta_Z(\mu,\nu)[4\|h\|_\infty + BL_h(r)],$$

which is (2.6.). ♦

Proof of Theorem 1: For all $\psi:X\to R$ with the property $\|\psi\|_{BL} \leq 1$ there holds according to Lemma 1b)

$$|\int_X \psi\, d(\mu F^{-1} - \nu F^{-1})| = |\int_Z \psi\circ F\, d(\mu-\nu)|$$
$$\leq \inf\{\beta_Z(\mu,\nu)[4 + BL_{\psi\circ F}(r)] + 4\varepsilon_\mu(r-1) : r > 1\}$$

But

$$BL_{\psi\circ F}(r) \leq \|\psi\|_\infty + \|\psi\|_L \cdot L_F(r) \leq \max\{1, L_F(r)\},$$

which yields (2.3.). (2.4.) follows immediately by putting $r := 1+\varepsilon_\mu^{-1}(\beta_Z(\mu,\nu))$ in (2.3.), noting that $\varepsilon_\mu(\varepsilon_\mu^{-1}(\beta_Z(\mu,\nu))) \leq \beta_Z(\mu,\nu)$ by the right continuity of ε_μ. ♦

Now we are going to deal with quantitative continuity of generalized moments with respect to bounded Lipschitz distance. The following theorem is a slight improvement of [13, Thm.2.1]:

Theorem 2. Let $h:Z\to R$ be an unbounded locally Lipschitz continuous mapping. Then there holds for any two probability measures μ,ν on $B(Z)$ and all $p\leq 1$:

$$|\int_Z h\, d(\mu-\nu)| \leq \beta_Z(\mu,\nu)^{1-(1/p)} \cdot [|h(0)| + 3(\|\underline{BL}_h(|z|)\|_{p,\mu} + \|\underline{BL}_h(|z|)\|_{p,\nu})] \qquad (2.8.)$$

where we use the abbreviations

$$|z| := d_Z(z,0) \quad \text{and} \quad \|\underline{BL}_h(|z|)\|_{p,\sigma} := (\int_Z \underline{BL}_h(|z|)^p\, \sigma(dz))^{1/p} \quad (\sigma=\mu,\nu)$$

and $\underline{BL}_h$ denotes the right continuous modification of the function BL_h (note that $BL_h \leq \underline{BL}_h$).

Proof: Using Lemma 1a) we obtain, for any r>0, the following chain of inequalities:

$$|\int_Z h\, d(\mu-\nu)|$$
$$\leq \int_{Z-K_r} (|h|+B_h(r))(\mu+\nu)(dz) + BL_r(h)\, \beta_Z(\mu,\nu)$$
$$\leq 2\int_{Z-K_r} (BL_h(|z|))(\mu+\nu)(dz) + BL_r(h)\, \beta_Z(\mu,\nu)$$

The first summand may be estimated as follows:

$$\int_{Z-K_r} (BL_h(|z|))(\mu+\nu)(dz)$$
$$\leq \underline{BL}_h(r)^{1-p} \int_Z (\underline{BL}_h(|z|)^p(\mu+\nu)\,(dz),$$

hence results

$$\left|\int_Z h\, d(\mu-\nu)\right|$$
$$\le 2\, \underline{BL}_h(r)^{1-p}[(\|\underline{BL}_h(|z|)\|_{p,\mu})^p + \|(\underline{BL}_h(|z|)\|_{p,\nu})^p] + BL_h(r)\, \beta_Z(\mu,\nu) \tag{2.9.}$$

Now we put

$$r := \sup\{s \ge 0 : BL_h(s) \le [\|\underline{BL}_h(|z|)\|_{p,\mu} + \|\underline{BL}_h(|z|)\|_{p,\nu} + |h(0)|]\, \beta_Z(\mu,\nu)^{-1/p}\}$$

In view of the left continuity of BL_h and the right continuity of $\underline{BL}_h$ we get

$$BL_h(r) \le [\|\underline{BL}_h(|z|)\|_{p,\mu} + \|\underline{BL}_h(|z|)\|_{p,\nu} + |h(0)|]\, \beta_Z(\mu,\nu)^{-1/p} \tag{2.10.}$$

and

$$\underline{BL}_h(r) \ge [\|\underline{BL}_h(|z|)\|_{p,\mu} + \|\underline{BL}_h(|z|)\|_{p,\nu} + |h(0)|]\, \beta_Z(\mu,\nu)^{-1/p},$$

the latter inequality implying

$$\underline{BL}_h(r)^{1-p} \le \min\{(\|\underline{BL}_h(|z|)\|_{p,\mu})^{1-p}, (\|\underline{BL}_h(|z|)\|_{p,\nu})^{1-p}\}\, \beta_Z(\mu,\nu)^{1-(1/p)} \tag{2.11.}$$

Combining (2.10.) and (2.11.) with (2.9.), we arrive at (2.8.). ♦

Remark 1.a) In virtue of the estimate

$$BL_h(r) \le L_h(r)(r+1) + |h(0)| \quad (r\ge 0), \tag{2.12}$$

Theorem 2 is better than Thm. 2.1. in [13] in the sense that the finite moment condition $\int_Z (L_h(|z|)|z|)^p \mu(dz) < \infty$ required there guarantees finiteness of $\|\underline{BL}_h(|z|)\|_{p,\mu}$, but not vice versa. Indeed, consider the example $Z:=\mathbf{R}$, $h:=\sin(z^2)$, $\mu(dz):=(z^4+1)^{-1}dz$, $p:=2$. Then $BL_h(r)\le 2r+1$, hence $\|\underline{BL}_h(|z|)\|_{2,\mu} < \infty$, whereas $L_h(r)\ge 2(r-\pi)$ and thus $\int_Z (L_h(|z|)|z|)^2 \mu(dz) = \infty$

b) Obviously, the inequalities in Theorems 1 and 2 remain valid if ε_μ, L_F and BL_h, respectively, are replaced by upper estimates. If, e. g., μ is "of Gaussian type", i.e. obeys an estimate

$$\varepsilon_\mu(r) \le c_1 \exp(-c_2 r^2) \quad (r>0) \tag{2.13.}$$

and F is "of exponential type", i.e. obeys

$$L_F(r) \le k_1 \exp(k_2 r) \tag{2.14.}$$

then (2.4.) yields

$$\beta_X(\mu F^{-1}, \nu F^{-1}) \le \gamma_1 \exp(\gamma_2 |\log \beta_Z(\mu,\nu)|^{1/2})\, \beta_Z(\mu,\nu) \tag{2.15.}$$

(Note that, for all $\delta>0$, the r.h.s. of (2.15.) is $o(\beta_Z(\mu,\nu)^{1-\delta})$ for small $\beta_Z(\mu,\nu)$.)

If F has property (2.14.), then $h(z) := (d_X(F(z),x_o))^k$ (where x_o is some fixed element of X and $k\in\mathbf{N}$) admits an estimate

$$BL_h(r) \le \alpha_1 k \exp(\alpha_2 k r) \tag{2.16.}$$

If, in addition to (2.16.), μ and ν are of "Gaussian type" (2.13.), then Theorem 2 yields, for all $\delta>0$:

$$\left|\int_Z h\, d(\mu-\nu)\right| = o(\beta_Z(\mu,\nu)^{1-\delta}) \text{ for small } \beta_Z(\mu,\nu) \tag{2.17.}$$

We conclude this section by giving examples that at least the order of the estimates in Theorems 1 and 2 is optimal for small $\beta_Z(\mu,\nu)$):

Example 1. $X = Z = \mathbf{R}_+$, $k > 1$, $F(z) := z^k$. For $0<\alpha<2^{-1}$ we put

$$\mu_\alpha := 2^{-1}(\delta_0+\delta_{\alpha^{1/(1-k)}}),\quad \nu_\alpha := 2^{-1}(\delta_0+\delta_{(\alpha^{1/(1-k)}+\alpha)}).$$

Then $\beta_Z(\mu_\alpha,\nu_\alpha) \le \alpha$, $\varepsilon_\mu^{-1}(\beta_Z(\mu_\alpha,\nu_\alpha))) = \alpha^{1/(1-k)}$, $L_F(r) = kr^{k-1}$, hence the r.h.s. of (2.4.) is bounded from above by a constant. On the other hand one has $\beta_X(\mu_\alpha F^{-1}, \nu_\alpha F^{-1}) \ge 2^{-1}$.

Example 2. $X = Z = \mathbf{R}_+$, $k,p > 1$, $h(z) := z^k$. For $0<\alpha<1$ we put

$$\mu_\alpha := \delta_0,\quad \nu_\alpha := (1-\alpha)\,\delta_0 + \alpha\delta_{\alpha^{-1/kp}}.$$

Then $\beta_Z(\mu_\alpha,\nu_\alpha) \le \alpha$ and $BL_h(r) \le (k+1)r^k$, leading to

$$\|\underline{BL}_h(|z|)\|_{p,\mu_\alpha}=0 \text{ and } \|\underline{BL}_h(|z|)\|_{p,\nu_\alpha} \leq (\alpha(k+1)^p(\alpha^{-1/kp})^{kp})^{1/p} = k+1.$$

Hence the r.h.s. of (2.8.) is bounded from above by const$\cdot\alpha^{1-1/p}$. But on the other hand there holds $|\int_Z h\, d(\mu_\alpha-\nu_\alpha)| = \alpha^{1-1/p}$.

3. An application to approximate solutions of stochastic differential equations

Let us return to the integral equation (1.1.). Under the conditions

$$\begin{array}{l} f \text{ is locally Lipschitz continuous and satisfies a linear growth condition} \\ g \text{ has a bounded and locally Lipschitz continuous derivative} \end{array} \tag{3.1.}$$

the solution $x = S_1(z)$ of (1.1.) has the following representation [1, 16]

$$\begin{aligned} x(t) &= \phi(\xi_z(t), z(t)-z(0)) \\ \xi_z(t) &= x_o + {}_0\int^t \eta(\xi_z(s), z(s)-z(0))ds \\ \eta(\alpha,\beta) &= (\delta\phi/\partial\alpha)(\alpha,\beta)^{-1} f(\phi(\alpha,\beta)) \\ (\delta\phi/\partial\beta)(\alpha,\beta) &= g(\phi(\alpha,\beta)) \;;\; \phi(\alpha,0) = \alpha \end{aligned} \tag{3.2.}$$

Defining x as in (3.2.) even for any bounded , measurable $z:[0,1]\rightarrow\mathbf{R}$, one gets a mapping $S:z\rightarrow x$ which can be shown [14, Thm.1] to be continuous with respect to the norm

$$\|z\|_1 := |z(0)| + {}_0\int^1 |z(s)|ds$$

on $\{z : \|z\|_\infty \leq R\}$ for each $R > 0$, hence is a continuous extension of S_1 in this sense.
Obviously S maps the space C[0,1] (of continuous functions) and the space D[0,1] (of right continuous functions with left limits), respectively, into itself, and it can be shown [16, 14] that S is locally Lipschitz continuous w.r. to the sup-norm on C[0,1] and w.r. to the modified Skorokhod metric d_o on D[0,1] defined by

$$d_o(z_1,z_2) := \inf_{\lambda\in\Lambda} \max\{ \|z_1-z_2\circ\lambda\|_\infty , \sup_{0\leq s<t\leq 1} |\log|\lambda(t)-\lambda(s)|\,(t-s)^{-1}| \} \tag{3.3.}$$

where Λ is the set of all mappings λ from [0,1] onto [0,1] which are strictly monotonically increasing.
A mapping $z\in D[0,1]$ is said to have finite quadratic variation along some fixed sequence of partitions τ_n of [0,1] with mesh size tending to zero, if the weak limit ζ of the measures

$$\zeta_n := \sum_{t_i\in\tau_n} (z(t_{i+1})-z(t_i))^2\, \delta_{t_i}$$

exists (cf. [6]); the distribution function of ζ is denoted by $t\rightarrow\langle z\rangle(t)$. For $z\in D[0,1]$ of finite quadratic variation, $x=S(z)$ obeys the integral equation [14, Prop.1]

$$\begin{aligned} x(t) = x_o &+ {}_0\int^t f(x(s))ds + {}_0\int^t g(x(s-))dz(s) + 1/2\cdot {}_0\int^t (gg')(x(s-))d\langle z\rangle^c(s) \\ &+ \sum_{s\leq t} [\phi(x(s-),\Delta z(s))-x(s-) - g(x(s-)\Delta z(s)] \end{aligned} \tag{3.4.}$$

where

$${}_0\int^t g(x(s-))dz(s) \quad := \quad \lim_{n\rightarrow\infty} \sum_{t_i\in\tau_n, t_i<t} g(t_i)(z(t_{i+1}) - z(t_i))$$

and

$$\langle z\rangle(t) = \langle z\rangle^c(t) + \sum_{s\leq t} \Delta z(s)^2$$

is the decomposition of $\langle z\rangle$ into its continuous and jump part.

For $z \in C[0,1]$ and $\langle z\rangle(t) \equiv t$ (which is a property shared by almost every Wiener path), (3.4.) specializes to

$$x(t) = x_o + {}_0\!\int^t (f+(1/2)gg')(x(s))ds + {}_0\!\int^t g(x(s-))dz(s) \qquad (3.5.)$$

(which, for a Wiener input z, is an Itô stochastic differential equation with Stratonovich correction).
For a piecewise constant function z, (3.4.) specializes to

$$x(t) = x_o + {}_0\!\int^t f(x(s))ds + \sum_{s\leq t} [\phi(x(s-),\Delta z(s))-x(s-)] \qquad (3.6.)$$

For certain coefficient functions f and g obeying (3.1.), the growth of S (and hence also that of L_S) may be larger than exponential, as the following example shows:

Example 3. Put $f(\alpha) := \alpha$, $g(\alpha) := \sin(\alpha\pi)$ $(\alpha \in \mathbf{R})$. Then the function ϕ occuring in (3.2.) is given by

$$\phi(\alpha,\beta) = (2/\pi)\arctan[\tan(\alpha\pi/2)\exp(\pi\beta)] \qquad (3.7.)$$

Put $x_o := 1$, and define, for any $n \in \mathbf{N}$, a sequence (t_m) by

$$\exp(t_1) = 1+1/n$$
$$(m-1/n)\exp(t_m) = m+1/n \quad (m\geq 2)$$

It is easily checked that

$$t_1 < 1/n \ ; \ 1/(nm) < t_m < 2/[n(m-1)] \ \text{ for } m\geq 2 \qquad (3.8.)$$

Choose $C_n \in \mathbf{R}$ such that

$$\phi(1/n, C_n) = 1 - 1/n$$

The following chain of implications

$$\arctan[\tan(\pi/2n)\exp(\pi C_n)] = (\pi/2)(1-1/n)$$
$$\Rightarrow \tan(\pi/2n)\exp(\pi C_n) = \tan[(\pi/2)(1-1/n)]$$
$$\Rightarrow \exp(\pi C_n) = \cot^2(\pi/2n) \leq n^2$$

shows that

$$C_n \leq \log n \qquad (3.9.)$$

Put

$$T_m := t_1+\ldots+t_m \ (m\geq 1), \quad T_0 := 0,$$

define

$$\underline{z}_n(t) := \begin{cases} 0 & \text{for } T_{2(m-1)} \leq t < T_{2m-1} \\ C_n & \text{for } T_{2m-1} \leq t < T_{2m} \end{cases}$$

and let z_n be the restriction of $\underline{z}_n$ to [0,1]. By (3.9.) there holds

$$\|z_n\|_\infty \leq \log n \qquad (3.10.)$$

The solution $x_n := S(z_n)$ of equation (3.6.) with input z_n increases exponentially on any interval $[T_{m-1}, T_m)$ from $m-1/n$ to $m+1/n$, and jumps at any time point T_m from $m+1/n$ to $m+1-1/n$. In paricular, x_n is increasing and obeys

$$x_n(T_m) \geq m. \qquad (3.11.)$$

In virtue of (3.8.) we get the estimate

$$T_m \leq (3/n)\log m \qquad (3.12.)$$

Combining (3.11.) and (3.12.), one arrives at

$$x_n(1) \geq \exp(n/3) \qquad (3.13.)$$

which together with (3.10.) yields

$$\|S(z_n)\|_\infty \geq \exp((1/3)\exp(\|z_n\|_\infty))$$ ♦

Under the following conditions, however, L_S has only exponential growth:

Theorem 3.[14, Thm.3] Assume, in addition to (3.1.), that f is globally Lipschitz continuous and that $0 < m \leq |g| \leq M < \infty$ for some real constants m,M. Then there holds for suitable k_1, k_2

$$L_S(r) \leq k_1\exp(k_2 r) \qquad (r>0) \tag{3.14.}$$

in any of the following cases:

a) $(Z, d_Z) = (X, d_X) = (C[0,1]$, sup-distance$)$

b) $(Z, d_Z) = (X, d_X) = (D[0,1], d_o)$

c) $(Z, d_Z) = (M[0,1], d_s)$; $(X, d_X) = (D[0,1], d_s)$

where $M[0,1] := \{z \in D[0,1] : z$ is nondecreasing$\}$, and d_s is the Skorokhod distance defined by

$$d_s(z_1,z_2) := \inf_{\lambda\in\Lambda} \max\{ \|z_1 - z_2 \circ \lambda\|_\infty, \|\lambda - id\|_\infty\} \tag{3.15.}$$

If μ_W is Wiener measure on C[0,1], then ε_{μ_W} obeys (2.13.) ; hence follows by (2.15.) that, under the assumptions of Theorem 3a) there exist constants γ_1, γ_2 such that for all probability measures ν on C[0,1] there holds

$$\begin{aligned} &\beta_{C[0,1]}(\mu_W S^{-1}, \nu S^{-1}) \\ &\leq \gamma_1 \exp(\gamma_2 |\log \beta_{C[0,1]}(\mu_W,\nu)|^{1/2})\, \beta_{C[0,1]}(\mu_W,\nu) \end{aligned} \tag{3.16}$$

If μ_P is (unit mean) Poisson measure on M[0,1], then a simple estimate shows that

$$\varepsilon_{\mu_P}(r) \leq 1/\Gamma(r) \tag{3.17.}$$

Hence follows by (2.4.) that, under the assumptions of Theorem 3c) there exists, for any $\delta>0$, a constant c such that for all probability measures ν on M[0,1] there holds

$$\beta_{D[0,1]}(\mu_P S^{-1}, \nu S^{-1}) \leq c \cdot \beta_{M[0,1]}(\mu_W,\nu)^{1-\delta} \tag{3.18.}$$

Finally we mention convergence rates with respect to bounded Lipschitz distance of some approximations to Wiener resp. Poisson distribution:

Example 4. Let, for $n \in N$, $(Y_{n,j})_{j=1,\ldots,n}$ be a sequence of independent random variables, with

$$P[Y_{n,j}=1] = 1/n = 1 - P[Y_{n,j}=0] \quad (j=1,\ldots,n)$$

Put $$z_n(t) := \sum_{1\leq i\leq j} Y_{n,i} \quad \text{for } j/n \leq t < (j+1)/n \text{ ; } 0 \leq j \leq n.$$

Let μ_n be the distribution of z_n, and μ_P be standard Poisson measure on M[0,1].Then there holds according to [4, Thm.6.1.]

$$\beta_{M[0,1]}(\mu_P, \mu_n) = O(n^{-1}) \tag{3.19.}$$

Combining (3.18.) and (3.19.) one gets

$$\beta_{D[0,1]}(\mu_P S^{-1}, \mu_n S^{-1}) = O(n^{-1+\delta}) \quad \text{for all } \delta>0. \tag{3.20.}$$

Example 5.a) Let, for $n \in N$, $(Y_{n,j})_{j=1,\ldots,n}$ be a sequence of independent random variables, with

$$P[Y_{n,j}=n^{-1/2}] = P[Y_{n,j}=-n^{-1/2}] = 1/2 \quad (j=1,\ldots,n)$$

Put $$z_n(t) := \sum_{1\leq i\leq j} Y_{n,i} + (t-j/n)Y_{n,j+1} \quad \text{for } j/n \leq t \leq (j+1)/n,\ 0\leq j\leq n.$$

Let μ_n be the distribution of z_n, and μ_W be standard Wiener measure on C[0,1].Then one derives from [8, Thm.1]:

$$\beta_{C[0,1]}(\mu_W, \mu_n) = O(n^{-1/2}\log n) \qquad (3.21.)$$

Combining (3.16.) and (3.21.) one gets, for suitable $\gamma>0$,

$$\beta_{C[0,1]}(\mu_W S^{-1}, \mu_n S^{-1}) = O(\, n^{-1/2}\exp(\gamma(\log n)^{1/2})) \qquad (3.22.)$$

b) If w(t) is a standard Wiener process and $w_n(t)$ is a "polygonal approximation" of w(t) (coinciding with w in t = 0, 1/n, 2/n,...,1 and piecewise linear between these points), then it can be shown (cf.[15, Remark 2b)]) that

$$E[\|w-w_n\|_\infty{}^p]^{1/p} = O(n^{-1/2}(\log n)^{1/2}) \qquad (3.23.)$$

holds for all $p\geq 1$.

c) The convergence rate (3.23.) even holds true if w_n is the conditional expectation of w with respect to a certain discrete σ-algebra. More precisely, let $I_{n,1},\ldots, I_{n,m(n)}$ be disjoint subintervals of R, each having standard normal probability $m(n)^{-1}$. Put

$$A_n := \sigma(\{\underline{w}_j \in I_{n,i}\} : j=1,\ldots,n;\ i=1,\ldots,m(n)\})$$

where $\underline{w}_j := n^{1/2}(w(j/n)-w((j-1)/n)$. In [15, Thm.1] it is proved that $w_n := E[w \mid A_n]$ (which is a "polygonal approximation of w with finitely many relizations") has the convergence rate (3.23.), provided that $\sup\{n/m(n) : n\in N\}$ is finite.

d) If - in either of the cases b) and c) - μ_n denotes the distribution of w_n, then (3.23.) (with p=1) implies immediately that

$$\beta_{C[0,1]}(\mu_W, \mu_n) = O(n^{-1/2}(\log n)^{1/2}) \qquad (3.24.)$$

which is a slightly better convergence rate than (3.21.).

If ψ is a real valued mapping on C[0,1] and J is a convex majorant of $BL_{\psi \circ S}$ such that $\int J(\|z\|_\infty)^p \mu_W(dz)$ is finite for all p>1 (a function J with these properties exists, e. g., for $\psi(x) = \|x\|_\infty{}^k$ $(k \in N)$ under the assumptions of Theorem 3, cf. (2.16.)), then Jensen's inequality guarantees that

$$\int J(\|z\|_\infty)^p\, \mu_n(dz) \leq \int J(\|z\|_\infty)^p\, \mu_W(dz) < \infty \quad (n\in N)\,;$$

together with Theorem 2 then follows for all $\delta>0$

$$\left| \int \psi(x)\, \mu_n S^{-1}(dx) - \int \psi(x)\, \mu_W S^{-1}(dx) \right| = O(n^{-(1/2)+\delta}).$$

References:

[1] Doss, H.: Liens entre équations différentielles stochastiques et ordinaires. Ann. Inst. Henri Poincaré, Sect. B.XIII, 99-125 (1977).

[2] Dudley, R.M.: Probabilities and Metrics. Lecture Notes Series No. 45. Aarhus Universitet 1976.

[3] Dudley, R.M.: Speeds of metric probability convergence. Z. Wahrscheinlichkeitsth. verw. Geb. 22, 323-332 (1972).

[4] Dudley, R.M.: Distances of probability measures and random variables. Ann. Math. Statist. 39, 1563-1572 (1968).

[5] Engl, H.W., Wakolbinger, A.: Continuity properties of the extension of a locally Lipschitz continuous map to the space of probability measures. Mh. Math. 100, 85-103 (1985).

[6] Föllmer, H.: Calcul d' Itô sans probabilités, in: Séminaire de probabilités XV. Lecture Notes Math. 850, 143-150. Berlin-Heidelberg-New York: Springer 1981.

[7] Ikeda, N., Watanabe,S.: Stochastic differential equations and diffusion processes. Amsterdam, Tokyo: North Holland/Kodansha 1981.

[8] Komlos, J., Major, P., Tusnády, G.: An approximation of partial sums of independent random variables, and the sample distribution function, II. Z. Wahrscheinlichkeitsth. verw. Geb. 34, 33-58 (1976).

[9] Konecny, F.: On Wong-Zakai approximation of stochastic differential equations. J. Multivariate Anal. 13, 605-611 (1983).

[10] Kubilius, K., Mikulevicius, R.; On the rate of convergence of semimartingales. Preprint, to appear.

[11] Marcus, S.I.: Modeling and approximation of stochastic differential equations driven by semimartingales. Stochastics 4, 223-245 (1981).

[12] Protter, Ph.: Approximations of solutions of stochastic differential equations driven by semimartingales. Ann. Probab. 13, 716-743 (1985).

[13] Römisch, W., Wakolbinger, A.: Obtaining convergence rates for approximations in Stochastic Programming. To appear in: J. Guddat (ed.), Parametric Optimization and Related Topics, Berlin: Akademie-Verlag.

[14] Römisch, W., Wakolbinger, A.: On Lipschitz dependence in systems with differentiated inputs. Math. Ann. 272, 237-248 (1985).

[15] Römisch, W., Wakolbinger, A.: Convergence rates of approximate solutions of stochastic differential equations. Technical report, Linz 1983.

[16] Sussmann, H. On the gap between deterministic and stochastic ordinary differential equations. Ann. Probab. 6, 19-44 (1978).

[17] Topsœ, F.: Preservation of weak convergence under mappings. Ann. Math. Statist. 38, 1661-1665 (1967).

[18] Whitt, W.: Preservation of rates of convergence under mappings. Z. Wahrscheinlichkeitsth. verw. Geb. 29, 39-44 (1974).

[19] Zolotarev, V.M.: Metric distances in spaces of random variables and their distributions, Math. USSR Sbornik 30, 373-401 (1976).

ON THE JOINT DISTRIBUTION OF THE BROWNIAN LOCAL AND OCCUPATION TIMES

Paavo Salminen

Åbo Akademi, Mathematical Institute,
SF-20500 Åbo, Finland

Abstract

Let $\mathbf{P}_x$ be the probability measure associated with a standard Brownian motion B when started at x. In a paper by Karatzas and Shreve the joint $\mathbf{P}_0$−distribution of the local time at zero, L_t, the occupation time Γ_t, of $[0, +\infty)$ and B_t is derived. They notice that this distribution coincides in the case $B_t < 0$ with the joint $\mathbf{P}_0$−distribution of the maximum $M_t := \sup_{s \le t} B_s$, the time point of the maximum $T := \inf\{s : B_s = M_t\}$ and B_t. Here we give a probabilistic explanation for this rather curious phenomenon.

1. Introduction

Let $B = \{B_t : t \geq 0\}$ be a standard Brownian motion, and $\mathbf{P}_x$ with it associated probability measure when B is started at x. Denote with $n_x(t, y)$ the $\mathbf{P}_x$−density of the first passage time $\tau_y := \inf\{t : B_t = y\}$ i.e.

$$n_x(t, y) = \frac{|y - x|}{\sqrt{2\pi t^3}} \exp - \frac{(y - x)^2}{2t}.$$

Let L_t^x be the jointly continuous version of the local time at x normalized so that for all Borel sets A

$$\int_0^t 1_A(B_s) ds = 2 \int_A L_t^x dx.$$

Further introduce the occupation measure, Γ_t, of $[0, \infty)$ i.e. $\Gamma_t := \text{meas}\{s \leq t : B_s \geq 0\}$, where "meas" means the Lebesgue measure of the set in the braces.

In [5] Karatzas and Shreve compute the joint $\mathbf{P}_0$-distribution of $L_t := L_t^0$, Γ_t, and B_t. Their result is

$$(1.1) \qquad \mathbf{P}_0(L_t \in dm, \Gamma_t \in ds, B_t \in dx) = \begin{cases} n_0(s, m) n_x(t - s, m) 2dx\, dm\, ds, & x < 0, \\ n_0(t - s, m) n_{-x}(s, m) 2dx\, dm\, ds, & x > 0. \end{cases}$$

In [5] this distribution is needed when computing the transition probabilities of the optimal process in a stochastic control problem (see also [1]).

Let now $M_t := \sup_{s \le t} B_s$, and $T := \inf\{s : B_s = M_t\}$. In [8] Lévy gives the joint $\mathbf{P}_0$−distribution of M_t, T, and B_t (see also [9], and [13]). Later on this distribution has been computed for various diffusions (see [4], [10], and [12]). Finally in [2] the general formula is given.

However we focus on the Brownian case where this distribution takes the form ($x \le m, s < t$)

(1.2) $$\mathbf{P}_0(M_t \in dm, T \in ds, B_t \in dx) = n_0(s,m)n_x(t-s,m)2dx\,dm\,ds.$$

Comparing (1.1) and (1.2) it is seen that the distributions coincide in the case $x < 0$. This fact is noticed in [5], where it is also stated that an explanation based on a path decomposition argument will be given later. Being not aware whether this has been done or not* we offer here an explanation based on the excursion theory (and path decompositions). We consider (1.2) as a known fact and deduce (1.1) from it. Note that the case $x > 0$ follows from the case $x < 0$ by symmetry.

Before starting this program we state

(1.3) **Corollary.** Let $\mathbf{P}^l_{00}$ be the law of a Brownian bridge from 0 to 0 with duration l, for short BB^l_{00}. Then the $\mathbf{P}^l_{00}$-distributions of (M_l, T) and (L_l, Γ_l) coincide.

Proof is immediate from (1.1) and (1.2).

2. Preliminaries

For a given $t > 0$ let $\lambda^t := \sup\{s \le t : B_s = 0\}$.

(2.1) **Proposition.** Given $\lambda^t = l$ the process $\{B_s : 0 \le s \le \lambda^t\}, B_0 = 0$, is a BB^l_{00}.

Proof. Consider the process $\hat{B}$ obtained from B by killing it exponentially. Then we may argue as in [9] Remark 2.5 (iii) to obtain that given $\lambda = l, \lambda := \sup\{s : \hat{B}_s = 0\}$, the process $\{\hat{B}_s : 0 \le s \le \lambda\}$ is a BB^l_{00}, from which the claim follows easily.

Next we recall a result from the Brownian excursion theory. Let X^1 and X^2 be two independent reflecting Brownian motions defined on the same probability space. Denote their local times (at zero) with ℓ^1 and ℓ^2, respectively, and set $A^1_a := \inf\{s : \ell^1_s > a\}$, and $A^2_a := \inf\{s : \ell^2_s > a\}$. A^1 and A^2 are independent Lévy processes with the Lévy measure $n(dr) = \frac{1}{\sqrt{2\pi r^3}}dr$. Then $A := A^1 + A^2$ is also a Lévy process and its Lévy measure is $\hat{n}(dr) = 2n(dr)$. By the independence, almost surely, A^1 and A^2 do not have common jumps. Therefore, if for a given a $A_{a-} < A_a$ then either $A^1_{a-} < A^1_a$ and $A^2_{a-} = A^2_a$ or $A^1_{a-} = A^1_a$ and $A^2_{a-} < A^2_a$. Further for a given $s > 0$ there exists an a such that either $s = A_a$ or $s \in [A_{a-}, A_a)$. Let

(2.3) $$X^*_s = \begin{cases} 0, & \text{if } s = A_a, \\ X^1_{v+A^1_{a-}}, & \text{if } s \in [A_{a-}, A_a) \text{ and } A^1_{a-} < A^1_a, \\ -X^2_{v+A^2_{a-}}, & \text{if } s \in [A_{a-}, A_a) \text{ and } A^2_{a-} < A^2_a, \end{cases}$$

* It has been done – see the note at the end of the paper

where $v = s - A_{a-}$.

(2.4) Proposition. The process X^* is a standard Brownian motion started at 0.

This follows from the characterization of a Brownian motion in terms of a Poisson point process and excursions (see [3] p. 123–131).

3. Proof of (1.1)

For the rest of the paper we fix a $t > 0$ and assume that $B_0 = 0$. Introduce $(\lambda := \lambda^t)$

(3.1)
$$\tilde{B} = \begin{cases} B^1_s, & s < T, \\ B^2_{s-T}, & T \le s < \lambda, \\ B^3_{s-\lambda}, & \lambda \le s \le t, \end{cases}$$

where $B^1_s := B_s$, $B^2_s := B_{\lambda - s}$, $s < \lambda$, $B^3_s := B_{\lambda+s}$, and $T := \inf\{s : B_s = M_\lambda\}$, $M_\lambda = \sup_{0 \le s \le \lambda} B_s$. It is well known that the time point when B attains the maximum value before the time λ is a.s. unique (see [14]). The following result is an easy corollary of Williams' path decomposition theorem (see [14], and [3] p. 136).

(3.2) Proposition. Given $\lambda = l, T = u$, and $M_\lambda = y$ the three parts of $\tilde{B}$ in (3.1) are independent. Under this condition $\{B^1_s : 0 \le s < T\}$ is identical in law with a Brownian motion killed when it hits y, for short ABM^y_0, conditioned by $\{\tau_y = u\}$. Similarly $\{B^2_s : 0 \le s < \lambda - T\}$ is identical in law with a ABM^y_0 conditioned by $\{\tau_y = l - u\}$, and $\{B^3_s : 0 \le s \le t - \lambda\}$ is identical in law with a Brownian meander of length $t - l$.

Next set

(3.3)
$$\bar{B}_s = \begin{cases} M^1_s - B^1_s, & s < T, \\ -(M^2_{s-T} - B^2_{s-T}), & T \le s < \lambda, \\ B^3_{s-\lambda}, & \lambda \le s \le t, \end{cases}$$

where $M^1_s = \sup_{0 \le v \le s} B^1_v$ for $s < T$, and $M^2_s = \sup_{0 \le v \le s} B^2_s$ for $s \le \lambda - T$. To analyze this process recall the well known result of Lévy which states that $M_s - B_s$ is a reflecting Brownian motion, for short RBM_0, and M_s is its local time i.e. P_0-a.s.

(3.4)
$$M_s = \lim_{\epsilon \downarrow 0} \frac{1}{2\epsilon} \int_0^s 1_{[0,\epsilon)}(M_u - B_u) du.$$

For a given RBM_0 we denote its local time (at zero) with ℓ and set $A_b = \inf\{s : \ell_s > b\}$. Clearly a.s. $b \mapsto A_b$ is right continuous with left limits. Using (3.2) we obtain

(3.5) **Proposition.** Given $\lambda = l, T = u$, and $M_\lambda = y$ the three parts of $\bar{B}$ in (3.3) are independent. Under this condition the first part is identical in law with a RBM_0 on $[0, \infty)$ conditioned by $\{A_{y-} = u\}$. Similarly the second part is identical in law with a RBM_0 on $(-\infty, 0]$ conditioned by $\{A_{y-} = l - u\}$.

Next we put the first and the second part of $\bar{B}$ together. This is done as in (2.3) by setting

$$X_s^1 = M_s^1 - B_s^1, 0 \le s < T, \qquad X_s^2 = -(M_s^2 - B_s^2), 0 \le s < \lambda - T,$$

$$A_a^1 = H_a^1 := \inf\{s : M_s^1 > a\}, \quad A_a^2 = H_a^2 := \inf\{s : M_s^2 > a\}.$$

Using the conditional independence it is seen that H^1 and H^2 do not have common jumps. We denote the resulting process with B^*.

Let

$$\Gamma_\lambda^* := \text{meas}\{s \le \lambda : B_s^* \ge 0\} \quad \text{and} \quad L_t^* := \lim_{\epsilon \downarrow 0} \frac{1}{4\epsilon} \int_0^t 1_{(-\epsilon,\epsilon)}(B_s^*) ds.$$

Obviously, by the construction, $\Gamma_\lambda^* = T$, and, by (3.4)

$$L_t^* = \lim_{\epsilon \downarrow 0} \frac{1}{4\epsilon} \Big(\int_0^T 1_{[0,\epsilon)}(M_u^1 - B_u^1) du + \int_0^{\lambda - T} 1_{(-\epsilon,0]}(-(M_u^2 - B_u^2)) du \Big) = \frac{1}{2}(M_T^1 + M_{\lambda-T}^2) = M_\lambda.$$

Consequently

$$\mathbf{P}_0(\Gamma_\lambda^* \in ds, L_t^* \in dm) = \mathbf{P}_0(T \in ds, M_\lambda \in dm), \tag{3.6}$$

and, because (see (2.1))

$$\mathbf{P}_0(\Gamma_\lambda^* \in ds, L_t^* \in dm) = \int_0^t \mathbf{P}_{00}^l(\Gamma_l^* \in ds, L_l^* \in dm) \mathbf{P}_0(\lambda \in ds),$$

where $\mathbf{P}_{00}^l$ is the law of a BB_{00}^l, we obtain

$$\mathbf{P}_{00}^l(\Gamma_l^* \in ds, L_l^* \in dm) = \mathbf{P}_{00}^l(T \in ds, M_l \in dm).$$

The conditional independence at λ gives

$$\mathbf{P}_0(\Gamma_\lambda^* \in ds, L_t^* \in dm, B_t^* \in dx) = \mathbf{P}_0(T \in ds, M_\lambda \in dm, B_t \in dx).$$

But $\Gamma_\lambda^* = \Gamma_t^*$, $M_\lambda = M_t$ if $B_t^* = B_t < 0$, and, therefore, we have (1.1) when the following proposition is verified.

(3.7) **Proposition.** The processes $\{B_s^* : 0 \le s \le t\}$ and $\{B_s : 0 \le s \le t\}$ are identical in law.

Proof. It is quite obvious from (3.5) and (3.6) that the processes

$$A_a^* := H_a^1 + H_a^2 \quad \text{and} \quad A_a := \inf\{s \le t : L_s > a\}$$

are identical in law. For given $0 < s_1 < \cdots < s_n \leq t$, Borel sets $S_1, \ldots, S_n$, and a path $a \mapsto A_a^* = \omega_a$ we have

$$\mathbf{P}_0(B_{s_1}^* \in S_1, \ldots, B_{s_n}^* \in S_n | A_\cdot^* = \omega_\cdot) = \mathbf{P}_0(B_{s_1} \in S_1, \ldots, B_{s_n} \in S_n | A_\cdot = \omega_\cdot).$$

This follows from (3.5) and the excursion characterization of Brownian motion (see [3] p. 123–131). Because A^* and A are identical in law the proof is complete.

Note. Since this work was done I learned about a technical report [6] and a forthcoming paper [7] by Karatzas and Shreve, where an explanation for the coincidence of the distributions (1.1) and (1.2) is given. Their approach is, however, opposite to the one presented here i.e., roughly speaking, a Brownian path is dissected, and re-assembled in such a way that local time becomes maximum, and occupation time becomes location of maximum. Although the basic path transformation (3.1) above and the one used by Karatzas and Shreve are essentially the same the further developments are very different. Therefore the approach presented above might have some interest.

References

[1] Beneš, V.E., Shepp, L.A., Witsenhausen, H.S.: Some solvable stochastic control problems. *Stochastics* 4(1), 1980, 39–84.

[2] Csáki, E., Földes, A., Salminen, P.: On the joint distribution of the maximum and its location for a linear diffusion. To appear in *Annales de l'I.H.P.*

[3] Ikeda, N., Watanabe, S.: *Stochastic Differential Equations and Diffusion Processes.* North-Holland, Amsterdam-Oxford-New York. Kodansha, Tokyo, 1981.

[4] Imhof, J.-P.: Density factorizations for Brownian motion, meander and three-dimensional Bessel process. *J. Appl. Prob.* 21, 1984, 500–510.

[5] Karatzas, I., Shreve, S.E.: Trivariate density of Brownian motion, its local and occupational times, with application to stochastic control. *Ann. Prob.* 12, 1984, 819–828.

[6] Karatzas, I., Shreve, S.E.: A decomposition of the Brownian path. Technical report, Department of Statistics, Columbia University, July 1984.

[7] Karatzas, I., Shreve, S.E.: A decomposition of Brownian path. To appear in *Statistics and Probability Letters.*

[8] Lévy, P.: *Processus Stochastiques et Mouvement Brownien.* Gauthier-Villers, Paris, 1948.

[9] Louchard, G.: Mouvement Brownien et valeurs propes du Laplacien. *Ann. Inst. de l'I.H.P.* IV, 1968, 331–342.

[10] Louchard, G.: Kac's formula, Lévy's local time and Brownian excursion. *J. Appl. Prob.* 21, 1984, 479–499.

[11] Salminen, P.: Brownian excursions, revisited. *Seminar on Stochastic Processes, 1983.* eds. Çinlar, E., Chung, K.-L., Getoor, R.K., Birkhäuser, Boston, 1984.

[12] Shepp, L.A.: The joint density of the maximum and its location for a Wiener process with drift. *J. Appl. Prob.* 16, 1979, 423–427.

[13] Vincze, I.: Einige zweidimensionale Verteilungs- und Grenzverteilungssätze in der Theorie der geordneten Stickproben. *Publ. Math. Inst. Hungar. Acad. Sci.* 11, 1957, 183–203.

[14] Williams, D.: Path decomposition and continuity of local time for one-dimensional diffusions. *Proc. London Math. Soc.* 28, 1974, 738–768.

LARGE DEVIATIONS ESTIMATES FOR SEMILINEAR STOCHASTIC EQUATIONS

W. Smoleński
Institute of Mathematics
Technical University of Warsaw
Warsaw, Poland

R. Sztencel
Department of Mathematics
University of Warsaw, PKiN
Warsaw, Poland

J. Zabczyk
Institute of Mathematics
Polish Academy of Sciences
Warsaw, Poland

In the note large deviations estimates of Freidlin-Wentzell are extended to stochastic, semilinear equations on Banach spaces.

1. Introduction

Let W_t , $t \in [0,T]$ be a standard m-dimensional Wiener process defined on a probability space $(\Omega, \mathcal{F}, \mathbb{P})$. Consider the following Ito stochastic equation on the state space $E = R^n$:

$$(1) \qquad dX = A(X)dt + \varepsilon BdW_t \ , \qquad X(0) = x$$

where A is a Lipschitz transformation from E into E, B a linear mapping from $E_0 = R^m$ into E, and ε a positive parameter. The equation (1) has a unique continuous solution $X^{x,\varepsilon}$. Let $y^{x,u}$ be the unique solution of a deterministic, controlled equation:

$$(2) \qquad \dot{y} = A(y) + Bu \ , \qquad y(0) = x$$

where u is an arbitrary E_0-valued, square integrable function on $[0,T]$. Let moreover $K(x,r)$, $x \in E$, $r > 0$ stand for the set of functions $y^{x,u}$ corresponding to all u satisfying $\int_0^T |u(s)|^2 ds \le r^2$.

Large deviations estimates of Freidlin-Wentzell, see [4], [5] and [1], can be formulated as follows.

Theorem 1. (i) For arbitrary $\delta > 0$, $\eta > 0$ and $c > 0$ there exists $\varepsilon_0 > 0$ such that for all $x \in E$, u satisfying $\int_0^T |u(s)|^2 ds \le c$ and $0 < \varepsilon < \varepsilon_0$:

$$(3) \quad \mathbb{P}\left(\sup_{t \in [0,T]} |X^{x,\varepsilon}(t) - y^{x,u}(t)| < \delta\right) \ge \exp - \frac{1}{2\varepsilon^2}\left(\int_0^T |u(s)|^2 ds + \eta\right)$$

(ii) For arbitrary $\delta > 0$, $\eta > 0$ and $r_0 > 0$ there exists $\varepsilon_0 > 0$ such that for arbitrary $x \in E$, $0 < r < r_0$ and $0 < \varepsilon < \varepsilon_0$:

$$(4) \quad \mathbb{P}(\text{distance}(X^{x,\varepsilon}, K(x,r)) > \delta) \le \exp - \frac{1}{2\varepsilon^2}(r^2 - \eta)$$

In the present note we show that estimates (3), (4) hold true when E_0 and E are separable Banach spaces and A is a semilinear operator. More specifically we assume that:

$$(5) \quad A(x) = \tilde{A}x + F(x), \qquad x \in D(\tilde{A}),$$

where $\tilde{A}$ is the infinitesimal generator of a C_0-semigroup $S(t)$, $t \ge 0$, densely defined on its domain $D(\tilde{A})$ and F is a Lipschitz mapping from E into E.

The estimates will be derived from inequalities of a similar type valid for Abstract Wiener Spaces, which can be found in [2], [11], [15] and [17]. We will present in the paper a different proof based on estimates for Gaussian measures proved in [3], [7] and [14]. To apply the abstract results we describe in detail the reproducing kernel corresponding to the solutions of the linear equation (1) with $A(x) = \tilde{A}x$, $x \in D(\tilde{A})$ and $F \equiv 0$. At this point an identification Lemma 1 plays a crucial role. In the proof of Lemma 1 we rely on some ideas due to Kanter [12]. The general case follows easily from the linear one because solutions of (1) are Lipschitz images of solutions to the linear equation with $F \not\equiv 0$.

Some large deviations estimates for solutions of stochastic equations in Banach spaces have been obtained in [10] and [18]. However results of [10] and [20] are of more special character. In particular in [20] the Wiener process W_t, $t \in [0,T]$ had to be finite dimensional.

The paper is a rewritten version of the report [18].

2. Gaussian measures and their reproducing kernels.

Let $\mathcal{B}(E)$ denote the σ-field of Borel subsets of E and let E' be the conjugate space of E. A probability measure γ on $(E, \mathcal{B}(E))$ is symmetric and Gaussian if for every $e' \in E'$, e' is a symmetric Gaussian random variable on $(E, \mathcal{B}(E), \gamma)$. It is well known, see [14], that there exists a unique Hilbert space H_γ continuousely embedded in E, with the scalar product $\langle \cdot , \cdot \rangle_\gamma$ and the norm $|\cdot|_\gamma$ such that for arbitrary finite family $e_1, e_2, \ldots, e_n \in E'$ the covariance matrix of $(e_1, \ldots, e_n)$ considered as a random vector on $(E, \mathcal{B}(E), \gamma)$ has n-dimensional, symmetric Gaussian distribution on R^n with the covariance matrix $(\langle e_i, e_j \rangle_\gamma)$. The space H_γ is called the reproducing kernel - RKHS(γ). If γ is full (i.e. if the closed linear support of γ equals E, then H_γ is dense in E. Thus we can identify E' with a subspace of H_γ', and (always) H_γ with H_γ'. The triple $E' \subset H_\gamma' = H_\gamma \subset E$ together with the Gaussian measure γ is called an Abstract Wiener Space. At the same time E' can be regarded as a subspace of $L^2(\gamma) = L^2(E, \mathcal{B}(E), \gamma)$ and H_γ' as the closure of E' in $L^2(\gamma)$. If γ is not full then there are elements of E' which are γ - a.s. zero and we are not able to identify E' with a subspace of H_γ'. However we can still introduce $|\cdot|_\gamma$ on E' be $|e'|_\gamma = \sup(e'(e) : e \in H_\gamma , |e|_\gamma \leq 1)$. Of course $e' = 0$ γ - a.s. iff $|e'|_\gamma = 0$.

It is not difficult to see that γ is the probability law of a random variable $\sum_{i=1}^{+\infty} e_i \xi_i$,

$$\mathcal{L}\Big(\sum_{i=1}^{+\infty} e_i \xi_i\Big) = \gamma \quad ,$$

where (e_i) is an arbitrary, orthonormal basis in H_γ and (ξ_i) a sequence of independent, identically distributed, standard, Gaussian random variables. Every $h \in H_\gamma$ extends to a random variable on E with $\mathcal{L}(h)$ being normal, with mean 0 and covariance h :

$$\mathcal{L}(h) = N(0, |h|_\gamma) .$$

In particular for $h \in H_\gamma$, $\langle h, \cdot \rangle_\gamma$ is γ-almost surely well defined.

Let E be endowed with a norm $\|\cdot\|$. Then the space $C_0([0,T],E)$ of E-valued continuous functions F on $[0,T]$ such that $F(0) = 0$ is a separable Banach space under the norm

$$|||F||| = \sup_{t\in[0,T]} \|F(t)\| \ .$$

For every $0 \leq t \leq T$ and every $e' \in E'$ let $\delta_t \otimes e'$ denote the element of $(C_o(0,T,E))'$ given by $(\delta_t \otimes e')(F) = e'(F(t))$. Let $A = \mathrm{span}(\{\delta_t \otimes e'\} ; t \in [0,T], e' \in E')$.

Let Γ be a probability measure on $C_o([0,T],E)$ and let H_Γ be a separable Hilbert space under a norm $|\cdot|_\Gamma$. Moreover, we assume that $H_\Gamma \subset C_o([0,T],E)$ and the embedding is continuous. For $\varphi \in (C_o(0,T,E))'$ let $|\varphi|_\Gamma$ be defined as $|\varphi|_\Gamma = \sup\{\varphi(f): f \in H_\Gamma, |f|_\Gamma \leq 1\}$.

<u>Lemma 1.</u> If for every $a \in A$, $\Gamma \cdot a^{-1} = N(0,|a|_\Gamma)$, then $\Gamma \cdot \varphi^{-1} = N(0, |\varphi|_\Gamma)$ for every $\varphi \in (C_o(0,T,E))'$. Consequently Γ is a Gaussian measure on $C_o(0,T,E)$ and $\mathrm{RKSH}(\Gamma) = (H_\Gamma, |\cdot|_\Gamma)$.

<u>Proof.</u> Let U be the unit ball of E and let S be a dense countable subset of $E \setminus U$. For every $s \in S$ let s' be an element of E' of norm one such that $s'(s) = \|s\|$.

The unit ball of $C_o(0,T,E)$ is equal to

$$\bigcap_{s\in S} \bigcap_{t\in Q\cap[0,T]} \{F \in C_o([0,T],E) : |s'(F(t))| \leq 1\} ,$$

where Q denotes the set of rational numbers.

Thus the σ-algebra generated by A coincides with the Borel σ-algebra of $C_o([0,T],E)$.

Let X be a $C_o([0,T],E)$-valued random variable with $\mathcal{L}(X) =$ and let Y be an independent copy of x. If $\alpha^2 + \beta^2 = 1$, then for every $a \in A$

$$\mathcal{L}(a(\alpha X + \beta Y)) = N(0, \alpha|a|_\Gamma) * N(0, \beta|a|_\Gamma) =$$

$$= N(0,|a|_\Gamma) = \mathcal{L}(a(X)) \ .$$

Therefore $\mathcal{L}(X) = \mathcal{L}(\alpha X + \beta Y)$ as cylinder measures on $C_o([0,T],E)$ endowed with the locally convex topology generated by A. Hence, $\mathcal{L}(X)$ and $\mathcal{L}(\alpha X + \beta Y)$ coincide as Radon measures on $C_o([0,T],E)$.

Now, take $\varphi \in (C_o([0,T],E))'$ and α, β, X, Y as above. Then $\mathcal{L}(\alpha\varphi(X) + \beta\varphi(Y)) = \mathcal{L}(\varphi(\alpha X + \beta Y)) = \mathcal{L}(\varphi(X))$. In other words, $\Gamma \cdot \varphi^{-1}$ is a symmetric 2-stable distribution on R, hence Gaussian.

It remains to prove that $(C_0([0,T],E))'$ and A have the same closure in $L_2(\Gamma)$.

Indeed, suppose that $\varphi \notin \bar{A}$, then there is $\varphi_0 \in \bar{A}$ such that $\psi = \varphi - \varphi_0 \perp \bar{A}$. Since $(C([0,T],E))'$ is a linear set of symmetric Gaussian r.v., ψ is independent of the σ-algebra generated by A. It follows that $\psi = 0$, i.e. $\varphi = \varphi_0$ Γ-almost everywhere. A contradiction completes the proof.

3. Large deviations estimates for Gaussian measures.

With the notation of Section 2 define for $r > 0$, $a \in E$, $B \in \mathcal{B}(E)$

$$K_r = \{ x \in H_\gamma \; ; \; |x|_\gamma \le r \}, \quad d(a,B) = \inf \{ \| a-x \| , \; x \in B \} .$$

Let moreover γ_ε be the probability law of

$$\varepsilon \sum_{i=1}^{+\infty} e_i \xi_i \quad : \quad \mathcal{L}(\varepsilon \sum_{i=1}^{+\infty} e_i \xi_i) = \gamma_\varepsilon \; .$$

The following theorem holds, compare [11].

Theorem 2. (i) For arbitrary $\delta > 0$, $\eta > 0$ and $c > 0$ there exists $\varepsilon_0 > 0$ such that for every $h \in H_\gamma$ with $|h|_\gamma \le c$ and $0 < \varepsilon < \varepsilon_0$

$$\gamma_\varepsilon(\{ x \in E \; ; \; \| x-h \| \le \delta \}) \ge \exp(-\frac{1}{2\varepsilon^2}(|h|_\gamma + \eta))$$

(ii) For arbitrary $\delta > 0$, $\eta > 0$ and $r_0 > 0$ there exists $\varepsilon_0 > 0$ such that for $0 < r < r_0$ and $0 < \varepsilon < \varepsilon_0$:

$$\gamma_\varepsilon(\{ x \in E; \; d(x,K_r) \ge \delta \}) \le \exp(-\frac{1}{2\varepsilon^2}(r^2 - \eta)) \; .$$

Proof (i). For $a \in E$ and $r > 0$ let $K(a,r) = \{ x \in E; \; \| x-a \| \le r \}$.

We shall prove the following estimate which is sharper than the required one: for $h \in H_\gamma$ and $r > 0$

$$(6) \qquad \gamma(K(h,r)) \ge \gamma((K(0,r)) \cdot \exp(-\tfrac{1}{2}|h|_\gamma^2) \; .$$

By the Cameron-Martin (cf. [14]) and the symmetry of γ we have:

$$\gamma((K(h,r)) = (\gamma * \delta_{-h})(K(0,r)) =$$

$$= \int_{K(0,r)} e^{(-\frac{1}{2}|h|_\gamma^2 - \langle h,x\rangle_\gamma)} \gamma(dx) =$$

$$= e^{-\frac{1}{2}|h|_\gamma^2} \cdot \int_{K(0,r)} e^{-\langle h,x\rangle_\gamma} \gamma(dx) =$$

$$= e^{-\frac{1}{2}|h|_\gamma^2} \cdot \int_{K(0,r)} \frac{1}{2}(e^{-\langle h,x\rangle_\gamma} + e^{\langle h,x\rangle_\gamma})\gamma(dx)$$

$$e^{-\frac{1}{2}|h|_\gamma^2} \cdot \gamma((K(0,r)) .$$

This completes the proof of (6) and (i) .

Proof of (ii). Let us remark that:

$$\gamma_\varepsilon(\{x\in E;\ d(x,K_s) > \delta\}) \le \mathbb{P}(\varepsilon\cdot\sum_{i=1}^{n} e_i\,\xi_i \notin K_s) +$$

$$+ \mathbb{P}(\|\ \varepsilon\cdot\sum_{i=n+1}^{\infty} e_i\,\xi_i\| > \delta) = \mathbb{P}(\sum_{i=1}^{n} \xi_i^2 > \frac{s^2}{\varepsilon^2}) +$$

$$+ \mathbb{P}(\|\sum_{i=n+1}^{\infty} e_i\,\xi_i\| > \frac{\delta}{\varepsilon}) \le (1-2\lambda)^{-\frac{n}{2}} \exp(-\frac{\lambda s^2}{\varepsilon^2}) +$$

$$+ \mathbb{E}\exp(r\|\sum_{n+1}^{\infty} e_i\,\xi_i\|^2)\exp(-\frac{r\delta^2}{\varepsilon^2}) = S_1 + S_2 .$$

We will estimate S_1 and S_2 separately.

Putting $\lambda = \frac{1-\alpha}{2}$ into S_1 we get

$$S_1 = \alpha^{-\frac{n}{2}} \exp(-\frac{1}{2\varepsilon^2}(s^2 - \alpha s^2)) =$$

$$= \alpha^{-\frac{n}{2}} \exp(-\frac{\alpha s^2}{\varepsilon^2})\exp(-\frac{1}{2\varepsilon^2}(s^2 - 2\alpha s^2)) .$$

Then for $\alpha = \frac{\eta}{2s^2}$

$$S_1 \le (\frac{2s_o^2}{\eta})^{\frac{n}{2}} \exp(-\frac{\eta}{2\varepsilon^2})\exp(-\frac{1}{2\varepsilon^2}(s^2-\eta))$$

$$\frac{1}{2}\exp(-\frac{1}{2\varepsilon^2}(s^2-\eta)) \quad \text{for} \quad \varepsilon < \xi(\eta, s_o),\ s < s_o .$$

To go further we need the following proposition which proof is implicitely contained in [3].

<u>Proposition 1.</u> Let E be a separable Banach space and let (X_n) be a sequence of symmetric Gaussian E-valued r.v.

If $\|X_n\| \to 0$ in probability, then for every $r \in R$

$$\lim_{n\to\infty} \mathbb{E}\exp(r\|X_n\|^2) = 1 .$$

<u>Proof</u>. If $r \le 0$ there is nothing to prove. For $r > 0$, fix $\varepsilon > 0$. Choose δ such that $e^{r\delta^2} = 1 + \varepsilon$.

Then $\mathbb{E}\exp(r\|X_n\|^2) \le 1 + \varepsilon + \int_{\|X_n\|>\delta} e^{r\|X_n\|^2} d\mathbb{P} = 1 + \varepsilon +$

$+ 2r\int_{\delta}^{\infty} te^{rt^2}\mathbb{P}(\|X_n\| > t)dt$.

By Fernique inequality (cf. [12]); if X is an E-valued Gaussian symmetric r.v. and $t \ge 4s \ge 0$, then

$$\mathbb{P}(\|X\| > t) \le \exp(\frac{t^2}{16s^2}\ln\alpha), \quad \text{where} \quad \alpha = \frac{P(\|X\| > s)}{P(\|X\| \le s)} .$$

Hence, to get $\mathbb{E}\exp(r\|X_n\|^2) \le 1 + 2\varepsilon$ it is enough to take $s = \frac{\delta}{4}$ and n such that

$$|\frac{\ln\alpha}{16s^2}| = |\frac{\ln\alpha}{\delta^2}| \quad \text{is sufficiently large.}$$

This completes the proof of the proposition.

By Proposition 1 :

$$S_2 = c_{r,n}\cdot\exp(-\frac{r\delta^2}{2}) ,$$

where $c_{r,n} \to 1$ as $n \to \infty$ and r is fixed.

Now fix δ, η and s_o. Choose $n = n(s_o, \delta)$ such that for $r = \frac{s_o^2}{2\delta^2}$, $c_{r,n} \leq 2$.

Then

$$S_2 \leq 2\exp(-\frac{s_o^2}{2\varepsilon^2}) \leq 2\exp(-\frac{s^2}{2\varepsilon^2})$$

$$\leq \exp(-\frac{1}{2\varepsilon^2}(s^2 - \eta)) \cdot 2\exp(-\frac{\eta}{\varepsilon^2}) .$$

Observe that $2\exp(-\frac{\eta}{\varepsilon^2}) \leq \frac{1}{2}$ for $\varepsilon < \varepsilon_o(\eta)$, hence $S_2 \leq \frac{1}{2}\exp(-\frac{1}{2\varepsilon^2}(s^2 - \eta))$ for $\varepsilon < \varepsilon_o(\eta, \delta)$ and finally,

$$S_1 + S_2 \leq \exp(-\frac{1}{2\varepsilon^2}(s^2 - \eta)) \quad \text{for } \varepsilon < \varepsilon_o(\delta, \eta, s_o) \text{ and } s < s_o .$$

This completes the proof of (ii).

4. The estimates for semilinear equations.

Let γ be a symmetric full Gaussian measure on a separable Banach space E_o and H its reproducing kernel. By the Kolmogoroff's extension theorem there exists an E_o-valued process W_t on $[0,T]$ with the following properties:

(a) $W_o = 0$

(b) $\mathcal{L}(h^{-\frac{1}{2}}(W_{t+h} - W_t)) = \gamma$, for $t, t+h \in [0,T]$

(c) W_t has independent increments

In particular, $\mathcal{L}(W_1) = \gamma$. Thus will be called the unit distribution of W_t. The existence of a version with continuous paths is a consequence of the following

Proposition 2. (see [6]). Let X_t, $t \in [a,b]$ be a process with values in a complete metric space (E_o, δ). If there exist positive constants c, α, β such that for $s,t \in [a,b]$:

$$E(\rho(X_t, X_s))^\beta \leq c|t-s|^{1+\alpha} ,$$

then X_t has a version with continuous sample paths.

Let E be also a separable Banach space and let F be a Lipschitz transformation from E into E. We will need the following lemma in the formulation of which $S(t)$, $t \geq 0$ stands for a C_0-semigroup of linear operators on E.

Lemma 2. For arbitrary $\xi \in C[0,T;E]$ there exists a unique function $y = \mathcal{K}(\xi) \in C[0,T;E]$ such that

$$(7) \qquad y(t) = \xi(t) + \int_0^t S(t-s)F(y(s))ds, \qquad t \in [0,T]$$

Moreover there exists a constant $C_T > 0$ such that for arbitrary $\xi, \eta \in C[0,T;E]$:

$$\sup_{t\in[0,T]} \| \mathcal{K}(\xi)(t) - \mathcal{K}(\eta)(t)\| \leq C_T \sup_{t\in[0,T]} \|\xi(t) - \eta(t)\|$$

Proof. The proof is classical. We show, for example the validity of the latter part of the lemma. There exist positive constants $K > 0$, $M > 0$, $\omega > 0$ such that

$$\|F(x)-F(y)\| \leq K\|x-y\| \quad \text{and} \quad \|S(t)\| \leq M \exp t\omega, \quad x,y \in E,\ t \geq 0.$$

If $z = \mathcal{K}(\eta)$ then:

$$e^{-\omega t}\|y(t)-z(t)\| \leq e^{-\omega t}\|\xi(t)-\eta(t)\| + MK\int_0^t e^{-\omega s}\|y(s)-z(s)\|ds$$

Consequently, by Gronwall's lemma:

$$e^{-\omega t}\|y(t)-z(t)\| \leq e^{MKt} \sup_{s\leq t} \|\xi(s)-\eta(s)\|, \quad t \in [0,T]$$

and therefore as C_T one can take $\exp(MK+\omega)T$.

By solutions $X = X^{x,\varepsilon}$ and $y = y^{x,u}$ of the infinite dimensional versions of the equations (1) and (2) we will understand solutions of the following integral equations:

$$(8) \quad X(t) = S(t)x + \int_0^t S(t-s)F(X(s))ds + \varepsilon\int_0^t S(t-s)BdW_s,$$

$$(9) \quad y(t) = S(t)x + \int_0^t S(t-s)F(y(s))ds + \int_0^t S(t-s)Bu(s)ds, \quad t \in [0,T].$$

To give a precise meaning to the equation (8) we need a definition of the stochastic integral

$$Z_t = \int_0^t S(t-s)BdW_s \ , \qquad t \in [0,T] \ .$$

An example by Yor [19] shows that even for a deterministic bounded Banach space-valued integrand and one dimensional Brownian motion the stochastic integral may fail to exist. Let us notice that the stochastic integral is certainly well defined if E_o and E are Hilbert spaces, see [16]. Let us observe that if we define

$$(10) \qquad \int_0^t f(s)dW_s = \sum_{k=0}^{n-1} a_k'(W_{t_{k+1}} - W_{t_k})$$

for a E_o'-valued step function $f(s) = \sum a_k' \chi_{]t_k,t_{k+1}]}(s)$

$0 = t_o < t_1 < \dots < t_n = t$, then

$$\mathbb{E}(\int_0^t f(s)dW_s)^2 = \int_0^t |f(s)|_\gamma^2 \, ds \quad .$$

Thus the definition can be extended to $f \in L_2([0,t],H_\gamma)$. So it is reasonable to say that $\int_0^t S(t-s)BdW_s$ exists if there exists an E-valued random variable Z_t such that for every $e' \in E'$

$$e'(Z_t) = \int_0^t B'S'(t-s)e'dW_s$$

Let us observe that for $f \in L_2([0,t],H_\gamma)$

$$\mathcal{L}(\int_0^t f(s)dW_s) = N(0, \int_0^t |f(s)|_\gamma^2 \, ds)$$

Thus, for every t, Z_t is a Gaussian E-valued r.v.

With the above explanations in mind we impose the following condition:

(11) The stochastic integrals

$$Z_t = \int_0^t S(t-s)BdW_s \ , \quad t \in [0,T]$$

exist and Z_t is a continuous E-valued process.

<u>Remark</u>. Even if E_o is a Hilbert space the continuity of Z_t is,

for general semigroups S(t), still an open problem see [13]. Some sufficient conditions for continuity are given in [13].

If the condition (11) is satisfied then

$$X^{x,\varepsilon}(t) = \mathcal{K}(S(\cdot)x + \varepsilon Z.)(t), \quad t \in [0,T]$$

and if $u(\cdot)$ is a square integrable function with values in $H_\gamma \subset E_0$ then

$$y^{x,u}(t) = \mathcal{K}(S(\cdot)x + \int_0^{\cdot} S(\cdot - s)Bu(s)ds)(t), \quad t \in [0,T] .$$

Here is the infinite dimensional version of Theorem 1.

Theorem 3. Assume that (5) and (11) hold.

(i) For arbitrary $\delta > 0$, $\eta > 0$ and $c > 0$ there exists $\varepsilon_0 > 0$ such that for all $x \in E$, u satisfying $\int_0^T |u(s)|_\gamma^2 ds \leq c$ and $0 < \varepsilon < \varepsilon_0$

$$\mathbb{P}\left(\sup_{t \in [0,T]} \| X^{x,\varepsilon}(t) - y^{x,u}(t)\| < \delta\right) \geq \exp - \frac{1}{2\varepsilon^2}\left(\int_0^T |u(s)|_\gamma^2 ds + \eta\right)$$

(ii) For arbitrary $\delta > 0$, $\eta > 0$ and $r_0 > 0$ there exists $\varepsilon_0 > 0$ such that for arbitrary $x \in E$, $0 < r < r_0$ and $0 < \varepsilon < \varepsilon_0$:

$$\mathbb{P}(\text{distance}(X^{x,\varepsilon}, K(x,r)) > \delta) \leq \exp - \frac{1}{2\varepsilon^2}(r^2 - \eta)$$

where $K(x,r)$ stands for the set of functions $y^{x,u}$ corresponding to all u satisfying

$$\int_0^T |u(s)|_\gamma^2 \, ds \leq r^2 .$$

Proof. The Lipschitzianity of the transformation $\mathcal{K}$ allows to reduce the proof to the linear case ($F \equiv 0$) and to zero initial condition, see [5]. Taking into account Theorem 2 it is enough to show that the distribution Γ of the process Z_t, $t \in [0,T]$ is a Gaussian measure on $C_0([0,T],E)$ and the reproducing kernel H_Γ consists of all functions $y(t) = \int_0^t S(t-s)Bu(s)ds$ $t \in [0,T]$ corresponding to $u \in L^2[0,T;H_\gamma]$ with the norm $|y|_\Gamma$:

$$|y|_\Gamma = \inf\{\int_0^T |u(s)|_\gamma^2\, ds \;;\; y(t) = \int_0^t S(t-s)Bu(s)ds, \quad t \in [0,T]\}.$$

We will use lemma 1.

Let $a = \sum_{i=1}^n e_i' \otimes \delta_{t_i}$ be an element of A. Then

$$\Gamma\circ a^{-1} = \mathcal{L}(\sum_{i=1}^n \int_0^{t_i} B'S'(t_i-s)e'dW_s).$$

We can assume that $t_1 < t_2 < \ldots < t_n$. Now

$$\Gamma\circ a^{-1} = \mathcal{L}(\int_0^{t_1} \sum_{i=1}^n B'S'(t_i-s)e_i'\, dW_s + \ldots + \int_{t_{n-1}}^{t_n} B'S'(t_n-s)e_i'dW_s)$$

$$= \mathcal{L}(\int_0^{t_1} \sum_{i=1}^n B'S'(t_i-s)e_i'dW_s) * \ldots * \mathcal{L}(\int_{t_{n-1}}^{t_n} B'S'(t_n-s)e_i'dW_s$$

$$= N(0, \int_0^{t_1} |\sum_{i=1}^n B'S'(t_i-s)e_i'|_\gamma^2\, ds) * \ldots * N(0, \int_{t_{n-1}}^{t_n} |B'S'(t_n-s)e_i'|_\gamma^2 ds)$$

$$= N(0, \int_0^T |\sum_{i=1}^n B'S'(t_i-s)e_i'\chi_{[0,t_i]}(s)|_\gamma^2\, ds)$$

On the other hand $|a|_\Gamma = \sup\{a(y) : y \in H_\Gamma, \quad |y|_\Gamma \le 1\}$.

Let $y(t) = \int_0^t S(t-s)Bu(s)ds$, $t \in [0,T]$. Then

$$a(y) = \sum_{i=1}^n \int_0^{t_i} B'S'(t_i-s)e_i'\, u(s)ds = \int_0^T (\sum_{i=1}^n B'S'(t_i-s)e_i'\chi_{[0,t_i]}(s))\cdot$$

$$\cdot\, u(s)ds.$$

It is elementary that

$$|a|_\Gamma = \int_0^T |\sum_{i=1}^n B'S'(t_i-s)e_i'\,\chi_{[0,t_i]}(s)|_\gamma^2\, ds.$$

Thus, by the lemma 1 Γ is Gaussian and $RKHS(\Gamma) = H_\Gamma$, which completes the proof.

5. Comments. The Lipschitz condition imposed on F is natural if one studies the exit problem from a bounded neighbourhood of an equilibrium point, see [20]. However for some other applications it is too restrictive, see [10] . The assumption that F is everywhere defined is not satisfied in some important cases, see [9] . It would be therefore desirable to extend Theorem 3 to the case of holomorphic semigroups S(t), $t \geq 0$ and F locally Lipschitz transformations defined on the domain $D(-\tilde{A})^{\alpha}$ of a fractional power $(-\tilde{A})^{\alpha}$ of the generator $\tilde{A}$, (α a positive number), see [9] .

References

[1] R.G.Azencott, Sur les grand deviations, Ecole d'Ete de Probabilite, Saint Flour (1978), LNiM 774.

[2] R.M.Dudley, I.Feldman and L.LeCam, On semi-norms and probabilities, and abstract Wiener spaces, Ann. Math. 93(1971), 390-408.

[3] M.X.Fernique, Integrabilité des vecteurs gaussiens. C.R. Acad. Sc. Paris, Serie A, 270(1970), 1698-1699.

[4] M.Freidlin and A.Wentzell, On small random perturbations of dynamical systems, Russian Math. Surveys, 25(1970), 1-55.

[5] M.Freidlin and A.Wentzell,Random perturbations of dynamical systems, Springer, Berlin New York, 1984.

[6] I.I.Gikhman and A.W.Skorokhod, Stochastic Processes, vol.1, Nauka, Moscow 1973.

[7] L.Gross, Measurable functions on Hilbert space, TAMS 105(1962), 372-390.

[9] D.Henry, Geometric theory of semilinear parabolic equations, Springer-Verlag, 1981.

[10] G.Jona-Lasinio and W.G.Faris, Large fluctuations for a non-linear heat equation with noise, Journal of Physics, A. Math. Gen., 15(1982), 441-459

[11] G.Kallianpur and H.Oodaira, Freidlin-Wentzell type estimates for Abstract Wiener Spaces; Sankhya, vol. 40, 1978, Series A, Pt. 2, 116-137.

[12] M.Kanter, Linear sample spaces and stable processes, J. Functional Analysis, 9(19), 441-459.

[13] P.Kotelenez, A submartingale type inequality with applications to stochastic evolution equations, Stochastics, 8(1982), 139-151

[14] H.H.Kuo, Gaussian measures in Banach spaces, LNiM 463, 1975.

[15] R.Le Page, Loglog law for Gaussian processes, ZW, 25(1973), 103-108.

[16] M.Metivier and J.Pellaumail, Stochastic Integration, Academic Press, 1980.

[17] H.Oodaira, Note on Freidlin-Wentzell Type estimates for stochastic processes, Proceedings, Measure Theory Applications to Stochastic Analysis, Eds. G.Kallianpur and D.Kölzow, LNiM 695, 1978, 145-154.

[18] W.Smoleński, R.Sztencel and J.Zabczyk, Large deviations estimate for Banach valued Wiener process, Preprint of the Institute of Mathematics, Polish Academy of Sciences, Warsaw, 1986.

[19] M.Yor, Sur les integrales stochastiques a valeurs dans un espace de Banach, C.R. Acad. Sc. Paris, t. 277, (17 septembre 1973).

[20] J.Zabczyk, Exit problem for infinite dimensional systems, Proceedings of a Workshop on Stochastic PDE's and Applications, Eds. G.Da Prato and L.Tubara, Trento, Italy, 1985.

CONTINUOUS DEPENDENCE FOR ITO EQUATIONS WITH RESPECT TO THE DRIFT INVOLVING LIE BRACKETS

C. VARSAN

Department of Mathematics, INCREST, Bd. Păcii 220, 79622 Bucharest, ROMANIA

Introduction

The problem we are concerned is a nonstandard continuous dependence for stochastic differential equations with respect to the drift coeffients. Roughly speaking it can be stated as follows. We are given a finite set of smooth functions $g_i(t,x):[0,T]xR^n\to R^n$, $i=1,\dots,m$, and denote $L(g_1,\dots,g_n)$ the Lie algebra generated by them, where $[g_i,g_j](t,x)= =((\partial g_j/\partial x)g_i-(\partial g_i/\partial x)g_j(t,x)$. Take $h_1,\dots,h_l \varepsilon L(g_1,\dots,g_m)$ and denote $y(\bullet)$ the solution of the Ito equation

$$(*)\quad dy=f(t,y)dt + [\sum_{i=1}^{\ell} u_i(t)h_i(t,y)]dt + \sum_{k=1}^{d}\sigma_k(t,y)dw_k(t), y(0)=x_0, t\varepsilon[0,T],$$

where f, $\sigma_k:[0,T]xR^n\to R$ are fixed. Along with (*) we consider

$$(**)\quad dx=f(t,x)dt+[\sum_{i=1}^{m} v_i(t)g_i(t,x)]dt+\sum_{k=1}^{d}\sigma_k(t,x)dw_k(t), x(0)=x_0, t\varepsilon[0,T]$$

where f, g_i, σ_k are the given functions.

The problem we answer is to approximate the solution in (*) by solutions in (**) using the usual metric $d(x(\bullet), y(\bullet))=(E \max_{t\varepsilon[0,T]} |x(t)-y(t)|^2)^{1/2}$. It can be done by defining an appropriate sequence of functions $\{v_i^h(\bullet)\}$ such that the corresponding solutions $x^h(\bullet)$ in (**) fulfils the goal; the sequence $\{v_i^h(\bullet)\}_{h>0}$ is unbounded with respect to h and since the pointwise convergence of the drift term in (**) to the drift term in (*) is meaningless we need a nonstandard approach.

This result is connected with the controllability properties of deterministic control systems as it appears in [1] and [2] and the techniques used here originate in [3]. In the stochastic case it completes the result in [4] by considering Lie brackets in the drift part.

In particular, it follows that if $g_j=\sigma_i$, $i=1,\dots,m$, $m=d$, then the support of the measure π_u on $C([0,T];R^n)$ generated by the solution $y_u(\bullet)$ in (*) is an invariant under the transformations of the drift f performed in (*); it equals the support of the measure π_0 generated by the solution in (*) which corresponds to $U_i=0$, $i=1,\dots,l$. It can be be seen using Girsanov's transformation of the probability measure in (**).

In deterministic case ($\sigma_k=0$, $k=1,\dots,d$) it gives the possibility to study controllability of the system (**) along a fixed trajectory via the enlarged system (*'). It can be stated more precisely as follows. Denote $\tilde{x}(t)$, $t\in[0,T]$, the solution in (**) and (*) which corresponds to $u_i=0$, $i=1,\dots,\ell$, and $v_j=0$, $i=1,\dots,m$, respectively. Suppose that dim span $\{\tilde{ad}^{(k)} f(h)\ (0,x_0): h\in L(g_1,\dots,g_m),\ k=0,1,2,\dots\}=n$ where $\tilde{ad}\ f(h)(t,x)=[f,h]\ (t,x)+\partial h/\partial t\ (t,x)$ and $[f,h]$, $L(g_1,\dots,g_m)$ are defined as above. Then for each $t\in(0,T]$ there exists a sphere $S(\tilde{x}(t),\rho_t)$ centered at $\tilde{x}(t)$, such that the initial point x_0 is steered to any point in $S(\tilde{x}(t),\rho_t)$ in time t by using bounded controls $u(t)\in U$ and trajectories in (*); the same property holds for the reduced system (**) but the control we have to use cannot be restricted to belong to the same set U. In our setting the controllability of the system (*) along $\tilde{x}(\bullet)$ at time t=T is preserved even if we restrict ourselves to the class of controls fulfilling $u_j(0)=u_i(T)$, $i=1,\dots,l$, $v_i(0)=v_j(T)$ $j=1,\dots,m$. The result in Theorem remains the same in the case that the Wiener process $w(\bullet)$ in (*) and (**) is replaced by a continuous square integrable martingale for which the quadratic variation matrix $V(t)=\langle M(t),\ M(t)\rangle$ has the form $V(t)=\int_0^t H(s,\omega)ds$ with H a bounded measurable matrix valued process. The invariance of the support of the measure π_u can be proved under the assumption that H is nonsingular and $H^{-1}(s,\omega)$ is bounded and measurable.

1. Formulation of the problem and main result

Denote $C_b^{\ell,p}([0,T\ xR^n)$ the space consisting of real functions which are continuously differentiable up to order ℓ with respect to $t\in[0,T]$, up to order p with respect to $x\in R^n$ and are bounded along with all their derivatives; if the boundedness condition is omitted we denote it by $C^{\ell,p}([0,T]xR^n)$. We are given $f,g_i,\sigma_k:[0,T]xR^n\to R^n$ which are continuous and $g_i\in C^{1,\infty}([0,T]xR^n)$, $f,\sigma_k\in C^{0,2}([0,T]xR^n)$, $i=1,\dots,m$ $k=1,\dots,d$. For $i_o,i_1\in\{1,\dots,m\}$ and $I=\{i_o,i_1\}$ define $|I|=2$, $g_I(t,x)=[g_{i_o},g_{i_1}](t,x)$, where $[g_i,g_j]/(t,x)$ denote the Lie bracket with respect to $x\in R^n$; generally for i_o, $i_1,\dots,i_L\in\{1,\dots,m\}$ and $I=\{i_o,\ i_1,\dots,i_L\}$, define $|I|=L+1$ and $g_I(t,x)=[g_{i_o},g_{I_1}]/(t,x)$, where $I_1=\{i_1,\dots,i_L\}$, $L\in N$. For each $u_i\in C([0,T];R)$, $i=1,\dots,m$, $u_I\in C^1([0,T];R)$, $2\le|I|\le L+1$, we consider the following stochastic differential equation

$$1)\quad dy=[f(t,y)+\sum_{i=1}^{m} u_i(t)g_i(t,y)+\sum_{|I|=2}^{L+1} u_I(t)g_I(t,y)]dt+\sum_{k=1}^{d}\sigma_k(t,y)dw_k(t)$$

$$y(0)=x_o,\ x_o\varepsilon R^n,\ t\varepsilon[0,T],$$

where $w(t)$, $t\varepsilon[0,T]$, is a standard Wiener process over the filtered probability space $\{\Omega,F,P;F_t\}$. We associate with (1) the following stochastic differential equation

$$2)\quad dx=[f(t,x)+\sum_{i=1}^{m} u_i(t)g_i(t,x)+\sum_{i=1}^{m} v_i(t)g_i(t,x)]dt+\sigma(t,x)dw(t)$$

$$x(0)=x_o,\ t\varepsilon[0,T],$$

where $\sigma=(\sigma_1,\dots,\sigma_d)$, f,g_i,σ_k,u_i are as in (1) and $v_i\varepsilon C^1([0,T];R)$, $v_i(0)=v_i(T)=0$.

We need the following conditions to be fulfilled

C_1) $\partial f/\partial x_j$, $\partial\sigma_k/\partial x_j\varepsilon C_b^{0,1}([0,T]xR^n)$, $g_i\varepsilon C_b^{1,L+2}([0,T]xR^n)$

C_2) $(\partial g_I/\partial x)f^*$, $(\partial g_I/\partial x)\sigma_k^*\varepsilon C_b([0,T]xR^n)$, $(\partial^2 g_I/\partial x_i\partial x_j)A_{ij}\varepsilon C_b([0,T]xR^n)$

for any $1\le|I|\le L$, $k=1,\dots,d$ $i,j\varepsilon\{1,\dots,m\}$, where $A=\sigma\sigma^*$, "v^*" is the transposed of v, and a vector or a matrix belogs to $C_b^{1,p}$ if all their components fulfil it.
In the case $f,\sigma_k\varepsilon C_b^{0,2}$ and $g_i\varepsilon C_b^{1,L+2}$ then the conditions (C_1) and (C_2) are fulfilled.

<u>Theorem</u>

Assume (C_1) and (C_2) are fulfilled for (1) and let $y(\cdot)$ be the solution in (1) corresponding to $u_i(\cdot),u_I(\cdot),2\le|I|\le L+1$, and $y(0)=x_0$. Then there exists a sequence $\{x^h(\cdot)\}_{h>0}$ of solutions in (2) such that

$$\lim_{h\to 0} E \max_{t\varepsilon[0,T]} |x^h(t)-y(t)|^2=0, \text{ uniformly with respect to}$$

x_o, $u_i(\cdot)$ and $u_I(\cdot)$ in bounded sets.

Remark 1

If we consider $u_i(\cdot)$ and $u_I(\cdot)$ in (1) as functions of (t,y) and fulfilling $u_i(\cdot)\varepsilon C_b^{0,1}[0,T]xR^n)$, $u_I(\cdot)\varepsilon C_b^{1,2}([0,T]xR^n)$ then the theorem remains unchanged except that in this case $v_i^h(\cdot)$ in (2) which determine $x^h(\cdot)$ are functions of (t,x) fulfilling

$v_i^h(\bullet)\varepsilon C_b^{1,1}([0,T]xR^n)$, $v_i^h(0,x)=v_i^h(T,x)=0$, $x\varepsilon R^n$.

Remark 2

If we relax the hypotheses in the Theorem by neglecting (C_2) and replacing $g_j\varepsilon C_b^{1,L+2}([0,T]xR^n)$ in (C_1) by $\partial g_j/\partial x_k\varepsilon C_b^{1,L+1}([0,T]xR^n)$ then using a standard argument of truncation it follows that there exists a sequence $\{x^h(\bullet)\}_{h>0}$ of solutions in (2) such that with probability one

$$\lim_{h\to 0}\max_{t\varepsilon[0,T]} |x^h(t)-y(t)|=0$$

uniformly with respect to x_o, $u_i(\bullet)$, $u_I(\bullet)$ in bounded sets.

Remark 3

Since $v_i^h(\bullet)$ and $u_i(\bullet)$ in (2) are continuous the theorem has an analogous version for the corresponding backward parabolic equations associated with (1) and (2).

2 Some auxiliary results and proof of the Theorem. We associate with (1) the maximal number L of the Lie brackets contained in (1) and call it the order of the system. To prove Theorem we need to approximate the solution in (1) by one determined by a system which has an order less than L.

It is done in the next Lemma by using the following approximate equation.

Denote $\tilde{f}(t,y)=f(t,y)+\sum_{i=1}^{m}u_i(t)g_i(t,y)+\sum_{|I|=2}^{L}u_I(t)g_I(t,y)$ and (1) is rewritten as

$$3)\quad dy=\{\tilde{f}(t,y)+\sum_{i=1}^{m}\sum_{j=1}^{\tilde{m}}U_{ij}(t)[g_i,b_j](t,y)\}dt+\sigma(t,y)dw(t),$$

$y(0)=x_o$, $t\varepsilon[0,T]$, where the second term in the right hand side of (3) is the same with the third term in (1).

Let N be a natural number. We consider a partition π_o of $[0,T]$ determined by the intervals $[kh,(k+1)h)$ $k=0,1,\dots,N-1$, with $|\pi_o|=h=T/N$. For each $k\varepsilon\{0,1,\dots,N-1\}$, let A_{ij}^k, $i=1,\dots,m$, $J=1,\dots,\tilde{m}$ be a partition of $[kh,(k+1)h)$ with $|A_{ij}^k|=h_1=h/m\tilde{m}$. Denote P^o the space consisting of scalar polynomial functions defined on $[0,1]$ and fulfilling $\int_o^1 p(t)dt=0$. Let $p_1(\bullet),p_2(\bullet)\varepsilon P^o$ be such that $p_i(0)=p_i(1)=0=dp_i/dt(0)=dp_i/dt(1)$ and $\int_o^1 p_2(t)\tilde{p}_1(t)dt=1$, where $\tilde{p}(t)=\int_o^t p(s)ds$. These functions will be fixed in the sequel and the could be chosen as polynomials of fifth and forth degree respectively. Let $p_i^k(t,h):[kh_1(k+1)h]\to R$, i=1,2, be defined by

$$p_i^k(t,h)=p_i(t-kh/h_1) \quad t\epsilon A_{11}^k=[kh,\ kh+h_1],\dots,$$
$$p_i^k(t,h)=p_i(t-(kh+(m\tilde{m}-1)h_1)/h_1), \quad t\epsilon A_{m\tilde{m}}^k=[kh,(k+1)h)$$
$$k=0,1,\dots,N-1.$$

Obviously $p_i^k(\bullet)\epsilon C^1([kh_1(k+1)g];R)$ and $p_i^k(kh)=p_i^k((k+1)h)=0$.
With (3) and partition π_o we associate the following differential equation of order L-1

$$4)\ dx=\{\tilde{f}(t,x)+\sum_{k=0}^{N-1}\ \sum_{i=1}^{m}\ \sum_{j=1}^{m}\ (m\tilde{m}/\sqrt{h_1})\ [p_1^k(t,h)U_{ij}(t)g_i(t,x)+p_2^k(t,h)b_j(t,x)]\,I_{ij}^k(t)\}dt$$
$$+\sigma(t,x)dw(t),\quad x(0)=x_o,\quad t\epsilon[0,T],\ \text{where } I_{ij}^k=\inf A_{ij}^k$$

Let $x^h(t)$, $t\epsilon[0,T]$, be the solution in (4) and denote

$$5)\quad M^h(t)=\int_o^t\{\sigma(s,x^h(s))-\sigma(s,y(s))+\sum_{i=1}^{m}\ \sum_{j=1}^{m}[\tilde{p}_1(s,h)\,(\partial/\partial x)\,(U_{ij}g_i)\sigma(s,x^h(s))$$
$$+\tilde{p}_2(s,h)\,(\partial b_j/\partial x)\sigma(s,x^h/s)]\}dw(s),\quad t\epsilon[0,T]$$

Let $\eta(r)$ be a random or a deterministic vector fulfilling $(E/\eta(r)/^2)^{1/2}\leq$ $\leq Cr$ for some fixed constant C>0.

Lemma 1

Assume that (C1) and (C2) are fulfilled. Let $x^h(\bullet)$ be the solution in (4) and $y(\bullet)$ fulfils (3). Then there exists a martingale $M^h(t)$, $t\epsilon[0,T]$ (see (5)) such that

$$x^h(t'')-x^h(t')=y(t'')-y(t')+M^h(t'')-M^h(t')+(t''-t')\eta(\sqrt{h})$$
$$t',t''\epsilon\{0,h,2h,\dots,(N-1)h=T\},\ t'<t'',\ \text{where}$$
$$(E|M^h(t'')-M^h(t')|^2)^{1/2}\leq\sqrt{(t''-t')}\eta(\sqrt{h})$$

uniformly with respect ot x_o, $u_i(\bullet)$, $u_I(\bullet)$, $2\leq|I|\leq L+1$ in bounded sets.

Proof

By definition

$$x^h(h_1)=x_0+\int_o^{h1}\tilde{f}(t,x^h(t))dt+(m\tilde{m}/\sqrt{h_1})\int_0^{h1}[p_1^0(t,h)u_{11}(t)g_1(t,x^h(t))+$$
$$+p_2^0(t,h)b_1(t,x^h(t)]dt+\int_0^{h1}\sigma(t,x^h(t))dw(t)=x_0+T_1+T_2+T_3.$$

By hypothesis $\tilde{f}$ and σ are Lipschitz continuous with respect to $x\in R^n$ and computation shows

$$6)\quad (E \max_{t\in[0,h]} |x^h(t)-y(t)|^2)^{1/2} \le (1-Ch)^{-1}\eta(\sqrt{h})$$

where $C>0$ is the Lipschitz constant for $\tilde{f}$ and σ. Using (6) we get

$$7)\quad T_1 = \int_0^{h1}\tilde{f}(t,y(t))dt + \int_0^{h1}[\tilde{f}(t,x^h(t))-\tilde{f}(t,y(t))]dt =$$

$$= \int_0^{h1}\tilde{f}(t,y(t))dt + h_1\eta(\sqrt{h})$$

$$8)\quad T_3 = \int_0^{h1}\sigma(t,y(t))dw(t) + \int_0^{h1}[\sigma(t,x^h(t))-\sigma(t,y(t))]dw(t) =$$

$$= \int_0^{h1}\sigma(t,y(t))dw(t) + M_1'(h_1), \text{ where } E|M_1'(h_1)|^2 \le h_1\eta(h).$$

Denote $\tilde{x}(s)=x^h(sh_1)$, $s\in[0,1]$, $A=\sigma\sigma^*$,

$$(Lu)(t,x) = [(\partial/\partial t + \sum_{i=1}^n \tilde{f}_i(t,x)/\partial/\partial x_i) + 1/2\sum_{i,j=1}^n A_{ij}(t,x)(\partial^2/\partial x_i\partial x_j)]u(t,x)$$

With these notations and using $p_i(\bullet)\in P^0$, $i=1,2$ we get

$$9)\quad T_2 = \tilde{m}\sqrt{h_1}\int_0^1[p_1(s)u_{11}(sh_1)g_1(sh_1,\tilde{x}(s)) + p_2(s)b_1(sh_1,\tilde{x}(s))]ds$$

$$= \tilde{m}h_1^{3/2}[\int_0^1 p_1(s)ds\int_0^s L(u_{11}g_1)(s_1h_1,\tilde{x}(s_1))ds_1 +$$

$$+ \int_0^1 p_2(s)ds\int_0^s (Lb_1)(s_1h_1,\tilde{x}(s_1))ds_1] +$$

$$+ \tilde{m}h_1[\int_0^1 p_1(s)ds\int_0^s (p_1(s_1)(\partial)/(\partial x)(u_{11}g_1)(s_1h_1,\tilde{x}(s_1)) +$$

$$+ p_2(s_1)(\partial)/(\partial x)(u_{11}g_1)b_1(s_1h_1,\tilde{x}(s_1))ds_1 +$$

$$+ \int_0^1 p_2(s)ds\int_0^s (p_1(s_1)(\partial b_1)/(\partial x)(u_{11}g_1)(s_1h_1,\tilde{x}(s_1)) +$$

$$+ p_2(s_1)(\partial b_1)/(\partial x)b_1(s_1h_1,\tilde{x}(s_1)))ds_1] +$$

$$+ (\tilde{m})/(\sqrt{h_1})\int_0^{h1}[\tilde{p}_1^0(t,h)(\partial/\partial x)(u_{11}g_1)\sigma(t,x^h(t)) + \tilde{p}_2^0(t,h)(\partial b_1)/(\partial x)\sigma(t,x^h(t))]dw(t) =$$

$$= T_2' + T_2'' + M_1''(h_1).$$

By hypothesis (see ((2)) we have

$$10)\quad |L(u_{11}g_1)(t,x^h(t))| + |Lb_1(t,x^h(t))| \le C_1, \quad t\in[0,T],$$ where $C_1>0$ is a constant wich doesn't depend on h, and using (10) in (9) we obtain

11) $T_2'=h_1\eta(\sqrt{h}), E|M_1''(h_1)|^2 \le h_1\eta(h)$

Since $\int_o^1 p_2(s)\tilde{p}_1(s)ds=-\int_o^1 p_1(s)\tilde{p}_2(s)ds=1$ and

$\int_o^1 p_i(s)(\tilde{p}_i(s))^j ds=0, i,j=1,2$ we get

12) $T_2''=m\tilde{m}h_1 u_{11}(0)[g_1,b_1](0, x_0)+h_1\eta(\sqrt{h})=$

$\int_o^h u_{11}(t)[g_1, b_1](t, y(t))dt+h_1\eta(\sqrt{h})$

Using (11) and (12) in (9) it follows

13) $T_2=\int_o^h u_{11}(t))[g_1, b_1](t, y(t))dt+h_1\eta(\sqrt{h})+M_1''(h_1)$

and from (7), (8) and (13) we get

14) $x^h(h_1)=x_0+\int_o^{h1}\tilde{f}(t,y)(t))dt+\int_o^{h1}u_{11}(t)[g_1, b_1](t,y(t))dt+$

$+\int_o^{h1}\sigma(t,y)(t)dw(t)+h_1\eta(\sqrt{h})+M_1(h_1)$ where

$M_1(h_1)=M_1'(h_1)+M_1''(h_1)$ fulfils

$E|M_1(h)^2|\le h_1\eta(h)$

On the next interval $[h_1, 2h_1]$ we repeat the computations for $[0,h_1]$
By definition

15) $x^h(2h_1)=x^h(h_1)+\int_{h_1}^{2h_1}\tilde{f}(t,x^h(t))dt+(m\tilde{m})/(\sqrt{h_1})\int_{h_1}^{2h_1}[p_1^0(t,h)u_{12}(t)g_1(t,x^h)(t))+$

$+p_2^0(t,h)b_2(t,x^h(t))dt+\int_{h_1}^{2h_1}\sigma(t,x^h(t))dw(t)=x^h(h_1)+\tilde{T}_1+\tilde{T}_2+\tilde{T}_3$

and we get easily

16) $\tilde{T}_1=\int_{h_1}^{2h_1}\tilde{f}(t,y)(t))dt+\int_{h_1}^{2h_1}[\tilde{f}(t,x^h(t))-\tilde{f}(t,y))]dt=$

$=\int_{h_1}^{2h_1}f(t,y(t))dt+h_1\eta(\sqrt{h})$

17) $\tilde{T}_3=\int_{h_1}^{2h_1}\sigma(t,y(t))dw(t)+\int_{h_1}^{2h_1}[\sigma(t,x^h(t))-\sigma(t,y(t))]dw(t)=$

$=\int_{h_1}^{2h_1}\sigma(t,y(t))dw(t)+M_2'(h_1)$

where

$$E|M_2'(h_1)|^2 \le h_1\eta(h), M_2'(h_1)=\int_{h_1}^{2h_1}[\sigma(t,x^h(t))-\sigma(t,y(t))]dw(t)$$

Similarly, repeating the computations in (9)-(11) we get

18) $\tilde{T}_2=\tilde{T}_2'+\tilde{T}_2''+M_2''(h_1)$

where

19) $\tilde{T}_2'=h_1\eta(h)$,

$$M_2''(h_1)=(m\tilde{m})/(\surd h_1)\int_{h_1}^{2h_1}[\tilde{p}_1^0(t,h)(\partial)/(\partial x)(u_{11}g_1)\sigma(t,x^h(t))+$$

$$+\tilde{p}_2^0(t,h)(\partial b_2)/(\partial x)\sigma(t,x^h(t))\; dw(t)$$

and $E|M_2''(h_1)|^2 \le h_1\eta(h)$.

Also, we have

20) $$\tilde{T}_2''=m\tilde{m}h_1u_{12}(h_1)[g_1,\; b_2](h_1,x^h(h_1))+h_1\eta(\surd h)=$$

$$=m\tilde{m}h_1u_{12}(0)[g_1,\; b_2](0,\; x_0)\; +\; h_1\eta(\surd h)=$$

$$=\int_0^h u_{12}(t)\,[g_1,\; b_2](t,y(t))dt+h_1\eta(\surd h)$$

Denote $M_2(h_1)=M_2'(h_1)+M_2''(h_1)$ and using (16)-(20) in (15) we get

21) $$x^h(2h_1)=x_0+\int_{h_1}^{2h_1}\tilde{f}(t,y(t))dt+\sum_{j=1}^{2}\int_0^h u_{ij}(t)[g_1,\; b_j](t,y(t))dt+$$

$$+\int_0^{2h_1}\sigma\,(t,y(t))dw(t)+2h_1\eta(\surd h)+M_1(h_1)+M_2(h_1)$$

where $M_1(h_1)$ is defined in (14), and

22) $E|M_1(h_1)+M_2(h_1)|^2=E|M_1(h_1)|^2+E|M_2(h_1)|^2 \le 2h_1\eta(h)$

Finally, for t=h, we get $M_i(h_1), i=1,\ldots,m\tilde{m}$, such that

23) $$x^h(h)=x^h(m\tilde{m}h_1)=x_0+\int_0^h\tilde{f}(t,y(t))dt+\sum_{i=1}^{m}\sum_{j=1}^{\tilde{m}}\int_0^h u_{ij}(t)[g_i,b_j](t,y(t))dt+$$

$$+\int_0^h\sigma(t,y(t))dw(t)+h\eta(\surd h)+M_1(h)=y(h)+h\eta(\surd h)+M_1(h)$$

where

$$24)\quad M_1(h)=\sum_{i=1}^{m\tilde{m}} M_i(h_1)=\int_o^h[\sigma(t,x^h(t))-\sigma(t,y(t))]dw(t)+$$

$$+\sum_{i=1}^{m}\sum_{j=1}^{\tilde{m}}(m\tilde{m}/\surd h_1)\int_o^h[\tilde{p}_1^0(t,h)(\partial)/(\partial x)(u_{ij}g_i)\sigma(t,x^h)(t))+$$

$$+\tilde{p}_2^0(t,h)(\partial)/(\partial x)(b_j)\sigma(t,x^h(t))]dw(t)$$

and $E|M_1(h)|^2\le h\eta(h)$.

Lemma was proved for $t''=h,t'=0$.

For the next interval $[h,2h]$ we have to repeat the computations done on $[0,h]$. Using (23) we get

$$25)\quad (E\max_{t\varepsilon[h,2h]}|x^h(t)-y(t)|^2)^{1/2}\le[(E|x^h(h)-y(h)|^2)^{1/2}+\eta(\surd h)\ (1-Ch)^{-1}\le$$

$$\le\eta(\surd h)(1+h+\surd h)(1-Ch)^{-1}$$

where $C>0$ is the Lipschitz constant for $\tilde{f}$ and σ.

For $t=2h$ we get a similar representation as in $t=h$. Namely

$$26)\quad x^h(2h)=x^h(h)+\int_h^{2h}\tilde{f}(t,y(t))dt+\sum_{i=1}\sum_{j=1}\int_h^{2h}u_{ij}(t)[g_i,\ b_j](t,y(t))dt+$$

$$+\int_h^{2h}\sigma(t,y(t))dw(t)+h\eta(\surd h)+M_2(h)=y(2h)+2h\eta(\surd h)+M_1(h)+M_2(h)$$

where $M_1(h)$ is defined in (24) and

$$27)\quad M_2(h)=\int_h^{2h}[\sigma(t,x^h(t))-\sigma(t,y(t))]dw(t)+$$

$$+\sum_{i=1}^{m}\sum_{j=1}^{\tilde{m}}(m\tilde{m})/(\surd h_1)\int_h^{2h}[\tilde{p}_1^1(t,h)(\partial)/(\partial x)(u_{ij}g_j)\sigma(t,x^h(t))+$$

$$+\tilde{p}_2^1(t,h)(\partial b_j)/(\partial x)\sigma(t,x^h)(t))]dw(t)$$

fulfil

$$28)\quad E|M_2(h)|^2=h\eta(h),E|M_1(h)+M_2(h)|^2=E|M_1(h)|^2+E|M_2(h)|^2=2h\eta(h)$$

By using an induction argument we get (see (25), (26))

$$29)\quad (E\max_{t\varepsilon[kh,(k+1)h]}|x^h(t)-y(t)|^2)^{1/2}\le\eta(\surd h)(1+kh+\surd kh)(1-Ch)^{-1}=\eta(\surd h)$$

and $M_1(h),\ldots,M_k(h)$ such that

30) $x^h(kh)=y(kh)+kh\eta(\sqrt{h})+\sum_{i=1}^{k} M_i(h)=y(kh)+kh\eta(\sqrt{h})+M^h(kh), k=0,1,\dots,N-1,$

where $M^h(t)$ is defined by

31) $M^h(t)=\int_o^t[\sigma(s,x^h(s))-\sigma(sy(s))]dw(s)+\sum_{i=1}^{m}\sum_{j=1}^{\tilde{m}}(m\tilde{m})/(\sqrt{h_1})$

$\int_o^t[\tilde{p}_1(s,h)(\partial)/(\partial x)(u_{ij}g_j)\sigma(s,x^h(s))+\tilde{p}_2(s,h)(\partial b_j)/(\partial x)\sigma(s,x^h(s))]dw(s))$

and fulfil

31') $E|M^h(k''h)-M^h(k'h)|^2=E|\sum_{i=k'}^{k''}M_i(h)|^2=\sum_{i=k'}^{k''}E|M_i(h)|^2=$

$=(k''-k')h\eta(h), k'<k'', k', k''\varepsilon\{0,1,\dots,N\}.$

From (30-(31') we get the conclusion. The proof is complete.

The approximation equation (4) has some coefficients $u_I(t,h)$ depending on h being unbounded with respect to h. These functions $u_I(t,h)$ are of class C^1 in $t\varepsilon[0,T]$, and with respect to h they fulfil the following condition

$hu_I(t,h)=\eta(\sqrt{h})u_I(t,h), h^2(\partial u_I)/(\partial t)(t,h)=\eta(\sqrt{h})v_I^1(t,h)$

where $u_I(\bullet,h)$, $v_I^1(\bullet,h)$ are uniformly bounded with respect to h.

These properties are essential in order to reduce the order of a system which has unbounded coefficients with respect to the parameter h. Now we consider the following stochastic equation

S) $dy=[f(t,y)+\sum_{i=1}^{m}u_i(t,h)g_i(t,y)+\sum_{|I|=2}^{L+1}u_I(t,h)g_I(t,y)]dt+\sigma(t,y)dw(t)$

$y(0)=x_0, t\varepsilon[0,T],$

where x_0,f,g_i,σ are as in (1) and $u_i(\bullet,h)\varepsilon C([0,T];R)$, $u_I(\bullet,h)\varepsilon C^1([0,T];R)$. With respect to the parameter h we assume that there exist $r(h)>0(r(h)=$ $=\eta(h))$ and a partition π_r of $[0,T]$ with intervals of the length r such that

a) $ru_I(t,h)=\eta(\sqrt{h})v_I(t,h)$, $1\le|I|\le L+1$, $t\varepsilon[0,T]$

b) $r^2(\partial u_I/\partial t)(t,h)=\eta(\sqrt{h})v_I^1(t,h)$, $t\varepsilon[0,T]$, $2\le|I|\le L+1$ where $v_I(\bullet)$, $v_I^1(\bullet)$ are uniformly bounded with respect to h.

Definition

A system S of order L for which there exists $r(h)>0$ such that $u_I(\bullet)$, $1\le|I|\le L+1$, fulfil (a) and (b) is called of index (L,r).

Lemma 2

Assume (C_1) and (C_2) fulfilled. Let (S) be a system of index (L,r), where $r=\eta(h)$. Then there exists a system (S_1) of index $(L-1, r_1)$ with $r_1=T/MK^4$, $M=\text{card}\,\{I:|I|=L+1\}$, $K=T/r$, such that the corresponding solutions $y(\bullet)$ in (S) and $y_1(\bullet)$ in (S_1) with $y(0)=y_1(0)=x_o$, fulfil

*) $y_1(t'')-y_1(t')=y(t'')-y(t')+(t''-t')\eta(\sqrt{h})+M^h(t'')-M^h(t')$

for $t'<t''$, $t',t''\in\{0,\tilde{r}_1,2\tilde{r}_1,\ldots,K^3\tilde{r}_1,\ldots,2K^3\tilde{r}_1,\ldots,K\tilde{r}_1^4=T\}$,

$\tilde{r}_1=T/K^4=Mr_1$

**) $M^h(t)$ is a martingale and $E|M^h(t'')-M^h(t')|^2=(t''-t')\eta(h)$

***) coefficients $u_I(\bullet)$ in (S_1) fulfil (a) and (b) with r replaced by r_1.

The proof of Lemma 2 repeats the same general scheme as in Lemma 1 and it is omitted.

Proof of the Theorem

Suppose L=1. By hypothesis the conditions in Lemma 1 are fulfilled and denote $x^h(\bullet)$ the solution in the approximate system (4). By definition we have

$$x^h(t)-y(t)=\int_o^t[f(s,x^h(s))-f(s,y(s))]ds+\int_0^t[\sigma(s,x^h(s))-\sigma(s,x^h(s,)(s))]dw(s),$$

$$+\sum_{i=1}^m\int_o^t v_i^h(s)g_i(s,x^h(s))ds-\sum_{|I|=2}\int_o^t u_I(1)g_I(s,y(s))ds$$

and

$$u_t=\sum_{i=1}^m\int_o^t v_i^h(s)g_i(s,x^h(s))ds=\sum_{i=1}^m\int_o^{m_1\tilde{r}_1}v_i^h(s)g_i(s,x^h(s))ds+$$

$$=\sum_{i=1}^m\int_{m_1\tilde{r}_1}^t v_i^h(s)g_i(s,x^h(s))ds$$

where $m_1\tilde{r}_1$ is the nearest node to t which is smaller than t in partition π_o.

Using Lemma 1 we get

$$u_t=x^h(m_1\tilde{r}_1)-x_o-\int_o^{m_1\tilde{r}_1}f(s,x^h(s))ds-\int_o^{m_1\tilde{r}_1}\sigma(s,x^h(s))dw(s)+$$

$$+\eta(\sqrt{h})=y(m_1,\tilde{r}_1)-x_o+m_1\tilde{r}_1\eta(\sqrt{h})-\int_o^{m_1\tilde{r}_1}f(s,x^h(s))ds-$$

$$-\int_o^{m_1\tilde{r}_1}\sigma(s,x^h(s))dw(s)+\sigma(\sqrt{h})=\int_o^{m_1\tilde{r}_1}[f(s,y(s))-f(s,x^h(s))]ds+$$

$$+\int_o^{m_1\tilde{r}_1}[\sigma(s,y(s))-\sigma(s,x^h(s))]dw(s)+\sum_{|I|=2}\int_o^t u_I(s)g_I(s)(s))ds+4\sigma(\sqrt{h})$$

Since

$$|\int_o^{m_1\tilde{r}_1}[\sigma(r,y(s))-\sigma(s,x^h(s))]dw(s)|^2\leq \max_{\tau\leq t}/\int_o^\tau[\sigma(s,y(s))-\sigma(s,x^h(s))]dw(s)|^2$$

it follows

$$E \max_{t\leq v}|x^h(t)-y(t)|^2\leq C\int_o^v E(\max_{s\leq t}|x^h(s)-y(s)|^2)dt+4\eta(\sqrt{h})$$

and $E \max_{t\varepsilon[0,T]} |x^h(t)-y(t)|^2\leq C_1h$ for some constant $C_1>0$.

Generally, for L≥1, applying Lemma 1 to (1) we get a system (S) of index (L-1, h|M), where h=T/N, M=card {I:|I|=L+1}.

In order to finish the proof we need to know that the previous estimate holds true in the case (1) is replaced by (S). It is enough to consider only the case when the order of the system (S) is equal to one.

By hypothesis (S) fulfils the conditions in Lemma 2, and applying Lemma 2 we get $y_1^h(\bullet)$ the solution of a system of order zero such that

$$y_1^h(t)-y^h(t)=\int_o^t[f^h(s,y_1^h(s))-f^h(s,y^h(s))]ds+\int_o^t[\sigma(s,y_1^h(s))-\sigma(s,y^h(s))]dw(s)+$$

$$+\sum_{i=1}^m\int_o^t v_i(s,h)g_i(s,y_1^h(s))ds-\sum_{|I|=2}\int_o^t u_I(s,h)g_I(s,y^h(s))ds$$

where $y^h(\bullet)$ is the solution in (S), and

$$U_t=\sum_{i=1}^m\int_o^t v_i(s,h)g_i(s,y_1^h(s))ds-\sum_{i=1}^m\int_o^{m_1\tilde{r}_1} v_i(s,h)g_i(s,y_1^h(s))ds+$$

$+\sum_{i=1}^m\int_{m_1\tilde{r}_1}^t v_i(s,h)g_i(s,y_1^h(s))ds$, where $m_1\tilde{r}_1\leq t$ is as before, fulfils

$$U_t=y_1^h(m_1\tilde{r}_1)-x_o+m_1\tilde{r}_1\eta(\sqrt{h})-\int_o^{m_1\tilde{r}_1}f^h(s,y_1^h(s))ds-$$

$$-\int_o^{m_1\tilde{r}_1}\sigma(s,y_1^h(s))dw(s))+\eta(\sqrt{h})=\int_o^{m_1\tilde{r}_1}[f^h(s,y^h(s))-f^h(s,y_1^h(s))]ds+$$

$$+\int_o^{m_1\tilde{r}_1}[\sigma(s,y^h(s))-\sigma(s,y_1^h(s))]dw(s)+\sum_{|I|=2}\int_o^t u_I(s,h)g_I(s,y^h(s)ds+4\eta(\sqrt{h}).$$

It follows

$$|y_1^h(t)-y^h(t)|^2 \le C\{|\int_{m_1\tilde{r}_1}^{t}[f^h(s,y_1^h(s))-f^h(s,y^h(s))]ds|^2+$$

$$+\max_{\tau\le t}|\int_0^{\tau}[\sigma(s,y_1^h(s))-\sigma(s,y^h(s))]dw(s)|^2+\eta(h)\}$$

for some constant $C>0$.

Since $t\varepsilon[m_1\tilde{r}_1,(m_1+1)\tilde{r}_1)$ and $\tilde{r}_1|f^h(s,y_1)-f^h(s,y)|\le\eta(\sqrt{h})|y_1-y|$ for any $s\varepsilon[0,T]$, $y_1,y\varepsilon R^n$, we obtain

$E\max_{t\varepsilon[0,T]}|y_1^h(t)-y^h(t)|^2\le C_2h$, for some constant $C_2>0$ and the proof is complete.

References

1. H.Kunita, Supports of Diffusion Processes and Controllability Problems, Intern.Symp.S.D.E. Kyoto, 1976, pp.163-185.
2. H.W.Knobloch, K.Wagner, On Local Controllability of Non-Linear Systems, Dynamical Systems and Microphisics Control Theory and Mechanics, 1984, pp.243-285.
3. C.Vârsan, On Local Controllability for Non-Linear Control Systems, Rev.Roum.Math.Pures Appl., vol.29, Nr.10, 1984, pp.907-919.
4. C.Vârsan, Continuous Dependence and Time Change for Ito Equations, J.Diff.Eq.vol.58, Nr.3, 1985, pp.295-306.

Stochastic Control Theory

A STOCHASTIC MAXIMUM PRINCIPLE

H. Becher

Technische Hochschule Carl Schorlemmer Leuna-Merseburg
Sektion Mathematik
DDR-4200 Merseburg
Otto-Nuschke-Str.

Let $(\Omega, \underline{\underline{F}}, P)$ be a complete probability space equipped with a filtration $(\underline{\underline{F}}_t)_{t \geq 0}$ satisfying the usual conditions of right continuity and completeness.

Further on let be:

- H, K, D : Real separable Hilbert spaces.
- V : Real separable reflexive Banach space with its dual V^*.
- $V \subset H \subset V^*$: V is continuously and densely embedded into H. Identifying H with its dual, H is dense in V^*.
- $\langle v, v^* \rangle$: Value of $v^* \in V^*$ at $v \in V$.
- $(w(t), \underline{\underline{F}}_t)$: Wiener process with values in K and a positive selfadjoint nuclear covariance operator Q.
- $\mathcal{L}_Q$: Space of the linear (not necessarily bounded) operators B with $B : Q^{1/2}K \longrightarrow H$ and $BQ^{1/2}$ is Hilbert-Schmidt operator.
- $\mathcal{B}_{[0,T]}$: σ-algebra of Borel subsets of $[0,T]$.
- $\tilde{D}$: Closed bounded convex subset in D.
- A : $\Omega \times [0,T] \times V \times D \longrightarrow V^*$.
- B : $\Omega \times [0,T] \times V \times D \longrightarrow \mathcal{L}_Q$.
- f : $\Omega \times H \longrightarrow R^1$.
- g : $\Omega \times [0,T] \times V \times D \longrightarrow R^1$.
- x_0 : $\underline{\underline{F}}_0$-measurable variable with values in H and $E\|x_0\|_H^2 < \infty$.
- $(z(t), \underline{\underline{F}}_t)$: Square integrable martingale with values in H and $z(t) = \int_0^t B_1(s) w(ds)$, $B_1 \in \mathcal{L}_Q$.

Let $(A(t,v,u))_{t \in [0,T]}$, $(B(t,v,u))_{t \in [0,T]}$ be progressively measurable processes for every $v \in V$, $u \in D$. Suppose that the functions A, B are hemicontinuous, monotonuos, coercitive and restricted on growth, i.e. there are positive constants c_1, c_2 and a non-negative process $\bar{q}(t)$ with $E\int_0^T \bar{q}(t)dt < \infty$ such that for every $\omega \in \Omega$, $t \in [0,T]$ and v, v_1, $v_2 \in V$ the following conditions are satisfied uniformly in $u \in D$:

(i) $\langle v, A(\cdot, v_1 + \lambda v_2, \cdot)\rangle$ is continuous with respect to real λ,

(ii) $2\langle v_1 - v_2, A(\cdot, v_1, \cdot) - A(\cdot, v_2, \cdot)\rangle + \| B(\cdot, v_1, \cdot) - B(\cdot, v_2, \cdot)\|^2_{\mathcal{L}_Q} \leq$
$$\leq c_1 \| v_1 - v_2 \|^2_H ,$$

(iii) $2\langle v, A(\cdot, v, \cdot)\rangle + \| B(\cdot, v, \cdot)\|^2_{\mathcal{L}_Q} + c_2 \| v \|^2_V \leq \bar{q} + c_1 \| v \|^2_H$,

(iv) $\| A(\cdot, v, \cdot)\|_{V^*} \leq \bar{q}^{1/2} + c_1 \| v \|_V$.

Moreover, suppose that the functions A, B are Lipschitz-continuous in $u \in D$ and Fréchet-differentiable with respect to $v \in V$.

Let f, g be convex functions which are continuously Fréchet-differentiable with respect to $v \in V$ and $(v,u) \in V \times D$, respectively. In addition, we assume that f, g and their Fréchet differentials satisfy growth conditions of the following types:

(v) $|g(\cdot, v, u)| \leq \mathrm{const}(\| v \|^2_H + \| u \|^2_D + 1)$,

(vi) $|\nabla_v g(\cdot, v, u)| \leq \mathrm{const}(\| v \|_H + \| u \|_D + 1)$.

Furthermore, g is assumed to be continuous in $(t,v,u) \in [0,T] \times V \times D$ so that for each sequence $(v_n) \subset V$ with $\lim_{n\to\infty} \| v_n - v \|_H = 0$ we have $\lim_{n\to\infty} g(t, v_n, u) = g(t, v, u)$.

Let $\tilde{U}$ be the class of all progressively measurable processes $(U(t))$ with values in $\tilde{D}$.

We consider the control problem

$$(1) \quad X(t) = X_0 + \int_0^t A(s, X(s), U(s))ds + \int_0^t B(s, X(s), U(s))w(ds) + z(t) ,$$

$$(2) \qquad J(U) := Ef(X(T)) + \int_0^T g(t, X(t), U(t))dt \overset{!}{=} \min_{U \in \tilde{U}}$$

where the functional J is assumed to be convex.

Remark 1. As solutions of (1) we obtain H-solutions in the sense of [5] which are continuously depended on $(U(t))$.

Remark 2. One easily finds that there exists an optimal control $(U^o(t)) \in \tilde{U}$.

Applying a duality theorem given in [1] we will present a maximum principle for the control problem (1), (2).

We introduce the following sets:

$$\mathcal{L}^2 := \left\{ (X(t)) \ : \ X(t) = X_0 + \int_0^t v^*(s)ds + \int_0^t b(s)w(ds) + z(t) \right\} ,$$

$$\bar{\mathcal{L}}^2 := \left\{ (\bar{P}(t)) \ : \ \bar{P}(t) = \bar{P}_0 + \int_0^t \bar{v}^*(s)ds + \int_0^t \bar{b}(s)w(ds) + M(t) \right\}$$

where the processes $X(t)$, $\bar{P}(t)$ take values in H which belong to V almost surely. The processes $v^*(t)$, $\bar{v}^*(t)$ and $b(t)$, $\bar{b}(t)$ take values in V^* and $\mathcal{L}_Q$, respectively. Furthermore, $X(t)$ and $\bar{P}(t)$ satisfy the assumptions of theorem I.3.2 in [5]. $M(t)$ is a H-valued square integrable martingale which is independent of $z(t)$. In addition, we assume that $(w(t),k_i)_K(M(t),h_j)_H$ are martingales for all $i,j = 1,2,\ldots$ where (k_n) and (h_n) are orthonormal bases in K and H, respectively.

Moreover, let L be a function defined on $\Omega \times [0,T] \times V \times V^* \times \mathcal{L}_Q$ as follows

$$L(t,v,v^*,b) := \begin{cases} \min g(t,v,u) & \text{if there exists } u \in \tilde{D} \text{ such} \\ & \text{that } v^* = A(t,v,u),\ b = B(t,v,u) \\ +\infty & \text{otherwise} \end{cases} .$$

We assume that the function L is convex.

Now let us introduce the functional

$$F_{fL}(X) := \begin{cases} Ef(X(T)) + E\int_0^T L(t,X(t),v^*(t),b(t))dt & \text{if } (X(t)) \in \mathcal{L}^2 \\ +\infty & \text{otherwise} \end{cases}$$

on $L_V^2(\Omega \times [0,T],\ \underline{\underline{F}} \times \mathcal{B}_{[0,T]},\ dP \times dt)$.

Using results of convex analysis we obtain

Lemma 1. The control problem (1), (2) is equivalent to the minimization problem

$$(3) \qquad \min \left\{ F_{fL}(X) \ : \ X \in L_V^2(\Omega \times [0,T],\ \underline{\underline{F}} \times \mathcal{B}_{[0,T]},\ dP \times dt) \right\} .$$

Therefore, from theorem 6.10 in [1] it follows

Lemma 2. $(U^o(t)) \in \tilde{U}$ is an optimal control of the problem (1), (2) if and only if there exist processes $(X^0(t)) \in \mathcal{L}^2$ and $(\bar{P}(t)) \in \bar{\mathcal{L}}^2$ such that the following subdifferential conditions hold:

$$(4) \qquad (\bar{v}^*(t), \bar{P}(t), \bar{b}(t)) \in \partial L(t, X^0(t), v^*(t), b(t)) \quad \text{a.s.,}$$

$$(5) \qquad -\bar{P}(T) \in \partial f(X^0(T)) \quad \text{a.s.}$$

By virtue of (4), (5) we obtain the main result:

Theorem. If $(U^0(t), X^0(t))$ is an optimal control of the problem (1), (2) then there exists a process $(\bar{P}(t)) \in \bar{\mathcal{L}}^2$ such that the following relations are verified:

$$(6) \qquad \bar{P}(t) = \bar{P}_0 - \int_0^t \nabla^*_{X^0(s)} A(s, X^0(s), U^0(s))(\bar{P}(s))ds$$
$$- \int_0^t \nabla^*_{X^0(s)} B(s, X^0(s), U^0(s))(\bar{b}(s))ds$$
$$+ \int_0^t \nabla_{X^0(s)} g(s, X^0(s), U^0(s))ds + \int_0^t \bar{b}(s)w(ds) + M(t) \quad \text{a.s.,}$$

$$(7) \qquad -\bar{P}(T) = \nabla_{X^0(T)} f(X^0(T)) \quad \text{a.s.,}$$

$$(8) \qquad \langle A(t, X^0(t), U^0(t)), \bar{P}(t)\rangle$$
$$+ \Big(B(t, X^0(t), U^0(t)), \bar{b}(t)\Big)_{\mathcal{L}_Q} - g(t, X^0(t), U^0(t))$$
$$= \max_{u \in \tilde{D}} \Big\{ \langle A(t, X^0(t), u), \bar{P}(t)\rangle$$
$$+ \Big(B(t, X^0(t), u), \bar{b}(t)\Big)_{\mathcal{L}_Q} - g(t, X^0(t), u) \Big\} \quad \text{a.s.}$$

Remark 3. The maximum principle which is given in above theorem includes the case of random functions A, B and, in particular, takes into consideration perturbations of state and control by Wiener process.

Remark 4. If $\underline{\underline{F}}_t$ is generated by $w(t)$ then in (6) we have $M \equiv 0$.

Remark 5. It's clear that the adjoint equation (6) is not of the type (1). Therefore, the $\underline{\underline{F}}_t$-measurability of the solution of (6) is

not obvious. But we can prove it by the following:
We approximate the control problem (1), (2) by a sequence of finite-dimensional control problems where we take the Galerkin approximations X_n of the state X. Then the adjoint equations of the finite-dimensional problems are also finite-dimensional. By using Bismut's theory (see [2]) we can show that the solutions $\bar{P}_n$ of the finite-dimensional adjoint equations are $\underline{\underline{F}}_t$-measurable. For them it's important to have the existence of inverse matrix of the fundamental solution for a finite-dimensional system of Ito equations.
For the Galerkin approximations $\bar{P}_n$ of the adjoint state it holds:

$$E \sup_t \| \bar{P}_n(t) \|_H^2 + E\int_0^T \| \bar{P}_n(t) \|_V^2 \, dt < \mathrm{const}(T, c_1, c_2) .$$

Hence we find a subsequence of $(\bar{P}_n)$ which weakly converges toward $\bar{P}$. Even we can show that the whole sequence $(\bar{P}_n)$ strongly converges. Consequently, the solution of adjoint equation (6) is $\underline{\underline{F}}_t$-measurable (see [3] for details).

Remark 6. If we analyse the maximum principle then there are problems to state processes $\bar{b}(t)$. However, the construction of ε-optimal strategies is possible (see [3]).

Remark 7. In general, the maximum principle gives only a necessary condition of optimality. Whereas, for instance, if (1) is linear and g separated with respect to v, u then the optimality condition is also sufficient (see [4]).

Remark 8. Let $\underline{\underline{G}}_t$ be a non-decreasing family of sub-σ-fields of $\underline{\underline{F}}_t$. If the admissible controls $(U(t))$ are $\tilde{D}$-valued functions with

$E\int_0^T \| U(t) \|_D^2 \, dt < \infty$ and adapted to $\underline{\underline{G}}_t$ then the $\underline{\underline{G}}_t$-measurability of optimal control will be guaranteed by introducing conditional expectation with respect to $\underline{\underline{G}}_t$ into (8).

Example. Let $V \subset H \subset V^*$ be a triple of Hilbert spaces, $D = K = H$, $A(.,X(.),U(.)) = AX(.)$, linear, and $\underline{\underline{F}}_t := \sigma\{w(s) : s \leq t\}$. Furthermore, we define the domain of A by $D(A) := \{x \in V : Ax \in H\}$.
Now we consider the problem

$$X(t) = X_0 + \int_0^t (AX(s) + U(s))ds + w(t) ,$$

$$J(U) = E \| X(T) - z_o \|_H^2 \stackrel{!}{=} \min_{U \in \tilde{U}} , \quad z_o \in H .$$

Then the following relations are verified:

$$\max_{u \in \tilde{D}} \left\{ \langle AX^o(t), E\{q(t) \mid \underline{\underline{F}}_t\} \rangle + \left(u, E\{ q(t) \mid \underline{\underline{F}}_t\} \right)_H \right\}$$

$$= \langle AX^o(t), E\{q(t) \mid \underline{\underline{F}}_t\} \rangle + \left(U^o(t), E\{q(t) \mid \underline{\underline{F}}_t\} \right)_H \quad \text{a.s.,}$$

$$\frac{dq(t)}{dt} = -A^* q(t) , \quad -q(T) = 2(X^o(T) - z_o) \quad \text{a.s.}$$

where $(q(t))$ is a process defined by $q(t) := -2\, \varrho^*(T-t)(X(T) - z_o)$; $\varrho(t)$ denotes the strongly continuous semigroup generated by A with its adjoint $\varrho^*(t)$.

In particular, if $V = \overset{o}{W}{}_2^1(G)$, $H = L^2(G)$, $G \subset R^1$, $\tilde{D} = [-1,1]$ then we have

$$U^o(t) = \operatorname{sgn} \left[E\{q(t) \mid \underline{\underline{F}}_t\} \right] .$$

References

[1] Becher, H.: Ein Beitrag zur Steuerung parabolischer Ito-Gleichungen, Dissertation A, Merseburg 1985.

[2] Bismut, J.M.: Convex functions in optimal stochastic control, J. Math. Anal. Appl., 44 (1973), 384-404.

[3] Grecksch, W.: Stochastische Evolutionsgleichungen und deren Steuerung, Teubner-Texte zur Mathematik, to appear 1987.

[4] Grecksch, W., Becher, H.: Anwendung der Dualitätstheorie in der stochastischen Steuertheorie, Math. Nachr., 117 (1984), 57-74.

[5] Krylov, N.V., Rozovskij, B.L.: About stochastic evolution equations. Results of science and engineering, ser. actual problems of mathematics, 14, Moscow 1979.

LINE INTEGRALS; STABLE SPACES OF MARTINGALES; COMPACTIZATION PROBLEMS IN OPTIMAL CONTROL

R.J. Chitashvili, M.G. Mania
Tbilisi Razmadze Mathematical Institute of the Acad. of Sci. of the Georgian SSR, 150 a Plekhanov Ave., Tbilisi, 380012, USSR

INTRODUCTION

The problem of optimal absolutely continuous change of measure involves the following initial objects: the probability space (Ω,F,G,P) with filtrations $G=(G_t,t\geq 0)$, $F=(F_t,t\geq 0)$, $G_t\subset F_t$, $t\geq 0$ satisfying the usual conditions and the family of probabilities $\mathbf{P}=(P^a,a\in A)$ on F_∞ locally equivalent to P, where F_t consists of the events which describe the behaviour of the controlled system up to the moment t, G_t forms the observable events up to the t-th moment and the parameter a implies decision.

The possibility of decision change with respect to the accumulated information leads to the extension of the class $\mathbf{P}$ by the introduction of controls. The simplest class of piecewise constant controls consists of G-predictable processes $u=(u_t(\omega),t\geq 0)$ of the following form $u_t(\omega) = \sum_i a_i I_{]\!]\tau_i,\tau_{i+1}]\!]}$, $a_i\in A$ where τ_i, $0<i\leq N$ is a sequence of stopping times $\tau_1<\tau_2<\ldots<\tau_{N-1}<\tau_N=\infty$ w.r.t. G. Denote such class by U_d.

The corresponding set of probabilities is defined by a set $\mathbf{R}_d$ of local densities $\mathbf{P}_d=(\rho^u\cdot P,\ \rho^u\in\mathbf{R}_d)$, $\rho^u_t=dP^u_t/dP_t$ where ρ^u_t is constructed according to the bifurcation principle

$$\rho^u_t = \prod_i \left(\rho^{a_i}_{t\wedge\tau_{i+1}} / \rho^{a_i}_{t\wedge\tau_i}\right),\quad u\in U_d \tag{1}$$

$\rho^a_t=dP^a_t/dP_t$ is a local density P^a w.r.t. P, and

$P^a_t=P^a|F_t, P_t=P|F_t$ are restrictions of the measures P^a,P on F_t.

The problem is to describe the elements of completion of the class of controls U_d induced by the closure (strong, weak, convex) of the set of measures $\mathbf{P}_d$ and, in particular, to derive an assertion on the optimal control existence in the maximization problem for the expected value

$$E^u \eta \longrightarrow \max_u$$

of F_{∞}-measurable random variables η .

In order to formalize this problem we introduce a class of local martingales $\mathbf{M}=(M^a, a \in A)$, $M^a \in \mathcal{M}_{loc}(F)$, by means of which the local densities from $\mathbf{R}=(\rho^a, a \in A)$ can be represented as exponential martingales $\rho^a=\mathcal{E}(M^a), a \in A$ which are the solutions of the equation $X_t=1+\int_0^t X_{s-}dM_s^a$, so that the closure problems for the multiplicative expressions (1) could be reduced to additive expressions.

Indeed, it can be easily shown that $\mathbf{R}_d=\mathcal{E}(\mathbf{M}_d) = (\mathcal{E}(M^u), M^u \in \mathbf{M}_d)$ where the class $\mathbf{M}_d$ consists of local martingales

$$M_t^u = \sum_i \left[M^{u\tau_i}_{t\wedge\tau_{i+1}} - M^{u\tau_i}_{t\wedge\tau_i}\right] , \quad u \in U_d. \tag{2}$$

For the beginning we shall describe the closures of $\mathbf{M}_d$ and to specifiy the problem we shall make some assumptions.

We assume that $\mathbf{M} \subset \mathcal{M}^2_{loc}$ (F,P) are locally square integrable martingales w.r.t. the measure P and the flow F; an increasing G-predictable process K exists which dominates the square characteristics $\langle M^a \rangle$. Denote

$$\varphi(a,b)=(\varphi_t(a,b)=d\langle M^a,M^b\rangle_t/dK_t, \quad t\geqslant 0, \quad a,b \in A).$$

Let ν be Dolean's measure $\nu(dt,d\omega)=dK_tP(d\omega)$ on σ-algebra $\mathcal{P}(F)$ of F-predictable subse ts of $[0,\infty]\times\Omega$. We also assume that G-stopping times $\tau_n, \tau_n<\tau_{n+1}, \tau_n \uparrow \infty$ exist such that $\sup_a \varphi(a,a)\cdot K_{\tau_n} \leqslant c_n$ (Here and in the forthcoming we use the notation $X\cdot S = (\int_0^t X_s dS_s)_{t\geqslant 0}$ for stochastic integrals w.r.t. semimartingales).

Let $H^2_{loc}(\tau)$ be a countable normed Hilbert subspace of $\mathcal{M}^2_{loc}$ of martingales localized by the moments $(\tau_n, n\geqslant 1)$ with a sequence of scalar products $(m,m')_n=Em_{\tau_n}m'_{\tau_n}$, $m,m' \in H^2_{loc}(\tau)$, $n\geqslant 1$.

It can be easily observed, applying the inequality $E\,\mathcal{E}^2_{\tau_n}(M^u)\leqslant e^{c_n}$ that $\mathbf{M}_d$ and $\mathbf{R}_d$ are bounded subsets of $H^2_{loc}(\tau)$. It enables us to consider the case $\sup_a \varphi(a,a)\cdot K_{\infty} \leqslant c$ in what follows. The remarks concerning the case of local conditions will be given below.

§ 1. DESCRIPTION OF THE CLOSURE OF $\mathbf{M}_d$, LINE INTEGRAL; STABLE SPACE OF MARTINGALES

It can be easily seen that the class $\mathbf{M}_d=(M^u, u \in U_d)$ is characterized as the minimal set of H^2 containing $\mathbf{M}=(M^a, a\in A)$ and stable w.r.t. the bifurcation of elements, i.e. $m^1, m^2 \in \mathbf{M}_d$ implies $(m^1_{t\wedge\tau} + m^2_{t\wedge\tau} - m^2_\tau)_{t\geq 0} \in \mathbf{M}_d$ where τ is an arbitrary G-stopping time. Hence, the desired strong, weak and convex closure of $\mathbf{M}_d$ denoted as $\mathbf{M}_s$, $\mathbf{M}_w$, $\mathbf{M}_g$, respectively, are subsets of the linear closure of $\mathbf{M}_d$: the stable space of martingales $\mathcal{L}^2(\mathbf{M})$ generated by the class $\mathbf{M}=(M^a, a \in A)$ (the minimal subspace of H^2 containing $\mathbf{M}$ and stable w.r.t. integration of G-predictable processes, i.e. $m \in \mathcal{L}^2(\mathbf{M})$ and $X \in \mathcal{P}(G)$ with $E(X^2 \cdot \langle m\rangle_{\infty}) < \infty$ imply $X \cdot m \in \mathcal{L}^2(\mathbf{M})$ [1]).

Hence, the problem we are interested in is associated to the description of the space $\mathcal{L}^2(\mathbf{M})$. To describe the space $\mathcal{L}^2(\mathbf{M})$ in "predictable terms" consider a space of processes which depend on the parameter $a \in A$ (classes of equivalence w.r.t. the measure ν)

$$H_0 = \left\{H(a,m),\ m \in H^2,\ H(a,m) = {}^G(d\langle M^a, m\rangle/dK)\right\}$$

${}^{G,P}X$ denotes a G-predictable projection of the process X w.r.t. the measure P.

H_0 can be supplied with a scalar product in the following way. Any martingale $m \in H^2$ can be expanded as a sum $m=m/\mathcal{L}^2(\mathbf{M})+m'$ where $m/\mathcal{L}^2(\mathbf{M})$ is a projection of m on $\mathcal{L}^2(\mathbf{M})$ and m' is orthogonal to the space $\mathcal{L}^2(\mathbf{M})$ (i.e. $\langle m', N\rangle=0$ for all $N \in \mathcal{L}^2(\mathbf{M})$).

Put

$$(H(.,m),\ H(.,N))_{H_0} = E\langle m/\mathcal{L}^2(\mathbf{M}),\ N/\mathcal{L}^2(\mathbf{M})\rangle.$$

Proposition 1 The spaces H_0 and $\mathcal{L}^2(\mathbf{M})$ are isometric.

Proof. It is known (see [1]) that $m' \perp\!\!\!\perp \mathcal{L}^2(\mathbf{M})$ is equivalent to $m' \perp\!\!\!\perp M^a$, $a \in A$. Therefore $H(a,m)=H(a,N)$ a.e. w.r.t. the measure ν implies $m/\mathcal{L}^2(\mathbf{M})=N/\mathcal{L}^2(\mathbf{M})$. □

In this context H_0 is defined in terms of $\mathcal{L}^2(\mathbf{M})$ itself. The Hilbert space H_0, however, admits analytical description in terms of the generating kernel $\varphi(a,b)$.
Introduce the following definition of the Hilbert space H^φ of G-predictable (w.r.t. (ω,t)) functions $h=h(a)=h_t(a), t \geq 0$ of a satisfying the following properties (of the scalar product $(.,.)_\varphi$):

1) ${}^G\varphi(a,.) \in H^\varphi$ for all $a \in A$;
2) If $h,g \in H^\varphi$ and $B \in \mathcal{P}(G)$, then $I_B h \in H^\varphi$ and $(I_B h, g)_\varphi = (h, I_B g)_\varphi$;

3) If $h=h(a)\in H^{\varphi}$, then $(h, {}^{G}\varphi(a,.))_{\varphi} = h(a)d\nu$ for any $B\in\mathcal{P}(G)$.

When φ does not depend on (ω,t) this definition coincides with the Hilbert space with a reproducing kernel (see, e.g. [2]).

Reproducing property 3), actually, reproduces the element h in the sense that $d/d\nu\,(hI_{(.)}, {}^{G}\varphi(a,.))_{\varphi} = h(a)$, $(.)\in\mathcal{P}(G)$.

<u>Proposition 2.</u> There exists a unique space H^{φ} with properties 1)-3) and $H^{\varphi} = H_o$.

<u>Proof.</u> First we show that the space H_o satisfies properties 1), 2) and 3). Properties 1), 2) immediately follow from the definition of H_o.

If $h\in H_o$, then $h=H(.,m)$ for some $m\in H^2$ and

$$(h, {}^{G}\varphi(.,a))_{H_o} = (H(.,m), {}^{G}\varphi(.,a))_{H_o} = E\langle m/\mathcal{L}^2(\mathbf{M}), M^a\rangle =$$

$$= H(a,m).\nu = \int h(a)d\nu.$$

Now let H^{φ} be some Hilbert space of G-predictable functions satisfying 1),2),3). Then for $B\in\mathcal{P}(G)$, evidently, $I_B\, {}^{G}\varphi(a,.) = H(.,I_B\cdot M^a)\in H_o$ and property 3) implies

$$(I_B{}^{G}\varphi(a,.),\ I_C{}^{G}\varphi(b,.) = \int I_{BC}{}^{G}\varphi(a,b)d\nu = (H(.,I_B\cdot M^a),\ H(.,I_C\cdot M^b))_{H_o}$$

It is sufficient to show that the closure of the linear hull of the elements $I_B{}^{G}\varphi(a,.)$, $B\in\mathcal{P}(G)$, $a\in A$, coincides with H^{φ}. But 2), 3) imply that any $h\in H^{\varphi}$ orthogonal to $I_B\,\varphi(a,.)$, $B\in\mathcal{P}(G)$, $a\in A$, equals zero ν-a.s. because

$$0=(h,I_B\,{}^{G}\varphi(a,.))_{\varphi} = (I_Bh, {}^{G}\varphi(a,.))_{\varphi} = \int I_Bh(a)d\nu \quad \text{for all } B\in\mathcal{P}(G) \quad .\Box$$

The problem of the description of $\mathcal{L}^2(\mathbf{M})$ of another character can be illustrated by the set $\mathbf{M}_d\subset\mathcal{L}^2(\mathbf{M})$. In particular, if $m\in\mathbf{M}_d$, then $m=M^u$, $u\in U_d$ and

$$H_t(a,m) = H_t(a,M^u) = \varphi_t(a,u_t) \tag{3}$$

Hence, the subset H^{φ}, corresponding to $\mathbf{M}_d$, is described w.r.t. formula (3) by the elements of U_d. Now we shall try to extend formula (3) to the whole space $\mathcal{L}^2(\mathbf{M})(H^{\varphi})$ which enables us to describe the extension of the set of controls U_g corresponding to the closure of $\mathbf{M}_d$. For the linear manifold

$$\ell(\mathbf{M}_d) = \{X\cdot M^u,\ u\in U_d,\ EX^2\cdot\langle M^u\rangle = EX^2\,\varphi(u,u)\cdot K_{oo} < \infty\}\subset\mathcal{L}^2(\mathbf{M})$$

it can be easily seen that

$$H_t(a,m) = \int_A \varphi_t(a,b)\,\mu_t(db) = \sum_i \varphi_t(a,a_i)X_t\, I_{]\!]\tau_i,\tau_{i+1}]\!]}$$

where for $m=X\cdot M^u$ μ is a G-predictable signed measure on $(A,\mathcal{A})$ concentrated at a finite number of points $\mu_t(\{a\}) = X_tI_{(u_t=a)}$,

$$\mu \in L_0^{\varphi} = \{\mu : \mu_t(\{a\}) = X_t I_{(u_t=a)}, t \geq 0,\ X \cdot M^u \in \ell(\mathbf{M}_d)\} \qquad (4)$$

The equality

$$(X^1 \cdot M^{u_1}, X^2 \cdot M^{u_2})_L = E \int_0^{\infty} \int_{A\times A} \varphi(a,b)\, \mu_t^1(da)\mu_t^2(db)dK_t \qquad (5)$$

with $\mu_t^i(\{a\}) = X_t^i I_{(u_t^i=a)}$ defines the scalar product in the linear manifold L_0^{φ} of all such measures and establishes isometry among $\mathcal{L}^2(\mathbf{M})$, H^{φ} and L^{φ}, where L^{φ} is a completion of L_0^{φ} corresponding to scalar product (5). In the case of a finite set $\mathbf{M}$ the description of the class $\mathcal{L}^2(\mathbf{M})$ in terms of L^{φ} was proved (for a particular case) in [3] and (for the general case) in [4], [1]. (For finite $\mathbf{M}$ the space L_0^{φ} becomes complete $L_0^{\varphi} = L^{\varphi}$.)

The space $\mathcal{L}^2(\mathbf{M})$ can be represented now as $\mathcal{L}^2(\mathbf{M}) = \{M^{\mu}, \mu \in L^{\varphi}\}$ where the element M^{μ} is uniquely defined either constructively as a limit in H^2 $M^{\mu} = \lim_{n\to\infty} M^{\mu_n}$, $\mu_n \in L_0$ and $\mu_n \to \mu$ in L^{φ} with $M_t^{\mu_n} = \int_0^t \int_A M(dt,a)\mu_n(t,da)$, or non-constructively by

$$H(a,M^{\mu}) = \lim_{n\to\infty} H(a,M^{\mu_n}) = \lim_{n\to\infty} \int_A \varphi(a,b)\mu^n(db) = \frac{d}{d\nu}(\delta_a I_{(.)}, \mu)_L.$$

in the sense of the convergence in H^{φ}, where $\delta_a \in L^{\varphi}$ is a measure concentrated at a single point $a \in A$.

An effective description of the closures of $\mathbf{M}_d$ and U_d in $\mathcal{L}^2(\mathbf{M})$ and L^{φ}, correspondingly, can be obtained in the case of a metric compact space (A,τ) with the metric $\tau(a,b), a,b \in A$ and the kernel $\varphi(a,b)$ continuous in a,b at ν-a.e. (ω,t).

Introduce the subsets $U_d \subset U_s \subset U_w \subset U_g \subset L^{\varphi}$ where U_s contains all G-predictable processes taking values in $A, (u: \Omega\times R^+ \to A)$. U_g contains all G-predictable probabilities on A, $u: \Omega \times R^+ \to Q(A)$ where $Q(A)$ is a set of all probabilities on A and the elements of U_s as well as the elements of U_g can be considered as degenerate distributions.

U_w contains all G-predictable processes having the form

$$u = I_{B(\nu)}u^1 + I_{B^c(\nu)}u^2, u^1 \in U_s,\ u^2 \in U_g \quad \text{and } B(\nu) \text{ is a union of } \nu\text{-}$$

atoms.

<u>Proposition 3 ([5],[6]).</u> If $\varphi(a,b)$ almost in all (ω,t) w.r.t. the measure ν is continuous w.r.t. the couple (a,b), A is a compact, then $\mathbf{M}_S = \{M^u, u \in U_S\}$, $\mathbf{M}_W = \{M^u, u \in U_W\}$, $\mathbf{M}_g = \{M^u, u \in U_g\}$.

In [5], [6] the case F=G was considered, but the proof of the assertion about the closure of $\mathbf{M}_d$ remains unchanged for $G \subset F$. Hence the strong closure of $\mathbf{M}_d$ is described in terms of "usual" controls from U_s and the convex closure - by "generalized" (relaxed) controls from U_g, where the martingale M^u is uniquely defined by the elements $u \in U_g$ according to the formula

$$H_t(a,M^u) = \int_A \varphi_t(a,b)u_t(db) \qquad (3')$$

and has the sense of a line integral w.r.t. the set $\mathbf{M}$ along the generalized curve and has a stronger continuity property: the convergence $u^n \to u$ ν-a.e. on (ω,t) in the sense of a weak convergence of values (Levy-Prokhorov's metric) in the space of probabilities $Q(A)$ on A implies the convergence $M^{u^n} \longrightarrow M^u$ in H^2.

In applications it is convenient to use the definition of M^u by the equality (see [5])

$$\langle M^u, m\rangle_t = \int_0^t \int_A H_s(m,a)u_s(da)dK_s, \qquad m \in H^2 \qquad (6)$$

and, finally, for the martingales $M^a = aM$, $a \in R^1$ (for $u \in U_s$) linear w.r.t. the parameter a M^u coincides with the usual (linear) integral $M^u = u\cdot M$ defined by (6) according to Kunita-Watanabe's scheme (evidently, $H_t(m,a) = a(d\langle M,m\rangle_t/dK)$ in this case). The stochastic line integral was introduced in [12] for φ continuous uniformly w.r.t. (ω,t) in a.

§ 2. CLOSURE OF $\mathbf{M}_d$ IN THE CASE OF A DISCONTINUOUS KERNEL

The extent of the enrichment of the class U_g (extension of the concept of control) for discontinuous φ can already be seen in the simple case when the kernels can be represented as $\varphi_t(a,b) = \varphi_t(a)\,\varphi_t(b)$ with G-predictable $\varphi_t(a)$ without the condition of ν-a.e. continuity in a. Note that the kernel φ has this form, particularly, when the martingales M^a can be represented as stochastic integrals $M^a = \xi(a)\cdot M$, $a \in A$.

Introduce ν-a.e. upper- and lower-continuous functions

$$\overline{\varphi}_t(a) = \lim_n \sup_{i:\tau(a_i,a)\leq 1/n} \varphi_t(a_i)$$

$$\underline{\varphi}_t(a) = \lim_n \inf_{i:\,\tau(a_i,a)\leq 1/n} \varphi_t(a_i)$$

where $A_0 = (a_i,\ i \geq 1)$ is an everywhere dense subset of the compact set A with the metric $\tau(a,b)$, $a,b \in A$.

Suppose that $\varphi_t(a)$ is stochastically continuous in a. Then by virtue of Doob's theorem a separable modification of the function φ exists.

The main result is the following assertion.

Proposition 4. $\mathbf{M}_g = \{M^{u,\gamma}, u \in U_g, \gamma \in \Gamma\}$, where Γ is a set of all G-predictable processes taking values in $[0,1]$ and $M^{u,\gamma}$ is defined by

$$H_t(M^{u,\gamma},a) = \varphi_t(a)\int_A(\overline{\varphi}_t(b)\gamma_t + \underline{\varphi}_t(1-\gamma_t))u_t(db) \qquad (8)$$

In other words, the convex closure of U_d in L^φ consists of elements $\mu=(u,\gamma)$ of G-predictable distributions on the expanded set $\tilde{A}=A\times\{1,2\}$ with the kernel $\tilde{\varphi}(\tilde{a},\tilde{b})=\tilde{\varphi}(\tilde{a})\tilde{\varphi}(\tilde{b})$ where $\tilde{\varphi}(\tilde{a})=\overline{\varphi}(a)$ for $\tilde{a}=(a,1)$ and $\tilde{\varphi}(\tilde{a})=\underline{\varphi}(a)$ for $\tilde{a}=(a,2)$ and with the scalar product

$$((u,\gamma),(u',\gamma'))_L = E\int_0^\infty \int_A\int_A(\overline{\varphi}(a)\gamma+\underline{\varphi}(a)(1-\gamma))(\overline{\varphi}(b)\gamma' +$$

$$+\underline{\varphi}(b)(1-\gamma')u(da)u'(db)dK_t.$$

The proof of Proposition 4 is based on the following lemma, which is interesting by itself.

First we shall introduce several objects.

Let Q(A) be a family of probability measures on the Borel σ-algebra $\mathcal{A}$ of the subsets A and $\tau^Q(q,q'), q,q' \in Q(A)$ denotes Levy-Prokhorov's metric.

Introduce a class of martingales $\mathbf{M}^q = \{M^q, q \in Q(A)\}$. The martingale $M^q, q \in Q(A)$ ([5]) is defined (for a constant generalized control $u \in U_g$) as a unique element $\mathcal{L}^2(\mathbf{M})$ for which

$$\langle M^q, m\rangle = \int_A \langle M^a, m\rangle\, q(da), \quad m \in \mathcal{L}^2(\mathbf{M})$$

and has the following properties (similar to those of M^a)

$$M^q \in \mathcal{M}^2_{loc}(P,F),\ \langle M^q\rangle \ll K,\quad d\langle M^q, M^{q'}\rangle/dK = \phi(q)\,\phi(q')$$

$$E\sup_q \phi^2(q)\cdot K_\infty < \infty \quad \text{where } \phi_t(q) = \int_A \varphi_t(a)q(da).$$

Note that Q(A) is also compact in the metric τ^Q and for the dense subset we have the set $Q_0(A)$ of all discrete measures from Q(A) concentrated at a finite number of points from A with rational values of measure for each point.

It can be easily shown that

$$\overline{\phi}(t,q) = \int_A \overline{\varphi}(t,a)q(da), \quad \underline{\phi}(t,q) = \int_A \underline{\varphi}(t,a)q(da)$$

and $\overline{\phi}(\underline{\phi})$ is the minimal (maximal) function ν-a.e. upper (lower) semi-continuous for which ν-a.e.

$$\underline{\phi}(q) \le \phi(q) \le \overline{\phi}(q), \quad q \in Q(A) \qquad (9)$$

for any separable modification $\phi(q)$.

Lemma 1. a) (Generalization of Philopov's lemma). If $(u_n, n\geq 1)\in U_g$ is a sequence such that $\phi(u_n)=\int_A \varphi_t(\omega,a)u_t^n(\omega,da)$ converges a.e. to some G-predictable function $Y=(Y_t(\omega),t\geq 0)$ then $u\in U_g$ and $\gamma\in\Gamma$ can be found such that ν-a.e.

$$Y_t = \overline{\Phi}_t(u_t)\gamma_t + \underline{\Phi}_t(u_t)(1-\gamma_t).$$

b) For any $\gamma\in\Gamma$ and $u\in U_g$ a sequence $(u_n, n\geq 1)\in U_g^o\subset U_g$ can be found (U_g^o is a class of controls u taking values in $Q_o(A)$) such that for any modification of ϕ

$$\phi(u_n)\longrightarrow \overline{\phi}(u)\gamma + \underline{\phi}(u)(1-\gamma)$$

Proof. a) To prove this lemma the assertion of Lemma 2 from [7] is used, which states the following.

If $g_n(x)$ is a sequence of measurable mappings of $(X,\mathcal{X})$ into a compact metric space Y, then a sequence of random variables $n_j(x)$, $j\geq 1$, $n_j: X\rightarrow N=(1,2,\dots)$ exists such that $n_j(x)\leq n_{j+1}(x)$, $\lim_{j\rightarrow\infty} n_j(x)=\infty$ and the sequence $(g_{n_j(x)}, j\geq 1)$ converges for all $x\in X$.

Hence there is a sequence $n_j(\omega,t)$ for which $u_{n_j(\omega,t)}$ converges to some $u\in U_g$ and, obviously,

$$\lim_{j\rightarrow\infty} \phi_t(u_{n_j}) = \lim_{n\rightarrow\infty} \phi_t(u_n) = Y_t$$

On the other hand,

$$\underline{\Phi}_t(u(t))\leq \lim_j \underline{\Phi}_t(u_{n_j})\leq Y_t\leq \lim_j \overline{\Phi}_t(u_{n_j})\leq \overline{\Phi}_t(u(t)).$$

Now if we determine the measurable function $\gamma(\omega,t)= =(\overline{\phi}(u)-Y)/(\overline{\phi}(u)-\underline{\phi}(u))$ we shall obtain the desired representation.

b) For a fixed $u\in U_g$ we can define the sequence $N_n(\omega,t)$, $n\geq 1$ as

$$N_n(\omega,t)=\min\left\{i:\ \phi(\omega,t,q_i)\geq \sup_{j:\ \tau^Q(q_j,u_t(\omega))\leq 1/n} \phi(\omega,t,q_j) - 1/n\right.$$

where $(q_1,q_2,\dots)$ is dense in $Q(A)$. Evidently,

$$\overline{\phi}(\omega,t,u(\omega,t))\geq \sup_{j\ :\ \tau^Q(q_j,u(t,\omega))\leq 1/n} \phi(\omega,t,q_j)$$

and if we define $u_n(\omega,t)=q_{N_n(\omega,t)}$, then

$$\phi(\omega,t,u_n(\omega,t))\geq \sup_{j:\ \tau^Q(q_j,u(\omega,t))\leq 1/n} \phi(\omega,t,q_j) - 1/n\ .$$

This together with (9) implies $\phi(u_n)\longrightarrow\overline{\phi}(u)$ ν-a.e. (Note that since u_n takes a countable number of values in $Q(A)$ the expression

$\phi(u_n)$ does not depend on the choice of the modification of ϕ). Similarly, we can show the existence of the sequence $v_n \in U_g^o$ for which $\phi(v_n) \longrightarrow \underline{\phi}(u)$. Obviously, for $\bar{u}_n = \gamma u_n + (1-\gamma)v_n$

$$\phi_t(\bar{u}_n) = \int_A \varphi(t,a)\bar{u}_n(t,da) = \gamma(\omega,t)\,\phi_t(u_n) + (1-\gamma(\omega,t)\,\phi_t(v_n)$$

$$\longrightarrow \gamma(\omega,t)\,\bar{\phi}_t(u) + (1-\gamma(\omega,t))\,\underline{\phi}_t(u),\ n \longrightarrow \infty .$$

Now it remains to approximate $\gamma(\omega,t)$ by simple functions $\gamma_n(\omega,t)$ so that the control $\tilde{u}_n = \gamma_n(\omega,t)u_n + (1-\gamma_n(\omega,t))v_n$ should belong to the class U_g^o $\square$.

<u>Proof of Proposition 4.</u> Lemma 1 b) implies that for any $M^{u,\gamma}$ $u \in U_g$, $\gamma \in \Gamma$ a sequence $(u_n, n \geq 1) \in U_g^o$ exists for which $E\langle M^{u,\gamma} - M^{u_n}\rangle \to 0$, $n \to \infty$. On the other hand, by Theorem 1.6 [6] implies that any element of $\{M^u, u \in U_g^o\}$ can be strongly approximated by the elements of $Co(\mathbf{M}_d)$. Hence $\mathbf{M}^{u,\gamma}$ belongs to the convex closure of $\mathbf{M}_d$.

Now we can show that the set $\{M^{u,\gamma},\ u \in U_g, \gamma \in \Gamma\}$ is strongly closed. Since the elements of $\mathbf{M}^{u,\gamma}$ are strongly approximated by the elements of $\{M^u,\ u \in U_g^o\}$ it is sufficient to show that for any sequence $\{u_n,\ n \geq 1\} \in U_g^o$ for which M^{u_n} strongly converges to some $M \in \mathcal{L}^2(\mathbf{M})$, $\gamma \in \Gamma$ and $u \in U_g$ can be found such that $M = M^{u,\gamma}$.

Therefore (taking into account that u_n takes a countable number of values in $Q(A)$) we have

$$E\langle M^{u_n} - M^{u_j}\rangle = E(\phi(u_n) - \phi(u_j))^2 \cdot K \to \infty,\ n \to \infty,\ j \to \infty$$

and a subsequence n' can be selected such that a.e. w.r.t. the measure γ

$$(\phi(u_{n'}) - \phi(u_{j'}))^2 \longrightarrow 0,\ n' \longrightarrow \infty,\ j' \longrightarrow \infty. \qquad (10)$$

For convenience we preserve the notation $(u_n, n \geq 1)$ for the selected subsequence.

Obviously, M^{u_n} weakly converges to M and hence for all $a \in A$ we have $E\langle M^{u_n}, M^a\rangle = E\,\phi(u_n)\,\varphi(a) \cdot K \to E\langle M, M^a\rangle$.

(10) implies that the sequences $(\phi(u_n)\,\varphi(a), n \geq 1)$ γ-a.e. converges to some G-predictable process H_t^a depending on the parameter $a \in A$. But by virtue of Lemma 1 a) $\gamma \in \Gamma$ and $u \in U_g$ exist such that γ-a.e.

$$H_t^a = \varphi_t(a)(\gamma_t \bar{\phi}_t(u) + (1-\gamma_t)\,\underline{\phi}_t(u)).$$

Hence

$$E\langle M, M^a\rangle = \lim_n E\langle M^{u_n}, M^a\rangle = E\int_0^\infty \varphi(a) \int_A (\bar{\varphi}(b)\gamma + \underline{\varphi}(b)(1-\gamma)u(db)\,dK =$$

$= E\langle M^{u,\gamma}, M^{a}\rangle$

which implies $M=M^{\gamma,u}$. □

For the general form of kernels φ the discription of the closure $\mathbf{M}_d$ is more sophisticated.

§ 3. CLOSURE OF A CLASS OF DENSITIES

We restrict ourselves by the consideration of a metric compact space A, a bounded ν-a.e. continuous kernel φ and a deterministic increasing process K. The notation $\mathbf{R}_s$, $\mathbf{R}_w$, $\mathbf{R}_g$ is used for the strong, weak and convex closures of the class $\mathbf{R}_d$ in H^2. We shall also use the index G which indicates G-predictable controls, e.g. $\mathbf{R}_d(G),\dots,\mathbf{R}_g(G)$, $U_d(G),\dots,U_g(G)$, in contrast to the notation for the classes $\mathbf{R}_d(F),\dots,U_g(F)$ corresponding to F-predictable controls.

Proposition 5. ([5],[6]) $\mathbf{R}_s(F)=\{\mathcal{E}(M^u),\ u\in U_s(F)\}$, $\mathbf{R}_w(F)=\{\mathcal{E}(M^u),\ u\in U_w(F)\}$, $\mathbf{R}_g(F)=\{\mathcal{E}(M^u),\ u\in U_g(F)\}$, $\mathbf{R}_g(F)$ is weakly compact in H^2_w.

Hence in the case of complete observation G=F the generalized controls from $U_g(F)$ enable us to describe the convex closure of $\mathbf{R}_d(F)$. In the case of partial observation $G\subset F$, however, the class of controls $U_g(G)$ is not sufficient for the description of $\mathbf{R}_g(G)$ and this is due to the fact that the class $\{\mathcal{E}(M^u), u\in U_g(G)\}=\mathbf{R}^o_g(G)$ is not convex now. At the same time since $\mathbf{R}_g(G)\subset\mathbf{R}_g(F)$ some subset $U_g(G,F)$, $U_g(G)\subset U_g(G,F)\subset U_g(F)$ exists such that $\mathbf{R}_g(G)=\{\mathcal{E}(M^u),$ $u\in U_g(G,F)\}$, but the elements of this class are not adapted to the flow G and, hence, are "unrealizable". The introduction of randomized controls enables us to describe the elements of the class $U_g(G,F)$ in "realizable" terms.

First we introduce spaces of programming controls. Let D_d, D_s, D_w, D_g denote subsets of the classes $U_d(G)$, $U_s(G)$, $U_w(G)$, $U_g(G)$ respectively, which consist of deterministic (independent of ω) functions. Let D_g be equipped with the topology of convergence $\alpha_n\to\alpha$, $n\to\infty$, $\alpha_n,\alpha\in D_g$ in the sense that

$$\int_0^\infty\int_A f_t(a)\alpha^n_t(da)dK_t\longrightarrow\int_0^\infty\int_A f_t(a)\,\alpha_t(da)dK_t \tag{11}$$

for all real functions which are bounded, measurable w.r.t. t and continuous in a, where the elements α of the classes D_d, D_s, D_w are indentified with the functions from D_g, which at every point t (at the points of discontinuity of the function K) take for their values the degenerate distribution $q_t \in Q(A)$ with $q_t(C) = I_{[\alpha_t \in C]}$, $C \in \mathcal{A}$.

We denote by Π a set of all probabilities π on the product $(\tilde{\Omega},\tilde{F}_{oo})=(\Omega,F_{oo}) \times (D_g,\mathcal{D}_{oo})$ with a Borel σ-algebra $\mathcal{D}_{oo}$ of subsets D_g such that $\Pi(B\times D_g) = P(B)$, $B \in F_{oo}$. The topology on Π will be defined by the convergence $\pi_n \to \pi$, $n \to oo$ which means

$$\int_{\Omega}\int_{D_g} f(\omega,\alpha)\,\pi^n(d\omega,d\alpha) \to \int_{\Omega}\int_{D_g} f(\omega,\alpha)\,\pi(d\omega,d\alpha) \tag{12}$$

for all real bounded functions f which are measurable w.r.t. ω and continuous in α (in D_g).

Consider a flow of σ-algebras $\tilde{F}=(\tilde{F}_t=F_t\times\mathcal{D}_t)_{t\geq 0}$ where $\mathcal{D}_t=\sigma(\alpha_s, s\leq t)$ is a σ-subalgebra of $\mathcal{D}_{oo}$. Introduce the subsets $\Pi_d(G)$, $\Pi_s(G)$, $\Pi_w(G)$, $\Pi_g(G)$ of the space Π of measures π concentrated on the subsets D_d, D_s, D_w, D_g, respectively, satisfying the conditions

$$\begin{aligned} &1)\ E^{\pi}\int_0^{oo} f_t(\alpha)dV_t = E^{\pi}\int_0^{oo} {}^{G,P}(f_t(\alpha))dV_t, \\ &2)\ E^{\pi} f_t(\alpha)\eta = E^{\pi} f_t(\alpha)E(\eta/F_t) \end{aligned} \tag{13}$$

for any function $f=f_t(\omega,\alpha)$ which is $\tilde{F}_t$-measurable for every t, an F_{oo}-measurable random variable η and any G-predictable increasing process V. (${}^{G,P}(.)$ denotes a G-predictable projection w.r.t. the measure P; and σ-algebras G_t, F_t are identified with the σ-algebras $G_t\times D_g$, $F_t \times D_g \subset \tilde{F}_t$.)

The elements π of the spaces $\Pi_d(G),\dots,\Pi_g(G)$ or the corresponding conditional distributions $\Pi(./\omega)$ ($\Pi(d\omega,d\alpha) = \Pi(d\alpha/\omega)\,P(d\omega)$) will be called the randomized controls. Property 1), in a certain sense, expresses the G-predictability of the controls, and Property 2) is equivalent to the condition $E^{\pi}(E^{\pi}(./F_{oo})/\tilde{F}_t) = E^{\pi}(./F_t)$, $t \geq 0$ and denotes the conditional independence of σ-algebras F_{oo} and $\tilde{F}_t$ for a given F_t, which, in its turn, is equivalent to the condition $M(F,P) \subset M(\tilde{F},\pi)$ (F-martingales are $\tilde{F}$-martingales w.r.t. the measure π) ([1]). In particular, the processes $(M^a, a \in A)$, $(M^{\alpha}, \alpha\in D_g)$ which are elements of $\mathbf{M}_d(G)$ for every $\alpha \in D_g$ are $\tilde{F}$-martingales.

The randomized controls define the class of probabilities $\tilde{\mathbf{P}}^{\Pi}_g(G) = \{\tilde{P}^{\pi}(d\omega,d\alpha)= \mathcal{E}_{oo}(M^{\alpha})\,\pi(d\omega,d\alpha),\ \pi\in\Pi_g(G))$ on $(\tilde{\Omega},\tilde{F})$, their restrictions $\mathbf{P}^{\Pi}_g(G) = \{P^{\pi} = \tilde{P}^{\pi}/F_{oo},\ \tilde{P}^{\pi}(B)=E^{\pi}I_B\,\mathcal{E}_{oo}(M^{\alpha}),\ B\in F_{oo}, \pi\in\Pi_g(G)$ on (Ω,F) and a class of densities $\mathbf{R}^{\Pi}(G) = \{\rho^{\pi}, \rho^{\pi}_t = E^{\pi}(\mathcal{E}_t(M^{\alpha})/F_t)$, $\pi \in \Pi_g(G)$, $t\geq 0\}$ w.r.t. the basic measure P (the subclasses $\tilde{P}^{\Pi}_d(G)$, $\tilde{P}^{\Pi}_s(G)$, $P^{\Pi}_d(G)$, $P^{\Pi}_w(G)$, $\mathbf{R}^{\Pi}_d(G)$, $\mathbf{R}^{\Pi}_s(G)$ can also be considered).

A study of randomized controls of this kind is natural since the classes $U_d(G),\dots,U_g(G)$ can be imbedded in the classes $\Pi_d(G),\dots,\Pi_g(G)$ if the elements u are considered as functions defined on Ω and taking for their values the degenerate distribution on the classes of programming controls D_d, D_s, D_w, D_g, respectively ($\omega \rightsquigarrow \Pi(C/\omega) = I_{[u(\omega,.)\in C]}, C \in \mathcal{D}_{oo}$). Obviously, with this imbedding $\mathcal{S}^{\pi} = \mathcal{S}^{u}$. Properties 1), 2) for $\mathbf{R}_g(G)$ are directly verified and

$$\mathbf{R}_d(G) \subset \mathbf{R}^{\Pi}_d(G),\dots,\mathbf{R}_g(G) \subset \mathbf{R}^{\Pi}_g(G)$$

We can also note that a more direct approach to the conversion of the class $\mathbf{R}_g(G)$ into a convex set by the immediate introduction of randomization on the class of controls $U_g(G)$, actually, leads to an excessive expansion of controls, Indeed, if μ is some probability on the Borel subsets of the set $U_g(G)$ with the topology of convergence $u^n \longrightarrow u$ which denotes convergence of $E \int_0^{oo} \int_A f_t(a) u^n_t(da) dK_t \longrightarrow E \int_0^{oo} \int_A f_t(a) u_t(da) dK_t$ for all bounded F-adapted functions f continuous in a, then we have

$$\mathcal{S}^{\mu}_t = \int_{U_g(G)} \mathcal{S}^u_t \, \mu(du) = \mathcal{S}^{\pi}_t = \int_{D_g} \mathcal{S}^{\alpha}_t \, \Pi^{\mu}(d\alpha/\omega)$$

$\pi^{\mu}(./\omega)$ is defined by

$$\pi^{\mu}(C|\omega) = \int I_{[u(\omega,.)\in C]} \, \mu(du), \; C \in \mathcal{D}_{oo}. \tag{14}$$

It can be seen that the measure π^{μ} with $\pi^{\mu}(d\omega, d\alpha) = \pi^{\mu}(d\alpha/\omega) \, P(d\omega)$ belongs to $\Pi_g(G)$, but on the other hand, it follows from (14) that the measure π^{μ} non-uniquely defines the measure μ. The definition of measure π^{μ} only determines the distribution of "coordinates" of $u(\omega,.)$ for fixed $\omega \in \Omega$.

Now we can show that the randomized controls form a compact space induced by simple controls from $U_d(G)$. First we must prove that the class $\mathbf{R}_g(G)$ can be imbedded in $\mathbf{R}_g(F)$. We shall associate with each probability $\pi \in \Pi_g(G)$ the measure ν^{π} on $\Omega \times D_g \times [0,oo[$ which is a restiction of the measure $\tilde{\nu}^{\pi}(d\omega, d\alpha, dt) = \pi(d\omega, d\alpha) dK_t$ to σ-algebra $\mathcal{P}(F) \otimes \mathcal{D}_{oo}$ ($\mathcal{P}(F)$ is σ-algebra of F-predictable sets on $[0,oo[\times\Omega$). Let ν^{π}_t denote F-predictable distribution on D_g defined by $\nu^{\pi}(d\omega, d\alpha, dt) = \nu(d\omega, dt) \, \nu^{\pi}_t(d\alpha)$. Condition 1) implies that $\nu^{\pi}_t(.)$ is, actually, G-predictable.

<u>Proposition 6.</u> For $\pi \in \Pi_g(G)$ we have $\mathcal{S}^{\pi} = \mathcal{S}^{u^{\pi}}$ where u^{π} is defined by the equality

$$u^{\pi}_t(C) = \int_{D_g} \mathcal{E}_{t-}(M^{\alpha}) \alpha_t(C) \, \nu^{\pi}_t(d\alpha) \Big/ \int_{D_g} \mathcal{E}_{t-}(M^{\alpha}) \, \nu^{\pi}_t(d\alpha), \; C \in \mathcal{A}. \tag{15}$$

<u>Proof</u> Since by Condition 2) $\rho^{\alpha}=\mathcal{E}(M^{\alpha})$ is an element of $\mathcal{M}^2(\pi,\tilde{F})$ and ρ^{π} is a filter of ρ^{α} w.r.t. the flow $(F_t,\ t\geq 0)$ general filtration formulas imply that the following equality $\langle\rho^{\pi},m\rangle=\langle\rho^{\alpha},m\rangle^{\pi,F}$ is valid ($(.)^{\pi,F}$ is an F-predictable dual projection w.r.t. the measure π) for any $m\in\mathcal{M}^2(P,F)$. Indeed, the difference $Z_t=\rho^{\alpha}_t m_t-\langle\rho^{\alpha},m\rangle_t$ is an $\tilde{F}$-martingale w.r.t. the measure π, hence the filter $\hat{Z}_t=\rho^{\pi}_t m_t-E^{\pi}(\langle\rho^{\alpha},m\rangle_t/F_t)$ will be an element of $\mathcal{M}^2(F,P)$. But by the definition of the dual projection the process $E^{\pi}(\langle\rho^{\alpha},m\rangle_t/F_t)-\langle\rho^{\alpha},m\rangle_t^{\pi,F}$ also belongs to $\mathcal{M}^2(P,F)$ and the equality $\langle\rho^{\pi},m\rangle=\langle\rho^{\alpha},m\rangle^{\pi,F}$ follows from the uniqueness of Doob-Meyer's decomposition. On the other hand, $\langle\mathcal{E}(M^{\alpha}),m\rangle=\int_0^t\mathcal{E}_{s-}(M^{\alpha})d\langle M^{\alpha},m\rangle=$ $=\int_0^t\mathcal{E}_{s-}(M^{\alpha})\int_A H_s(m,a)\,\alpha_s(da)dK_s$ and, hence,

$$\langle\rho^{\pi},m\rangle=\Big(\int_0^t\mathcal{E}_{s-}(M^{\alpha})\int_A H_s(m,a)\,\alpha_s(da)dK_s\Big)^{\pi,F}=\int_0^t\rho^{\pi}_{s-}\int_A H_s(m,a)u^{\pi}_s(da)dK_s$$

so the assertion is proved.

In what follows we shall need the following lemma.

On the product $(X,\mathcal{X})\times(Y,\mathcal{Y})$ of measurable spaces we consider a set $\mathfrak{M}$ of all probabilities μ such that $\mu(B\times Y)=\mu_0(B),B\in\mathcal{X}$ where μ_0 is a fixed probability on $\mathcal{X}$. Suppose σ-algebra $\mathcal{X}$ is countably generated and let $(\mathcal{X}_n,n\geq 1)$ be a sequence of nondecreasing finite algebras which generate $\mathcal{X}$. Besides let $\mathcal{Y}$ be a Borel σ-algebra of the metric compact Y. We consider the convergence $\mu^n\longrightarrow\mu$ which denotes the convergence of integrals

$$\int f(x,y)\,\mu^n(dx,dy)\longrightarrow\int f(x,y)\,\mu(dx,dy) \tag{16}$$

for any $f\in\mathcal{F}$ from the class of real bounded functions which are $\mathcal{X}$-measurable w.r.t. x and continuous in y. Let $\mathfrak{M}^0$ denote a subset $\mathfrak{M}$ of measures μ with degenerate conditional distributions $\mu(dy/x)$ of the form $\mu(\{y_i\}/x)=1$ for $x\in B_i$, $B_i\in\mathcal{X}_n$ $\sum_{i=1}^m B_i=X$, $y_1,y_2,\dots,y_m\in Y$.

<u>Lemma 2.</u> a) $\mathfrak{M}$ is a metric compact; b) if the measure μ_0 has no atoms then $\mathfrak{M}^0$ is dense in $\mathfrak{M}$; c) $\mathfrak{M}$ is a convex closure of $\mathfrak{M}^0$.

<u>Proof.</u> a) Let $g=(g_i,i\geq 1)$ be an orthonormal basis of the Hilbert space $L_2(X,\mathcal{X},\mu_0)$. For a sequence of numbers $(c_i,\ i\geq 1)$ with $c_i\geq 0$ $\sum_{i=1}^{\infty}c_i<\infty$ $\mu_0\{x:\sum_{i=1}c_ig_i^2(x)<\infty\}=1$ and therefore the mapping

$x,y\rightsquigarrow g(x),y$ transfers the measures $\mu\in\mathfrak{M}$ into the measures μ^g on the product $\Gamma\times Y$ where Γ is a Hilbert space of the sequences $z\in R^{(\infty)}$ with the norm $\|z\|_c=(\sum_{i=1}^{\infty}c_iz_i^2)^{1/2}$.

Let $(\mu^n,n\geq 1)$ be a sequence from $\mathfrak{M}$ and $(\mu^g_n,n\geq 1)$ be the corresponding images. The compactness of Y and the condition $\mu_n^g(dz\times Y)=\mu_0^g(dz)$ imply weak compactness of the sequence μ_n^g and let $\mu_{n'}^g$ be a subsequence weakly converging to some measure μ^g on $\Gamma\times Y$. Consider the measure μ on $\mathcal{X}\otimes\mathcal{Y}$ defined as $\mu(dx,dy) = \mu_0(dx)\,\mu^g(dy/g(x))$ where $\mu^g(./z)$ is a conditional distribution corresponding to the measure μ^g. We can show that the sequence $\mu_{n'}$ converges to μ in the sense of (16). The convergence of (16) takes place for the functions of the class $\mathcal{F}^g = \{f : f_\ell(x,y) = \sum_{k=1}^{\ell} g_k(x)c_k^f(y),$ $c_k^f(y) = \int_X f(x,y)g_k(x)\mu_0(dx),\ f\in\mathcal{F},\ 1\le \ell<\infty\}$. Indeed,

$$\int_{X\times Y} f(x,y)\mu_{n'}(dx,dy) = \int_{\Gamma\times Y} \bar{f}_\ell(z,y)\mu_{n'}^g(dz,dy) \longrightarrow \int_{\Gamma\times Y} \bar{f}_\ell(z,y)\,\mu^g(dz,dy) =$$

$$= \int_{X\times Y} f_\ell(x,y)\,\mu(dx,dy)$$

since the functions $\bar{f}_\ell(z,y) = \sum_{k=1}^{\ell} z_k c_k^f(y)$ are continuous on Y and uniformly integrable in n' w.r.t. the sequence $\mu_{n'}^g$.

Now it is sufficient to show that for every function $f\in\mathcal{F}$ a sequence $f_k\in\mathcal{F}^g$ can be found such that

$$\int_X \sup_y |f(x,y)-f_k(x,y)|^2\,\mu_0(dx) \longrightarrow 0,\ k\longrightarrow\infty . \tag{17}$$

For the functions from $\mathcal{F}$ which are simple in x and can be represented as $f(x,y) = \sum_{k=1}^{m} f_k(y)\varphi_k(x)$ with $\varphi_k(x)=I_{B_k}\in\mathcal{X}_n$, $\sum_{k=1}^{m} B_k = X$ with functions f_k continuous in y, this approximation follows from the estimator

$$|f(x,y)-f_\ell^f(x,y)|^2 = \Big[\sum_{k=1}^{m} f_k(y)(\varphi_k(x) - \sum_{i=1}^{\ell} g_i(x)c_i^{\varphi_k})\Big]^2 \le$$

$$\le c_m \max_{y,k} f_k^2(y) \sum_{k=1}^{m} (\varphi_k(x) - \sum_{i=1}^{\ell} g_i(x)c_i^{\varphi_k})^2 ,\ \text{where}$$

$$c_i^{\varphi_k} = \int_X \varphi_k(x)g_i(x)\,\mu_0(dx)$$

Finally, $f\in\mathcal{F}$ can be approximated by simple functions (w.r.t. x) from $\mathcal{F}$. Let $(h^i, i\ge 1)$ be a dense set in the space of continuous functions on Y. Put $\tau_k(x) = \min\{ i\ge 1 : \sup_y |f(x,y) - h^i(y)| \le \frac{1}{k}\}$ and let $f_{k,N}(x,y) = h^{\tau_k(x)}(y)\cdot I_{[\tau^k(x)\le N]}$, $k\ge 1$, $N\ge 1$. Then

$$|f(x,y)-f_{k,N}(x,y)| = |f(x,y)I_{[\tau_k(x)>N]} +$$

$$+ \sum_{i=1}^{N} I_{[\tau_k(x)=i]}(f(x,y)-h^i(y))| \le$$

$\leq \sup_{x,y} f(x,y)I_{[\tau_k(x) > N]} + \frac{1}{k}$ which proves the assertion since $\mu_o(\tau_k(x) > N) \longrightarrow 0$ as $N \longrightarrow \infty$. Besides it is sufficient to verify the convergence in (16) for a countable class of functions of the form $\sum_{k=1}^{m} h_k(y)I_{B_k}(x)$, $B_k \in \mathcal{X}_n$, $\sum_{k=1}^{m} B_k = X$, $n \geq 1$, which proves the metrizability of $\mathcal{M}$.

b) Let $Q_o(Y)$ denote a set of probabilities on Y concentrated at a finite number of point from Y. Since $Q_o(Y)$ is dense everywhere in the set of all probabilities on Y in Levy-Prokhorov's metric the set of measures μ with $\mu(./x)$ of the form $\mu(\{y_i\}/x) = \varrho_i(x)$ $\sum_{i=1} \varrho_i(x) = 1$ for some $y_1, y_2, \ldots, y_n$ Y is dense everywhere in $\mathcal{M}$. The approximation of such measures by the elements from $\mathcal{M}^o$ is shown in [6] (Appendix A).

c) For the proof it is sufficient to note that on the atoms of the measure μ_o $\mu(./x)$ can be considered constant in x and apply the density $Q_o(Y)$ in $Q(Y)$.

Proposition 7. a) The spaces D_g, $\Pi_g(G)$ are compact; b) if K is continuous and the probability P has no atoms then $D_g = D_w$ is a closure of D_d, $\Pi_g(G) = \Pi_w(G)$ is a closure of $U_d(G)$ when the latter is naturally imbedded in the space Π_g; c) D_g, $\Pi_g(G)$ are convex closures of D_d and $U_d(G)$, respectively.

Proof. a) It is sufficient to apply assertion a) of Lemma 2 first to the space D_g (with $X = [0,\infty[$, $Y = A$, $\mu_o dt = dK_t$) and then to the space $\Pi_g(G)(X = \Omega, Y = D_g, \mu_o = P)$. The fact of Π_g being closed can be obtained passing to the limit in relations (13), where the study may, evidently, be restricted by the functions f continuous in α; c) assertion b) of the lemma implies that the class of measures Π for which $\Pi(d\alpha/\omega)$ is representable as

$$\Pi(\{\alpha^i\}/\omega) = I_{B_i}(\omega), \quad \sum_{i=1}^{m} B_i = \Omega, \quad B_i \in F_\infty \tag{18}$$

for some $\alpha^1, \ldots, \alpha^n$ from the set D_d (dence in D_g) is dense in Π. But this measure is equivalent to the function $u_t^\pi(\omega) = \sum_i \alpha_t^i I_{B_i}(\omega)$, $t \geq 0$.

To prove the proposition it is sufficient to verify that the process $u_t^\pi(\omega)$ for $\pi \in \Pi_g(G)$ belongs to $U_d(G)$. But then it is sufficient to apply condition (13) to the measure π with $f_t(\alpha) = X_t I_{[\alpha_t = a]}$ for $a \in A$ and the F_t-predictable process X. Since the integration w.r.t. the measure π from (18) means the substitution of $\alpha = u^\pi$ in (13) these conditions require that the process $I_{[u_t^\pi = a]}$ should be G-predictable and, hence, $u^{\pi} \in U_d(G)$. Now assertion c) immediately follows from Lemma 2 c).

The derivation of compactness for the set $P^{\Pi}(G)$ is connected to the continuity condition for the exponent $\mathcal{E}(M^{\alpha})$ w.r.t. the parameter α.

Proposition 8. In almost all ω w.r.t. the measure P let the expression $Z^{\alpha}(\eta) = E(\mathcal{E}_{\infty}(M^{\alpha})\eta/G_{\infty})$ be continuous in α in D_g. Then the class $\mathbf{R}_g(G)$ is compact in H^2, $\mathbf{R}_g^{\Pi}(G)=\mathbf{R}_g(G)$. In the maximization problem $E^{\Pi}\mathcal{E}_{\infty}(M^{\alpha})\eta = E^{P^{\pi}}\eta$ for F_{∞}-measurable r.v. η with $E\eta^2 < \infty$ an optimal randomized control from $\Pi_g(G)$ exists. If K is continuous and P has no atoms, then $\mathbf{R}_g^{\Pi}(G)=\mathbf{R}_w^{\Pi}(G) = \mathbf{R}_w(G)=\mathbf{R}_g(G) = \{\mathcal{E}(M^u), u \in U_g(G,F)\}$ and the elements of the class $U_g(G,F)$ are defined by formula (15).

The proof follows from Proposition 7 and the possibility of representation of the expectation $E^{\pi}\mathcal{E}_{\infty}(M^{\alpha})\eta$ in the form $E^{\pi}\mathcal{E}_{\infty}(M^{\alpha})\eta = E^{\pi}E(\mathcal{E}_{\infty}(M^{\alpha})\eta/G_{\infty})$ which results from Property (13)1) of the measures $\pi \in \Pi_g$.

Consider the following important case of the continuity of $Z^{\alpha}(\eta)$

Let $(\Omega, F, G, P) = (X\times Y, \mathcal{X}\otimes\mathcal{Y}, \mathcal{X}, P)$ where $\mathcal{X}=(\mathcal{X}_t)_{t\geq 0}$, $\mathcal{Y}=(\mathcal{Y}_t)_{t\geq 0}$ be flows of σ-algebras of the subsets of X and Y and the flow $\mathcal{X}$ be identified with $\mathcal{X}\times Y$. Suppose that

$$M^a=M^a(x,y) = N^a(x,y) + L(x,y), \quad a \in A \tag{19}$$

where $N^a(x,y)$ is a $\mathcal{Y}$-martingale w.r.t. the conditional measure $P_x(.) = P(./x)$ for every $x \in X$, $[N^a, L]=0$, ($[.,.]$ denotes the joint quadratic variation) and let $E_x(\mathcal{E}(L)\eta)^2 < \infty$ (E_x denotes conditional expectation $E(./G_{\infty}) = E(./x)$), in particular, $Z_x^{\alpha}(\eta)=E_x\mathcal{E}(N^{\alpha})\tilde{\eta}\,(\tilde{\eta}=\mathcal{E}(L)\eta)$.

Proposition 9. $Z_x^{\alpha}(\eta)$ is continuous in α in D_g almost for all x.

Proof. If for every fixed $x \in X$ probabilities P_x^{α}, $\alpha \in D_g$ on $\mathcal{Y}_{\infty}$ are considered which are defined as $P_x^{\alpha}(dy)= \mathcal{E}(N^{\alpha}(x))\, P_x(dy)$ we have $Z_x^{\alpha}(\eta)-Z_x^{\beta}(\eta) = E_x^{\alpha}\tilde{\eta}(x)-E_x^{\beta}\tilde{\eta}(x)$ for $\alpha, \beta \in D_g$ where $N^{\alpha}(x)=N^{\alpha}(x,.)$ $\tilde{\eta}(x)=\tilde{\eta}(x,.)$. As it is shown in [6] the difference $Z^{\alpha}-Z^{\beta}$ can be represented as

$$Z_x^{\alpha}(\eta)-Z_x^{\beta}(\eta)=E_x\,\mathcal{E}_{\infty}(N^{\alpha}(x))\int_0^{\infty}\int_A H_s(m^{\beta}(x),a)(\alpha_s(da)-\beta_s(da))dK_s \tag{20}$$

where $m^{\beta}(x)$ is a martingale w.r.t. the measure P^x, $m^{\beta}(x) = S^{\beta}(x) - \langle S^{\beta}(x), N^{\beta}(x)\rangle$ and $S_t^{\beta}(x)=E_x(\tilde{\eta}(x)/\mathcal{Y}_t)$ (see also [8]).

To obtain the assertion it is sufficient to show that almost for all x the function $\gamma_t^{\alpha,\beta}(x,a) = E_x\,\mathcal{E}(N^{\alpha}(x))H_t(m^{\beta}(x),a)$ is continuous in a for every $\beta \in D_g$ uniformly in $\alpha \in D_g$ and $\int_0^{\infty}\sup_{a,\alpha}\gamma_t^{\alpha,\beta}(x,a)dK_t < \infty$.

Using the estimators

$$|H_t(m^\beta,a)|^2 = \left|\frac{d\langle m^\beta(x),M^a\rangle_t}{dK_t}\right|^2 \le \left|\frac{d\langle m^\beta(x)\rangle_t}{dK_t}\right| \left|\frac{d\langle N^a\rangle_t}{dK_t}\right| \le$$

$$\le c_1(d\langle m^\beta(x)\rangle_t/dK_t), \quad E_x\, \mathcal{E}^2(N^\alpha(x)) \le e^c$$

(valid by virtue of the conditions $\langle N^\alpha\rangle \le c$ and $d\,\langle N^a\rangle/dK =$ $= \varphi(a,a) \le c_1$ we have $\int_0^\infty (\sup_{a,\alpha} \gamma_t^{\alpha,\beta}(x,a))^2 dK_t \le e^c c_1 E_x\langle m^\beta(x)\rangle$. But [6] (Appendix 2) implies that $E_x\langle m^\beta(x)\rangle \le c_2 E_x\, \tilde{\eta}^2(x)$ where c_2 depends only on the constant c. Similarly we obtain

$$\left|\gamma_t^{\alpha,\beta}(x,a) - \gamma_t^{\alpha,\beta}(x,b)\right| \le e^c\, E_x \left|\frac{d\langle m^\beta(x)\rangle_t}{dK_t}\right| \left|\frac{d\langle N^a - N^b\rangle_t}{dK_t}\right| \le e^c\, E_x \left|\frac{d\langle m^\beta(x)\rangle_t}{dK_t}\right| \cdot$$

$$\cdot\left|\varphi_t(a,a) - 2\,\varphi_t(a,b) + \varphi_t(b,b)\right|$$ and the assertion follows from the boundedness and a.e. continuity of the kernel φ.

In the optimization problem for the set $\mathbf{M}$ from the countable normed Hilbert space $H^2_{loc}(\tau)$ one has only to specifiy the conditions for the r.v. η (not for every F_∞-measurable and even bounded η the expression $E^u\eta$ will be a bounded functional on $\mathbf{R}_g$). A sufficient condition under which the assertion about the optimal control existence holds is the F_{τ_n}-measurability of η for some $n \geq 1$ or, more generally, if $\lim_{t\to\infty} \sup_u E^u(\eta/F_t) = \eta$ P^u-a.s. for $u \in U_g(F)$ (see [6]).

As an example illustrating condition (19) a control problem can be considered for a partially observable diffusion type process

$$\begin{aligned} dy_t &= f_t(x,y,u_t)dt + b_t(x,y)dW_t \\ dx_t &= \varphi_t(x,y)dt + c_t(x)dV_t \end{aligned} \tag{21}$$

with bounded non-anticipative functions f, φ, b, c ($|b|, |c| > 0$) and independent Wiener processes W, V when for the measure P we have the distribution on $C^{(2)}_{[0,T]}$ of the solutions of a homogeneous system of equations ($f = \varphi = 0$) and the martingale M^a has the form

$$M^a_t(x,y) = \int_0^t \frac{f_s(x,y,a)}{b^2_s(x,y)}\, dy_s + \int_0^t \frac{\varphi_s(x,y)}{c^2_s(x)}\, dx_s = N^a(x,y) + L(x,y)$$

and it is important that the coefficient φ does not contain any control parameter a, otherwise the stochastic integral $\int_0^t \int_A \frac{\varphi_s(x,y,a)}{c_s(x)}\, \alpha_s(da)dx_s$ is not continuous in α (in D_g) for almost all x.

The randomized controls were introduced in [9] and further considered in [10], [11] for the diffusion Markov processes. The compactization of the space of measures Π_g in a topology which is stronger than that considered in the above papers - a topology of weak convergence on $X \times Y \times D_g$ - enables us, in particular, to avoid the conditions of continuity of the coefficients in (21) in x, y. Note, however, that these papers considered the case of degenerate diffusion too when the distribution of the non-observable component y is constructed on the basis of the strong solution.

We do not know any general assertion on the sufficiency of controls from $U_g(G)$. The fact that the usual controls from U_s are not sufficient even in the case of complete observations of the functionals of the general type $\eta = \eta(\omega, \alpha)$ is shown by an example due to Varadhan (see [9]). The fact that the randomized controls from the class $\Pi_s(G)$ concentrated on the usual programming controls are not sufficient can be seen from the following elementary example which, actually, belong to the problem of deterministic control (case $\eta = \eta(\omega)$).

Example. Let the controlled system be defined by the equations

$$dy_t = u_t dt + dW_t, \quad dx_t = dV_t, \quad 0 \le t \le 1,$$

for $A = \{1, -1\}$ and, hence, the controls from $\Pi_s(G)$ are concentrated on D_s which consists of the functions α taking values -1 and 1 and, evidently, for W and α are independent. In the minimization problem for the functional

$$I^\pi = E^\pi \mathcal{E}_1(\int \alpha_t dW_t) \int_0^1 W_t^2 dt = E^\pi \int_0^1 y_t^2 \, dt =$$

$$= E^\pi \Big[\int_0^1 (\int_0^t \alpha_s ds)^2 dt + \int_0^1 W_s^2 ds\Big] = E^\pi \int_0^1 (\int_0^t \alpha_s ds)^2 dt + \frac{1}{2}$$

for any function α from D_s it is evident that $\int_0^1 (\int_0^t \alpha_s ds)^2 dt > 0$ and, hence, $I^\pi > \frac{1}{2}$, $\pi \in \Pi_s(G)$. At the same time for $u \in U_g(G)$ with $u \equiv 0$ we have $I^\pi = \frac{1}{2}$.

REFERENCES

1. Jacod J. Calcul stochastique et problemes de martingales. Lect.Notes in Math. Vol. 714, Springer-Verlag: Berlin-Heidelberg-New York, 1979

2. Vakhania N.N., Tarieladze V.I., Chobanyan Yu. Probability distributions in Banach spaces. Moscow, Nauka, 1985 (in Russian).

3. Gal'chuk L.I. Structure of vector martingales. Proc. of the Seminar in the Theory of Random processes in Druskininkai,1974 (in Russian).

4. Meyer P.A. Integrales hilbertiennes S.P.S.XI Lect.Notes in Math, 581, 446-462. Springer Verlag:Berlin-Heidelberg-New York, 1977.

5. Chitashvili R.J., Mania M.G. On optimal controls in the problem of locally absolutely continuous change of measures. Proc. of IV Intern. Vilnius Conf. on Probab.Th.and Math.Stat.,Vilnius, 1985.

6. Chitashvili R.J.,Mania M.G. Optimal locally absolutely continuous change of measure in filtered space. Stochastics (to appear).

7. Chitashvili R.J., Toronjadze T.A. On one-dimensional stochastic differential equations with unit diffusion coefficient. Stochastics, 1981, vol. 4, pp. 281-315.

8. Chitashvili R.J. Martingale ideology in the theory of controlled stochastic processes, Lect.Notes in Math., Vol. 1021, 1983.

9. Fleming W.H., Pardoux E. Optimal control for partially observed diffusions, SIAM J. Control Optim., 20 (1982), 258-285.

10. Bismut J.M. Partially observed diffusions and their control, SIAM J. Control Optim., 20 (1982), 302-309.

11. Fleming W.H., Nisio M. On stochastic relaxed control for partially observed diffusions, Nagoya Math. J., Vol. 93 (1984), 71-108.

12. Gikhman I.I., Skorokhod A.V. Stochastic differential equations. Kiev: Naukova Dumka, 1982.

PARTIALLY OBSERVABLE CONTROL OF
DIFFUSIONS WITH CORRELATED NOISE

Nicole EL KAROUI
Laboratoire de Probabilités
Université de Paris VI:tour 46-56
4, Place Jussieu 75005 PARIS

INTRODUCTION

We are concerned with the control of diffusions under partial observations with correlated noise. The state and the observation processes are governed by the stochastic differential equations:

$$\begin{cases} dX_t = b(t,X_t,Y_t,U_t)\,dt + \sigma_j(t,X_t,Y_t,U_t)\,dB_t^j + \gamma_i(t,X_t,Y_t)\,dW_t^i \\ dY_t = h(t,X_t,Y_t)\,dt + dW_t \\ X_t = X_r \quad \text{and} \quad Y_t = Y_r \quad \text{for} \quad t \leq r, \quad \text{and} \quad P \circ X_r^{-1} = \mu \end{cases}$$

where (b, σ_j, γ_i) (resp h) are n-dimensional (resp d-dimensional) vector fields, and $(B^j)_1^k$ and $(W^i)_1^d$ are independent Brownian motions.

Of course, the control process U_t is required to be adapted to the filtration ($\underline{\underline{F}}_t^Y$) of observations. Under these conditions, we can calculate recursively the conditional distribution of X_t given $\underline{\underline{F}}_t^Y$, such that

$\pi_t(f) := E\left[f(X_t)/\underline{\underline{F}}_t^Y\right]$ by the Fujisaki-Kallianpur-Kunita formula:

$$\begin{cases} d\pi_t(f) = \pi_t(Lf(t,.,Y_t,U_t)\,dt + \{\pi_t(\Gamma f(t,.,Y_t) - \pi_t(f)\pi_t(h)(t,Y_t)\} \\ \qquad\qquad \{dY_t - \pi_t(h)(t,Y_t)\,dt\} \\ \pi_t = \pi_r \quad \text{for} \quad t < r \end{cases}$$

where $Lf(t,x,y,u)$ is the elliptic operator associated with the state process $X.$, and $\Gamma f(t,x,y)$ is a d-dimensional vector with coordinates

$$\Gamma f(t,x,y)_i = \sum_{j=1}^{n} \gamma_i^j(t,x,y)\frac{\partial f(t,x)}{\partial x_j} + h_i(t,x,y)\,f(t,x)$$

The problem is to find a control which maximizes a criterion in the form, $j(r,y,\mu,\mathcal{U}) := E\int_r^T k(s,Y_s,X_s,U_s)\,ds + g(Y_T,X_T)$.

where k is an instantaneous reward and g a terminal reward.

An important question is to show that an optimal control can be chosen as a function of the filter (π_t). It is known in the literature as a "separation theorem ".

We solve these problems by making use of the same arguments as in the non correlated noise case, as we have explained in [E.H.J.2]. The main difficulty is to give a good statement of a relaxed problem. Then, by the compactness of the distributions of controlled state, observation and filter processes and a "nice" dependence with respect to the initial conditions, we prove the existence of an optimal markovian filter.

I . STATEMENT OF THE PROBLEM, and DYNAMIC PROGRAMMING EQUATIONS

The problem of existence of controls, which are adapted to the observation filtration $\underline{\underline{F}}_t^Y$ is not easy: we must know the observation in order to determine the adapted control U_t, which is used to control the state X_t, which gives the observation Y_t.....and so on.

Let us weaken the measurability conditions on the control U_t and require the adaptation of U_t with respect to a filtration $\underline{\underline{F}}_t$ which is larger that $\underline{\underline{F}}_t^Y$ and smaller that $\underline{\underline{F}}_t^{X,Y} = \underline{\underline{G}}_t$.
The basic property that we require on the filtration $\underline{\underline{F}}_t$ is that the conditional distribution of X_t given $\underline{\underline{F}}_t$, π_t, is still given by the Fujisaki-Kallianpur-Kunita formula.

Moreover, in order to avoid the introduction of convexity assumptions on the coefficients, we use relaxed controls, i.e. $\underline{\underline{F}}_t$-adapted processes which take values in the space $\mathcal{P}(A)$ of probability measures on the metric compact space A.

A- Hypotheses and notations

Let A be a metric compact space and $\mathcal{P}(A)$ the set of all probability measures on A, endowed with the weak* topology. A generic point in $\mathcal{P}(A)$ is q and the integration with respect to q is denoted by $q(f) := \int f(a)\, q(da)$.

(A_1) The functions $\{b_k, \sigma_j^k, \gamma_i^k, h_k\}$, defined on $R^+ \times R^n \times R^d \times A$ are bounded and uniformly continuous with respect to (x,y,a).

(A_2) The elliptic operator associated with the state process (X_t) is defined on $C_b^{1,2}(R^+ \times R^n)$ by:

$$Lf(t,x,y,\alpha) := 1/2 \sum_{i,j} a_{i,j}(t,x,y,\alpha)\frac{\partial^2 f}{\partial x_i \partial x_j}(t,x) + \sum_j b_j(t,x,y,\alpha)\frac{\partial f}{\partial x_j}(t,x) + \frac{\partial f}{\partial t}(t,x)$$

where $a_{i,j}(t,x,y,\alpha) := \sum_{l=1}^{k} \sigma_i^l \sigma_j^l (t,x,y,\alpha) + \sum_{m=1}^{d} \gamma_i^m \gamma_j^m (t,x,y)$

The correlation between state and observation processes is described by the operator Γ from $C_b^1(R^+{}_\times R^n)$ into R^d, given by

$$\Gamma f(t,x,y) := \Big[\sum_{m=1} \gamma_i^m \frac{\partial f}{\partial x_m}(t,x) + h_i(t,x)\, f(t,x) \Big]_{i=1}^{d}$$

Notice that Γ is independent of controls.

The elliptic operator associated with the pair (signal-observation) processes is given by:

$$\Lambda f(t,x,y,\alpha) := Lf(t,x,y,\alpha) + 1/2 \sum_j \frac{\partial^2 f}{\partial y_j^2}(t,x,y) + \sum_{i=1}^{n} \sum_{m=1}^{d} \gamma_i^m \frac{\partial^2 f}{\partial x_i \partial y_m}(t,x,y)$$

$$+ \sum_{j=1}^{d} h_j(t,x,y) \frac{\partial f}{\partial y_j}(t,x,y)$$

(A_3) The functions k and g which characterize the instantaneous and terminal rewards are bounded on $R^+{\times}R^n{\times}R^d{\times}A$

We use also rewards which depend on the filter, given by bounded functions K and G on $R^+{}_\times R^d{}_\times \mathcal{P}(R^n)_\times A$.

For example, $K(t, y, m, \alpha) = \int k(t,x,y,\alpha)\, m(dx)$

$G(y,m) = \int g(x,y)\, m(dx)$

(B_1) Let $(\mathcal{X}, \underline{\underline{\mathcal{X}}}_t)$ be the canonical space of the state process $C([0,T], R^n)$, endowed with the right continuous filtration associated with $\underline{\underline{\mathcal{X}}}_t^o = \sigma(x_s; s \leq t)$. A generic point of $\mathcal{X}$ is denoted by $[x]$ or (x_t) .

In the same way, we introduce the canonical space of observation process $\mathcal{Y} = C([o,T], R^d)$, and the one associated with the filter process $\Pi = C([o,T], \mathcal{P}(R^n))$.

(B_2) Let V the compact space of positive Radon measures on $[0,T]{\times}A$ whose projection on $[0,T]$ is the Lebesgue measure, i.e. q belongs to V iff

$q(dt,da) = dt\; q(t,da)$ where $q(t,.)$ is a Borel kernel on A.

V is endowed with the weak (*) topology. It is the same as the stable topology, for which the mappings $q \to q(f)$ are continuous, when f is a bounded function, continuous with respect to a but not necessary in (t,a). The space V is endowed with its Borel σ-algebra $\underline{\underline{V}}$ and its natural filtration $\underline{\underline{V}}_t$ generated by $\{ 1_{[o,t]} \cdot q \;;\; q \in V \}$.

(B_3) Let $(\hat{\mathcal{X}}, \hat{\underline{\underline{\mathcal{X}}}}_t)$ be the space $\mathcal{Y}{\times}\Pi{\times}V{\times}\mathcal{X}$ with the filtration $\underline{\underline{\mathcal{Y}}}_t \times \underline{\underline{\Pi}}_t \times \underline{\underline{V}}_t \times \underline{\underline{\mathcal{X}}}_t$ and $(\bar{\mathcal{Y}}, \bar{\underline{\underline{\mathcal{Y}}}}_t)$ be the space $\mathcal{Y}{\times}\Pi{\times}V$ with the filtration $\underline{\underline{\mathcal{Y}}}_t \times \underline{\underline{\Pi}}_t \times \underline{\underline{V}}_t$.

We use frequently the term:

$C_t(f,[\hat{x}]) := f(t,x_t,y_t) - f(r,x_r,y_r) - \int_r^t \int_A \Lambda f(s,x_s,y_s,a)\, q(ds,da)$

We use also the notation $\int_r^t q_s(\Lambda f)(s,x_s,y_s)\, ds$ and $\Delta = \int_r^T q_s.K(s,y_s,\pi_s)ds + G(Y_T,\pi_T)$.

Since the control problem involves only the distribution of the controlled processes, we use a weak formulation to define a control .

Definition 1 : An admissible control is a system $\mathcal{U}(r,y,\mu) = (\Omega, \underline{\underline{G}}_t, \underline{\underline{F}}_t, Y_t, \pi_t, q_t, X_t, P)$ *such that :*

i) $\underline{\underline{F}}_t$ *and* $\underline{\underline{G}}_t$ *are two right-continuous filtrations such that* $\underline{\underline{F}}_t \subsetneq \underline{\underline{G}}_t$. ($\underline{\underline{F}}_t$ *is the observable filtration)*

ii) $(q_t(da))$ *is an* $\underline{\underline{F}}_t$*-adapted process, which takes its values in* $\mathcal{P}(A)$. *(It is a relaxed control) .*

iii) (Y_t) *and* (π_t) *are* $\underline{\underline{F}}_t$*-adapted continuous processes with values in* R^d *and* $\mathcal{P}(R^n)$ *respectively, so-called observation and filter processes, such that:*

— $Y_t - Y_r - \int_r^t \pi_s(h)\, ds$ *is a* $\underline{\underline{F}}_t$*- Brownian motion*

— *For* $\phi \in C_b^{1,2}(R^n)$ *and* $r \leqslant t$

[F.K.K.] or [K.S.] $$d\pi_t(\phi) = q_t.\pi_t(L\phi)(t,Y_t)\, dt + \{\pi_t(\Gamma\phi) - \pi_t(\phi)\pi_t(h)\,(t,Y_t)\}\{dY_t - \pi_t(h)(t,Y_t)\, dt\}$$

$$Y_t = Y_r = y \quad , \pi_t = \pi_r = \mu \qquad \text{for } t \leqslant r$$

iv) The state process X_t *and the observation process* Y_t *are connected by :* for $f \in C_b^{1,2}(R^+ \times R^n \times R^d)$

$$C_t^f = f(t,X_t,Y_t) - f(r,X_t,Y_t) - \int_r^t q_s(\Lambda f)(s,X_s,Y_s)\, ds$$

is a $\underline{\underline{G}}_t$ *- martingale for* $r \leqslant t$ *, and* $X_t = X_r$ *for* $t \leqslant r$

The distribution of X_r *is* μ .

v) π_t *is the conditional distribution of* X_t *given* $\underline{\underline{F}}_t$ *i.e.*

$$\pi_t(f) = E[f(X_t)/\underline{\underline{F}}_t] \qquad P.\ p.s. \quad \text{for } t \in R^+$$

The criterion associated with this control is

$$J(r,y,\mu,\mathcal{U}) = E\int_r^T q_s(K)(s,Y_s,\pi_s)\, ds + G(Y_T,\pi_T)$$

<u>REMARK</u>: When $K = m(k)$ and $G = m(g)$, we find again the partially observable control problem of the state (X_t). In the general case, we can also see the problem as a totally observable control problem of the pair (Y_t,π_t), where π_t is a measure-valued process, submitted to the constraint to be the filter process of the controlled state (X_t) . We have discussed in details these various points of view in [E.H.J. 2].

<u>COMMENTS</u>: a) It is well-known that the condition iii) implies that

(X_t, Y_t) satisfies a S.D.E., in the same form as in the introduction, with coefficients $q_t \cdot b(t,X_t,Y_t)$, $\overline{\sigma}_j(t,X_t,Y_t,q_t)$, $\gamma_i(t,X_t,Y_t)$ and $h(t,X_t,Y_t)$. Here $\overline{\sigma}(t,X_t,Y_t,q_t)$ is a square root of the matrix

$$\left[\int q_t(da)\, \Sigma\, \sigma_i^l\, \sigma_j^l\, (t, X_t, Y_t, a) \right]_{i,j}$$

b) Under the assumptions (A_1) and (A_2), there exists at least an admissible control.

A very important class of admissible controls is the class of controls defined on the canonical space. It is sufficiently large to characterize the value-function of the control problem, and sufficiently regular for beeing endowed with a nice topology.

Definition 2 : *A (P.O.) control rule with initial condition* (r, y, μ) *is a probability measure* R *on the space* $(\hat{\mathfrak{X}}, \hat{\mathfrak{X}}_t)$ *such that* $(\hat{\mathfrak{X}}, \overline{\mathcal{Y}}_t, \hat{\mathfrak{X}}_t, y_t, \pi_t, q, x_t, R)$ *is an admissible control.*

We denote by $\mathcal{R}(r,y,\mu)$ *the set of (P.O.) control rules with initial conditions* (r,y,μ) *and by* $J(r,y,\mu,R)$ *the associated reward.*

More generally, if the distribution of the pair (y,μ) *is the probability* M *on* $R^d \times \mathcal{P}(R^n)$, *we denote by* $\mathcal{R}(r,M)$ *the set of associated control rules.*

REMARK : We can associate with every admissible control, by transport, a (P.O.) control rule without changing the criterion.

The characterization of a control rule that we give below is an immediate consequence of the definition 1, since the adaptation of the processes $(y_t, \pi_t, q_t dt)$ holds.

Proposition 3: A probability measure R *on* $\hat{\mathfrak{X}}$ *is a control rule with initial conditions* (r,y,μ) *if and only if:*

i) $y_t - y_r - \int_r^t \pi_s(h)\, ds$ *is a* $\overline{\mathcal{Y}}_t$ *Brownian motion for* $t \geq r$ *and* (π_t) *satisfies the* [F.K.K.] *equation .*

ii) $C_t(f, [\hat{x}])$ *is a* $\hat{\mathfrak{X}}_t$ *-martingale for* $t \geq r$.

iii) For $r \leq t$, $x_t = x_r$, $y_t = y_r$, $\pi_t = \pi_r = \mu$

iv) For $H \in b\overline{\mathcal{Y}}_t$, *and* $t \geq 0$ $\quad R(H\, f(x_t)) = R(H\, \pi_t(f))$

Corollary 4 : The graph Q *of the multivalued mapping* $(r,y,\mu) \Longrightarrow \mathcal{R}(r,y,\mu)$ *is a Borel subset of* $R^+ \times R^d \times \mathcal{P}(R^n) \times \mathcal{P}(\hat{\mathfrak{X}})$

Proof: By definition, $Q = \{(r,y,\mu,R);\ R \in \mathcal{R}(r,y,\mu)\}$
The properties i),ii) iii),iv) of the Proposition 3, which express that a pair belongs to Q, can be checked from a denumbrable family of conditions : we require of f and ϕ to belong to denumbrable dense sets $\mathcal{H}$ and $\mathcal{K}$ of functions, and on times t to be ration ; we can also verify the martingale or conditionning properties from a denumbrable family of random variables, which generate the σ-fields $\hat{\underline{\mathcal{X}}}_t$ and $\overline{\underline{\mathcal{Y}}}_t$ for t rational.

Since the mappings $[\hat{x}] \Longrightarrow C_t(f, [\hat{x}])$, $[\hat{x}] \Longrightarrow (\phi(x_t), \pi_t(\phi))$ $[\hat{x}] \Longrightarrow \left[\int_r^t q_s \cdot \pi_s(\Lambda\phi)ds, \pi_t(\Gamma\phi) - \pi_t(\phi)\pi_t(h)\right]$ are Borel, (in fact continuous), the mappings:

$$R \Longrightarrow R[1_H(C_t(f) - C_s(f))] \qquad H \in \hat{\underline{\mathcal{X}}}_s \quad s,t \text{ rational}$$

$$R \Longrightarrow R(1_H(f(x_t)),\ R(1_H \pi_t(f)) \qquad H \in \overline{\underline{\mathcal{Y}}}_t \quad t \text{ rational}$$

are Borel. In the same way, we prove by using a martingale formulation that $\{R ; (y_t, \pi_t)$ satisfies i) Prop.3 $\}$ is Borel.

It is also obvious that the graph of the set of probabilities which satisfy the initial conditions iv) Prop.3 is measurable. Therefore, Q is a Borel set of the polish space $R^+ \times R^d \times \mathcal{P}(R^n) \times \mathcal{P}(\hat{\mathcal{X}})$. □

Theorem 5: Let $V(r,y,\mu)$ be the value-function associated with the control problem :

$$V(r,y,\mu) = \sup\left\{ R\int_r^T q_s(K)(s,y_s,\pi_s)\,ds + G(y_T,\pi_T) ; R \in \mathcal{R}(r,y,\mu)\right\}$$

For every probability measure on $R^d \times \mathcal{P}(R^n)$

$$\int V(r,y,\mu)\, M(dy,d\mu) = \sup\left\{R\int_r^T q_s(K)(s,y_s,\pi_s)\,ds + G(y_T,\pi_T) ;\ R \in \mathcal{R}(r,M)\right\}$$

Proof: We use the notation Δ, introduced in B_3.
The analytic functions theory proves that the value-function $V(r,y,\mu) = \sup\{R(\Delta) ; R \in \mathcal{R}(r,y,\mu)\}$ is analytic, because by the Corollary 4 the graph of $\mathcal{R}(r,y,\mu)$ is a Borel subset of polish product spaces.

For a probability measure M on $R^d \times \mathcal{P}(R^n)$, there exists two Borel functions v_1 and v_2, $v_1 \leqslant V \leqslant v_2$, and such that

$$M(v_1)(r) = M(V)(r) = M(v_2)(r)$$

The sections of the graph

$$Q_\varepsilon = \{(r,y,\mu,R) ; R(\Delta) + \varepsilon \geq v_1(r,y,\mu),\ R \in \mathcal{R}(r,y,\mu)\}$$

are not empty. By a selection theorem ([D.M.]), there exists a measurable kernel $R^\varepsilon_{(r,y,\mu)}$, which belongs to $\mathcal{R}(r,y,\mu)$ and such that :

$$M(V)(r) = M(v_1)(r) = \int M(dy,d\mu)\, R_{(r,y,\mu)}(\Delta) + \varepsilon \leqslant \sup\{R(\Delta); R \in \mathcal{R}(r,M)\} + \varepsilon$$ □

It remains to show that the control problem is strongly Markovian, i.e. that the set of control rules is stable by conditioning and concatenation to the stopping times with respect to the natural observable filtration $\overline{\mathcal{Y}}^{\circ}_{t}$.

Let R be a control rule and τ a $\overline{\mathcal{Y}}^{\circ}_{t}$ -stopping time. It is classical that the property i) Prop.3 holds between the times $\tau(\omega)$ and t, with initial conditions $(y_{\tau(\omega)}, \pi_{\tau(\omega)})$ for the conditional distribution of R given $\overline{\mathcal{Y}}_{\tau}$, and the property ii) Prop.3 holds also for the conditional distribution of R given $\hat{\underline{\mathcal{X}}}_{\tau}$, with initial conditions (x_{τ}, y_{τ}).

By the property iii) Prop.3 , π_{τ} is the conditional distribution of x_{τ} given $\overline{\mathcal{Y}}_{\tau}$ under the probability R . Hence, the property ii) Prop.3 holds for the conditional distribution of R given $\overline{\mathcal{Y}}_{\tau}$ but with initial conditions $(\tau, y_{\tau}\ \pi_{\tau})$.

By the same arguments, we can prove the stability by concatenation, ([E.H.J.]2 .Th 3.2.).

As in [Ek] , we can deduce from the exchange property between "sup" and "integration" established in Theorem 5 and from the strongly Markovian character of this control problem, the equations of dynamic programming.

For the simplicity of notations, we set, for each borelian bounded function F,

$$\Delta_r^t(F) := \int_r^t q_s \cdot K(s, y_s, \pi_s) ds + F(t, y_t, \pi_t) \quad ; \qquad \Delta = \Delta_r^t(G)$$

Theorem 6 : Let τ *be a* $\overline{\mathcal{Y}}^{\circ}_{t}$*-stopping time.*

i) Dynamic programming equations :

$$V(r,y,\mu) = \sup\{ R(\Delta_r^{\tau}(V));\ R \in \mathcal{R}(r,y,\mu) \}$$

ii) For $R \in \mathcal{R}(r,y,\mu)$, $\Delta_r^t(V)$ *is a R-supermartingale*

iii) A rule R^* *is optimal iff* $\Delta_r^t(V)$ *is a* R^**-martingale.*

The proof is the same as in ([E.H.J]2)

II- COMPACTNESS OF THE RULES AND EXISTENCE OF A MARKOVIAN OPTIMAL FILTER

We now prove the compactness of the set of control rules and the regularity of the value-function V, when the functions K and G are regular .

For the compactness, we use tightness criteria for probability measures on the canonical spaces $\mathcal{Y}, \Pi, \mathcal{X}$. We recall the one relating to the space Π, which is less known.

Lemma 7 : Let $\mathcal{N}$ be a family of probability - measures on Π
Let (f_k) be a countable family of functions, which is dense in $C_0(R^d)$. If for each k, the laws of the real process $(\pi_t(f_k))$ are tight, and for each $t \in Q^+$, the laws of π_t are tight in $\mathcal{P}(R^n)$, then $\mathcal{N}$ is precompact.

Proposition 8 : The multi-valued mapping $(r,y,\mu) \Longrightarrow \mathcal{R}(r,y,\mu)$ is compact-valued, with closed graph Q.

Proof: To show the precompactness of the set $\mathcal{R}(r,y,\mu)$, it suffices to prove that the marginal sets $\mathcal{R}_{|V}$, $\mathcal{R}_{|\mathcal{Y}}$, $\mathcal{R}_{|\Pi}$, $\mathcal{R}_{|\mathcal{X}}$, are precompact. (Here the symbol $\mathcal{R}|\mathcal{Y}$ denote the set of restrictions to $\mathcal{Y}$ of elements of $\mathcal{R}$).
But the processes x_t, y_t, $\pi_t(\phi)$ are semimartingales which satisfy the hypothesis of theorem 1.4.6. in ([S.V.]), because the functions which arise in their decompositions are bounded. Therefore the laws of these semimartingales verify the assumptions of the Lemma 7.

Now we prove that the graph Q is closed. The compactness of the set $\mathcal{R}(r,y,\mu)$ follows immediately.
So let (r_n, y_n, μ_n, R_n) be a sequence which converges to (r,y,μ, R). We want to show that if R_n belongs to $\mathcal{R}(r_n,y_n,\mu_n)$, then its limit R belongs to $\mathcal{R}(r,y,\mu)$.
It is sufficient to show that the properties i),ii),iii),iv) Prop3 are still true after passing to the limit; it is easy to verify except forthe property ii): we should use an appropriate martingale problem for the pair of semimartingales $(y_t, \pi_t(\phi))$

Remark : As in the first part of the proof, we show that the set $\mathcal{R}(K) := \{R;\ R \in \mathcal{R}(r,y,\mu);\ (r,y,\mu) \in K\}$ is also precompact for a given compact set K.

Theorem 9 : Assume the functions F and G to be uniformly u.s.c..
- *The value-function V is u.s.c.*
- *The set $\mathcal{R}^*(r,y,\mu)$ of optimal rules is non empty and the multi-valued mapping $(r,y,\mu) \Longrightarrow \mathcal{R}^*(r,y,\mu)$ is compact-valued with borelian graph.*
- *The family $\mathcal{R}^*(r,y,\mu)$ is stable by conditioning and concatenation*

Proof: We suppose K and G to be sufficiently regular, so that Δ is u.s.c. on $\hat{\mathcal{X}}$.

a) The mapping $R \to R(\Delta)$ is u.s.c. from the compact space $\overline{\mathcal{R}(K)}$ in $\mathcal{P}(\hat{\mathcal{X}})$. Then, $V(r,y,\mu) = \sup\{ R(\Delta);\ R \in \mathcal{R}(r,y,\mu)\}$ is also u.s.c., since the graph Q is closed.

b) The u.s.c. function $R(\Delta)$ reaches its maximum $R^*(\Delta)$ in the compact space $\mathcal{R}(r,y,\mu)$, and the set $\mathcal{R}^*(r,y,\mu)$ of optimal rules is a closed subset of this compact space.

Let Q^* be the graph of the multifunction $\mathcal{R}^*(r,y,\mu)$.

$$Q^* = Q \cap \{ (r,y,\mu,R);\ R(\Delta) = V(r,y,\mu)\}$$

is obvious a Borel subset of Q, which is closed if V is continuous.

c) Let R^* be an optimal rule, and τ a $\mathcal{Y}^\circ_\tau$ stopping time.

By the dynamic programming equation,

$$R^*\int_r^\tau q_s(K)(s,y_s,\pi_s)\,ds + V(\tau,y_\tau,\pi_\tau) = R^*\int_r^\tau q_s(K)(s,y_s,\pi_s)ds + \Delta_\tau^T(G)$$

Hence, $\quad R^*(V(\tau,y_\tau,\pi_\tau)) = R^*(\Delta_\tau^T(G))$

Given $A \in \mathcal{Y}^\circ_\tau$, let τ_A denote be the stopping time:

$$\tau_A = \tau \text{ on } A \quad\text{and}\quad \tau_A = T \quad\text{on}\quad A^c$$

$$V(\tau_A,y_{\tau_A},\pi_{\tau_A}) = 1_A\, V(\tau,y_\tau,\pi_\tau) + 1_{A^c}\, G(y_T,\pi_T) \qquad \text{and so}$$

$$R^*(1_A\, V(\tau,y_\tau,\pi_\tau)) = R^*(1_A\, \Delta^T(G))$$

The stability under conditioning follows.

In the same way, we prove the stability by concatenation.

Using Krylov's ideas, which are explained in Stroock and Varadhan, ([S.V.] ch 12), we prove the existence of a family of optimal control rules such that (y_t,π_t) is a strong Markov process . This arguments are already used in control problems in ([Ha], [E.H.J.] 1and2)

Theorem 10: Assume K and G to be uniformly uppersemicontinuous. There exists a (r,y,μ)-measurable family of optimal control rules such that the process (y_t,π_t) is a Strong Markov process with respect to its canonical filtration.

Proof: The proof is adapted from ([S.V.] p293) and is the same as in ([E.H.J.] 1).

Let (λ_n) be a dense subset of $[0,T]$ and (f_n) be a dense subset of $C_o(R^+ \times R^d \times \mathcal{P}(R^n))$ and let (t_m,ϕ_m) be an enumeration of $\{(\lambda_n,f_k);n\geq 1,k\geq 1\}$

We define by induction the following control problems:

$$\mathcal{R}_o^*(r,y,\mu) = \mathcal{R}^*(r,y,\mu) \qquad V_1(r,y,\mu) = V(r,y,\mu)$$

$$\mathcal{R}^*_{m+1}(r,y,\mu) = \{R\in\mathcal{R}^*_m(r,y,\mu);\ R(\phi_m(t_m,y_{t_m},\pi_{t_m}) = V_m(r,y,\mu)\}$$

and $\quad V_m(r,y,\mu) = \sup\{R(\phi_m(t_m,y_{t_m},\pi_{t_m}))\ ;\ R\in\mathcal{R}^*_m(r,y,\mu)\}$

As in Theorem 9, the sets $\mathcal{R}^*_m(r,y,\mu)$ are non-empty compact sets, stable under conditioning and concatenation, but the functions V_m are only analytic.
Therefore their intersection $\mathcal{R}^*_\infty(r,y,\mu) = \bigcap\mathcal{R}^*_m(r,y,\mu)$ is a non-empty compact set.
It remains to show that the restriction of $\mathcal{R}^*_\infty(r,y,\mu)$ to $\mathcal{Y}\times\Pi$ has at most an element for each (r,y,μ).

Let R_1 and R_2 be two points in $\mathcal{R}^*_\infty(r,y,\mu)$. Then, for each pair (p,n),

$$R_1(f_p(\lambda_n,y_{\lambda_n},\pi_{\lambda_n})) = R_2(f_p(\lambda_n,y_{\lambda_n},\pi_{\lambda_n})) \quad .$$

Therefore,we deduce from the density of ϕ_n and f_p and from the continuity of $t\to R(f_p(t,y_t,\pi_t))$ that

$$R_1(f(t,y_t,\pi_t)) = R_2(f(t,y_t,\pi_t)) \quad .$$

Denote by $Q^*_t f(r,y,\mu)$ this quantity. It is an analytic semi-group,because the graph of $\mathcal{R}^*_\infty(r,y,\mu)$ is analytic.
Using the stability under conditioning, we prove that R_1 and R_2 coincide on the family of random variables

$$Z = \prod_{i=1}^{n} f_i(t_i,y_{t_i},\pi_{t_i})$$

which generate the σ-field $\underline{\mathcal{Y}}_T\times\underline{\Pi}_T$.
The restriction of the set $\mathcal{R}^*_\infty(r,y,\mu)$ to $\mathcal{Y}_T\otimes\Pi_T$ is stable by conditioning and concatenation; moreover it has only one element.
It is exactly the strong Markov property.

III- COMPARISON BETWEEN THE DIFFERENT CONTROL PROBLEMS.

We come back to the initial problem of partially observable control,(cf introduction), in order to show that the value-function $V(r,y,\mu)$ of relaxed control problem is the same as the value- function of initial problem, where the relaxed controls are adapted to the observations filtration $\underline{F}^y_t$, enlarged with the initial conditions. This results require that weak uniqueness of controlled equations holds.

We use the Zakai transformation, in order to give a precise meaning to these notions. (For example, [Da]) .

We recall that, if $(\Omega, \underline{\underline{F}}_t, Y_t, \Pi_t, P)$ is a solution of (F.K.K.) (def 1) we introduce a new probability measure on $(\Omega, \underline{\underline{F}}_T)$ with $t\in[0,T]$ by:

$$\frac{dP^\circ}{dP} = \exp-\int_r^t \Pi_s(h)\, dY_s + \tfrac{1}{2}\int_r^t \Pi_s^2(h)\, ds = \sigma_t^{-1}(1,\Pi)$$

$\sigma_t(\Phi) = \sigma_t(1,\Pi)\Pi_t(\Phi)$ satisfies the Zakai equation :

(Z) $d\sigma_t(\Phi) = q_t \cdot \sigma_t(L\Phi)\, dt + \sigma_t(\Gamma\Phi)\, dY_t$, where (Y_t) is a P°-Brownian motion.

The relation between Zakai equation and (F.K.K.) equation is one-to-one. We refer to (F.K.K.)° equation for the solutions of (F.K.K.) under the probability measure P°, for which (Y_t) is a Brownian motion.

<u>Definition 11</u>: *Let Q be a probability measure on $\mathcal{Y}\times V$ such that (y_t) is a Brownian motion, with initial condition $y_t = y$ for $t\leq r$. If there exists at most one extension $\hat{Q}$ of Q to $\mathcal{Y}\times V\times\Pi$ such that:*

- *(y_t) is a $(\mathcal{Y}\times V\times\Pi)_t$ Brownian motion .*
- *(π_t) satisfies (F.K.K.)° (or equivalently $\sigma_t(1,\pi)\,\pi_t$ satisfies (Z)).*

we say that weak uniqueness holds for the (F.K.K.) equation.

<u>Remark</u>: When the S.D.E. of the pair (state- observation) has a unique weak solution, the (F.K.K.) equation has an unique strong solution, which is adapted to the filtration $\underline{\underline{F}}_t^Y$ enlarged with the initial conditions , if we suppose that the control (q_t) is Q.p.s. constant. In the general case, we have the following proposition:

<u>Proposition 12</u>: *Assume that weak uniqueness holds for (F.K.K.) equation.*

All solution of (F.K.K.) on $\mathcal{Y}\times V\times\Pi$ can be constructed from the filter process associated with a control rule .

<u>Proof</u>: Let Q be a probability measure on $\mathcal{Y}\times V$, and $\hat{Q}$ an extension of Q , solution of (F.K.K.) on $\mathcal{Y}\times V\times\Pi$. There exists a probability measure on an enlarged space $\bar{\mathcal{Y}}\times W\times\mathcal{X}\times\hat{R}$, and continuous processes B_t^j , $j=1\ldots k$, x_t , such that:

- B^j are k independent Brownian motions, which are independent of the filtration $\bar{\mathcal{Y}}_T$.
- (x_t) is a solution of the S.D.E.:

$$dx_t = (q_t.b + \gamma.h)(t,y_t,x_t)\, dt + q_t.\sigma_j(t,y_t,x_t)\, dB_t^j + \gamma_i(t,y_t,x_t)dy_t^i$$

Let $L_t := \exp\int_r^t h(s,x_s,y_s)\, dy_s - \tfrac{1}{2}\int_r^t |h(s,x_s,y_s)|^2 ds$ be an exponential martingale.

Denote r_t be the conditional distribution (under the probability measure $L_T.\hat{R}$) of x_t given $(\mathcal{Y}\times V)_t$.
Then, (y_t, r_t) satisfies (F.K.K.) for the probability $L_T.\hat{R}$, but also for the restriction of this probability measure to the filtration $(\mathcal{Y}\times V)_t$. This probability is given by $\sigma_t(1,r).Q$. So, by changing the probability, r_t is a solution of (F.K.K.)° for the probability Q .
By weak uniqueness for the (F.K.K.) equation, $\hat{Q}$ is the distribution on $\mathcal{Y}\times V\times\Pi$ of the process (r_t) , and so the canonical process (π_t) is $\hat{Q}$ .p.s. indistinguable from the process (r_t) which is $(\mathcal{Y}\times V)_t$-adapted by construction.

Remark: Under the hypothesis of weak uniqueness for (F.K.K.), the relaxed (P.O.) control problem can be viewed as a totally observable control problem of the solutions of (F.K.K.) without any constraint.

Subsequently, we approach an arbitrary relaxed control by step relaxed controls, which have a finite support. These approximations are very important in the works of Krylov ([Kr]) and Nisio ([Ni]). The proof is the same as in Prop .4.5 and 4.6 in [E.H.J.1] , if weak uniqueness of (F.K.K.) holds. She is too long,in order that we can it give here.

We now can prove the main result of this part.

Theorem 13: *Separation theorem*

Assume that:- the pathwise uniqueness holds for the (F.K.K.) equation associated with constant relaxed control.

- the weak uniqueness holds for (F.K.K.).

- the reward functions K and G are uniformly continuous and bounded.

Then, the original problem and the relaxed problem have the same value-functions $V(r,y,\mu)$. Moreover, the function V is continuous.

Proof: We proof this result in two steps.

-a- Let $\hat{Q}$ be a probability measure on $\mathcal{Y}\times V\times\Pi$, solution of (F.K.K.)°, and associated with a step control.
Denote by $\overline{\Pi}_t$ the right-continuous filtration generated by $(y_., \pi_.)$.
A $\overline{\Pi}_t$-adapted control is viewed as a control which is "feedback" with respect to the pair (observation-filter).
We denote by $q^f(s,da)$ an $\overline{\Pi}_s$-adapted probability on A, such that:

$$\text{for } g \geq 0 , \quad q^f(s,g) = \hat{Q}(q(s,g)|\overline{\Pi}_s) .$$

Notice that $q^f(s,.)$ is still a step control with respect to the time.

It follows immediately from (F.K.K.), that $q_t.\pi_t(L\phi)$ is $\bar{\Pi}_t$-adapted. Therefore, we can write q_s^f instead of q_s in (F.K.K.) equation without any change for π_t.
By the proposition 11, $(y_t, q_t^f, \pi_t, \hat{Q})$ can be associated with a control rule, $\hat{R}^f$. Notice that $J(r,y,\mu,\hat{R}^f) = J(r,y,\mu,\sigma_T(1,\pi).\hat{Q})$.

-b- It remains to prove that we can choose a version of (π_t) which is adapted to the filtration $\underline{\underline{F}}_t^Y$, since the control q^f is a step feedback control.
This result follows from the pathwise uniqueness of the solution of (F.K.K.) associated with constant controls, proved by Kunita in ([Ku]).
We suppose that the feedback control q^f is constant on the intervals $]t_i, t_{i+1}]$, with values $q_i([y]_{t_i}, [\pi]_{t_i})$. We construct by switching to times t_i, an unique strong solution of (F.K.K.) on the space $(\mathcal{Y}, \mathcal{Y}_t, Q)$. Here the filtration $\mathcal{Y}_t$ is enlarged with the initial conditions of the pair (y_t, π_t).
Since $y_.$ and $\pi_.$ are constant before r , we have $q_o(.,.) = q(y,\mu)$.
Let us denote by π_t^1 the solution of (F.K.K.) associated with $q(y,\mu)$. She is $\underline{\underline{F}}_t^Y \vee \sigma(\mu)$-measurable.
Define $q^1(.,.) = q_1([y]_{t_1}, [\pi]_{t_1}^1)$. This random variable is $\underline{\underline{F}}_{t_1}^Y \vee \sigma(\mu)$ measurable and is independent on $(y_{t+t_1} - y_{t_1})$.
Let π^2 be the solution of (F.K.K.), starting from $\pi_{t_1}^1$ at time t_1, associated with the control q^1, and the independent Brownian motion $(y_{t+t_1} - y_{t_1})$.
Let the process π_t defined by:

$$\pi_t = \mu \quad t \leq r \quad , \pi_t = \pi_t^1 \quad t \in]r, t_1]$$
$$\pi_t = \pi_t^2 \quad t \in]t_1, t_2]$$

(π_t) is a solution of (F.K.K.) on $[r, t_2]$, $\underline{\underline{F}}_t^Y \vee \sigma(\mu)$ adapted.
It remains to define inductively the process π_t .
-c- Under the assumptions of the theorem, the proof of the continuity of the value-function is classical.(For example, [E.H.J.1] or [E.H.J.2]).

REFERENCES :

[Da] M.H.DAVIS : Lectures on stochastic control and non linear filtering.
Tata Institue of fundamental research (1984). Springer Verlag.

[D.M.] C.DELLACHERIE, P.A.MEYER : Probabilités et potentiel
Chapitres I à IV. Herman 1979.

[Ek] N.EL KAROUI: Les aspects probabilistes du contrôle stochastique.
Ecole d'été de Saint-Flour, 1979.
Lectures notes in Math. (1981), p.74-239 .

[E.H.J.1] N.EL KAROUI,D.HUU NGUYEN,M,JEANBLANC-PICQUE : Compactification méthods in the control of degenerate diffusions: existence of an optimal control.
To appear in Stochastics .(1986) .

[E.H.J.2] THE SAME : Existence of an optimal markovian filter for the control under partial observations
To appear in Siam Journ. of Control (1986) .

[Fl] W.H.FLEMING : Measure-valued processes in the control of partially observable stochastic systems.
App.Math.Optim.6 (1980), p.271-285 .

[F.N.] W.H.FLEMING,N.NISIO : On stochastic relaxed control for partially observed diffusions.
Nagoya Math.J. Vol.93 (1984),p.71-108 .

[F.P.] W.H.FLEMING,E.PARDOUX : Optimal control for partially observed diffusions
Siam J.Control and Opt. 20, 2 (1982), p.261-285 .

[Ha] U.G.HAUSSMANN : Existence of optimal markovian controls for degenerate diffusions.
Proceedings of the third Bad-Honnef Conference (1985)
Lect.Notes in control and information sciences.

[J.M.] J.JACOD, J.MEMIN :Weak and strong solutions of stochastic differential equations: Existence and stability.
Proc.Durham 18 .Stochastic integral (1980)
Lect.Notes in Math n° 851 Springer Verlag.

[Kr] N.V.KRYLOV : Controlled diffusions processes
Springer Verlag (1980)

[Ku] H.KUNITA : Asymptotic behavior of the nonlinear filtering errors of Markov processes
Journal of Multi. Analy. 1 (1971), p.365-393 .

[Ni] M.NISIO : Stochastic control theory
Tata Institute of fundamental research
ISI Lect. Notes 9 .(1981)

[Pa] E.PARDOUX : Equations du filtrage non lineaire, de la prediction et du lissage.
Stochastics 6 (1982) 193-231 .

[S.V.] D.W.STROOCK, S.R.S. VARADHAN : Multidimensional diffusion processes.
Springer-Verlag (1979) .

SOME NEGATIVE PROPERTIES OF NASH-EQUILIBRIUM STRATEGIES IN STOCHASTIC DIFFERENTIAL GAMES

S. D. Gaidov
Centre of Mathematics, P. O. Box 325
BG - 7000 Rousse, Bulgaria

SUMMARY

Many-player nonzero-sum games in which the dynamics is described by Ito stochastic differential equations are considered. Some of the negative properties of the Nash-equilibrium strategies are studied. Examples, illustrating these properties are presented. Notions of solutions of games improving the Nash-equilibrium are introduced.

1. PRELIMINARIES

Let us consider the game

$$\Gamma = \langle I, \Sigma, \{\mathcal{U}_i\}_{i\in I}, \{J_i\}_{i\in I}\rangle.$$

Here $I = \{1,\dots,N\}$ is the set of the players participating in Γ. The evolution of the dynamic system Σ is described by a stochastic differential equation of the type

$$dx(t) = f(t,x(t),u_1,\dots,u_N)dt + b(t,x(t),u_1,\dots,u_N)dw(t)$$

where $t\in[t_o,T]$, $T>t_o\geq 0$, with initial condition $x(t_o) = x_o \in R^n$. The process $w = \{w(t),\ t\in[t_o,T]\}$ is a standard m - dimensional Wiener process, defined on some complete probability space (Ω , $\mathcal{F}$, P) and adapted to a family $F = \{\mathcal{F}_t,\ t\in[t_o,T]\}$ of nondecreasing sub - σ - algebras of $\mathcal{F}$. $x(t)\in R^n$ is the state vector process and $u_i\in U_i\subset R^{n_i}$ is the control of the i - th player, $i\in I$. Now let us make the following assumptions about the functions $f(t,x,u_1,\dots,u_N)$ and $b(t,x,u_1,\dots,u_N)$. Suppose

$$f : [t_o,T] \times R^n \times U_1 \times \dots \times U_N \longrightarrow R^n$$

and

$$b : [t_o,T] \times R^n \times U_1 \times \dots \times U_N \longrightarrow R^n \times R^m$$

have continuous partial derivatives and let $C > 0$ be a constant such that

$$|f(t,0,\dots,0)| + |b(t,0,\dots,0)| \leq C,$$

$$|f_x| + |b_x| + \sum_{i \in I} (|f_{u_i}| + |b_{u_i}|) \leq C,$$

where $|.|$ is a general symbol for the norm in the respective space.

Each player has perfect observations of the state process $x(t)$ at every moment $t \in [t_o,T]$ and constructs his strategy in the game Γ as an admissible feedback control (see [3]): $u_i(t) = u_i(t,x(t))$, where $u_i(.,.) : [t_o,T] \times R^n \longrightarrow U_i$ is a Borel function satisfying the following conditions:

(i) There exists a constant $M_i > 0$ such that
$|u_i(t,x)| \leq M_i(1 + |x|)$ for all $t \in [t_o,T]$, $x \in R^n$;

(ii) For each bounded set $B \subset R^n$ and $T^* \in (t_o,T)$ there exists a constant $K_i > 0$ such that for arbitrary $x,y \in B$, $t \in [t_o,T^*]$
$|u_i(t,x) - u_i(t,y)| \leq K_i |x - y|$.

Denote by $\mathcal{U}_i$ the set of strategies of the i - th player, $i \in I$, and let the vector of strategies $u = (u_1,\dots,u_N)$ be called for brevity simply a strategy.

The assumptions mentioned above imply the existence and sample path uniqueness of the solution $X = \{ x(t), t \in [t_o,T] \}$ of Ito's equation, corresponding to the control u. Let $\mathcal{A}(u)$ be the infinitesimal operator of the Markov process X (see [2]).

Next, consider the continuous functions L_i, Q_i satisfying the polynomial growth conditions:

$$|L_i(t,x,u_1,\dots,u_N)| \leq C_i(1 + |x| + \sum_{i \in I} |u_i|)^k,$$

$$|Q_i(t,x)| \leq C_i(1 + |x|)^k,$$

where C_i, k are positive constants. Introduce now the cost - function $J_i(u)$ of the i - th player

$$J_i(u) = E_{t_o,x_o} \{ Q_i(T,x(T)) + \int_{t_o}^{T} L_i(t,x(t),u_1,\dots,u_N)dt \} , \quad i \in I.$$

The object of each player in the game Γ is to minimize his own cost-function.

Let us consider also the linear - quadratic game Γ_{1q}, where the the dynamics Σ is described by

$$dx(t) = [A(t)x(t)+ \sum_{i\in I} B_i(t)u_i]dt+b(t)x(t)dw(t),\ x(t_o) = x_o,$$

the controls (strategies) are functions of the type $u_i = F_i(t)x(t)$ and the cost - functions are given as follows

$$J_i(u) = E_{t_o,x_o} \{x'(T)D_i x(T)+ \int_{t_o}^{T} [x'(t)M_i(t)x(t)+\sum_{j\in I} u_j' N_j^i(t)u_j]\ dt\},$$

$i\in I$. Here $x(t)$, x_o, $u_i \in R^n$ and w is onedimensional standard Wiener process. The matrices $A(t)$, $B_i(t)$, $b(t)$, $F_i(t)$, $M_i(t)$, $N_j^i(t)$ are continuous and with dimensions $n \times n$, and D_i are constant $n \times n$ - matrices. Also let D_i, $M_i(t)$, $N_j^i(t)$ be symmetric.

2. BASIC LEMMA

Let in the game Γ_{1q} the matrix $N_l^i(t)$ be negative definite and $|x_o| \neq 0$. Then for arbitrary strategy $u = (u_1,\dots,u_N)$ there exists a strategy $\bar{u}_l$ of the l - th player such that

$$J_i(u\|\bar{u}_l) < J_i(u),$$

where $(u\|\bar{u}_l) = (u_1,\dots,u_{l-1},\bar{u}_l,u_{l+1},\dots,u_N) = (u_l,u_{I\setminus l})$.

Sketch of proof : Consider the function

$$G_i(t,x,u) = W_t(t,x)+[A(t)x+ \sum_{j\in I} B_j(t)u_j]'W_x(t,x)+x'M_i(t)x+$$

$$+\frac{1}{2}\ x'b'(t)W_{xx}(t,x)b(t)x+tr[b(t)b'(t)W_{xx}(t,x)]+ \sum_{j\in I} u_j' N_j^i(t)u_j,$$

defined for $t\in[t_o,T]$, $x \in R^n$, $u_i\in R^n$, $i\in I$.

Step 1. Let $u = (u_1,\dots,u_N)$ be an arbitrary strategy, i.e. $u_j=F_j(t)x$, $j\in I$. Then the differential equation

$$G_i(t,x,F_1(t)x,\dots,F_N(t)x) = 0$$

with boundary condition $W(T,x) = x'D_i x$ has a unique solution of the form

$$W(t,x) = x'\Theta(t)x + r(t),$$

where the symmetric $n \times n$ - matrix $\Theta(t)$ is the unique continuous solution of the linear equation

$$\dot{\Theta}(t)+[A'(t)+\sum_{j\in I}F_j'(t)B_j'(t)]\Theta(t)+\Theta(t)[A(t)+\sum_{j\in I}B_j(t)F_j(t)]+$$

$$+b'(t)\Theta(t)b(t)+M_i(t)+\sum_{j\in I}F_j'(t)N_j^i(t)F_j(t)=0$$

with the boundary condition $\Theta(T) = D_i$ and for the scalar function $r(t)$ we have

$$r(t)=\int_t^T tr[b(s)b'(s)\Theta(s)]ds.$$

The application of Ito - Dynkin's formula for $W(t,x)$, arbitrary u and its corresponding solution of the linear stochastic equation $x(t)$ implies

$$W(t_o,x_o)=J_i(u).$$

Step 2. Once more consider $G_i(t,x,u)$, where $W(t,x) = x'\Theta(t)x + r(t)$. If $u_{I\setminus 1} = (u_1,\dots,u_{1-1},u_{1+1},\dots,u_N)$, where $u_j = F_j(t)x$, $j\in I\setminus 1$, then we have the representation

$$G_i(t,x,u_1,u_{I\setminus 1}) = x'S(t)x+u_1'B_1'(t)\Theta(t)+x'\Theta(t)B_1(t)u_1+u_1'N_1^i(t)u_1.$$

Here $x'S(t)x$ is the sum of all terms, which do not contain u_1. Since $N_1^i(t)<0$, there exists $\delta>0$ such that $u_1'N_1^i(t)u_1\leq -\delta u_1'u_1$, for all $t\in[t_o,T]$, $u_1\in\mathcal{U}_1$. Thus

$$G_i(t,x,u_1,u_{I\setminus 1})\leq x'S(t)x+u_1'B_1'(t)\Theta(t)+x'\Theta(t)B_1(t)u_1-\delta u_1'u_1.$$

Now take $\bar{u}_1 = \bar{c}x$, where $\bar{c}$ is a constant which will be determined later. Then

$$G_i(t,x,u\parallel\bar{u}_1)\leq x'\{S(t)+\bar{c}[B_1'(t)\Theta(t)+\Theta(t)B_1(t)]-\delta\bar{c}^2E_n\}x.$$

Denote by $x'P(t,\bar{c})x$ the quadratic form in the former inequality. The constant $\bar{c}$ can be chosen sufficiently great such that all main minors of $P(t,\bar{c})$ be as follows: $P_1(t,\bar{c})<0$, $P_2(t,\bar{c})>0$, $P_3(t,\bar{c})<0,\dots$, for each $t\in[t_o,T]$. Then Sylvester's criteria implies that the quadratic form $x'P(t,\bar{c})x$ is negative definite. Therefore

$$G_i(t,x,u\parallel\bar{u}_1)\leq x'P(t,\bar{c})x<0,\ \text{for all } t\in[t_o,T],\ x\in R^n,\ |x|\neq 0.$$

Step 3. Next we shall use a result given in [1] . Let $\bar{x}(t)$ be the solution of the linear stochastic equation corresponding to $u\parallel\bar{u}_1$, i.e.

$$d\bar{x}(t) = [A(t)\bar{x}(t)+\bar{c}B_1(t)\bar{x}(t)+\sum_{j\in I\setminus 1} B_j(t)F_j(t)\bar{x}(t)]dt+b(t)\bar{x}(t)dw(t)$$

where $t\in[t_o,T]$ and $\bar{x}(t_o) = x_o$. The drift and diffusion coefficients of this equation satisfy conditions (1.3), see Ch.V,§1. Moreover the process $\bar{x}(t)$ is regular, see Ch.III,§4. So we can use Proposition 2.3, see Ch.V,§2 and conclude that if $|x_o| \neq 0$ then $|\bar{x}(t)| \neq 0$ for each $t\in[t_o,T]$, i.e. the point $x = 0$ cannot be reached by the sample path of $\bar{x}(t)$. Therefore the restriction $|x| \neq 0$ in Step 2 is of no importance.

Step 4. Taking into consideration the results of the former two steps we have

$$G_i(t,\bar{x}(t),u\ \bar{u}_1) < 0 \text{ for each } t\in[t_o,T].$$

Next we use again Ito - Dynkin's formula for $W(t,x)$, but with $u\|\bar{u}_1$ and $\bar{x}(t)$ and we get $W(t_o,x_o) > J_i(u\|\bar{u}_1)$. Finally we have

$$J_i(u) = W(t_o,x_o) > J_i(u\|\bar{u}_1).$$

Remark 1. This result is established for fixed $i,l\in I$, but it holds for arbitrary $i,l\in I$.

Remark 2. An analogue of the Basic lemma for deterministic differential games was considered by prof. V. I. Zhukovskii in his lectures at the Summer School of Operations Research in Primorsko, September 23 - 30, 1984.

Remark 3. It is easy to verify the following similar to our lemma

Proposition: Let in the game Γ_{1q} the matrix $N_1^i(t)$ be positive definite and $|x_o| \neq 0$. Then for arbitrary strategy $u = (u_1,\dots,u_N)$ there exists a strategy $\bar{u}_1$ of the l - th player such that

$$J_i(u\|\bar{u}_1) > J_i(u) \quad \text{for all } i,l\in I.$$

3. NASH-EQUILIBRIUM. SOME NEGATIVE PROPERTIES

Definition: The strategy $u^n = (u_1^n,\dots,u_N^n)$ is called a Nash-equilibrium strategy for the game Γ, if for each $u_i\in \mathcal{U}_i$

$$J_i(u^n\|u_i) \geqq J_i(u^n),\ i\in I.$$

There is a great variety of publications on Nash-equilibrium strategies (see also [4], [5], [13]). This notion for a solution of a differential game has a number of positive properties. Let us compare it for example with the guaranteeing (minimax) strategies (see [7],

[13]). Nevertheless Nash-equilibrium strategies possess some negative properties.

<u>Property 1:</u> The definition implies that in the Nash-equilibrium situation the policy of each player is based on the actions of the partners, which are advantageous for him. Here the role of every player is passive.

It can turn to be active if we consider Berge-equilibrium strategies (see [9],[10]), studied for two - player games.

<u>Property 2:</u> The unilateral deviation from the Nash-equilibrium of one player implies not only the maximization of his cost - function, but it can also imply the maximization of the cost - functions of the other players.

For example, let us consider the class of linear - quadratic stochastic differential games Γ_{lq}. Remark 3 leads directly to the verification of Property 2. Indeed, in the conditions of the Proposition for each Nash-equilibrium strategy $u^n = (u_1^n,\dots,u_N^n)$ there exists a strategy $\bar{u}_l$ of the l - th player such that

$$J_i(u^n \| \bar{u}_l) > J_i(u^n) \quad \text{for all} \quad i,l \in I.$$

<u>Property 3:</u> Domination of Nash-equilibrium. Here we give some results from our paper [8].

<u>Definition:</u> The strategy u^d dominates the Nash-equilibrium startegy u^n in the game Γ, if

$$J_i(u^d) \leqq J_i(u^n), \; i \in I.$$

Consider the functions

$$G_i(t,x,u) = W_t^i(t,x) + \mathcal{A}(u)W^i(t,x) + L_i(t,x,u), \; i \in I,$$

defined for all $t \in [t_o,T]$, $x \in R^n$, $u = (u_1,\dots,u_N)$, $u_i \in U_i$, $i \in I$.

<u>Proposition:</u> If u^d is the solution of the system of algebraic inequalities

$$G_i(t,x,u^d) \leqq 0, \; i \in I,$$

then u^d dominates u^n.

Moreover, a class of linear - quadratic stochastic differential games with explicit form of the dominating strategy is given in [8].

Property 2 and Property 3 can be overcome if we pass to Pareto-optimum (see [4],[6]). But the coincidence of Pareto-optimum and Nash-equilibrium is nearly utopic. Thus we come to the consideration of Z-equilibrium (see [11]) and objection and counter - objection strategies (see [9], [12]).

Property 4: Nash-equilibrium does not exist for a sufficiently wide class of stochastic differential games.

Let us again consider the linear - quadratic game Γ_{lq}. For arbitrary $u = (u_1, \ldots, u_N)$ there exists a strategy $\bar{u}_i \in \mathcal{U}_i$ of the i - th player such that

$$J_i(u \| \bar{u}_i) < J_i(u), \ i \in I,$$

by the Basic lemma in its conditions, i.e. there exists no Nash-equilibrium for this game.

Property 4 shows once more that Nash-equilibrium possesses some essential negative properties.

Remark: A sufficiently complete list of the positive and negative properties of Nash-equilibrium strategies for deterministic differential games is given in [14].

Acknowledgement: The author expresses his gratitude to Dr. J. Stoyanov for the useful discussions on the topic and for his support.

REFERENCES

[1] Chasminskii, R. Z. Stability of Systems of Differential Equations with Random Perturbations of their Parameters (in Russian). Nauka, Moscow (1969).
English transl.: Stochastic Stability of Differential Equations. Sijthoff and Noordhoff, Alphen aan den Rijn (1980).

[2] Dynkin, E. B. Markov Processes (in Russian). Fizmatgiz, Moscow (1963).
English transl.: Springer - Verlag, Berlin (1965).

[3] Fleming, W. H., Rishel, R. W. Deterministic and Stochastic Optimal Control. Springer - Verlag, Berlin - Heidelberg - New York (1975).
Russian transl.: Mir, Moscow (1978).

[4] Gaidov, S. D. Basic Optimal Strategies in Stochastic Differential Games. C. R. Acad. Bulgare Sci., 37 (1984), 457 - 460.

[5] Gaidov, S. D. Nash-Equilibrium in Stochastic Differential Games. In: Computers and Mathematics with Applications. Special Issue dedicated to R. E. Bellman (Editors: G. Adomian and E. S. Lee). Pergamon Press, New York (1986), in print.

[6] Gaidov, S. D. Pareto-Optimality in Stochastic Differential Games. Problems of Control and Information Theory, 6 (1986), in print.

[7] Gaidov, S. D. Guaranteeing Strategies in Stochastic Differential Games. In: Mathematics and Education in Mathematics. Proceedings of the 15th Spring Conference of UBM. Bulg. Acad. Sci., Sofia (1986), 379 - 383.

[8] Gaidov, S. D. On the Domination of Nash-Equilibrium in Stochastic Differential Games. In: Differential Equations and Applications. Proceedings of the 3rd International Conference. Centre of Mathematics, Technical University, Rousse (1986), 655 - 658.

[9] Gaidov, S. D. Optimal Strategies in Two - Player Stochastic Differential Games. C. R. Acad. Bulgare Sci., 39 (1986), 33 - 36.

[10] Gaidov, S. D. Berge-Equilibrium in Stochastic Differential Games. Submitted to Serdica.

[11] Gaidov, S. D. Z-Equilibrium in Stochastic Differential Games (in Russian). In: Many Player Differential Games (in Russian). Centre of Mathematics, Technical University, Rousse (1984), 53 - 63.

[12] Gaidov, S. D. Objection and Counter - Objection Strategies for Stochastic Differential Games. Submitted to Stochastics.

[13] Gaidov, S. D. Equilibrium Strategies in Stochastic Differential Games. Ph. D. Thesis (in Bulgarian). Sofia (1986).

[14] Zhukovskii, V. I., Tynyanskii, N. T. Equilibria in Multicriterial Dynamic Systems (in Russian). Moscow State University, Moscow (1984).

FINITE DIMENSIONAL APPROXIMATION OF AN OPTIMAL CONTROL PROBLEM FOR STOCHASTIC PARTIAL DIFFERENTIAL EQUATIONS

W. Grecksch
Technische Hochschule Carl Schorlemmer Leuna-Merseburg
Sektion Mathematik
DDR-4200 Merseburg
Otto-Nuschke-Str.

Let K, H, D be separable Hilbert spaces. $(\Omega, \mathcal{F}, P, \mathcal{F}_t, w(t))$ is a probability space equipped with a filtration $\mathcal{F}_t$ and a Wiener process $(w(t))$ with values in K and a positive self adjoint nuclear covariance operator Q.

Further on let be:

- $\mathcal{D}$: Convex bounded closed nonempty set in D.
- $V \subseteq H \subseteq V^*$: A rigged Hilbert space, V be a separable reflexive Banach space, which is a dense subset of H, there exists a constant $\gamma > 0$ with $\|v\|_H \leq \gamma \|v\|_V$ for all $v \in V$. V^* is the dual space of V.
- $\langle v^*, v \rangle$: The value of the functional $v^* \in V^*$ for $v \in V$.
- $\mathcal{L}_Q$: Space of the linear operators C with $C : Q^{1/2}K \longrightarrow H$ and $CQ^{1/2}$ is a Hilbert Schmidt operator.
- $\|C\|_Q$: Hilbert Schmidt norm of $CQ^{1/2}$.
- $A : \Omega \times [0,T] \times V \times D \longrightarrow V^*$.
- $B : \Omega \times [0,T] \times V \times D \longrightarrow \mathcal{L}_Q$.
- $f : \Omega \times H \longrightarrow R^1$, on H continuous.
- $g : \Omega \times [0,T] \times H \times D \longrightarrow R^1$, on $[0,T] \times H \times D$ continuous and $|g(t,v,u)| \leq c(\|v\|_H^2 + \|u\|_D^2 + 1)$.
- E : Mathematical expectation.
- X_0: $\mathcal{F}_0$ measurable variable with values in H and $E\|X_0\|_H^2 < \infty$.
- $\varphi : \Omega \times [0,T] \to [0,\infty[$ with $E\int_0^T \varphi(\omega,t)dt < \infty$.
- μ : Lebesgue measure on $[0,T]$.
- $\tilde{U}$: The class of all progressive measurable processes $(U(t))$ with values in $\mathcal{D}$ and $E\int_0^T \|U(t)\|_D^2\, dt < \infty$.

The processes $(A(t,v,u))_{t\in[0,T]}$, $(B(t,v,u))_{t\in[0,T]}$ are progressive measurable for all $v \in V$, $u \in \mathcal{D}$.

The following properties are fulfiled for a constant $a_1 > 0$ and for all $\omega \in \Omega$, $t \in [0,T]$, $u \in \mathcal{D}$, $v, v_1, v_2 \in V$:

(V1) $\langle A(t, v_1 + \lambda v_2, u), v \rangle$ is continuous for $\lambda \in R^1$.

(V2) $2\langle A(t,v_1,u)-A(t,v_2,u), v_1-v_2\rangle + \|B(t,v_1,u)-B(t,v_2,u)\|_Q^2 \leq 0$.

(V3) $2\langle A(t,v,u),v\rangle + \|B(t,v,u)\|_Q^2 + a_1 \|v\|_V^2 \leq \varphi(t)$.

(V4) $\|A(t,v,u)\|_{V^*} \leq \varphi^{1/2}(t)$.

(V5) $\lim_{n\to\infty} (P\times\mu)\{(\omega,t) : \|A(t,v,U_n(t))-A(t,v,U(t))\|_{V^*} > \varepsilon\} = 0$,

$\lim_{n\to\infty} (P\times\mu)\{(\omega,t) : \|B(t,v,U_n(t))-B(t,v,U(t))\|_Q > \varepsilon\} = 0$

for all $\varepsilon > 0$, $(U(t))$, $(U_n(t)) \in \tilde{U}$ $(n=1,2,\dots)$ with

$\lim_{n\to\infty} (P\times\mu)\{(\omega,t) : \|U_n(t)-U(t)\|_D > \varepsilon\} = 0$.

We consider the problem of the optimal control

$$dX(t) = A(t,X(t),U(t))dt + B(t,X(t),U(t))w(dt),\ X(0)=X_0 \qquad (1)$$

$$J(U,X) := Ef(X(T)) + E\int_0^T g(t,X(t),U(t))dt \stackrel{!}{=} \min_{U\in\tilde{U}} . \qquad (2)$$

<u>Definition 1</u>. A H-solution of (1) is a H-valued $\mathcal{F}_t$-measurable continuous process $(X(t))$ with:

a) $X(t) \in V$ (a.e.), $E\int_0^T (\|X(t)\|_H^2 + \|X(t)\|_V^2)dt < \infty$.

b) It exists $\Omega' \subset \Omega$ with $P(\Omega') = 1$ and

$$(X(t)-X_0,v)_H = \int_0^t \langle A(t,X(t),U(t)),v\rangle dt + \int_0^t (v,B(t,X(t),U(t))w(dt))_H$$

for all $\omega \in \Omega'$, $t \in [0,T]$ and $v \in Y$, Y is dense in V.

<u>Theorem 1</u>.([2]) There is a H-solution with

$$E\sup_{t\in[0,T]} \|X(t)\|_H^2 + E\int_0^T \|X(t)\|_V^2\, dt \leq \mathrm{const}\cdot(E\|X_0\|_H^2 + E\int_0^T \varphi(t)dt).$$

<u>Remark 1</u>. Special cases of (1), (2) are the optimal control of parabolic Ito equations with boundary conditions of Cauchy [2] or Neumann [1].

<u>Lemma 1</u>. Given $(U(t))$, $(U_n(t)) \in \tilde{U}$ $(n=1,2,\dots)$ with

$\lim_{n\to\infty} (P\times\mu)\{(\omega,t) : \|U_n(t)-U(t)\|_D > \varepsilon\} = 0$ for all $\varepsilon > 0$.

Then $\lim_{n\to\infty} E\|X_n(t)-X(t)\|_H^2 = 0$ for all $t \in [0,T]$, where $X_n(\cdot)$ and $X(\cdot)$ are the H-solutions of (1) for the controls $(U_n(t))$ and $(U(t))$.

Proof. It follows from the Ito formula for $\| . \|_H^2$ and the properties of the Ito integral:

$$E\|X_n(t)-X(t)\|_H^2$$

$$= 2\, E \int_0^t \langle A(s,X_n(s),U_n(s))-A(s,X(s),U_n(s)),X_n(s)-X(s) \rangle \, ds$$

$$+ 2\, E \int_0^t \langle A(s,X(s),U_n(s))-A(s),X(s),U(s)),X_n(s)-X(s) \rangle \, ds$$

$$+ 2\, E \int_0^t \| B(s,X_n(s),U_n(s))-B(s,X(s),U_n(s)) \|_Q^2 \, ds$$

$$+ 2\, E \int_0^t \| B(s,X(s),U_n(s))-B(s,X(s),U(s)) \|_Q^2 \, ds.$$

Because of the Schwarz inequality and the condition (V2) we have

$$E\|X_n(t)-X(t)\|_H^2$$

$$\leq 2\, (E \int_0^t \|A(s,X(s),U_n(s))-A(s,X(s),U(s))\|_{V*}^2 \, ds)^{1/2} \times$$

$$\times (E \int_0^t \|X_n(s)-X(s)\|_V^2 ds)^{1/2} + 2E \int_0^t \|B(s,X(s),U_n(s))-B(s,X(s),U(s)\|_Q^2 ds.$$

The theorem 1 shows that $(E \int_0^t \|X_n(s)\|_V^2 \, ds)$ is bounded, therefore $\exists\, r > 0$ with $E \int_0^t \|X_n(s)-X(s)\|_V^2 \, ds \leq r$ for all n. It follows from the condition (V3) and (V4) that $\|A(s,X(s),U_n(s))-A(s,X(s),U(s))\|_{V*}^2$ and $\|B(s,X(s),U_n(s))-B(s,X(s),U(s))\|_Q^2$ have integrable majorants. Then we obtain with the condition (V5) and the Lebesgue convergence theorem $\lim_{n\to\infty} E\|X_n(t)-X(t)\|_H^2 = 0$ for all $t \in [0,T]$ ∎

Let (h_n) be orthonormal base in H with $h_n \in V$. For orthonormal bases $(e_n) \subset K$, $(d_n) \subset D$ we consider the following system of equations of Galerkin

$$dX_k^{rn}(t) = \langle A(t,X^{rn}(t),U^r(t)),h_k \rangle \, dt$$

$$+ \sum_{j=1}^{n} (B(t,X^{rn}(t),U^r(t))e_j,h_k)_H(w(dt),e_j)_K \tag{3}$$

$$X_k^{rn}(0) = (X_o,h_k)_H$$

(k=1,...,n; n=1,2,...; r=1,2,...), where $X^{rn}(t) := \sum_{j=1}^{n} X_j^{rn}(t)h_j$ and $U^r(t) := \sum_{j=1}^{r} (U(t),d_j)d_j$.

Remark 2. The systems (3) are finite dimensional Ito equations.

We introduce the problem

$$J(U^r, X^{rn}) \stackrel{!}{=} \min_{U^r \in \mathfrak{U}^r} , \qquad (4)$$

where $\mathfrak{U}^r$ is the set of all processes $\sum_{j=1}^{r} U_j(t)d_j$ with progressive measurable real processes $(U_j(t))$ and $E\int_0^T |U_j(t)|^2\, dt < \infty$.

Lemma 2. It is $\lim_{n\to\infty} E\|X^{rn}(t)-X^r(t)\|_H^2 = 0$ for all $t \in [0,T]$, where $(X^r(t))$ is the H-solution of (1) for $(U(t))=(U^r(t))$.

The proof follows from the Ito formula for $\|\cdot\|_H^2$ and the condition (V2).

Let $(Y_n(t))$, $(Y(t))$, and Y_n, Y denote random processes and random variables with values in H and $\lim_{n\to\infty} E\int_0^T \|Y_n(t)-Y(t)\|_H^2\, dt = 0$, $\lim_{n\to\infty} E\|Y_n-Y\|_H^2 = 0$. Suppose $(Y_n(t))$, $(Y(t))$, Y_n, Y satisfy the conditions

$\lim_{n\to\infty} E\int_0^T g(t,Y_n(t),U(t))dt = E\int_0^T g(t,Y(t),U(t))dt$ and

$\lim_{n\to\infty} Ef(Y_n) = Ef(Y)$. For instance this conditions are fulfiled for bounded functions f, g.

Suppose the problems (1), (2) and (3), (4) have solutions(U^o, X^o) and (U_o^{rn}, X_o^{rn}).

Lemma 3. It is $\lim_{r,n\to\infty} J(U_o^{rn}, X_o^{rn}) = J(U^o, X^o)$.

Proof. We denote with $X(U_o^{rn})$ the solution of (1) for $U=U_o^{rn}$.

Let $\varepsilon > 0$ be arbitrary. There exists numbers $r_o(\varepsilon)$, $n_o(\varepsilon)$ such that the following inequalities are satisfied for $r \geq r_o(\varepsilon)$, $n \geq n_o(\varepsilon)$:

$$|J(U_o^r, X(U_o^r)) - J(U_o^r, X^n(U_o^r))| < \frac{\varepsilon}{2} , \qquad (5)$$

where $U_o^r(t) := \sum_{k=1}^{r} (U^o(t), d_k)d_k$ and $X^n(U_o^r)$ is the solution of (3) for $U=U_o^r$.

$$| J(U_o^{rn}, X(U_o^{rn})) - J(U_o^{rn}, X_o^{rn}) | < \varepsilon \tag{6}$$

and

$$| J(U^o, X^o) - J(U_o^r, X(U_o^r)) | < \frac{\varepsilon}{2} \ . \tag{7}$$

(5), (6) follow from lemma 2 and the properties of f and g.

(7) follows from Lemma 1 and the properties of f and g.

Now we prove $|J(U^o,X^o) - J(U_o^{rn},X_o^{rn})| < \varepsilon$ for $r \geq r_o(\varepsilon)$, $n \geq n_o(\varepsilon)$.

For $J(U^o,X^o) - J(U_o^{rn},X_o^{rn}) > 0$ we get with (6)

$$0 < J(U^o,X^o) - J(U_o^{rn},X_o^{rn}) \leq J(U_o^{rn},X(U_o^{rn})) - J(U_o^{rn},X_o^{rn}) < \varepsilon .$$

For $J(U^o,X^o) - J(U_o^{rn},X_o^{rn}) < 0$ we get with (5), (7)

$$0 > J(U^o,X^o) - J(U_o^{rn},X_o^{rn}) \geq J(U^o,X^o) - J(U_o^r,X^n(U_o^r))$$

$$= [J(U^o,X^o) - J(U_o^r,X(U_o^r))] + [J(U_o^r,X(U_o^r)) - J(U_o^r,X^n(U_o^r))]$$

$$> -\frac{\varepsilon}{2} - \frac{\varepsilon}{2} = -\varepsilon \ \blacksquare$$

The main result is:

<u>Theorem 2.</u> There are numbers r, n so, that the control $(U_o^{rn}(t))$ is ε-optimal for (1), (2), that is

$$|J(U_o^{rn},X(U_o^{rn})) - J(U^o,X^o)| < \psi(\varepsilon) ,$$

where $X(U_o^{rn})$ is the H-solution of (1) for $U = U_o^{rn}$ and $\psi:]0,\infty[\to]0,\infty[$ with $\psi(\varepsilon) \downarrow 0$ for $\varepsilon \downarrow 0$. Here is $\psi(\varepsilon) = \varepsilon$.

<u>Proof.</u> Let $r_o(\varepsilon)$, $n_o(\varepsilon)$ be from the proof of lemma 3. Then we have

$$| J(U_o^{rn},X(U_o^{rn})) - J(U^o,X^o) |$$

$$\leq |J(U^o,X^o) - J(U_o^{rn},X_o^{rn})| + |J(U_o^{rn},X_o^{rn}) - J(U_o^{rn},X(U_o^{rn}))|$$

$$\leq \frac{\varepsilon}{2} + \frac{\varepsilon}{2} = \varepsilon \text{ for } r \geq r_o(\varepsilon),\ n \geq n_o(\varepsilon),$$

using lemma 3 and formula (6) $\blacksquare$

<u>Remark 3.</u> The question about ε-optimal controls is important for evolution equations with small nonlinearities too. Given $A(t,x,u) = A(t)x + \varepsilon A_1(t,x) + R_1(t)u$, $B(t,x,u) = B(t)x + \varepsilon B_1(t,x)$

+ $R_2(t)u$, where $A(t)(\cdot)$, $B(t)(\cdot)$, $R_1(t)(\cdot)$, $R_2(t)(\cdot)$ are linear and $-A(t)(\cdot)$ is coercitive. In [1] is proved with results of convex analysis in the case of the existence of optimal controls, that the optimal control of the linear problem ($\varepsilon=0$) is ε-optimal for (1), (2). If we determine a ε-optimal control for the linear system, with obove theorem, then this control is also ε-optimal for (1), (2).

Example. Let $V=\overset{\circ}{W}{}_2^1(G)$, $H=L^2(G)$, $V^*=W_2^{-1}(G)$, $G\subset R^1$, $K=R^1$, $D=H$,

$A=\frac{1}{2}\frac{\partial^2}{\partial x^2}$, $\mathcal{F}_t:=\sigma\{w(s):s\in[0,t]\}$, $(U(t))=(U(t;x))$ with $|U(t;x)|\leq 1$, $(h_j)\subset H$ orthonormal base with $h_j\in V$ and $n=r$. We consider for $\varepsilon>0$

$$dX(t;x)=\Big[\frac{1}{2}\frac{\partial^2 X(t;x)}{\partial x^2}+U(t;x)\Big]dt+\varepsilon\Big|\frac{\partial}{\partial x}X(t;x)\Big|\,w(dt)+w(dt)$$

$$X(0;x)=X_0(x)\,, \tag{8_ε}$$

$$J(U,X)=\frac{1}{2}E\int_G X^2(T;x)dx\overset{!}{=}\min_U\,. \tag{9_ε}$$

and

$$dX(t;x)=\Big[\frac{1}{2}\frac{\partial^2 X(t;x)}{\partial x^2}+U(t;x)\Big]dt+w(dt),X(0;x)=X_0(x), \tag{8_0}$$

$$J(U,X)=\frac{1}{2}E\int_G X^2(T;x)dx\overset{!}{=}\min_U\,. \tag{9_0}$$

In the example (8_0), (9_0) we get for (3), (4)

$$dX_k^{nn}(t)=\Big[-\sum_{j=1}^{n}\int_G\frac{d}{dx}h_j(x)\frac{d}{dx}h_k(x)\,X_j^{nn}(t)+\int_G U^n(t;x)h_k(x)dx\Big]dt$$

$$+\int_G h_k(x)dx\,w(dt),\ X_k^{nn}(0)=\int_G h_k(x)X_0(x)dx\ (k=1,\dots,n), \tag{10}$$

$$J(U^n,X^{nn})=\frac{1}{2}\sum_{k=1}^{n}E\int_G (X_k^{nn}(t))^2dt\overset{!}{=}\min_{U^n\in\mathfrak{U}^n}\,. \tag{11}$$

Let $\mathfrak{A}:=\left(\int_G\frac{d}{dx}h_j(x)\frac{d}{dx}h_k(x)\,dx\right)_{j,k=1,\dots,n}$. It follows from the maximum principle [1] that

$$U_o^{nn}(t)=\mathrm{sgn}\left\{ -\exp(\alpha(T-t)) \cdot \begin{pmatrix} E\{X_1^{nn}(T;x)\,|\,\mathcal{F}_t\}h_1(x) \\ \vdots \\ E\{X_n^{nn}(T;x)\,|\,\mathcal{F}_t\}h_n(x) \end{pmatrix} \right\}$$

is an optimal control for (10), (11). Let $n \geq n_o(\varepsilon)$. The Theorem 2 shows that $(U_o^{nn}(t))$ is ε-optimal for (8_o), (9_o). It follows from the remark 3 that this control is ε-optimal for (8_ε), (9_ε) too.

References.

[1] Grecksch, W. Stochastische Evolutionsgleichungen und deren Steuerung. Teubner-Texte zur Mathematik. To appear 1987.

[2] Krylov, N.V., Rozovskij, B.L. About stochastic evolution equations. Results of science and engineering, ser. actual problems of mathematics 14 (1979), 71-146.

[3] Melnik, S.A. ε-optimal control of a random system with distributed parameters. Theory of random processes (Kiev) 10 (1982), 58-64.

SOME EXAMPLES OF THE OPTIMAL CONTROL OF DIFFUSIONS WITH PARTIAL OBSERVATION AND NON-GAUSSIAN INITIAL CONDITION

U.G. Haussmann
Mathematics Department, University of British Columbia
121 - 1984 Mathematics Road
Vancouver, B.C.
Canada V6T 1Y4

1. Introduction

We consider a control system where the state X_t is an R^n- valued stochastic process satisfying

(1.1) $$dX_t = [A(t)X_t + B(t)u(t,Y) + a(t)]dt + C(t)dW_t + D(t)dY_t,$$

X_o is a given random variable, and the observation is an R^d-valued process satisfying

(1.2) $$dY_t = [H(t)X_t + G(t)u(t,Y) + h(t)]dt + dV_t,\ Y_o = 0.$$

Here $\{W_t : 0 \leq t \leq T\}$ and $\{V_t : 0 \leq t \leq T\}$ are R^m and R^d-valued (respectively) independent standard Wiener processes which are also independent of X_o, and T is fixed and finite. The controls u are suitably measurable functions mapping $[0,T] \times C(0,T;R^d)$ into U, a Borel subset of R^k. Here $C(0,T;R^d)$ denotes the set of continuous functions $[0,T] \to R^d$. A, B, C, D, H, G are bounded, continuous matrix-valued (of suitable dimensions) functions of t, and a, h are bounded, continuous vector-valued functions of t. We impose further constraints on X_o and the controls u below. The performance criterion is the cost J(u) given by

(1.3) $$J(u) = E\{\int_0^T \ell(t,X_t,u(t,Y)dt + c(X_T)\}$$

Then the problem is to find a control u* such that $J(u^*) = \inf J(u)$.

When $U = R^k$, ℓ and c are quadratic and X^o is a (possibly degenerate) Gaussian random variable then Wonham, [11], showed that among the Lipschitz continuous controls an optimal one could be found as the solution of a reduced completely observable control problem whose state is the conditional mean of X_t given $\underline{\underline{Y}}_t$, the σ-algebra generated by the past observations, Y_s, $o \leq s \leq t$. The fact that the problem can be solved by first solving a filtering problem, i.e. finding the conditional mean of X_t given $\underline{\underline{Y}}_t$, and then solving a control problem with complete information, is known as the separation principle. The same result can be found in [5] for a wider class of controls. In a more general setting the problem was attacked by Haussmann, [7], (the proof of [7], theorem 2.1, contains an error: the innovation ν may not be an $\{\underline{\underline{F}}'_t\}$ Brownian motion, but the reader may consult [4], Corollary 2, for a correct proof), and by Christopeit and Helmes, [3].

In [7] and [3] the Beneš-Hilborn predicted miss problem ($\ell = 0$, $c(x) = g(v.x)$ for suitable g) is also solved. However the assumption that X_o is Gaussian is always retained, because the classic Kalman filtering theory requires this. Recently Beneš and Karatzas, [1], Kolodziej and Mohler [9], and Makowski, [10], have been able to overcome this limitation. We use the method of Makowski as extended in [8] to solve both problems without the hypothesis that X_o be Gaussian.

The approach is again first to filter the state, but now we have two sufficient statistics : in addition to the conditional mean of X_t given $\underline{\underline{Y}}_t$ we require another statistic which is however independent of the control (given $\underline{\underline{Y}}_t$). The reduced completely observable control problem (i.e. the separated problem) is more difficult to solve now, but under suitable hypotheses we are still able to do so by using convexity as in [6]. The final step is then to show that the control which is optimal for the reduced problem is also optimal for the original one.

In section two we define the problem precisely, and in section three we carry out the filtering to obtain the sufficient statistics. In the last two sections we then treat the linear regulator and the predicted miss problem. Complete proofs can be found in a forthcoming article to appear in Stochastics.

2. The Problem

We begin by defining the admissible controls and the corresponding solutions of (1.1), (1.2). Let X_o be a given $\mathbf{R}^n$-valued random variable with mean m_o and let the distribution of

$$\xi_o := x_o - m_o$$

be P_o. We assume

(A_1) there exists $\varepsilon > 0$ such that

$$\int_{\mathbf{R}^n} \exp(\varepsilon|x|^2)P_o(dx) < \infty.$$

Let us write P_W^m for Wiener measure on $C(0,T;\mathbf{R}^m)$ and $\{\underline{\underline{B}}_t^m\}$ for the canonical Borel filtration of $C(0,T;\mathbf{R}^m)$. Moreover $\underline{\underline{B}}^n$ stands for the Borel σ-algebra on $\mathbf{R}^n$. Now we define a probability space with filtration,

$(\Omega,\underline{\underline{F}},\{\underline{\underline{F}}_t\},\mathring{P})$ by

$$\Omega = \mathbf{R}^n \times C(0,T;\mathbf{R}^m) \times C(0,T;\mathbf{R}^d),$$

$$\underline{\underline{F}}_t = \underline{\underline{B}}^n \times \underline{\underline{B}}_t^m \times \underline{\underline{B}}_t^d, \qquad \underline{\underline{F}} = \underline{\underline{F}}_T,$$

$$\mathring{P} = P_o \times P_W^m \times P_W^d.$$

The generic element of Ω is $\omega = (x, w, y)$ so that if $\xi_o(\omega) := x$, $W_t(\omega) := w(t)$, $Y_t(\omega) = z(t)$, then ξ_o has distribution P_o, W and Y are standard Wiener processes and the three variables are mutually independent.

We define

$$\underline{U}_o = \{u:\ [0,T] \times C(0,T;R^d) \to U,\ \{\underline{\underline{B}}^d_t\}\text{-progressively measurable},\ u(\cdot,y) \in L^2(0,T;U)\ \forall\ y\}.$$

Recall that the data are bounded and continuous.

2.1 Proposition. Assume A_1. Then for each u in $\underline{U}_o$, (1.1) has a unique strong solution, X^u, on $(\Omega, \underline{\underline{F}}, \{\underline{\underline{F}}_t\}, \overset{\circ}{P})$ with $X^u_o = \xi_o + m_o$.

Let us define for each t in [0, T] a random variable on $(\Omega, \underline{\underline{F}}, \overset{\circ}{P})$ by

$$\chi^u_t = \exp \{\int_o^t [H(s)X^u_s + G(s)u(s,Y) + h(s)]'\, dY_s - \frac{1}{2}\int_o^t |H(s)X^u_s + G(s)u(s,Y) + h(s)|^2 ds\}$$

where ' denotes transpose. Now define

$$\underline{U}_1 = \{u \in \underline{U}_o : \overset{\circ}{E}\ \chi^u_t = 1\}.$$

If we define the measure P^u by $dP^u = \chi^u_T\, d\overset{\circ}{P}$ then Girsanov's theorem states that P^u is a probability measure on $(\Omega, \underline{\underline{F}})$ and that if a stochastic process V^u_t is defined by

$$dV^u_t = dY_t - [H(t)X^u_t + G(t)u(t,Y) + h(t)]dt,\ V_o = 0,$$

then W_t, V^u_t are independent standard Wiener processes on $(\Omega, \underline{\underline{F}}, \{\underline{\underline{F}}_t\}, P^u)$.

2.2 Proposition. Assume A_1. Then for each u in $\underline{U}_1$, (X^u, Y) is a solution of (1.1), (1.2) on $(\Omega, \underline{\underline{F}}, \{\underline{\underline{F}}_t\}, P^u)$ when $V_t = V^u_t$, i.e. is a weak solution of (1.1), (1.2). Moreover it is the unique weak solution.

2.3 Notation. We let $R_{11}(t)$ be the unique symmetric positive semi-definite solution of

$$\dot{R}_{11} = C(t)\, C(t)' + A(t)R_{11}(t) + R_{11}(t)A(t)' - R_{11}H(t)'H(t)R_{11},\quad R_{11}(0)=0,$$

Φ_t: the fundamental matrix of A(t) with $\Phi_o = I$,

Ψ_t: the fundamental matrix of $A(t) - R_{11}(t)H(t)*H(t)$ with $\Psi_o = I$,

$$N(t) = \int_o^t \Phi_s' H(s)'H(s)\Phi_s ds$$

$$Q(t) = \int_o^t \Psi_s' H(s)'H(s)\Psi_s ds$$

$$R_{12}(t) = \Psi_t - \Phi_t,\quad R_{22}(t) = N(t) - Q(t),$$

$$R(t) = \begin{bmatrix} R_{11}(t) & R_{12}(t) \\ R_{12}(t)' & R_{22}(t) \end{bmatrix},$$

$$\xi_t = \Phi_t\ \xi_o$$

$$(\Lambda^u_t)^{-1} = \exp \{-\int_o^t [H(s)\xi_s]'dV^u_s - \frac{1}{2}\int_o^t |H(s)\xi_s|^2 ds\}.$$

Observe that given $\underline{\underline{F}}_o$, ξ_t is a continuous function, hence bounded on [0,T]. Since also V_t^u is independent of $\underline{\underline{F}}_o$, then

$$E^u (\Lambda_T^u)^{-1} = E^u E^u \{(\Lambda_T^u)^{-1} | \underline{\underline{F}}_o\} = E^u 1 = 1.$$

Thus $\bar{P}^u$ defined by $d\bar{P}^u = (\Lambda_T^u)^{-1} dP^u$ is a probability measure on $(\Omega, \underline{\underline{F}})$.

We can now define the admissible controls:

$$\underline{U} = \{u \in \underline{U}_1 : \int_o^T E^u |u(t,Y)|^2 dt < \infty, \int_o^T \bar{E}^u |u(t,Y)|^2 dt < \infty\}.$$

The requirement that u(t,Y) be square integrable under dt $d\bar{P}^u$ is needed to obtain the conditional distribution of X_t^u given Y_t. The following result is useful in this connection.

2.4 Proposition. Assume A_1. If u is in $\underline{U}_1$ and if

(2.1) $$\int_o^T E^u |u(t,Y)|^{2q} dt < \infty$$

with $q = 1 + (4\varepsilon)^{-1} [|N(T)| + (|N(T)|^2 + 8\varepsilon|N(T)|)^{1/2}]$, then

$$\int_o^T \bar{E}^u |u(t,Y)|^2 dt < \infty.$$

To end this section let us summarize the problem. We wish to find u^* in $\underline{U}$ such that

$$J(u^*) = \inf\{J(u) : u \in \underline{U}\}$$

where J is given by (1.3). Finally we add that as noted in [8], Remark 2.1, no generality is gained by placing a non-singular matrix coefficient in front of dV in (1.2).

3. The Filter

We require now the conditional distribution of X_t^u given $\underline{\underline{Y}}_t$. The method of Makowski [10] as presented in [8] allows us to prove the following results.

Let us define

$$\rho_o(t,\alpha) = \int_{R^n} \exp[\alpha'x - x'Q(t)x/2] P_o(dx),$$

$$\rho_1(t,\alpha) = \int_{R^n} x \exp[\alpha'x - x'Q(t)x/2] P_o(dx),$$

$$\rho_2(t,\alpha) = \int_{R^n} x\, x' \exp[\alpha'x - x'Q(t)x/2] P_o(dx).$$

$$\rho(t,\alpha) = \rho_1(t,\alpha)/\rho_o(t,\alpha)$$

$$\sigma(t,\alpha) = \rho_2(t,\alpha)/\rho_o(t,\alpha) - \rho(t,\alpha)\rho(t,\alpha)'$$

$$m_t^u = E^u(X_t^u | \underline{\underline{Y}}_t) \quad S_t^u = E^u\{(X_t^u - m_t^u)(X_t^u - m_t^u)' | \underline{\underline{Y}}_t\}$$

then it follows that

$$S_t^u = S(t,n_t^u) = R_{11}(t) + \Psi_t \sigma(t,n_t^u)\Psi_t'.$$

Next we introduce the innovation ν^u defined by

$$d\nu_t^u = dY_t - [H(t)m_t^u + G(t)u(t,Y) + h(t)]dt, \quad \nu_o^u = 0.$$

Then ν^u is a standard Wiener process on $(\Omega, \underline{\underline{F}}, \{\underline{\underline{Y}}_t\}, P^u)$. At this point we need some properties of the functions ρ and σ. Let us write ρ_α for $\partial\rho/\partial\alpha$ and $\dot{\rho}$ for $\partial\rho/\partial t$.

3.1 Proposition. Assume A_1. Then ρ, σ and ρ_α are locally bounded, and $\rho_\alpha = \sigma$.

3.2 Corollary.

$$(3.1) \qquad dm_t^u = [\bar{A}(t)m_t^u + \bar{B}(t)u(t,Y) + \bar{a}(t)]dt + \bar{S}(t,n_t^u)d\nu_t^u, \; m_0^u = m_0,$$

$$(3.2) \qquad dn_t^u = \Psi_t'H(t)'H(t)\Psi_t\rho(t,n_t^u)dt + \Psi_t'H(t)'d\nu_t^u, \; n_0^u = 0,$$

$$(3.3) \qquad dY_t = \left[H(t)m_t^u + G(t)u(t,Y) + h(t)\right]dt + d\nu_t^u, \; Y_0 = 0,$$

where $\bar{A} = A + DH$, $\bar{B} = B + DG$, $\bar{a} = a + Dh$, $\bar{S} = D + SH'$.

3.3 Theorem. If ϕ is a measurable function such that $\phi(X_t^u)$ is P^u-integrable, then

$$(3.4) \qquad E^u\{\phi(X_t^u)|\underline{\underline{Y}}_t\} = \frac{\int_{R^n} E\{\phi(\tilde{Z})\}\exp[x'n_t^u - x'Q(t)x/2]P_o(dx)}{\int_{R^n} \exp[x'n_t^u - x'Q(t)x/2]P_o(dx)}$$

where $\tilde{Z}$ is a random variable with distribution $N\left(m_t^u + \Psi_t(x - \rho(t,n_t^u)),\; R_{11}(t)\right)$.

We conclude this section with a discussion of (3.2). From Proposition 3.1 it follows that (3.2) has a unique strong solution up to an explosion time, since ρ is locally bounded and locally Lipschitz. However, we shall require a unique strong solution up to time T, hence we assume

(A_2) $\qquad \sigma$ is bounded on $[0,T] \times R^n$.

Let us mention two cases for which (A_1), (A_2) hold.

3.4 Example. If P_o has support inside $\{|x| \le M\}$ then

$$\int \exp(\varepsilon|x|^2 P_o(dx) \le \exp(\varepsilon M^2) < \infty,$$

$$|\rho(t,\alpha)| \le M, \; |\sigma(t,\alpha)| \le 2M^2.$$

3.5 Example. If P_o is the distribution $N(0, R_0)$ then

$$\int \exp(\varepsilon|x|^2)P_o(dx) < \infty$$

provided $\varepsilon < [2\,\lambda_{max}(R_0)]^{-1}$ where $\lambda_{max}(R_0)$ is the largest eigenvalue of R_0 (if it is zero then ε can be chosen arbitrarily). If we decompose R^n into $\Delta_+ + \Gamma$ where Δ is the range of R_0 and Γ is its orthogonal complement, i.e. its null space, and if we write R_+, Q_+ for the restrictions of R_0, $Q(t)$ to Δ then

$$\rho_1(t,\alpha) = \int_{\Delta} (z + \tilde{R}_+\alpha_+) \exp [-z'R_+^{-1}z/2]dz$$

where $\alpha = \alpha_+ + \alpha_0$ with α_+ in Δ, α_0 in Γ and $\tilde{R}_+ = (Q_+ + R_+^{-1})^{-1}$.

If we now define $\tilde{R}$ by $\tilde{R}\alpha = \tilde{R}_+\alpha_+$ then

$$\rho(t,\alpha) = \tilde{R}\alpha,\ \sigma(t,\alpha) = \tilde{R},\ S(t,\alpha) = R_{11}(t) + \Psi_t\tilde{R}\Psi_t'.$$

Thus σ is bounded. Moreover, as it should, S is independent of α so (3.1) does not depend on n_t^u, and (3.2) is not needed in this case.

4. The Linear Regulator

We assume $U = R^k$,

(4.1) $$J(u) = E\{\int_0^T [X_t^{u'} M(t)X_t^u + u(t,Y)' N(t)u(t,Y)]dt + X_T^{u'} F X_T^u\},$$

where M, N are bounded and measurable, M(t) and F are positive semi-definite and N(t) is positive definite. Since u is square integrable as is X_0 then so is X_t^u. Hence Theorem 3.3 implies

$$\begin{aligned} E^u\{X_t^{u'} M(t)X_t^u|\underline{\underline{Y}}_t\} &= E^u\{tr[M(t)X_t^{u'}|\underline{\underline{Y}}_t\} \\ &= tr[M(t)R_{11}(t)] + E^u tr\{M(t)[m_t^u m_t^{u'} + \Psi_t\sigma(t,n_t^u)\Psi_t']\} \\ &= tr[M(t)E^u S(t,n_t^u)] + E^u\{m_t^{u'} M(t)m_t^u\} \end{aligned}$$

where tr denotes trace. But the distribution of n_t^u under P^u is independent of u since the solution of (3.2) is strong and unique. It follows that

(4.2) $$J(u) = E^u\{\int_0^T [m_t^{u'} M(t)m_t^u + u(t,Y)'N(t)\ u(t,Y)]dt + m_T^{u'} F\ m_T^u\} + \Delta$$

where Δ is a constant. We define a function V on $[0,T] \times R^n$ as follows:

$$V(t,m) = m'P(t)m + 2\pi(t)'m + q(t)$$

$$\dot{P} + P\bar{A}(t) + \bar{A}(t)'P - P\ \bar{B}(t)\ N(t)^{-1}\bar{B}(t)'P + M(t) = 0,\ P(T) = F,$$

$$\dot{\pi} + \left(\bar{A}(t) - \bar{B}(t)N(t)^{-1}\bar{B}(t)'P(t)\right)'\pi + P(t)\bar{a}(t) = 0,\ \pi(T) = 0,$$

$$\dot{q} + 2\bar{a}(t)'\pi(t) - \pi(t)'\bar{B}(t)N(t)^{-1}\bar{B}(t)' + E^u tr[\bar{S}(t,n_t^u)'P(t)\bar{S}(t,n_t^u)] = 0,$$

$$q(T) = 0,$$

with P(t) positive semi-definite and symmetric. Finally define

(4.3) $$\bar{u}(t,m) = -N(t)^{-1}[B(t) + D(t)G(t)]'[P(t)m + \pi(t)].$$

4.1 Proposition. Assume A_1 and A_2. If there exists a control u* in $\underline{U}$ such that $u^*(t,Y) = \bar{u}(t,m_t^{u^*})$ a.s., then u* is optimal.

Proof: For any u in U

$$0 \le [u + N^{-1}\bar{B}'(Pm + \pi)]'N[u + N^{-1}\bar{B}'(Pm + \pi)]$$

with equality if $u = \bar{u}(t,m)$. Now Itô's lemma and (4.2) imply

$$V(0,\ m_o^u) \le E^u J(U) - \Delta$$

with equality if $u(t,Y) = \bar{u}(t,m_t^u)$. We have used the fact that $\bar{S}$ is bounded and m^u is square integrable $dt \times dP^u$. The result follows.///

It remains to show that u^* as in the proposition exists in $\underline{U}$. On $(\Omega,\overset{\circ}{P})$ let (m^*,n^*) be the unique strong solution of

$$dm_t = [\bar{A}(t)m_t + \bar{B}(t)\bar{u}(t,m_t) + \bar{a}(t)]dt + \bar{S}(t,n_t)[dY_t - (H(t)m_t + G(t)\bar{u}(t,m_t) + h(t)dt]$$

$$\begin{aligned}(4.4)\qquad &= \{[A(t) - S(t,n_t)H(t)'H(t) - (B(t) - S(t,n_t)H(t)'G(t))N(t)^{-1}\\ &\quad (B(t) + D(t)G(t))'P(t)]m_t + [a(t) - S(t,n_t)H(t)'h(t) -\\ &\quad (B(t) - S(t)H(t)'G(t))N(t)^{-1}(B(t) + D(t)G(t))'\pi(t)]\}dt +\\ &\quad (D(t) + S(t,n_t)H(t)')dY_t,\end{aligned}$$

$$dn_t = \Psi_t'H(t)'H(t)\Psi_t\rho(t,n_t)dt + \Psi_t'H(t)'[dY_t - (H(t)m_t + G(t)\bar{u}(t,m_t) + h(t))dt]$$

$$\begin{aligned}(4.5)\qquad &= \Psi_t'H(t)'\{H(t)\Psi_t\rho(t,n_t) - [H(t) - G(t)N(t)^{-1}(B(t)+D(t)G(t))'P(t)]m_t\\ &\quad + [h(t) - G(t)N(t)^{-1}(B(t) + D(t)G(t))'\pi(t)]\}dt + \Psi_t'H(t)'dY_t.\end{aligned}$$

Since the solutions are strong then we have progressively measurable functions $\bar{m}$, $\bar{n}$ such that $m_t^*(\omega) = \bar{m}(t, Y(\omega))$, $n_t^*(\omega) = \bar{n}(t, Y(\omega))$.
Now let $u^*(t,y)$, $= \bar{u}(t, \bar{m}(t,y))$. We shall show that u^* is optimal.

Using the linearity in m of $\bar{u}$ and Proposition 2.4 it can be shown that $u^*$is an element of $\underline{U}$. To show that it is optimal we apply Proposition 4.1, so we must only show that

$$m_t^* = E^{u^*}\{x_t^* | \underline{\underline{Y}}_t\} := m_t^{u^*}.$$

According to Corollary 3.2 on $(\Omega, \underline{\underline{F}}, P^u)$

$$(4.6)\qquad dm_t^{u^*} = [\bar{A}(t)m_t^{u^*} + \bar{B}(t)u^*(t,Y) + \bar{a}(t)]dt + \bar{S}(t,n_t^{u^*})[dY_t - (H(t)m_t^{u^*} + G(t)u^*(t,Y) + h(t))dt]$$

$$(4.7)\qquad dn_t^{u^*} = \Psi_t'H(t)'(t)\Psi_t\rho(t,n_t^{u^*})dt + \Psi_t'H(t)'[dY_t - (H(t)m_t^{u^*} + G(t)u^*(t,Y) + h(t))dt].$$

By the equivalence of $\overset{\circ}{P}$ and P^u, (4.6), (4.7) also hold on $(\Omega, \underline{\underline{F}}, \overset{\circ}{P})$; moreover since $\bar{u}(t,m_t^*) = u^*(t,Y)$ then (m^*,n^*) satisfy the same equations (replace $\bar{u}$ by u^* in (4.4), (4.5)) and hence by uniqueness $m^{u^*} = m^*$ a.s. $(\overset{\circ}{P})$ thus a.s. (P^{u^*}), and the proof is complete.

5. The Predicted Miss Problem

Now $J(u) = E^u(v \cdot X_T^u)^2$ for some given v in R^n and

$$U = \{u \in R^k : |u_i| \leq 1, i = 1, 2, \ldots, k\}$$

We assume that a = 0, h = 0, G = 0, A, B, C, D, H are continuous and bounded (in fact C need only be measurable and bounded). Let us define v_t by

(5.1) $\quad dv_t = -\bar{A}(t)'v_t dt,\ v_T = v,$ and set $\zeta_t^u = v_t' m_t^u$.

From Theorem 3.3 we obtain

5.1 Proposition. Assume A_1 and A_2. If u is in $\underline{U}$ then

(5.2) $\quad d\zeta_t^u = v_t' B(t)u(t,Y)dt + v_t'\bar{S}(t,n_t^u)d\nu_t^u,\ \zeta_0^u = v_0 \cdot m_0,$

(5.3) $\quad dn_t^u = \Psi_t'H(t)'H(t)\Psi_t\rho(t,n_t^u)dt + \Psi_t'H(t)'d\nu_t^u,\ n_0^u = 0,$

(5.4) $\quad E^u\{(v \cdot X_T^u)^2 | \underline{\underline{Y}}_T\} = (\zeta_T^u)^2 + v'S(T,n_T^u)v.$

5.2 Corollary. Assume A_1 and A_2. If u is in $\underline{U}$ then

(5.5) $\quad J(u) = E^u(\zeta_T^u)^2 + \Delta$

for some constant Δ.

Proof: (5.3) has a unique strong solution which is independent of u, hence

$$\Delta: = E^u\{v'S(T,n_T^u)v\}$$

is a constant. The result now follows from (5.4).///

Considering (5.2) and (5.5) leads us to believe that an optimal control should always push ζ towards 0 as much as possible. We define

(5.6) $\quad \bar{u}(t,\zeta) = -\text{sgn}\,[B(t)'v_t]\ \text{sgn}\zeta.$

Note that if v is a vector then so is sgn v and $(\text{sgn } v)_i : = \text{sgn } v_i$. The next result is analogous to Proposition 4.1 but is more difficult to prove. The proof relies on the pathwise uniqueness of solutions of (5.3), on the convexity of x^2 and on an invariance property of (5.2) when $u = \bar{u}$.

5.3 Proposition. Assume A_1, A_2 and

(A_3) $\quad P_o$ is symmetric,

(A_4) $\quad$ for each t there exits $\gamma(t)$ such that

$$v_t'[D(t) + S(t,n)H(t)']\gamma(t) > 0 \ \forall\, n.$$

If there exists a control u* in $\underline{U}$ such that

(5.8) $\quad u^*(t,Y) = \bar{u}(t,\zeta_t^{u^*})$ a.s.

then u* is optimal.

To prove that such a u* exists we proceed as in the last section, but since $\bar{u}$ is no longer linear we cannot deduce immediately that (4.4), (4.5) have a unique strong solution. Nevertheless a technical argument, cf. [2], shows that this is still true. The remainder of the proof remains the same. Thus

$$u^*(t,Y) = -\text{sgn}[B(t)'v_t]\ \text{sgn}[v_t'\, E^{u^*}(X_t^* \mid \underline{\underline{Y}}_t)]$$

is optimal.

REFERENCES

[1] V.E. Beneš and I. Karatzas, Estimation and control for linear, partially observed systems with non-Gaussian initial distribution, Stochastic Proc. Applic. 14 (1983), 233 - 248.

[2] N. Christopeit and K. Helmes, Optimal control for a class of partially observable systems, Stochastics 8 (1982), 17-38.

[3] __________________________, The separation principle for partially observed linear control systems : a general framework, in "Filtering and Control of Random processes", Lecture Notes in Control and Information Sciences 61 (1984), 36-60.

[4] R. Cohen, A solution to the partially observed control problem of linear systems with non-quadratic cost, in "Stochastic Differential System," Lecture Notes in Control and Information Science 78 (1986), 127-136.

[5] W.H. Fleming and R.W. Rishel, Deterministic and Stochastic Control, Springer-Verlag, New York, 1975.

[6] U.G. Haussmann, Extremal controls for completely observable diffusions, in "Advances in Filtering and Optimal Stochastic Control", Lecture Notes in Control and Information Sciences 42 (1982), 149 - 160.

[7] _____________________, Optimal control of partially observed diffusions via the separation principle, in "Stochastic Differential Systems", Lecture Notes in Control and Information Sciences 43 (1982), 302 - 311.

[8] U.G. Haussmann and E. Pardoux, A conditionally almost linear filtering problem with non-Gaussian initial condition, submitted to Stochastics.

[9] W.K. Kolodziej and R.R. Mohler, State estimation and control of conditionally linear systems, SIAM J. Control Opt. 24 (1986), 497 - 508.

[10] A.M. Makowski, Filtering formulae for partially observed linear systems with non-Gaussian initial conditions, Stochastics 16 (1986), 1-24.

[11] M.W. Wonham, On the separation theorem of stochastic control, SIAM J. Control 6 (1968), 312 - 326.

A PROBLEM OF NON-ZERO SUM STOPPING GAME

J.P. LEPELTIER AND E. ETOURNEAU
Département de Mathématiques
Université du Maine

B.P. 535 - 72017 LE MANS CEDEX
FRANCE

Introduction

If the zero-sum stopping game (still called Dynkin Game) has been intensively studied, the generalization to the non-zero sum case i.e the game with two payoffs

$$J^1(T_1,T_2) = E(X^1_{T_2} 1_{(T_1<T_2)} + Y^1_{T_2} 1_{(T_2\leq T_1)})$$

$$J^2(T_1,T_2) = E(X^2_{T_2} 1_{(T_2\leq T_1)} + Y^2_{T_1} 1_{(T_1<T_2)})$$

and where the first (resp. second)player chooses T_1 (resp. T_2) in view to maximize J^1 (resp. J^2) has been only studied by A. Bensoussan and A. Friedman [2] in the diffusion case with techniques of variational inequalities, and more recently by Y. Ohtsubo [5] and M. Morimoto [7] in the discrete time case.

In a first part we study the non-zero sum game in the setting of the General Theory of Processes (continuous time) and prove under the assumptions :

.X^1, X^2, Y^1, Y^2 are right-continuous and upper-left semicontinuous

(H) positive processes

.$X^1 \leq Y^1$, $X^2 \leq Y^2$

.Y^1 and Y^2 are supermartingales, $Y^1_\infty = Y^2_\infty = o$

that there exists an equilibrium point for the game. For this part similar results have been obtained independently by M. Morimoto [8] in a preprint.

In the second part, we use these results to study the Markov case i.e. the case with payoffs :

$$J^1(T_1,T_2) = E_x(f^1(X_{T_1}) 1_{(T_1<T_2)} + g^1(X_{T_2}) 1_{(T_2\leq T_1)})$$

$$J^2(T_1,T_2) = E_x(f^2(X_{T_2}) 1_{(T_2\leq T_1)} + g^2(X_{T_1}) 1_{(T_1<T_2)})$$

where $(\Omega, \underline{\underline{F}}, \underline{\underline{F}}_t, (X_t), \theta_t, P^x\ x\in E)$ is the canonic realization of a right Markov process. With the assumptions :

(H_m)
. f^1, f^2, g^1, g^2 are right-continuous, upper left-semicontinuous on the trajectories, positive
. $f^1 \le g^1$, $f^2 \le g^2$
g^1 and g^2 are excessive.

we prove that there exists an equilibrium Markov point, a.s. independent of the initial law of the process.

Finally, in this paper the proofs of the results are just briefly sketched. For more details we can see [4].

§1. The general case

Definition I.1. We call non-zero sum game on stopping times the game defined by $(\Omega, \underline{\underline{F}}, \underline{\underline{F}}_t, \underline{P}, J^1(T_1,T_2), J^2(T_1,T_2))$ where

- $(\Omega, \underline{\underline{F}}, \underline{\underline{F}}_t, P)$ is a filtered probability space with "usual conditions"
- $J^1(T_1,T_2) = E(X^1_{T_1} 1_{(T_1<T_2)} + Y^1_{T_2} 1_{(T_2 \le T_1)})$

(resp. $J^2(T_1,T_2) = E(X^2_{T_2} 1_{(T_2 \le T_1)} + Y^2_{T_1} 1_{(T_1<T_2)})$)

is the gain for the first (resp. second) player, when the $\underline{\underline{F}}_t$ stopping time T_1 (resp. T_2) is the strategy of the first (resp. second) player.

We are looking for an equilibrium point for the game i.e. a pair (T^*_1, T^*_2) of $\underline{\underline{F}}_t$ stopping times such that :

$$J^1(T^*_1,T^*_2) \ge J^1(T_1,T^*_2),\quad J^2(T^*_1,T^*_2) \ge J^2(T^*_1,T_2)$$

for all pair (T_1,T_2) of $\underline{\underline{F}}_t$ stopping times.

This result will be obtained under the assumptions (H) of the introduction. For this we set a sufficient condition for an equilibrium which is a simple generalization of theorem 1 of [6] relative to the Dynkin case.

Theorem I.2 The pair (T^*_1,T^*_2) is an equilibrium point for the game since there exists a pair (W^1,W^2) of right-continuous process such that

(1) $X^1 \le W^1$, $X^2 \le W^2$

(2) $W^1_{T^*_1} = X^1_{T^*_1}$, $W^2_{T^*_2} = X^2_{T^*_2}$

(3) $W^1_{T^*_2} = Y^1_{T^*_2}$, $W^2_{T^*_1} = Y^2_{T^*_1}$

(4) $W^2_{t\wedge T^*_1}$ and $W^1_{t\wedge T^*_2}$ are supermartingales

(5) $W^1_{t\wedge T^*_1\wedge T^*_2}$ and $W^2_{t\wedge T^*_1\wedge T^*_2}$ are martingales.

By proving the existence of W^1 and W^2 such that if T^*_1 (resp T^*_2) denotes the beginning of the set $(W^1 = X^1)$ (resp. $(W^2 = X^2)$) the properties (1), (3), (4), (5) are true, by theorem I.2 we shall deduce that (T^*_1,T^*_2) is an equilibrium.

The processes W^1 and W^2 will be defined in the following as fixed points of some applications.

Definition I.3 We denote by E_1 (resp. E_2) the set of lower right semicontinuous $\underline{\underline{F}}_t$ - adapted processes V^1 (resp. V^2) such that $X^1 \leq V^1 \leq Y^1$ (resp $X^2 \leq V^2 \leq Y^2$)

Definition I.4 For V^1 (resp V^2) in E_1 (resp E_2) the stopping time D_1 (resp D_2) is the beginning of $(V^1 = X^1)$ (resp $(V^2 = X^2)$). We define :

$$F_1(V^2) = R(X^1 1_{[0,D_2[} + Y^1 1_{[D_2\, +\infty\, [})$$

$$(\text{resp. } F_2(V^1) = R(X^2 1_{[0,D_1[} + Y^2 1_{[D_1,+\infty[})$$

where R denotes the operator of Snell's envelope

We have the following properties of F_1 and F_2

Theorem I.5 For $\{i,j\} = \{1,2\}$ F_i (resp. F_j) is a non-increasing application from E_j to E_i (resp. from E_i to E_j). Finally $F = F_1{}^\circ F_2$ has a fixed point W^1.

Proof Since X^i and Y^i are right-continuous, the process $X^i 1_{[0,D_j[} + Y^i 1_{[D_j,+\infty[}$ is also right-continuous, and also its Snell's envelope $f_i\ (V^j)$. Furthermore since $X^i \leq Y^i$ we have $X^i \leq X^i 1_{[0,D_j[} + Y^i 1_{[D_j,+\infty[} \leq Y^i$ and applying the Snell's envelope operator :

$X^i \leq R(X^i) \leq F_i\ (V^j) \leq R(Y^i)=Y^i$, since Y^i is a supermartingale, and $F_i(V^j)$ is in E_i.

The fact that F_i is non-increasing is obvious since $X^i \leq Y^i$. Now define $(V_i)_{i\varepsilon I}$ the set of lower right semicontinuous adapted processes such that $X^1 \leq V^i \leq Y^1$ and $F(V^i) \geq V^i$. Using the results of Dellacherie [3] we define :

$$W^1 = P\text{-ess.sup}_{i\varepsilon I}\ V^i$$

and prove that W^1 is a fixed point of $F = F_1{}^\circ F_2$ using the fact that if F_1 and F_2 are nonincreasing, then $F = F_1{}^\circ F_2$ is nondecreasing.

Now set $W^2 = F^2(W^1)$. We remark that W^1 and W^2 are right-continuous since they are Snell's envelopes of right-continuous processes. We have now the final result of this part.

Theorem I.6 If T_1^* (resp. T_2^*) is the beginning of the set $(W^1 = X^1)$ (resp $W^2 = X^2$)) then (T_1^*,T_2^*) is an equilibrium for the game.

Proof We verify successively the conditions (1) (3) (4) (5) of the theorem I.2.

(1) is verified since W^i is in E_i (i-1,2) which implies

$X^i \leq R(X^i) \leq W^i \leq Y^i$ (i = 1,2).

(3) since Y^j is a supermartingale we can remark that

$W^j = F_j(W^i)\, 1_{[0,T_i^*[} + Y^j\, 1_{[T_i^*,+\infty[}$ and finally

$W^j_{T_i^*} = Y^j_{T_i^*}$

(4) is obvious since W^j (j=1,2) are supermartingales.

(5) We remark that if Z denotes the process

$Z = X^j\, 1_{[0,T_i^*[} + Y^j 1_{[T_i^*,+\infty[}$, then $T^* = D(W^j = Z)$

is an optimal stopping time for the optimal stopping problem relative to Z, and consequently $W^j_{t\wedge T^*}$ is a martingale ([3]). Proving that $T_1^* \wedge T_2^* \leq T^*$ we deduce finally that $W^j_{t\wedge T^*\wedge T_1^*\wedge T_2^*} = W^j_{t\wedge T_1^*\wedge T_2^*}$ is also a martingale.

§2. The Markov Case

Now $(\Omega, \underline{\underline{F}}, \underline{\underline{F}}_t, (X_t), \theta_t, P^x, x \in E)$ is the canonic realization of a right markov process, valued in a lusinian space E. The gains J^1 and J^2 are given by :

$$J^1(T_1,T_2) = E_x(f^1(X_{T_1})\, 1_{(T_1 < T_2)} + g^1(X_{T_2})\, 1_{(T_2 \leq T_1)})$$

$$J^2(T_1,T_2) = E_x(f^2(X_{T_2})\, 1_{(T_2 \leq T_1)} + g^2(X_{T_1})\, 1_{(T_1 < T_2)})$$

where f^1, f^2, g^1, g^2 are $\underline{\underline{B}}_e$ measurable (i.e. with respect to the σ-algebra generated by the excessive functions). The results of this part are obtained with the assumptions (H_m) of the introduction.

Definition II-1 For i=1,2, A_i will be the set of all $\underline{\underline{B}}_e$-measurable functions V^i defined on E and such that $f^i \leq V^i \leq g^i$.

Now if V^i belongs to A^i we define D_i as the beginning of the set $(V^i = f^i)$. Finally for all x in E, and all $\underline{\underline{F}}_t$-stopping time T set :

$$K^2(T,V^1) = f^2(X_T)\, 1_{(T<D_1)} + g^2(X_T)\, 1_{(T\geq D_1)}$$

$$K^2(T,V^2) = f^1(X_T)\, 1_{(T<D_2)} + g^1(X_T)\, 1_{(T\geq D_1)}$$

$$\phi^2(V^1)(x) = \sup_T E_x(K^2(T,V^1))$$

$$\phi^1(V^2)(x) = \sup_T E_x(K^1(T,V^2)$$

We can prove the important result that the Snell's envelopes of $f^2(X)\, 1_{[0,D_1[} + g^2(X)\, 1_{[D_{11},+\infty[}$ and $f^1(X)\, 1_{[0,D_2[} + g^1(X)\, 1_{[D_{21},+\infty[}$ are markovian and a.s. independent of the initial law. More precisely we have the :

Theorem II.2

For all V^1 in A_1 (resp V^2 in A_2) the process $\Phi^2(V^1)(X)$ (resp $\Phi^1(V^2)(X)$) is for all law P^x, the Snell's envelope of $f^2(X)\, 1_{[0,D_1[} + g^2(X)\, 1_{[D_1,+\infty[}$

(resp. $f^1(X)\, 1_{[0,D_2[} + g^1(X)\, 1_{[D_{21},+\infty[}$

The proof of this result is not obvious and cannot be detailed here ; we use for this results on the Markov Processes Theory notably by Azema [1]. For more details see [4].

By Theorem II-2, for all process $V^1(X)$ (resp. $V^2(X)$) we can now associate the process $\Theta_2(V^1(X))$ (resp. $\Theta_1(V^2(X))$ which is for all law P^x the Snell's envelope of the process

$f^2(X)\, 1_{[0,D_1[} + g^2(X)\, 1_{[D_1,+\infty[}$ (resp. $f^1(X)\, 1_{[0,D_2[} + g^1(X)\, 1_{[D_{21},+\infty[}$ and we can prove the

Theorem II.3 The application $\Theta_1 o \Theta_2$ has a fixed point $W^1(X)$. Furthermore, if $W^2(X) = \Theta_2(W^1(X))$, the pair of $\underline{\underline{F}}_t$-stopping times (T_1^*, T_2^*) when T_1^* (resp. T_2^*) is the beginning of the set $(W^1(X) = f^1(X))$ (resp. $(W^2(X) = f^2(X))$ is an equilibrium point for the game, for all law P^x.

Proof

The applications Θ_1 and Θ_2 have the same properties as the applications ϕ_1 and ϕ_2 of the General Theory (Def. I-4 and Th. I-5). Then $\Theta_1 o \Theta_2$ has a fixed point $W^1 = \sup V_i\,(X) = V(X)$. The end of the proof is the same as in Theorem I-6.

BIBLIOGRAPHY

[1] J. AZEMA Théorie Générale des processus et retournement du temps. Annales Scientifiques de l'E.N.S., 4è série, T.6, fas. 4 (1973).

[2] A. BENSOUSSAN, A. FRIEDMAN Non zero sum stochastic differential games with stopping times and free boundary problem. Trans. AMS, vol. 231, n°. 2 (1977).

[3] C. DELLACHERIE Sur l'existence de certains ess. inf. et ess. sup. de familles de processus mesurables. Séminaire de Probabilités de Strasbourg. XII - Lecture Notes in Maths. Springer Verlag

[4] N. EL KAROUI Cours de St.-Flour sur le Contrôle Stochastique. Lecture Notes in Maths n° 576. Springer Verlag

[5] E. ETOURNEAU Résolution d'un problème de Jeu de somme non nulle sur les temps d'arrêt. Thèse 3ème cycle - Paris VI - Avril 1986.

[6] Y. OHTSUBO A non zero sun extension of the Dynkin's stopping problem - Preprint (1985).

[7] M. MORIMOTO Dynkin games and martingale methods. Stochastics 13 (1984) 213 - 228.

[8] M. MORIMOTO Non-zero-sum Discrete Stochastic Games with Stopping Times. Rob. Th. Rel. Fields 72, 155-160 (1986).

[9] M. MORIMOTO On non-cooperative n-player cyclic stopping games* Preprint (1986).

LIMIT THEOREMS OF PROBABILITY THEORY AND OPTIMALITY IN LINEAR CONTROLLED SYSTEMS WITH QUADRATIC COST

Petr Mandl

Department of Probability and Mathematical
Statistics, Charles University
Sokolovská 83, 186 00 Prague 8, Czechoslovakia

1. Introduction

Linear autonomous controlled systems satisfying

$$(1) \qquad dX_t = fX_t dt + gU_t dt + dW_t, \qquad t \geqq 0, \quad X_0 = x ,$$

will be dealt with. X_t is n-dimensional and $W = \{W_t, t \geqq 0\}$ is the n-dimensional Wiener process. $U = \{U_t, t \geqq 0\}$ is an m-dimensional control process depending nonanticipatively on the observation of X. We introduce the quadratic cost functional

$$C_T = \int_0^T (X_t' c X_t + |U_t|^2) dt, \qquad T \geqq 0 .$$

Prime denotes the trans position, c is a nonzero nonnegatively definite matrix. The pairs of matrices (f,g) and $(f, \sqrt{c})$ are assumed to be stabilizable. This implies (see [3]) that the steady state matrix Riccati equation

$$(2) \qquad wf + f w - wgg w + c = 0$$

has a unique nonnegatively definite solution.

Set

$$(3) \qquad k = - g w .$$

Then

$$(4) \qquad \hat{U}_t = kX_t , \qquad t \geqq 0 ,$$

is the optimal stationary control.

The average cost optimality of (4) is well known. Namely, setting

(5) $\theta = \text{trace } w$,

we have

(6) $\lim_{T\to\infty} C_T/T = \theta$ a.s. ,

and this is the best result achievable. (6) can be regarded as optimality of (4) with respect to the Law of Large Numbers. In fact, $\hat{U}$ is optimal with respect to all principal limit theorems. The case of the Central Limit Theorem was dealt with in [4] , the main result will be restated. In the present paper we concentrate on the Arcsine Law and on the Law of the Iterated Logarithm.

Let us recall the stochastic ordering of random variables. ξ is stochastically greater or equal η if

$$P(\xi \leq y) \leq P(\eta \leq y) , \qquad y \in (-\infty,\infty) .$$

We shall use stochastic ordering in the asymptotic sense. Let $F(y)$ be a distribution function, and let ξ_T, $T>0$, be a family of random variables. The relation

(7) $\xi_T \succ F(y)$

means

(8) $\limsup_{T\to\infty} P(\xi_T \leq y) \leq F(y)$

in all continuity points of F. The convergence in distribution, i.e.,

$$\lim_{T\to\infty} P(\xi_T \leq y) = F(y)$$

in all continuity points of F, will be expressed by

$$\xi_T \asymp F(y) .$$

(8) implies

$$\liminf_{T\to\infty} E\, h(\xi_T) \geq \int_{-\infty}^{\infty} h(y)\, dF(y)$$

for cach nondecreasing, continuous and bounded function $h(y)$.

Let $\Phi(y)$ denote the distribution function of the normal distribution $N(0,1)$. In [4] we proved the following result.

<u>Proposition 1.</u> Let the control U in (1) be arbitrary but such that

(9) $$\lim_{t\to\infty} E|X_t|^2/\sqrt{t} = 0.$$

Then

$$(C_T - \theta T)/\sqrt{T} \succeq \Phi(y/\sqrt{\Delta}),$$

with

(10) $$\Delta = \text{trace } v,$$

where

v is the unique nonnegatively definite solution of the matrix equation

(11) $$v(f + gk) + (f + gk)'v + 4w^2 = 0.$$

If, in addition to (9),

(12) $$\operatorname*{p\,lim}_{T\to\infty} \frac{1}{\sqrt{T}} \int_0^T |U_t - kX_t|^2\, dt = 0,$$

then

(13) $$(C_T - \theta T)/\sqrt{T} \asymp \Phi(y/\sqrt{\Delta}).$$

p lim refers to convergence in probability. According to Proposition 1, if U_t approaches to $\hat{U}_t$ so that (12) holds, then the limit distribution of $(C_t - \theta T)/\sqrt{T}$ is $N(0,\Delta)$, and this cannot be improved in the sense of asymptotic stochastic ordering provided that (9) holds.

The principal tool in proving Proposition 1 is the relation

(14) $$C_T - \theta T + X_T'wX_T - x'wx - \int_0^T |U_t - kX_t|^2\, dt = 2\int_0^T X_t'wdW_t,\quad T\geq 0,$$

which will be also exploited in this pap er. To demonstrate (14) we employ the Itô formula to get

$$C_T' + X_T'wX_T - x'wx = C_T + \int_0^T d(X'wX) = \int_0^T (X'cX + |U|^2)\, dt +$$

$$+ 2\int_0^T (X'wfX + X'wgU)\, dt + 2\int_0^T X'wdW + \int_0^T \operatorname{tr} w\, dt,$$

and rewrite the result using

$$2X'wfX + X'cX = |kX|^2 ,$$

and (3), (5).

The stochastic integral on the right-hand side of (14) can be represented by means of a random time change in a Wiener process $\mathcal{W} = \{\mathcal{W}_t, t \geq 0\}$ (see [2]),

$$(15) \qquad 2\int_0^T X_t' w dW_t = \mathcal{W}_{V_T} , \qquad V_T = 4\int_0^T |wX_t|^2 \, dt , \qquad T \geq 0 .$$

V_T is a quadratic functional, and (10), (11) lead to a relation similar to (14),

$$(16) \qquad V_T - \Delta T + X_T' v X_T - x'vx - 2\int_0^T X_t' v g(U_t - kX_t) \, dt = 2\int_0^T X_t' v dW_t , \; T \geq 0.$$

2. The Arcsine Law

Let χ denote the indicator function. The average

$$(17) \qquad \frac{1}{T}\int_0^T \chi\{C_t > \theta t\} \, dt$$

is the proportion of time spent by C_t above θt. We shall apply the Arsine Law for the Wiener process (see [1]) to establish the asymptotic lower bound for the distribution of (17).

<u>Proposition</u> 2. Let U be such that

$$(18) \qquad \lim_{t \to \infty} E|X_t|^2/\sqrt{t} = 0.$$

Then

$$(19) \qquad \frac{1}{T}\int_0^T \chi\{C_t > \theta t\} \, dt \succeq \frac{2}{\pi} \text{ arc sin } \sqrt{y} , \qquad y \in [0,1] .$$

Proof. Introduce the function

$$k_b(x) = 1 \quad \text{for} \quad x > b, \quad k_b(x) = 0 \quad \text{for} \quad x \leq 0,$$

and denote

$$A_T = \int_0^T |U_t - kX_t|^2 \, dt .$$

Using (14),(15) write

$$\frac{1}{T}\int_0^T k_0(C_t - \theta t)\,dt = \frac{1}{T}\int_0^T k_0(\mathcal{W}_{V_t} - X_t' w X_t + x' w x + A_t)\,dt =$$

$$= \int_0^1 k_0\Big(\frac{1}{\sqrt{T}}\,\mathcal{W}_{V_{Tz}} - X_{Tz}' \, w X_{Tz} + x' w x + A_{Tz}\Big)\Big)\,dz =$$

$$= \int_0^1 k_{\varepsilon/\sqrt{\Delta}}\Big(\frac{1}{\sqrt{\Delta}}\ {}^T\mathcal{W}_{\Delta z}\Big)\,dz - \int_0^1 \Big(k_\varepsilon({}^T\mathcal{W}_{\Delta z}) -$$

$$- k_0\Big({}^T\mathcal{W}_{\frac{1}{T} V_{Tz}} - \frac{1}{\sqrt{T}}\, X_{Tz}' \, w \, X_{Tz} + \frac{1}{\sqrt{T}}\, x' w x + \frac{1}{\sqrt{T}}\, A_{Tz}\Big)\Big)\,dz ,$$

where $\varepsilon > 0$, and ${}^T\mathcal{W}_u = \frac{1}{\sqrt{T}}\,\mathcal{W}_{Tu}$, $u \geqq 0$, is a Wiener process.

For $y \in [0,1]$ and $\delta > 0$ arbitrary we have then with obvious shortenning of the denotation

$$P\Big(\frac{1}{T}\int_0^T k_0(C_t - \theta t)\,dt \leqq y\Big) \leqq P\Big(\int_0^1 k_{\varepsilon/\sqrt{\Delta}}\,dz \leqq y + \delta\Big) +$$

(20)

$$+ P\Big(\int_0^1 (k_\varepsilon - k_0)\,dz > \delta\Big).$$

Note that the before last term does not depend on T and that

$$\lim_{\varepsilon,\delta \to 0} P\Big(\int_0^1 k_{\varepsilon/\sqrt{\Delta}}\,dz \leqq y + \delta\Big) = \frac{2}{\pi}\ \text{arc sin}\ \sqrt{y} .$$

Thus, it remains to prove the negligibility of the last term in (20) as $T \to \infty$.

With regard to (16) we have for T sufficiently large

$$k_\varepsilon - k_0 \leqq \chi\{\frac{1}{\sqrt{T}} X'_{Tz} wX_{Tz} > \frac{\varepsilon}{4}\} + \chi\{\frac{1}{\sqrt{T}} X'_{Tz} vX_{Tz} > \frac{\varepsilon}{4}\} +$$

(21) $$+\chi\{ {}^T\mathcal{W}_{\Delta z} - {}^T\mathcal{W}_{\frac{1}{T}V_{Tz}} \geqq \frac{\varepsilon}{2} + \frac{1}{\sqrt{T}} A_{Tz} ; |\frac{1}{T} V_{Tz} - \Delta z| \leqq$$

$$\leqq \frac{\varepsilon}{2\sqrt{T}} + \frac{2}{T} |\int_0^{Tz} X'_s \, v \, g(U_s - kX_s) ds| + |\frac{2}{T}\int_0^{Tz} X'_s \, v \, d\, W_s|\}.$$

It holds

$$\frac{2}{T} |\int_0^{Tz} X' vg(U - kX) \, ds| \leqq 2 \, |v||g| \sqrt{\frac{\int_0^{Tz} |X|^2 ds}{T^{3/2}}} \sqrt{\frac{A_{Tz}}{\sqrt{T}}} .$$

Hence, for $\gamma > 0$, the last term in (21) is majorized by

$$\chi\{ 2|v||g| \sqrt{\frac{\int_0^{Tz} |X|^2 ds}{T^{3/2}}} > \gamma\} + \chi\{\frac{2}{T} |\int_0^{Tz} X' vdW| > \frac{\gamma}{T^{1/4}}\} +$$

(22)

$$+\chi\{\sup_{|\Delta z - t| \leqq \frac{\varepsilon}{2\sqrt{T}} + \frac{\gamma}{T^{1/4}} + \gamma\sqrt{\frac{A_{Tz}}{\sqrt{T}}}} ({}^T\mathcal{W}_{\Delta z} - {}^T\mathcal{W}_t) \geqq \frac{\varepsilon}{2} + \frac{A_{Tz}}{\sqrt{T}}\} .$$

Let us denote by R_z the right-hand side of (21) augmented by replacing its last term by (22). Then

(23) $$\int_0^1 (k_\varepsilon - k_0) \, dz \leqq \int_0^1 R_z \, dz$$

We shall prove that

(24) $$\limsup_{T\to\infty} E \int_0^1 R_z \, dz = \limsup_{T\to\infty} \int_0^1 E \, R_z dz \leqq \varphi(\gamma)$$

with

$$\varphi(\gamma) = 2\Phi(-\frac{\varepsilon}{10\sqrt{\gamma}}) + 2\sum_{j=1}^{\infty}\Phi(-\frac{\sqrt{j}}{5\sqrt{\gamma}}) .$$

(23) and (24) imply

$$\limsup_{T\to\infty} P(\int_0^1 (k_\varepsilon - k_0)\,dz > \delta) \leq \limsup_{T\to\infty} P(\int_0^1 R_z\,dz > \delta) \leq \frac{\varphi(\gamma)}{\delta} .$$

Since

$$\lim_{\gamma\to 0} \varphi(\gamma) = 0 ,$$

the desired negligibility of the last term in (20) will be then cstablished.

ER_z is the sum of the probabilities of the respective events in (21), (22). To demonstrate (24) note first that (18) implies

$$(25) \qquad P(\frac{1}{\sqrt{T}} X'_{Tz}\, w\, X_{Tz} > \frac{\varepsilon}{4}) \to 0 , \quad P(\frac{1}{\sqrt{T}} X'_{Tz}\, v\, X_{Tz} > \frac{\varepsilon}{4}) \to 0 .$$

Furthermore,

$$E\int_0^T |X|^2\,ds \,/\, T^{3/2} \to 0 ,$$

and hence

$$(26) \qquad P(2|v||g|\sqrt{\frac{\int_0^{Tz} |X|^2\,ds}{T^{3/2}}} > \gamma) \to 0 .$$

Similarly,

$$\frac{1}{T^{3/2}}\, E(\int_0^{Tz} X'v\,dW)^2 = \frac{1}{T^{3/2}} \int_0^{Tz} EX'v^2X\,ds \to 0,$$

and thus

$$(27) \qquad P(\frac{2}{T}|\int_0^{Tz} X'v\,dW| > \frac{\gamma}{T^{1/4}}) \to 0 .$$

Finally, denote by M the last event in (22). For T suficiently large

$$P(M) \leq \sum_{j=0}^{\infty} P(M;\ j\sqrt{T} \leq A_{Tz} < (j+1)\sqrt{T}) \leq$$

$$\leq \sum_{j=0}^{\infty} P(\sup_{|\Delta z - t| \leq 3\gamma\sqrt{j+1}} (\mathcal{W}_{\Delta z} - \mathcal{W}_t) \geq \frac{\varepsilon}{2} + j) \leq$$

$$\leq \sum_{j=0}^{\infty} P(\sup_{|\Delta z - t| \leq 3\gamma\sqrt{j+1}} (\mathcal{W}_{\Delta z - 3\gamma\sqrt{j+1}} - \mathcal{W}_t) \geq \frac{\varepsilon}{4} + \frac{1}{2} j) =$$

$$= \sum_{j=0}^{\infty} 2\Phi\left(\frac{-\varepsilon/4 - j/2}{\sqrt{6\gamma\sqrt{j+1}}}\right) \leq \varphi(\gamma) .$$

From here and from (25), (26), (27) we conclude that (24) holds. □

The lower bound in Proposition 2 is attained under same conditions as in Proposition 1.

<u>Proposition 3.</u> Let U be such that (18) and

(28) $$\operatorname*{p\,lim}_{T\to\infty} \frac{1}{\sqrt{T}} \int_0^T |U_t - kX_t|^2 \, dt = 0$$

hold. Then

(29) $$\frac{1}{T} \int_0^T \chi\{C_t > \theta t\} \, dt \asymp \frac{2}{\pi} \arcsin \sqrt{y}, \qquad y \in [0,1] .$$

Proof. With regard to (19) we have to verify

(30) $$\liminf_{T\to\infty} P(\frac{1}{T} \int_0^T k_0(C_t - \theta t) \, dt \leq y) \geq \frac{2}{\pi} \arcsin\sqrt{y}, \ y\in[0,1] .$$

Analogously to (20) we deduce the inequalities

$$P(\frac{1}{T} \int_0^T k_0(C_t - \theta t) \, dt \leq y) \geq P(\int_0^1 k_{-\varepsilon/\sqrt{\Delta}} \, dz \leq y - \delta) -$$

$$- P(\int_0^1 (k_0 - k_{-\varepsilon}) \, dz \geq \delta) .$$

To estimate the last probability we have

$$\int_0^1 (k_0 - k_{-\varepsilon})\,dz = \int_0^1 R_z^* \,dz \;,$$

where R_z^* differs from R_z only in the last term of (22)which is to be modified to

$$(32) \qquad \chi\Big\{ \sup_{|\Delta z - t| \le \frac{\varepsilon}{2\sqrt{T}} + \frac{\delta^2}{T^{1/4}} + \delta\sqrt{\frac{A_{Tz}}{\sqrt{T}}}} ({}^T\mathcal{W}_t - {}^T\mathcal{W}_{\Delta z}) \ge \frac{\varepsilon}{2} - \frac{A_{Tz}}{\sqrt{T}} \Big\}.$$

Since (28) says that

$$\operatorname*{p\,lim}_{T\to\infty} \; A_{Tz}/\sqrt{T} = 0 \;,$$

the probability of the event in (32)tends to 0 as $T\to\infty$. From this and from (25), (26), (27) we deduce the negligibility of the last term in (31). The before last term is independent of T and converges to $\frac{2}{\pi}$ arcsin $\sqrt{y}$ as $\varepsilon, \delta \to 0$. We conclude that (30) holds. □

Propositions 2 and 3 can be generalized to curves of the form

$$(33) \qquad l(t) = \theta t + \nu\sqrt{(\Delta t)} + o(\sqrt{t}) \;, \qquad t \in [0,\infty).$$

Let $F_\nu(y)$ denote the distribution function of the time which a trajectory of the Wiener process spends above ν during the time interval $[0,1]$. I.e., for $y \in [0,1)$,

$$F_0(y) = \frac{2}{\pi} \arcsin \sqrt{y} \;,$$

$$F_\nu(y) = \int_0^y F_0\Big(\frac{y-t}{1-t}\Big) \frac{t^{-3/2}}{\sqrt{2\pi}} e^{-\frac{\nu^2}{2t}}\,dt \;, \qquad \nu < 0 \;,$$

$$F_\nu(y) = \int_0^{1-y} F_0\Big(\frac{y}{1-t}\Big) \frac{t^{-3/2}}{\sqrt{2\pi}} e^{-\frac{\nu^2}{2t}}\,dt + 2\Phi\Big(\frac{\nu}{\sqrt{1-y}}\Big) - 1, \qquad \nu > 0 \;.$$

<u>Proposition 4.</u> Let (18) hold, and let $l(t)$ be as in (33). Then

$$\frac{1}{T}\int_0^T \chi\{C_t > l(t)\}\, dt \succ F_\nu(y) .$$

If (18), (28) are fulfilled then

$$\frac{1}{T}\int_0^T \chi\{C_t > l(t)\}\, dt \asymp F(y) .$$

3. The Law of the Iterated Logarithm

The Law of the Iterated Logarithm provides bounds for the fluctuation of C_T around θT. The bounds are attained under the optimal stationary control (4) and cannot be improved within the class of controls defined in the next proposition. We abbreviate almost surely to a.s.

Proposition 5. Let U be such that for an $\varepsilon > 0$

$$\lim_{t\to\infty} |X_t|^{2+\varepsilon}/\sqrt{t} = 0 \quad \text{a.s.} \tag{34}$$

Then almost surely the following inequalities hold,

$$\limsup_{T\to\infty} (C_T - \theta T)/\sqrt{(2T \log\log T)} \geqq \sqrt{\Delta} , \tag{35}$$

$$\liminf_{T\to\infty} (C_T - \theta T)/\sqrt{(2T \log\log T)} \geqq -\sqrt{\Delta}. \tag{36}$$

If together with (34)

$$\lim_{T\to\infty} \int_0^T |U_t - kX_t|^2\, dt/\sqrt{(T \log\log T)} = 0 \text{ a.s.}, \tag{37}$$

then equality holds in (35), (36) almost surely .

Proof. Consider again (14), (15), (16). Denote

$$A_T = \int_0^T |U_t - kX_t|^2\, dt, \qquad B_T = 2\int_0^T X_t'\, v\, g\, (U_t - kX_t)\, dt ,$$

$$f(x) = \sqrt{(2 x \log \log x)} .$$

Note that

$$(38) \qquad f'(x) \leqq f(x)/x \qquad \text{for} \quad x \geqq 6 ,$$

and that

$$|B_T| \leqq 2|v||g| \sqrt{\int_0^T |X|^2 \, dt} \sqrt{A_T} .$$

In virtue of (34) we have for T sufficiently large

$$(39) \qquad |B_T| \leqq T^{3/4-\delta} \sqrt{A_T}$$

with some $\delta > 0$. Here, as well as in the sequel, we omit the statement almost surely. (34) implies also

$$2 \int_0^T X' v \, dW = o(T) , \qquad T \to \infty$$

The time change to the Wiener process and the Law of the Iterated Logarithm can be used to prove this.

Write according to (16)

$$(40) \qquad \Delta T \ldots = \Delta T + o(T) = \Delta T - X_T' v X_T + x' v x + 2 \int_0^T X' v dW = V_T - B_T .$$

From (14) and from (34) follows

$$(41) \qquad \frac{C_T - \theta T}{f(\Delta T \ldots)} = \frac{\mathcal{W}_{V_T}}{f(V_t)} + Z_T + o(1) , \qquad T \to \infty ,$$

where

$$(42) \qquad Z_T = \frac{\mathcal{W}_{V_T} + A_T}{f(\Delta T \ldots)} - \frac{\mathcal{W}_{V_T}}{f(V_T)} .$$

Assume first

$$\lim_{T\to\infty} V_T = \infty .$$

Applying the Law of the Iterated Logarithm to the first term on the right in (41) we obtain

$$\limsup_{T\to\infty} \frac{1}{\sqrt{\Delta}} (C_T - \theta T)/f(T) \geqq 1 + \liminf_{T\to\infty} Z_T ,$$

$$\liminf_{T\to\infty} \frac{1}{\sqrt{\Delta}} (C_T - \theta T)/f(T) \geqq -1 + \liminf_{T\to\infty} Z_T .$$

Consequently, to demonstrate (35), (36) it suffices to show that

(43) $$\liminf_{T\to\infty} Z_T \geqq 0 .$$

(42) Can be rewritten as

$$Z_T = \frac{\mathcal{W}_{V_T}}{f(V_T)} \left(\frac{f(\Delta T\ldots+B_T)}{f(\Delta T\ldots)} - 1\right) + \frac{A_T}{f(\Delta T\ldots)} .$$

Since

$$\limsup_{T\to\infty} |\mathcal{W}_{V_T}| / f(V_T) = 1 ,$$

(43) will be proved by verifying

(44) $$\liminf_{T\to\infty} \left(\mp \frac{f(\Delta T\ldots+B_T)}{f(\Delta T\ldots)} \pm 1 + \frac{A_T}{f(\Delta T\ldots)}\right) \geqq 0 ,$$

where the upper signs are in vigour for $B_T > 0$, the lower ones for $B_T < 0$.

Let T be suficiently large and let $B_T > 0$. Then from (38), (39) follows

$$- \frac{f(\Delta T\ldots+B_T)}{f(\Delta T\ldots)} + 1 + \frac{A_T}{f(\Delta T\ldots)} \geqq \frac{-B_T f\ (\Delta T\ldots)}{f(\Delta T\ldots)} + \frac{A_T}{f(\Delta T\ldots)} \geqq$$

$$\geqq - \frac{B_T}{\Delta T\ldots} + \frac{A_T}{f(\Delta T\ldots)} = - o(1) \sqrt{\frac{A_T}{f(\Delta T\ldots)}} + \frac{A_T}{f(\Delta T\ldots)} \geqq - \frac{o(1)^2}{4} .$$

This establishes (44) with the upper signs.

Consider

$$(45) \qquad \frac{f(\Delta T\ldots+B_T)}{f(\Delta T\ldots)} - 1 + \frac{A_T}{f(\Delta T\ldots)}$$

(45) is nonnegative if $A_T \geqq f(\Delta T\ldots)$. Thus, let

$$(46) \qquad A_T < f(\Delta T\ldots) .$$

Then in virtue of (39)

$$|B_T| \leqq T^{3/4-\delta} \sqrt{f(\Delta T\ldots)} = o(T) .$$

Consequently, the first term in (45) approaches 1 as $T \to \infty$ provided that (46) holds. Hence, (44) is valid.

Assume now

$$(47) \qquad \lim_{T\to\infty} V_T < \infty .$$

From (14), (15) then follows

$$\liminf_{T\to\infty} (C_T - \theta T)/f(T) \geqq 0 ,$$

and hence (36). The nonvalidity of (35) would require

$$\limsup_{T\to\infty} A_T/f(T) < \sqrt{\Delta} .$$

But this together with (39) implies $B_T = o(T)$, which with regard to (40) yields $V_T \to \infty$ in contradiction to (47).

To prove the second assertion of Proposition 5 note that from (37), (39), (40) follows

$$A_T/f(\Delta T\ldots) = o(1),\ B_T = o(T),\ V_T = \Delta T + o(T) .$$

From here and from (42) we conclude that Z_T in (41) is negligible, and consequently equalities in (35), (36) hold. □

4. References

[1] K. Itô, H. P Mc Kean, Jr.: Diffusion Processes and their Sample Paths. Springer - Verlag, Berlin - Heidelberg 1965.

[2] H. P. Mc. Kean, Jr.: Stochastic Integrals Academic Press. New York - London 1969.

[3] V. Kučera: A review of the matrix Riccati equation. Kybernetika (Prague) 9 (1973), 42 - 61.

[4] P. Mandl: Asymptotic ordering of probability distributions for linear controlled systems with quadratic cost. In Stochastic Differential Systems (N. Christopeit, K. Helmes, M. Kohlmann editors), Lecture Notes in Control and Inf. Sc., Springer - Verlag Berlin-- Heidelberg 1986, 277 - 283.

Special Problems in Martingale Theory and Stochastic Calculus

A MINIMAL FLUCTUATION PROPERTY FOR COIN TOSSING AND LOCALLY SYMMETRIC MARTINGALES

J.M.C. Clark
Department of Electrical Engineering
Imperial College
London SW7 2BT

H. Föllmer
Mathematikdepartement
ETH-Zentrum
CH-8092 Zurich

1. Introduction

Let $\{F_n : n \in N\}$ be a non-decreasing family of σ-algebras in a probability space. We consider conditions under which an $\{F_n\}$-adapted convergent process $\{S_n\}$ converges to its limit as fast (with probability one) as any other adapted process.

The speed of convergence of an adapted process $\{Y_n\}$ will be measured by a normed modulus of error $\overline{\lim}_{n\to\infty}|S-Y_n|/a_n$. Comparisons between the speeds of convergence of different processes based on such a normed modulus will generally depend on the choice of norming sequence $\{a_n\}$ (consider, for instance, the difficulty of deciding which of two numerical sequences $\{x_n\}$, $\{y_n\}$ converges faster to zero if $\underline{\lim}_n\, x_n/y_n < 1 < \overline{\lim}_n x_n/y_n$). So, even though S_n converges as fast to S as any other adapted process (or has the "minimal fluctuation" property, for short) with respect to a particular norming sequence, it may fail to do so for a different norming sequence. It turns out, however, that if the distributions of $\{S_n\}$ obey a local symmetry condition, then $\{S_n\}$ has a minimal fluctuation property that holds for a wide variety of norming sequences.

In this paper we illustrate the idea of the minimal fluctuation property by a simple example based on coin tossing, and for this we give a complete proof. In the final section we state some further results for a large class of processes satisfying various symmetry conditions. Proofs, and some applications in the approximation of stochastic integrals and stochastic differential equations, will appear elsewhere.

2. A coin tossing example

Let $\{X_k\}(k=1,2,\ldots)$ be a sequence of independent binary random variables on (Ω,F,P), with $P[X_k=+1] = P[X_k=-1]=1/2$. Consider the random variable

$$M = \sum_{k=1}^{\infty} 2^{-k} X_k$$

and the associated martingale

$$M_n = E[M|F_n] = \sum_{t=1}^{n} 2^{-k} X_k \qquad (n = 1,2,\ldots)$$

where $F_n = \sigma(X_1,\ldots,X_n)$.

Theorem 1: For any adapted process $\{Y_n\}$ $(n=1,2,\ldots)$, and for any adapted positive norming process $\{A_n\}$ $(n=1,2,..)$

(1) $$\overline{\lim}_n \frac{|Y_n-M|}{A_n} \geq \overline{\lim}_n \frac{|M_n-M|}{A_n} \qquad \text{a.s.}$$

Proof: 1) For $\beta \in (0,1)$ take $N = N(\beta)$ large enough so that $\sum_{k>N} 2^{-k} < \beta$. For any $\alpha > 0$ let T_n $(n=0,1,..)$ be the sequence of stopping times defined by

(2) $$T_0 = 0$$

$$T_{n+1} = \min\{n>T_n+N \mid \frac{|Y_n-M_n|}{A_n} + \frac{2^{-n}}{A_n}(1-\beta) \geq \alpha\}.$$

Let

(3) $$B^+_{n+1} = \{X_{T_n+1} = \ldots = X_{T_n+N} = 1\} \cap \{\frac{Y_{T_n}-M_{T_n}}{A_n} \leq 0\} \cap \{T_n < \infty\}.$$

$$B^-_{n+1} = \{X_{T_n+1} = \ldots = X_{T_n+N} = -1\} \cap \{\frac{Y_{T_n}-M_{T_n}}{A_n} > 0\} \cap \{T_n < \infty\}$$

and $B_{n+1} = B^+_{n+1} \cup B^-_{n+1}$ $(n=0,1,\ldots)$. Since $B_{n+1} \in F_{T_{n+1}}$ and

$$P[B_{n+1}|F_{T_n}] = 2^{-N} I_{\{T_n<\infty\}} ,$$

the generalized Borel-Cantelli lemma implies

(4) $$\sum_n I_{B_{n+1}} = \infty \qquad \text{a.s. on}$$

$$\{\sum_n P[B_{n+1}|F_{T_n}] = \infty\} = \bigcap_n \{T_n < \infty\}.$$

2) Since

$$M_{T_n} - M \geq (1-\beta)2^{-T_n} \qquad \text{on } B^-_{n+1}$$

and $$M_{T_n} - M \leq -(1-\beta)2^{-T_n} \qquad \text{on } B^+_{n+1}$$

we obtain

$$\frac{|Y_{T_n}-M|}{A_{T_n}} = \left| \frac{Y_{T_n}-M_{T_n}}{A_{T_n}} + \frac{2^{-T_n}}{A_{T_n}} \frac{M_{T_n}-M}{2^{-T_n}} \right|$$

$$\geq \frac{|Y_{T_n}-M_{T_n}|}{A_{T_n}} + \frac{2^{-T_n}}{A_{T_n}}(1-\beta)$$

$$\geq \alpha \qquad \text{on } B_{n+1} .$$

Thus, (4) implies

$$\overline{\lim}_n \frac{|Y_n-M|}{A_n} \geq \alpha \qquad \text{a.s.} \quad \text{on} \bigcap_n \{T_n < \infty\};$$

hence

(5) $$\overline{\lim}_n \frac{|Y_n-M|}{A_n} \geq \overline{\lim}_n \left(\frac{|Y_n-M_n|}{A_n} + \frac{2^{-n}}{A_n}(1-\beta)\right) \qquad \text{a.s.}$$

On $\{\overline{\lim}_n \frac{2^{-n}}{A_n} = \infty\}$ we obtain $\overline{\lim}_n \frac{|Y_n-M|}{A_n} = \infty$, on $\{\overline{\lim}_n \frac{2^{-n}}{A_n} < \infty\}$ we can let β tend to 0, and in both cases we get

(6) $$\overline{\lim}_n \frac{|Y_n-M|}{A_n} \geq \overline{\lim}_n \left(\frac{|Y_n-M_n|}{A_n} + \frac{2^{-n}}{A_n}\right) \qquad \text{a.s.}$$

Note that (6) is in fact an equality, since $|M_n-M| \leq 2^{-n}$ implies

$$\frac{|Y_n-M|}{A_n} \leq \frac{|Y_n-M_n|}{A_n} + \frac{2^{-n}}{A_n} .$$

In particular we obtain for $Y_n = M_n$ the relation

(7) $$\overline{\lim}_n \frac{|M_n-M|}{A_n} = \overline{\lim}_n \frac{2^{-n}}{A_n} .$$

But (5) and (7) imply (1) and the proof is complete.

<u>Remark</u>. Suppose that $\lim_n \frac{2^{-n}}{A_n}$ exists. Then we can state, more precisely than in (1):

(8) $$\overline{\lim}_n \frac{|Y_n-M|}{A_n} = \overline{\lim}_n \frac{|Y_n-M_n|}{A_n} + \overline{\lim}_n \frac{|M_n-M|}{A_n} \qquad \text{a.s.}$$

This follows from (7) and the equality in (6).

3. Locally symmetric process

Suppose $\{F_n\}$ is a nondecreasing filtration in a probability space (Ω,F,P).

Definition A process $\{X_n\}$ adapted to $\{F_n\}$ is locally symmetric if for all $m,n>0$ and all Borel $B \in R^n$

$$P[(X_{n+1} \ldots X_{n+m}) \in B|F_n] = P[(-X_{n+1} \ldots -X_{n+m}) \in B|F_n].$$

If a process $\{S_n\}$ has locally symmetric differences - that is, the process $X_n = S_n - S_{n-1}$ is locally symmetric - and it is integrable, then clearly it is a martingale. The coin-tossing process M_n of the previous section is a martingale of this sort.

Integrability, however, is not necessary for the minimal fluctuation property to hold:

Theorem 2 Suppose $S_n \to S$ a.s. as $n \to \infty$ and $\{S_n\}$ has locally symmetric differences. Then for any adapted process $\{Y_n\}$ and for any positive numerical sequence $\{a_n\}$

$$\overline{\lim}_n |S-Y_n|/a_n \geq \overline{\lim}_n |S-S_n|/a_n \qquad \text{a.s.}$$

Theorem 1 established the minimal fluctuation property for random normings, as long as these were adapted. The following is a partial extension of this for locally symmetric processes:

Theorem 3 Suppose the difference process $\{X_n\}$ of $\{S_n\}$ is locally symmetric. Suppose $\{A_n\}$ is any adapted positive norming process satisfying, for some fixed m, $\overline{\lim}_n A_{n+m}/A_n < 1$,. Then, for any adapted process $\{Y_n\}$

$$\overline{\lim}_n|S-Y_n|/A_n \geq \overline{\lim}_n|S-S_n|/A_n \qquad \text{a.s.}$$

on the set where the right-hand member is finite.

In many applications the local symmetry condition holds only asymptotically. This is true of certain non-symmetric martingale transforms of Gaussian martingales. It turns out that the minimal fluctuation property still holds, though for a more limited class of numerical norming sequences. The following theorem illustrates this.

Suppose $\{W_n\}$ is a sequence of independent normal random variables, with $EW_n = 0$ and $EW_n^2 = 1$, adapted to $\{F_n\}$. Suppose $\{H_n\}$ is an $\{F_n\}$-predictable process (i.e. H_n is F_{n-1}-measurable) and $\sum_1^\infty EH_n^2 < \infty$. Then the transform martingale $S_n = \sum_1^n H_k W_k$ converges a.s. to a limit S. Let $\phi(x)$ denote $\sqrt{(2x \log \log x^{-1})}$.

Theorem 4 If $\{H_n\}$ satisfies

i) $\overline{\lim}_n \, EH_n^2/\sum_{n+1}^{\infty} EH_k^2) < \infty$

ii) $\lim_n |H_n|/(EH_n^2)^{1/2} = \gamma > 0$ a.s.

and $\{a_n\}$ is any norming sequence of positive constants such that

iii) $0 < \underline{\lim}_n \, a_n/\phi(\sum_{n+1}^{\infty} EH_k^2) \leq \overline{\lim}_n \, a_n/\phi(\sum_{n+1}^{\infty} EH_k^2) < \infty.$

then for any adapted sequence $\{Y_n\}$

$$\overline{\lim}_n |S-Y_n|/a_n \geq \overline{\lim}_n |S-S_n|/a_n > 0 \quad \text{a.s.}$$

If $a_n \sim \phi(\sum_{n+1}^{\infty} EH_k^2)$ the right-hand member is γ a.s.

Further generalizations are possible. Heyde [1] has given a law of the iterated logarithm for the tails $M_\infty - M_n$ of convergent martingales $\{M_n\}$, in which the norming sequence is $\phi(\langle M\rangle_\infty - \langle M\rangle_n)$, where $\langle M\rangle_n$ is the predictable conditional variance process of $\{M_n\}$. It can be shown that martingales have the minimal fluctuation property with respect to this norming sequence under condiitons similar to those of Heyde. These results will appear elsewhere.

[1] **C.C. Heyde.** On central limit and iterated logarithm supplements to the martingale convergence theorem. J. Appl. Prob. 14 758-775, 1977.

THE FUNCTIONAL LAW OF THE ITERATED LOGARITHM FOR LÉVY'S AREA PROCESS*

by

Kurt Helmes, Bruno Remillard, and Radu Theodorescu

University of Kentucky, Laval University, Laval University

Let $\{\frac{1}{2}L(t): t \geq 0\}$ be Lévy's area process and define a random sequence $\{f_n: n \geq 1\}$ in $C[0,1]$ by $f_n(t) = L(nt)/nl(n)$, $t \in [0,1]$, $n \geq 1$, where $l(n)=1$ for $n=1,2$ and $l(n)=\log\log n$ for $n \geq 3$. It is shown that, with probability 1, the set of limit points of $\{f_n: n \geq 1\}$ is $H = \{x: x(t) = \int_0^t (z_1(u)\dot{z}_2(u) - \dot{z}_1(u)z_2(u))du,\ (z_1,z_2) \in K,\ t \in [0,1]\}$, where $K = \{(z_1,z_2): z_j(0)=0,\ z_j$ absolutely continuous, $j=1,2$, $\frac{1}{2}\int_0^1 (|\dot{z}_1(u)|^2 + |\dot{z}_2(u)|^2 du \leq 1\}$.

AMS 1980 Subject Classifications: Primary 60F17, 60G17; Secondary 60J65, 58G32.

Key words and phrases: Law of the iterated logarithm, stochastic area process.

1. INTRODUCTION

Let $(\Omega,\mathbf{F},P)$ be a complete probability space, let $\mathbf{W} = \{W(t): t \geq 0\}$ be a Wiener process on $(\Omega,\mathbf{F},P)$ with values in $\mathbb{R}^2$, and let $\mathbf{F}_t$ denote the completion of $\sigma\{W(s): 0 \leq s \leq t\}$, $t \geq 0$. Furthermore set $\mathbf{L} = \{L(t): t \geq 0\}$, where $L(t) = \int_0^t \langle JW(s), dW(s)\rangle$, $t \geq 0$, $J = \begin{pmatrix} 0 & -1 \\ 1 & 0 \end{pmatrix}$, and $\langle\cdot,\cdot\rangle$ denotes the scalar product in $\mathbb{R}^2$. Then $\frac{1}{2}\mathbf{L}$ is the area process introduced by Lévy [9] (see also [7, p.168] and [8, p.385]) as a stochastic analogue of the formula which in the deterministic case yields the area enclosed by a smooth curve and the chord connecting the origin with a point on the curve. In recent years this process has found applications in analysis (see, e.g., [4]) as well as in statistical inference (see, e.g., [3] and [11, p.212]).

The law of the iterated logarithm for the area process and related processes has been proved to hold at infinity [2] and at 0 [5]. In the present paper we prove the functional law of the iterated logarithm (compact LIL in the sense of [1]). However the compact set we derive is quite different from that one occuring when one considers a Wiener process.

Let $L_n(t) = L(nt)$, $t \geq 0$, $n \geq 1$, and define the random sequence $\{f_n: n \geq 1\}$ in C (C denotes the space of all real-valued continuous functions on $[0,1]$, vanishing at 0, with the supremum norm $\|\cdot\|$) by $f_n(t) = L_n(t)/nl(n)$, $t \in [0,1]$, $n \geq 1$, where $l(n)=1$ for $n=1,2$ and

* Work supported by the Natural Sciences and Engineering Research Council of Canada, by the Fonds F.C.A.R. of the Province of Quebec, and by the Deutsche Forschungsgmeinschaft.

$l(n)=\log\log n$ for $n\geq 3$. We prove that there is a compact set H such that $P(\lim_{n\to\infty} d(f_n,H)=0)=1$, where $d(x,A)=\inf_{y\in A}\|x-y\|$, and $P(C(f_n:n\geq 1)=H)=1$; here $C(a_n:n\geq 1)$ is the set of limit points of $\{a_n:n\geq 1\}$ in C, $H=\{x=\rho(z):\ \rho(z)(t)=\int_0^t \langle Jz(u),\dot z(u)\rangle du,\ z\in K,\ t\in[0,1]\}$, where $K=\{z=(z_1,z_2):\ z_j(0)=0,\ z_j$ absolutely continuous, $j=1,2$, $\frac{1}{2}\int_0^1 |\dot z(u)|^2du\leq 1\}$, $|\cdot|$ stands for the usual norm in $\mathbb{C}$, $\dot z=(\dot z_1,\dot z_2)$, and $\dot z_j$ is the derivative of z_j, $j=1,2$.

We tacitly use the following properties of **L**: (i) (symmetry) $-\frac{1}{2}\mathbf{L}$ is an area process; (ii) (scaling) for any $c>0$, $\frac{1}{2}\{L(ct)/c:t\geq 0\}$ is an area process; (iii) **L** is a mean square integrable martingale (see [5]).

2. RELATIVE COMPACTNESS

We start by proving the following approximation lemma:

Lemma 1. *For any $\epsilon>0$, there is a $c_1(\epsilon)>1$ such that*

$$P\left(\max_{n_j<n\leq n_{j+1}}\|L_n-L_{n_j}\|>\epsilon\phi(n_j)\ \text{i.o.}\right)=0,$$

where $n_j=1+[c^j]$, $j\geq 1$, $1<c<c_1$, $\phi(n)=nl(n)$, $n\geq 1$, and $[x]$ is the integral part of x.

Proof. For any $c>1$, let $m>1$ be an integer to be specified later, and set $n_j=1+[c^j]$, $j\geq 1$. We have

$$\max_{n_j<n\leq n_{j+1}}\|L_n-L_{n_j}\|\leq \max_{n_j<n\leq n_{j+1}}\left\{\max_{1\leq k\leq m}\left\{\sup_{\frac{k-1}{m}\leq t\leq\frac{k}{m}}|L_n(t)-L_{n_j}(t)|\right\}\right\}$$

$$\leq 2\max_{1\leq k\leq m}\left\{\sup_{\frac{k-1}{m}n_j\leq t\leq\frac{k}{m}n_{j+1}}\left|L(t)-L_{n_j}\left(\frac{k-1}{m}\right)\right|\right\}.$$

Hence

$$P\left(\max_{n_j<n\leq n_{j+1}}\|L_n-L_{n_j}\|>\epsilon\phi(n_j)\right)\leq\sum_{k=1}^{m}P\left(\sup_{\frac{k-1}{m}\leq t\leq\frac{k}{m}n_{j+1}}\left|L(t)-L_{n_j}\left(\frac{k-1}{m}\right)\right|>\frac{\epsilon}{2}\phi(n_j)\right)$$

$$\leq \exp\{-2l(n_j)\}\sum_{k=1}^{m}E\left(\exp\left\{\frac{4}{\epsilon n_j}\left|L_{n_{j+1}}\left(\frac{k}{m}\right)-L_{n_j}\left(\frac{k-1}{m}\right)\right|\right\}\right) \tag{1}$$

by virtue of Doob's submartingale inequality. Now

$$E\left(\exp\left\{\frac{4}{\epsilon n_j}\left|L_{n_{j+1}}\left(\frac{k}{m}\right)-L_{n_j}\left(\frac{k-1}{m}\right)\right|\right\}\right)\leq 2E\left(\exp\left\{\frac{8}{\epsilon n_j}\left(L_{n_{j+1}}\left(\frac{k}{m}\right)-L_{n_j}\left(\frac{k}{m}\right)\right)\right\}\right)^{\frac{1}{2}}\times$$

$$\times E\left(\exp\left\{\frac{8}{\epsilon n_j}\left(L_{n_j}\left(\frac{k}{m}\right)-L_{n_j}\left(\frac{k-1}{m}\right)\right)\right\}\right)^{\frac{1}{2}}=:2AB\,.$$

Let us look at each factor. First, in view of Proposition 2.3 [5], for B we have

$$E\left\{\exp\left\{\frac{8}{\epsilon n_j}\left(L_{n_j}\left(\frac{k}{m}\right)-L_{n_j}\left(\frac{k-1}{m}\right)\right)\right\}\right\}$$

$$=E\left\{\exp\left\{\frac{8}{\epsilon}\left(L\left(\frac{k}{m}\right)-L\left(\frac{k-1}{m}\right)\right)\right\}\right\}=\left(\cos\frac{8}{m\epsilon}-\frac{8(k-1)}{m\epsilon}\sin\frac{8}{m\epsilon}\right)^{-1}$$

$$\leq\left(\cos\frac{8}{m\epsilon}-\frac{8}{\epsilon}\sin\frac{8}{m\epsilon}\right)^{-1}\leq 2$$

for $m>m_1(\epsilon)$. Next A converges when $j\to\infty$ to

$$\left(\cos\frac{8k(c-1)}{\epsilon m}-\frac{8k}{\epsilon m}\sin\frac{8k(c-1)}{m\epsilon}\right)^{-1}\leq\left(\cos\frac{8(c-1)}{\epsilon}-\frac{8}{\epsilon}\sin\frac{8(c-1)}{\epsilon}\right)^{-1}\leq 2$$

for $1<c<c_1=1+m_1(\epsilon)$. Hence the last expression in (1) is bounded above by const./j^2 for $j>j_1(c)$ and the conclusion now follows from the Borel-Cantelli lemma.

By making use of Lemma 1 we can now prove the following relative compactness property:

Theorem 2. *The sequence* $\{f_n : n\geq 1\}$ *is relatively compact with probability 1.*

Proof. It follows from the Arzelà-Ascoli Theorem that we only have to prove that, for any $\epsilon>0$, there is $\delta>0$ such that

$$P\left(\sup_{|t-s|<\delta}|f_n(t)-f_n(s)|>\epsilon \text{ i.o.}\right)=0;$$

note that $P(f_n(0)=0,\ n\geq 1)=1$. In virtue of Lemma 1, it suffices to show that, for any $\epsilon>0$ and any $c>1$, there is an integer $m_2(\epsilon)$ such that $m>m_1(\epsilon)$ implies that

$$P\left(\sup_{|t-s|<1/m}|f_{n_j}(t)-f_{n_j}(s)|>\epsilon \text{ i.o.}\right)=0,$$

where $n_j=1+[c^j]$, $j\geq 1$. Now (see the proof of Lemma 1),

$$P\left(\sup_{|t-s|<1/m}|f_{n_j}(t)-f_{n_j}(s)|>\epsilon\right)\leq\sum_{k=1}^{m}P\left(\sup_{\frac{k-1}{m}\leq t\leq\frac{k}{m}}\left|f_{n_j}(t)-f_{n_j}\left(\frac{k-1}{m}\right)\right|>\frac{\epsilon}{3}\right)\leq\frac{\text{const.}}{j^2}$$

for $m>m_1(\epsilon)$ (so $m_2(\epsilon)=m_1(\epsilon)$). Hence, by virtue of the Borel-Cantelli lemma,

$$P\left(\sup_{|t-s|<1/m}|f_{n_j}(t)-f_{n_j}(s)|>\epsilon \text{ i.o.}\right)=0$$

for $m>m_1(\epsilon)$.

3. THE STRUCTURE OF THE DERIVED SET

We shall prove first an analytic result for the set H. Denote by C' the topological dual of C.

Lemma 3. *Let* $\xi\in C'$ *be fixed. Then* $\sup_{x\in H}\xi(x)=\theta(\xi)$, *where*

$$\theta(\xi)=\inf\{\lambda>0: E(\exp\{\xi(L)/\lambda\})<\infty\}.$$

Proof. First $\theta(\xi)\le 2\|\xi\|/\pi$ since $E(\exp\{\|L\|/\lambda\})<\infty$ for $\lambda>2/\pi$. Next, we shall prove that $\sup_{x\in H}\xi(x)\le\theta(\xi)$. Let $z\in K$ be fixed and define $\tilde{P}$ by its Radon-Nikodym derivative with respect to P,

$$\frac{d\tilde{P}}{dP}=\exp\left\{\int_0^1 <\dot{z}(u),dW(u)> - \tfrac{1}{2}\int_0^1 |\dot{z}(u)|^2\,du\right\}.$$

Further, put

$$V(t)=\int_0^t <JW(u),\dot{z}(u)>du+\int_0^t <Jz(u),dW(u)>$$

and

$$\tilde{V}(t)=\int_0^t <J\tilde{W}(u),\dot{z}(u)>du+\int_0^t <Jz(u),d\tilde{W}(u)>,$$

where $\tilde{W}(t)=W(t)-z(t)$, $t\in[0,1]$. We recall that $\tilde{\mathbf{W}}=\{\tilde{W}(t):t\in[0,1]\}$ is a Wiener process with respect to $\tilde{P}$. For $\lambda>\theta(\xi)$ fixed, we have

$$E\left(\exp\left\{\xi(L)/\lambda\right\}\right)=E\left(\exp\left\{(\xi(\tilde{L})+\xi(\tilde{V})+\xi(\rho(z)))/\lambda\right\}\right)$$

$$=\exp\left\{\xi(\rho(z))/\lambda\right\}E\left(\frac{d\tilde{P}}{dP}\exp\left\{\xi(\tilde{L}+\tilde{V})/\lambda-\int_0^1<\dot{z}(u),d\tilde{W}(u)>-\tfrac{1}{2}\int_0^1|\dot{z}(u)|^2du\right\}\right)$$

$$=\exp\left\{\xi(\rho(z))/\lambda-\tfrac{1}{2}\int_0^1|\dot{z}(u)|^2du\right\}E\left(\exp\left\{\xi(L+V)/\lambda-\int_0^1<\dot{z}(u),dW(u)>\right\}\right),$$

where $\tilde{\mathbf{L}}$ is the area process corresponding to $\tilde{\mathbf{W}}$. Now let

$$Y=\exp\{\xi(V)/\lambda-\int_0^1<\dot{z}(u),dW(u)>\}.$$

Then

$$E(\exp\{\xi(L)/\lambda\})=2E(\exp\{\xi(L)/\lambda\}1_{\{Y\ge 1\}})$$

and

$$E(\exp\{\xi(L)/\lambda\}Y)=E(\exp\{\xi(L)/\lambda\}(Y+Y^{-1})1_{\{Y\ge 1\}}).$$

Since $\lambda>\theta(\xi)$ and $(x+x^{-1})/2\ge 1$ for $x\ge 1$, we have

$$\exp\left\{\xi(\rho(z))/\lambda-\tfrac{1}{2}\int_0^1|\dot{z}(u)|^2du\right\}\le 1.$$

Therefore $\xi(x)\le\lambda$ for all $x\in H$ and $\lambda>\theta(\xi)$. Thus $\sup_{x\in H}\xi(x)\le\theta(\xi)$.

We next prove that $\theta(\xi)\le\sup_{x\in H}\xi(x)$. Set $\psi(\varsigma)=E(\exp\{i\varsigma\xi(L)\})$, $\varsigma\in\mathbb{C}$. Then ψ is analytic in $|\mathrm{Im}\,\varsigma|<1/\theta(\xi)$. Using the fact that $\xi(L)=\int_0^1 F(u)<JW(u),dW(u)>$, where $F(t)=\xi(1_{[t,1]}(\cdot))$, $t\in[0,1]$, we conclude from Theorem 1 [6, p.183] and the linearization of the Ricatti equation that, for $|\mathrm{Im}\,\varsigma|<1/\theta(\xi)$, $\psi(\xi)=1/h_\varsigma(0)$, where h_ς is the solution of the following boundary problem: h_ς continuously differentiable, $\ddot{h}_\varsigma(t)=\varsigma^2F(t)h_\varsigma(t)$, a.e. $t\in(0,1)$, $h_\varsigma(1)=1$, $\dot{h}_\varsigma(1)=0$. Hence $1/\theta(\xi)=\inf\{\lambda>0:h_{i\lambda}(0)=0\}$. Let $\eta=i/\theta(\xi)$; then $h_\eta(0)=0$. Further set $z(t)=\exp\{\eta G(t)\}h_\eta(t)$, $t\in[0,1]$, where $\dot{G}=F$ and $G(0)=0$. Then $z(0)=0$, z is continuous in $[0,1]$, and z is absolutely continuous on $[0,1]$. Next set

$$x(t) = \frac{1}{2i}\int_0^t (\bar{z}(u)\dot{z}(u) - z(u)\dot{\bar{z}}(u))du = \rho(\text{Re } z, \text{Im } z)(t), \quad t\in[0,1].$$

Then

$$\xi(x) = \frac{1}{2i}\int_0^t F(u)(\bar{z}(u)\dot{z}(u) - z(u)\dot{\bar{z}}(u))du = \frac{1}{\theta(\xi)}\int_0^1 F^2(u)h_\eta^2(u)du,$$

$$\int_0^1 |\dot{z}(u)|^2 du = \int_0^1 \dot{h}_\eta^2(u)du + \frac{1}{\theta^2(\xi)}\int_0^1 F^2(u)h_\eta^2(u)du$$

and

$$\frac{1}{\theta^2(\xi)}\int_0^1 F^2(u)h_\eta^2(u)du = -\int_0^1 h_\eta(u)\ddot{h}_\eta(u)du = \int_0^1 \dot{h}_\eta^2(u)du.$$

Hence $\xi(x) = \frac{\theta(\xi)}{2}\int_0^1 |\dot{z}(u)|^2 du$, and we conclude that $\theta(\xi) \le \sup_{x\in H}\xi(x)$. Thus $\sup_{x\in H}\xi(x) = \theta(\xi)$.

Next, by adapting a result of J. Kuelbs ([10, p.240]), we shall prove

Theorem 4. *$P(C(f_n : n\ge 1)\subset H) = 1$ and $P(\lim_{n\to\infty} d(f_n, H) = 0) = 1$.*

Proof. Since $\{f_n : n\ge 1\}$ is relatively compact with probability 1 (cf. Theorem 2), it is only necessary to prove that, for any $\xi\in C'$,

$$P(\limsup_{n\to\infty}\xi(f_n)\le\theta(\xi)) = 1.$$

In the light of Lemma 1, we need only show that, for any $\epsilon > 0$ and any $c > 1$,

$$P(\xi(f_{n_j}) > (1+\epsilon)\theta(\xi) \text{ i.o.}) = 0,$$

where $n_j = 1 + [c^j]$, $j\ge 1$. But

$$P(\xi(f_{n_j}) > (1+\epsilon)\theta(\xi))$$
$$\le \exp\{-(1+\epsilon)^{1/4} l(n_j)\} E(\exp\{(1+\epsilon)^{-1/4}\xi(L)/\theta(\xi)\}) \le \text{const. } j^{-(1+\epsilon)^{1/4}}, \quad j\ge 1,$$

and our result follows.

Theorem 5. *$P(C(f_n : n\ge 1)\supset H) = 1$.*

Proof. Since H is separable, it suffices to show that, for any $\epsilon > 0$ and any $x\in H$, we have $P(\|f_n - x\| < \epsilon \text{ i.o.}) = 1$.

Let $m > 1$ be an integer to be specified later. Then

$$\|f_n - x\| \le \max_{1\le k\le m}\left|f_n\left(\frac{k}{m}\right) - x\left(\frac{k}{m}\right)\right| + \tau\left(f_n, \frac{1}{m}\right) + \tau\left(x, \frac{1}{m}\right),$$

where $\tau(y,\delta) = \sup_{|t-s|<\delta}|y(t) - y(s)|$. Theorem 2 implies that there is an integer $m_3(\epsilon)\ge 1$ such that $\tau(x, 1/m) < \epsilon/4$ and $P(\tau(f_n, 1/m) > \epsilon/4 \text{ i.o.}) = 0$ for $m\ge m_3(\epsilon)$. Hence

$$P(\|f_n - x\| < \epsilon \text{ i.o.}) \geq P\Big(\max_{1\leq k\leq m} \big|f_n\big(\frac{k}{m}\big) - x\big(\frac{k}{m}\big)\big| < \frac{\epsilon}{2} \text{ i.o.}\Big)$$

$$\geq P\Big(\max_{1\leq k\leq m} \big|f_{d^n}\big(\frac{k}{m}\big) - x\big(\frac{k}{m}\big)\big| < \frac{\epsilon}{2} \text{ i.o.}\Big)$$

for any integer $d > 1$.

Next there is $m_4(\epsilon) \geq m_3(\epsilon)$ such that

$$P\Big(|L(m^n)| > \frac{\epsilon}{4}\, m^{n+1}\, l(m^{n+1}) \text{ i.o.}\Big) = 0$$

for $m > m_4(\epsilon)$. Hence

$$P(\|f_n - x\| < \epsilon \text{ i.o.})$$

$$\geq P\Big(\max_{1\leq k\leq m} \big|L(m^n k) - L(m^n) - x\big(\frac{k}{m}\big) m^{n+1}\, l(m^{n+1})\big| < \frac{\epsilon}{4}\, m^{n+1}\, l(m^{n+1})\Big)$$

for $m > m_4(\epsilon)$.

Moreover

$$L(m^n k) - L(m^n) = \tilde{\tilde{L}}(m^n(k-1)) + \langle JW(m^n), \tilde{\tilde{W}}(m^n(k-1))\rangle\,,$$

where $\tilde{\tilde{\mathbf{W}}} = \{\tilde{\tilde{W}}(t) = W(t+m^n) - W(m^n):\ t \geq 0\}$ is a Wiener process independent of $\mathbf{F}_{m^n}$ and $\tilde{\tilde{\mathbf{L}}}$ is the area process associated with $\tilde{\tilde{\mathbf{W}}}$. Next we establish that $\langle JW(m^n), \tilde{\tilde{W}}(m^n(k-1))\rangle / m^{n+1}\, l(m^{n+1})$ is arbitrarily small for m sufficiently large. First

$$P(|\langle JW(m^n), \tilde{\tilde{W}}(m^n(k-1))\rangle > \frac{\epsilon}{8}\, m^{n+1}\, l(m^{n+1}))$$

$$\leq 2\exp\{-2l(m^{n+1})\}\, E\Big[E\Big(\exp\{\frac{16}{\epsilon\, m^{n+1}} \langle JW(m^n), \tilde{\tilde{W}}(m^n(k-1))\rangle\} \,|\, \mathbf{F}_{m^n}\Big)\Big]$$

$$\leq 2\exp\{-2l(m^{n+1})\}\, E\Big(\exp\{\frac{128}{\epsilon^2 m} |W(1)|^2\}\Big)$$

$$= 2\exp\{-2l(m^{n+1})\}\Big(1 - \frac{256}{\epsilon^2 m}\Big)^{-1} \leq \frac{\text{const.}}{(n+1)^2}$$

for $m > m_5(\epsilon) \geq m_4(\epsilon)$. In view of the Borel-Cantelli Lemma, we obtain

$$P\Big(\max_{1\leq k\leq m} |JW(m^n), \tilde{\tilde{W}}(m^n(k-1))\rangle / \phi(m^{n+1})| > \frac{\epsilon}{8} \text{ i.o.}\Big) = 0\,.$$

Taking into account the above inequalities, we find that

$$P(\|f_n - x\| < \epsilon \text{ i.o.}) \geq P\Big(\max_{1\leq k\leq m} \big|(\tilde{\tilde{L}}(m^n(k-1))/\phi(m^{n+1})) - x\big(\frac{k}{m}\big)\big| < \frac{\epsilon}{8} \text{ i.o.}\Big)$$

for $m > m_5(\epsilon)$. By virtue of the Borel-Cantelli Lemma, $P(\|f_n - x\| < \epsilon \text{ i.o.}) = 1$ if the series

$$\sum_{n=1}^{\infty} P\Big(\max_{1\leq k\leq m} \big|(\tilde{\tilde{L}}(m^n(k-1))/\phi(m^{n+1})) - x\big(\frac{k}{m}\big)\big| < \epsilon\Big)$$

diverges for each $\epsilon > 0$ and m sufficiently large. The relation

$$P\Big(\max_{1\leq k\leq m} \big|(\tilde{\tilde{L}}(m^n(k-1))/\phi(m^{n+1})) - x\big(\frac{k}{m}\big)\big| < \epsilon\Big) = P\Big(\max_{1\leq k\leq m} \big|f_{m^{n+1}}\big(\frac{k-1}{m}\big) - x\big(\frac{k}{m}\big)\big| < \epsilon\Big)$$

holds, so $P(\|f_n - x\| < \epsilon \text{ i.o.}) = 1$ if the series $\sum_{n=1}^{\infty} P(\|f_{m^n} - x\| < \epsilon)$ diverges for each $\epsilon > 0$ and each integer $m > 1$. Indeed, let $z \in K$ be such that $x = \rho(z)$. Then, proceeding as in Lemma 3, we obtain

$$P(\|f_k - x\| < \epsilon)$$
$$= E\Big(\exp\{\int_0^1 <\dot{y}_k(u), dW(u)> - \tfrac{1}{2}\int_0^1 |\dot{y}_k(u)|^2\,du\}$$
$$\times \exp\{-\int_0^1 <\dot{y}_k(u), d\tilde{W}(u)> - \tfrac{1}{2}\int_0^1 |\dot{y}_k(u)|^2\,du\}\, 1_{\{\|L+\tilde{V}\| < \epsilon l(k)\}}\Big),$$

where $\tilde{W}(t) = W(t) - y_k(t)$ and $y_k(t) = z(t)(l(k))^{1/2}$, $t \in [0,1]$. It follows that

$$P(\|f_k - x\| < \epsilon) = \exp\left\{-\tfrac{1}{2} l(k) \int_0^1 |\dot{z}(u)|^2\,du\right\} p_k\,,$$

where

$$p_k = E\left[\exp\left\{-(l(k))^{1/2}\int_0^1 <\dot{z}(u), dW(u)>\right\} 1_{\{\|L+V\| < \epsilon l(k)\}}\right]$$

and $\liminf_{k \to \infty} p_k \geq \frac{1}{2}$. Consequently

$$\sum_{n=1}^{\infty} P(\|f_{m^n} - x\| < \epsilon) \geq \text{const.} \sum_{n=1}^{\infty} \frac{1}{n}\, p_{m^n} = \infty$$

for each $\epsilon > 0$ and each integer $m > 1$. Therefore $P(\|f_n - x\| < \epsilon \text{ i.o.}) = 1$ for each $\epsilon > 0$ and each $x \in H$.

Combining Theorem 4 and Theorem 5, we finally obtain

Theorem 6. *$P\big(\lim_{n \to \infty} d(f_n, H) = H) = 0\big) = 1$ and $P(C(f_n : n \geq 1) = H) = 1$.*

An immediate consequence of Theorem 6 and the proof of Lemma 3 is the following result:

Theorem 7. *Let $\xi \in C'$. Then*

$$P(\limsup_{n \to \infty} \xi(L(n)) / \theta(\xi) n l(n) = 1 \text{ i.o.}) = 1,$$

where $1/\theta(\xi)$ is the smallest eigenvalue of the following boundary problem: v continuously differentiable, $\ddot{v}(t) = \mu^2 F(t) v(t)$, a.e. $t \in (0,1)$, $v(0) = \dot{v}(1) = 0$, where $F(t) = \xi(1_{[t,1]}(\cdot))$, $t \in [0,1]$.

From Theorem 7 we obtain the ordinary law of the iterated logarithm:

Theorem 8. $P(\limsup_{n \to \infty} L(n) / (2/\pi) n l(n) = 1 \text{ i.o.}) = 1$.

Proof. In view of Theorem 7 we must show that $\theta(\delta_1) = 2/\pi$, where δ_1 is the Dirac measure at 1. This follows by computing the smallest eigenvalue of the problem specified in Theorem 7. (Another way to find $\theta(\delta_1)$ is to use the relation $E(\exp\{\nu L(1)\}) = 1/\cos\nu$, $\nu < \pi/2$.)

4. EXTENSIONS

By making use of the same technique as above, we can also prove the functional law of the iterated logarithm for Berthuet's [2] type processes as well as for those considered by Schott [12].

Comment. It was brought to our attention that P. Baldi, in a forthcoming paper to appear in Probability Theory and Related Fields, proved the functional law of the iterated logarithm for a class of diffusion processes containing those considered in this paper. He essentially uses large deviations techniques in his proofs. A summary of his results already appeared in C.R. Acad. Sci. Paris Sér. I, 1984, 299, 463-466.

REFERENCES

[1] Acosta, A. de and Kuelbs, J., Some results on the cluster set $C(\{S_n/a_n\})$ and the LIL. *Ann. Probability,* 11 (1983), 102-122.

[2] Berthuet, R., Loi du logarithme itéré pour certains intégrales stochastiques. *C.R. Acad. Sci. Paris. Sér. A,* 289 (1979), 813-815.

[3] Dugué, D., Pour réunir les tests de Cramér-von Mises, Kolmogoroff-Smirnoff, Paul Lévy et Laplace-Gauss. *C.R. Acad. Sci. Paris. Sér. I,* 296 (1983), 89-91.

[4] Gaveau, B., Principe de moindre action, propagation de la chaleur et estimées sous elliptiques sur certains groupes nilpotents. *Acta. Math.,* 139 (1977), 95-153.

[5] Helmes, K., The "local" law of the iterated logarithm for processes related to Lévy's stochastic area process. *Studia Math.,* 84 (1986).

[6] Helmes, K. and Schwane, A., Lévy's stochastic area formula in higher dimensions. *J. Functional Anal.,* 54 (1983), 177-192.

[7] Hida, T., *Brownian Motion.* Springer (1983), New York.

[8] Ikeda, N. and Watanabe, S., *Stochastic Differential Processes and Diffusion Processes.* Elsevier North-Holland (1981), New York.

[9] Lévy, P., Le mouvement brownien plan. *Amer. J. Math.,* 62 (1949), 487-550.

[10] Kuelbs, J., The law of the iterated logarithm and related strong convergence theorems for Banach space valued random variables. *Lecture Notes in Mathematics,* No. 539 (1979), 224-314.

[11] Lipster, R.S. and Shiryaev, A.N., *Statistics of Random Processes.* Vol. 2, Springer (1977), New York.

[12] Schott, R., Une loi du logarithme itéré pour certaines intégrales stochastiques. *C.R. Acad. Sci. Paris. Sér. I,* 292 (1983), 295-298.

K. Helmes,
Department of Mathematics
University of Kentucky
Lexington, KY 40508

B. Remillard and R. Theodorescu
Département de mathématiques, statistique et actuariat
Université Laval
Québec, Canada G1K 7P4

ITO-VENTZEL'S FORMULA FOR SEMIMARTINGALES, ASYMPTOTIC PROPERTIES OF MLE AND RECURSIVE ESTIMATION

N.L.Lazrieva, T.A.Toronjadze

Tbilisi Razmadze Mathematical Institute of the Academy of Sciences of the Georgian SSR, 150 a Plekhanov Ave., Tbilisi, 380012

0°. Introduction. The present paper deals with the following problems: 1° gives a generalization of Ito-Ventzel's formula for the general form of semimartingales; 2° is devoted to asymptotic properties of the maximum likelihood estimator (MLE) in a general scheme of statistical models, the condition being expressed in terms of predictable characteristics of some transformations of martingales defining the density; in 3° an algorithm is proposed for the construction of recursive estimators (RE), based on the analysis of a stochastic differential equation for the MLE obtained by Ito-Ventzel's formula, examples are given and on the basis of the results of 2° asymptotic properties of recursive estimators are studied in a particular case. Finally, in 4° a short review of the results obtained in this field is presented. In what follows the notations of [1] will be used.

1°. Ito-Ventzel's Formula. Let on some stochastic basis $(\Omega, \underline{\underline{F}}, \underline{F}, P)$ with filtration $\underline{F} = \{\underline{\underline{F}}_t\}_{0 \le t \le T}$, $(\underline{\underline{F}}_s \subset \underline{\underline{F}}_t \subset \underline{\underline{F}},\ s \le t,\ \underline{\underline{F}} = \underline{\underline{F}}_T)$ satisfying the usual conditions a family of semimartingales $F(x) = \{F(t,x) = M(t,x) + A(t,x), 0 \le t \le T\}, x \in R^1$, where $M(\cdot,x) \in \mathcal{M}^2(\underline{F}, P)$, $A(\cdot,x) \in \mathcal{A}(\underline{F}, P)$ and the semimartingale $\xi(t) = m(t) + a(t)$ with $m \in \mathcal{M}^2(\underline{F}, P), a \in \mathcal{A}(\underline{F}, P)$ be given.

Our aim is to derive a formula for the differential of the composite function $F(t, \xi(t))$. This formula contains the so called stochastic line integral, the construction scheme for which will be given below.

Let $\mathcal{L}^2 = \mathcal{L}^2\{M(\cdot,x), x \in R^1\}$ denote the minimal stable subspace of the space H^2 containing stochastic integrals $-\{h \cdot M(\cdot,x), h \in (f \in \mathcal{P}: Ef^2 \circ \langle M(\cdot,x)\rangle_T < \infty, x \in R^1\}$. Suppose the following conditions are satisfied:

1) a process $K \in \mathcal{A}^+_{loc}$ exists such that for all $x \in R^1$ $[M(\cdot,x)] \ll K$;

2) the derivative $\varphi(\cdot,x,y)=d[M(\cdot,x),M(\cdot,y)]/dK$ is continuous w.r.t. a couple of variables (x,y) μ^K- a.s. for all (t,ω) where μ^K is Dolean's measure of the process K ;

3) $\sup_x \varphi(\cdot,x,x)\circ K_t<\infty$ P-a.s , $t\in[0,T]$;

An element of $\mathcal{L}^2 - \int_0^{\cdot} M(ds,u_s)$ such that for any $m\in\mathcal{L}^2$

$$E[\int_0^{\cdot} M(ds,u_s),m]_T = E\, f(u,m)\circ[m]_T \equiv E\int_0^T f(s,u(s),m)\,d[m]_s,$$

where $f(\cdot,x,m)$ is a modification of the derivative $d[M(\cdot,x),m]/dK$ continuous w.r.t. x ,(which exists by virtue of condition 1)) is called the line integral w.r.t. a family of martingales $\{M(\cdot,x),\ x\in R^1\}$ along a predictable curve $u=(u(t,\omega),\ 0\le t\le T)$ (denoted as $u\circ M$).

Under the above conditions such an element exists and is unique.

Further assume that: 4) a function $a(\cdot,x)$ continuous in x μ^K-a.s. for all (t,ω) exists such that $A(t,x)=a(\cdot,x)\circ K_t$.

The line integral w.r.t. a family of semimartingales $F(x), x\in R^1$, along the curve u (denoted as $u\circ F$) will be a sum

$$u\circ F = u\circ M + u\circ A$$

where

$$u\circ A_t = \int_0^t a(s,u_s)\,dK_s$$

under the condition that the latter integral exists.

We shall list some properties of the line integral $u\circ M$:

a) for any $m\in\mathcal{L}^2$ $[u\circ M,m]=f(u,m)\circ[m]$

b) $\Delta(u\circ M)_t = M(t,u(t))-M(t-,u(t))\equiv\Delta M(t,u(t))$; c) for any stopping time τ $(u\circ M)^\tau = u\circ M^\tau$; d) $[u_1\circ M-u_2\circ M]=g(\cdot,u_1,u_2)\circ K$ where $g(t,x,y)=\varphi(t,x,x)-2\varphi(t,x,y)+\varphi(t,y,y)$; e) if for any $C>0$ $E\sup_{|x|\le C}\varphi(\cdot,x,x)\circ K_T<\infty$ then $\lim_{n\to\infty} E\langle u_n\circ M-u\circ M\rangle_T=0$ when $\lim_{n\to\infty} u_n=u$ μ^K- a.s.

In the space of semimartingales $S=\{S:\ S=m+a,\ m\in\mathcal{M}^2(F,P),\ a\in\mathcal{A}(F,P)\}$ a norm $\|S\|=E^{1/2}[m]_T+E\operatorname{var}_{[0,T]} a$ is introduced.

Now we shall establish the conditions under which Ito-Ventzel's formula is valid:

(i) The mapping $x\to F(x)$ is twice continuously differentiable w.r.t. x in the sense of the norm $\|\cdot\|$, the second derivative $\ddot F(t,x)$ is continuous in x for all (t,ω) , the functions $F(t,0), \dot F(t,0)$ $\sup_{|x|\le K}|\ddot F(t,x)|$ are locally bounded.

Let $\Phi(t,x)$ denote one of the processes $F(t,x)$ or $\dot F(t,x)$. By virtue of (i) $\Phi(t,x)$ can be represented as

$$\Phi(t,x)=M_\Phi(t,x)+A_\Phi(t,x),$$

where $[M_\Phi(\cdot,x)]=\mathcal{G}_\Phi(\cdot,x,x)\circ K$, $A_\Phi(t,x)=a_\Phi(\cdot,x)\circ K_t$.

(ii) The mappings $\mathcal{G}_\Phi(t,x,x)$ and $a_\Phi(t,x)$ are continuous in x μ^K- a.s. for all (t,ω) and for any $c>0$ $E\sup_{|x|\le c}\mathcal{G}_\Phi(\cdot,x,x)\circ K_T<\infty$, $E\sup_{|x|\le c}|a_\Phi(\cdot,x)|\circ K_T<\infty$.

Theorem 1.1. Let conditions (i) and (ii) be satisfied. Then

$$F(t,\xi(t))-F(0,\xi(0))=\int_0^t F(ds,\xi(s-))+\int_0^t \dot F(s-,\xi(s-))d\xi_s+\frac{1}{2}\int_0^t \ddot F(s-,\xi(s-))d\langle\xi^c\rangle_s+$$
$$+\Big[\int_0^\cdot \dot F(ds,\xi(s-)),\xi\Big]_t+\sum_{s\le t}\big(F(s-,\xi(s))-F(s-,\xi(s-))-\dot F(s-,\xi(s-))\Delta\xi_s\big)+ \quad (1)$$
$$+\sum_{s\le t}\big(\Delta F(s,\xi(s))-\Delta F(s,\xi(s-))-\Delta\dot F(s,\xi(s-))\Delta\xi_s\big),$$

where $\Delta F(s,\xi(s))=\Delta F(s,x)|_{x=\xi(s)}$. Formula (1) can be rewritten as

$$dF(\xi)=d(\xi_-\circ F)+d(\dot F_-(\xi_-)\cdot\xi)+\frac{1}{2}d(\ddot F_-(\xi_-)\circ\langle\xi^c\rangle)+d[\xi_-\circ\dot F,\xi]+$$
$$+dS(F_-(\xi)-F_-(\xi_-)-\dot F_-(\xi_-)\Delta\xi)+dS(\Delta F(\xi)-\Delta F(\xi_-)-\Delta\dot F(\xi_-)\Delta\xi).$$

Remark 1.1. Conditions of Theorem 1.1 guarantee the existence of all terms which enter formula (1).

We shall give a general description of the proof of Theorem 1.1 Using the standard technique (see, e.g. [1]) we can reduce the general case to the consideration of the continuous bounded semimartingale ξ and of the bounded $\dot F$ and $\ddot F$. Further if $\{\tau_i^n\}$ is some sequence of finer partitionings of the interval $[0,T]$, then the difference $F(t,\xi(t))-F(0,\xi(0))$ can be written as an identity:

$$F(t,\xi(t))-F(0,\xi(0))=\sum_i\big(F(\tau_{i+1}^n,\xi(\tau_{i+1}^n))-F(\tau_i^n,\xi(\tau_i^n))\big)=$$
$$=\sum_i\big(F(\tau_{i+1}^n,\xi(\tau_{i+1}^n))-F(\tau_i^n,\xi(\tau_{i+1}^n))\big)+ \quad (2)$$
$$+\sum_i\big(F(\tau_i^n,\xi(\tau_{i+1}^n))-F(\tau_i^n,\xi(\tau_i^n))\big)\equiv I_1^n+I_2^n$$

and it is shown that $I_2^n\xrightarrow{P}\dot F_-\cdot\xi+\frac{1}{2}\ddot F_-\circ\langle\xi\rangle$, I_1^n can be written as

$$I_1^n=\sum_i\big(F(\tau_{i+1}^n,\xi(\tau_i^n))-F(\tau_i^n,\xi(\tau_i^n))\big)+\Big\{\sum_i\big(F(\tau_{i+1}^n,\xi(\tau_{i+1}^n))-F(\tau_{i+1}^n,\xi(\tau_i^n))\big)+$$
$$+\sum_i\big(F(\tau_i^n,\xi(\tau_{i+1}^n))-F(\tau_i^n,\xi(\tau_i^n))\big)\Big\}\equiv I_{1,1}^n+I_{1,2}^n$$

and on the basis of the properties of the line integral we prove that

$$I_{1,1}^n\xrightarrow{P}\xi\circ F,\quad I_{1,2}^n\sim\sum_i\big(\dot F(\tau_{i+1}^n,\xi(\tau_i^n))-\dot F(\tau_i^n,\xi(\tau_i^n))\big)\big(\xi(\tau_{i+1}^n)-\xi(\tau_i^n)\big)\xrightarrow{P}$$
$$\xrightarrow{P}\langle\xi\circ\dot F,\xi\rangle.$$

2°. Local Asymptotic Normality (LAN) of the Family of Distributions and Asymptotic Properties of the MLE. Consider a sequence of general statistical models

$$\mathcal{E}_n = (\Omega^n, \underline{F}^n, \underline{F}^n, P^n_\theta, P^n),\ n \geq 1,\ \theta \in \Theta \subset R^1,$$

where for any $n \geq 1$, $(\Omega^n, \underline{F}^n, \underline{F}^n, P^n)$ is a stochastic basis with the filtration $\underline{F}^n = (\underline{F}^n_t)_{0 \leq t \leq T}$ satisfying the usual conditions; for all θ $P^n_\theta \sim P^n$; $P^n_\theta \neq P^n_{\theta'}$ when $\theta \neq \theta'$; $P^n_\theta | \underline{F}^n_0 = P^n | \underline{F}^n_0$.

Let $P^n_\theta(t)$ and $P^n(t)$ denote restrictions of the measures P^n_θ and P^n on σ-algebra $\underline{F}^n_t$, respectively, and let $\rho^n_\theta(t) = dP^n_\theta(t)/dP^n(t)$. It is well-known (see, e.g., [1]) that a local martingale $M^n_\theta = \{M^n_\theta(t), 0 \leq t \leq T\} \in \mathcal{M}_{loc}(\underline{F}^n, P^n)$ exists such that $\rho^n_\theta(t) = \mathcal{E}_t(M^n_\theta)$ where $\mathcal{E}(\cdot)$ is Dolean's exponential curve.

Suppose the following regularity conditions hold: 1) for any $n \geq 1$ and $\theta \in \Theta$, $M^n_\theta \in \mathcal{M}^2(\underline{F}^n, P^n)$; 2) the family of martingales $\{M^n_\theta, \theta \in \Theta\}$ is continuously differentiable in the sense of the norm $\|\cdot\|$ (if $m \in \mathcal{M}^2$, then $\|M\| = E^{1/2}\langle M\rangle_T$); 3) for any $n \geq 1$ and $\theta \in \Theta$ we have $|\Delta \dot{M}^n_\theta(t) \Delta M^n_\theta(t)(1 + \Delta M^n_\theta(t))^{-1}| \leq C^n_t$ where $E^n_\theta \sum_{s \leq t} C^n_s < \infty$.

If conditions 1)-3) are satisfied, then it can be easily seen that

$$\frac{d}{d\theta} \ln \rho^n_\theta = L(\dot{M}^n_\theta, M^n_\theta),$$

where $d/d\theta \ln \rho^n_\theta$ denotes a derivative of $\ln \rho^n_\theta$ in the sense of the semimartingale norm, introduced in 1° and

$$L(m, M) \overset{def}{\equiv} m - \langle m^c, M^c\rangle - \sum \Delta m \Delta M (1 + \Delta M)^{-1}.$$

By virtue of the generalized Girsanov's theorem $L(\dot{M}^n_\theta, M^n_\theta) \in \mathcal{M}_{loc}(\underline{F}^n, P^n_\theta)$. Further we assume that 4) the Fisher information $I^n_\theta \overset{def}{=} E^n_\theta \langle L(\dot{M}^n_\theta, M^n_\theta)\rangle_T$ is positive and finite for all $n \geq 1$ and $\theta \in \Theta$. Let $\varphi^{-2}_n(\theta) = I^n_\theta$. Denote $\mathcal{U}_{n,\theta} = \varphi^{-1}_n(\theta)(\Theta - \theta)$ and for any $u \in \mathcal{U}_{n,\theta}$ define the normed likelihood relation $Z^n_\theta(u) = dP^n_{\theta + u\varphi_n(\theta)}/dP^n_\theta$. It can be easily seen that $Z^n_\theta(u) = \mathcal{E}(L)$, where $L = L(M^n_{\theta + u\varphi_n(\theta)} - M^n_\theta, M^n_\theta)$.

<u>Theorem 2.1.</u> Let the following conditions hold: uniformly w.r.t. θ on any compact $K \subset \Theta$

(a) $\lim_{n \to \infty} \varphi_n(\theta) = 0$; (b) $P^n_\theta - \lim_{n \to \infty} \varphi^2_n(\theta) \langle L(\dot{M}^n_\theta, M^n_\theta)\rangle = 1$;

(c) $P^n_\theta - \lim_{n \to \infty} \{|x|^2 I_{\{|x| > \varepsilon\}} * \nu^n_\theta\} = 0$ for any $\varepsilon > 0$ where ν^n_θ is a compensator of the jump measure of the process $\varphi_n(\theta) L(\dot{M}^n_\theta, M^n_\theta)$ w.r.t. the measure P^n_θ;

(d) $\lim_{n \to \infty} \sup_{\theta \in K} \sup_{|\theta - y| < \varphi^\delta_n(\theta)} \varphi^2_n(\theta) E^n_\theta [L(\dot{M}^n_y - \dot{M}^n_\theta, M^n_\theta)] = 0$ for any $\delta \in (0,1)$.

Then the family of measures $\{P^n_\theta, \theta \in \Theta\}$ satisfies the uniform LAN condition, i.e.

$$\ln Z^n_\theta(u) = \Delta^n_\theta \cdot u - u^2/2 + \psi^n_\theta(u) \tag{2}$$

where on any compact $K \subset \Theta$ uniformly w.r.t. θ

$$\lim_{n\to\infty} \mathcal{L}(\Delta_\theta^n | P_\theta^n) = \mathcal{N}(0,1), \quad P_\theta^n - \lim_{n\to\infty} \Psi_\theta^n = 0.$$

Remark 2. In the proof of the theorem $\varphi_n(\theta) L(\dot{M}_\theta^n, M_\theta^n)$ stands for Δ_θ^n and the sum $J_n^1 + J_n^2$ stands for $\Psi_\theta^n(u)$, where $J_n^1 = L(M_{\theta+u\varphi_n(\theta)}^n - M_\theta^n, M_\theta^n) - u\cdot\Delta_\theta^n$, $J_n^2 = \ln Z_\theta^n(u) - L(M_{\theta+u\varphi_n(\theta)}^n - M_\theta^n, M_\theta^n) + u^2/2$.

The convergence of the distribution of Δ_θ^n to the normal one immediately follows from the conditions (a),(b)and (c) and the functional central limit theorem for martingales. The partitioning of $\Psi_\theta^n(u)$ into the sum $J_n^1 + J_n^2$ is determined by purely technical considerations since it is more convenient to prove the convergence to zero of J_n^1 and J_n^2 separately.

Asymptotic properties of the MLE are described by the following theorem.

Theorem 2.2. Let the following conditions be satisfied:

1) the family of measures $\{P_\theta^n, \theta \in \Theta\}$ satisfies the uniform LAN condition,

2) a constant $C_0 > 0$ exists such that for any $n \geq 1, \theta, \theta_1, \theta_2 \in \Theta$

$$\varphi_n^2(\theta_1) E_\theta^n \langle L(\dot{M}_{\theta_2}^n, M_\theta^n)\rangle < C_0,$$

3) constants $\gamma>0, \chi>0, C>0$ exist such that for any $\theta \in K, n>0, K \subset \Theta$ K- compact $h_\theta^n(u) \geq C|u|^\gamma = cg(u)$, where $h_\theta^n(u) = -\ln E_\theta^n\{\exp -\chi[\langle L^c\rangle + S(\ln^p K - {}^pK + 1) - S^p((1+\Delta L)^\beta - 1 - \beta\Delta L]\}$ where $L = L(M_{\theta+u\varphi_n(\theta)}^n - M_\theta^n, M_\theta^n)$, $0<\beta<1$, $K = (1+\Delta L)^p$, $S^p(\cdot)$ is a dual predictable projection of the sum $S(\cdot)$ and ${}^pK$ is a predictable projection of the process K.

Then uniformly w.r.t. θ on any compact $K \subset \Theta$ we have

1. $P_\theta^n - \lim_{n\to\infty} \hat{\theta}^n = \theta$ 2. $\mathcal{L}(\varphi_n^{-1}(\theta)(\hat{\theta}^n - \theta) | P_\theta^n) \to \mathcal{N}(0,1)$;

3. all moments of the r.v. $\varphi_n^{-1}(\theta)(\hat{\theta}^n - \theta)$ converge to corresponding moments of the standard normal distribution.

If, in addition the condition

4) $\lim_{n\to\infty} \sup_{|\theta-\theta_0| < \varphi_n^\delta(\theta)} |\varphi_n(\theta)\varphi_n^{-1}(\theta_0) - 1| = 0, \quad \delta \in (0,1)$

is satisfied, then the MLE $\hat{\theta}^n$ is asymptotically efficient, i.e.

4. $\lim_{\delta\to 0} \lim_{n\to\infty} \sup_{|\theta-\theta_0|<\delta} E_\theta^n \varphi_n^{-2}(\theta_0)(\hat{\theta}^n - \theta)^2 = 1.$

Remark 2.2. If the process L is left quasi-continuous, then $K=1$.

3^0. Recursive Estimation. a) Principle Underlying the Establishment of the Equation for the Recursive Estimator. The MLE is obtained from

$$L(\dot{M}_\theta, M_\theta) = 0. \quad (3)$$

Denote $F(t,\theta)=L_t(\dot{M}_\theta, M_\theta)$. Assume that $\hat{\theta}$ is a semimartingale. Applying Ito-Ventzel's formula to $F(t,\hat{\theta}_t)$, (3) will, formally, give the following implicit differential equation for $\hat{\theta}$

$$d\hat{\theta}_t = (\tilde{F}(t-,\hat{\theta}_{t-})-1)^{-1}\{F(dt,\hat{\theta}_{t-})+\bar{F}(t-,\hat{\theta}_{t-})d\hat{\theta}_t-\tfrac{1}{2}\ddot{F}(t-,\hat{\theta}_{t-})d\langle\theta^c\rangle_t- \\ - dS_t(F(t-,\hat{\theta}_t)-F(t-,\hat{\theta}_{t-})-\dot{F}(t-,\hat{\theta}_{t-})\Delta\hat{\theta}_t)- \\ - dS_t(\Delta F(t,\hat{\theta}_t)-\Delta F(t,\hat{\theta}_{t-})-\Delta\dot{F}(t,\hat{\theta}_{t-})\Delta\hat{\theta}_t)- \\ - d\langle\hat{\theta}_-\cdot\dot{F},\hat{\theta}^c\rangle_t - \Delta\dot{F}(t,\hat{\theta}_{t-})\Delta\hat{\theta}_t)\}, \tag{4}$$

where $\dot{F}(t,\theta)=\tilde{F}(t,\theta)-1+\bar{F}(t,\theta)$, $\tilde{F}(t,\theta)=-[L(\dot{M}_\theta,M_\theta)]_t$, $\bar{F}(t,\theta)=L_t(\ddot{M}_\theta,M_\theta)$ and $\tilde{F}-1\leq -1$.

We propose the following equation for the recursive estimator $\tilde{\theta}$:

$$d\tilde{\theta}_t = (\tilde{F}(t-,\tilde{\theta}_{t-})-1)^{-1}\{-F(dt,\tilde{\theta}_{t-})+dS((\tilde{F}(t,\tilde{\theta}_{t-})-1)^{-1}\Delta F(t,\tilde{\theta}_{t-})\Delta\tilde{F}(t,\tilde{\theta}_{t-}))\}, \tag{5}$$

which can be derived from (5) according to the following principle: we remind that $\mathcal{Y}_t^{-2}(\theta)=E_\theta\langle L(\dot{M}_\theta,M_\theta)\rangle_t$. Evidently, the investigation of (4) should be carried out under the same assumptions, when the asymptotic properties of $\hat{\theta}_t$ are known for $t\to\infty$. Therefore we assume that $\mathcal{Y}_t^2(\theta)[L(\dot{M}_\theta,M_\theta)]_t\to 1$ as $t\to\infty$ (in the case of i.i.d. $\mathcal{Y}_n^{-1}\sim\sqrt{n}$), $[L(M_\theta^{(i)},M_\theta)]_t/[L(M_\theta^{(j)},M_\theta)]_t\to \mathrm{const}$, $t\to\infty$, $i,j=1,2,3$, where $M_\theta^{(k)}$ is the k-th strong derivative of the martingale M_θ. Besides it is evident that $(\hat{\theta}_t-\theta)$ when $t\to\infty$ has the order of an infinitesimal equal to $\mathcal{Y}_t^{-1}(\theta)$.

Under these assumptions we can calculate this order for the terms which enter (4) when $t\to\infty$. Therefore we calculate $d\hat{\theta}_t^c$ and $\Delta\hat{\theta}_t$. We have

$$d\hat{\theta}^c=(\tilde{F}(t-,\hat{\theta}_{t-})-1)^{-1}(-M^c(dt,\hat{\theta}_{t-})-\bar{F}(t-,\hat{\theta}_{t-})d\hat{\theta}_t^c), \\ \Delta\hat{\theta}_t=(\tilde{F}(t-,\hat{\theta}_{t-})-1)^{-1}(-\Delta F(t,\hat{\theta}_{t-})-\bar{F}(t,\hat{\theta}_{t-})\Delta\hat{\theta}_t-\delta_t(\ddot{F},\Delta\hat{\theta}_t^2)). \tag{6}$$

Putting (6) into (4) we see that only the terms entering the right-hand side of (5) have the order of an infinitesimal $\mathcal{Y}_t(\theta)$, while this order for the remaining terms of (4) is $\mathcal{Y}_t^2(\theta)$, hence, they can be omitted.

Now the asymptotic properties of the recursive estimator b can be studied in the following manner: instead of the direct investigation of (5), as it has been done earlier, we can prove that the solutions of (4) and (5) are "approaching" and then on the basis of the properties of the MLE $\hat{\theta}$ stated in 2° we can conclude that RE $\tilde{\theta}$ has the same properties.

b) <u>Examples.</u> 1. Let $X_1,X_2,\cdots,X_n$, $n\geq 1$, be independent observations of the random variable X with the distribution density $f(x,\theta)$, $\theta \in R^1$, and the unknown parameter θ is to be estimated.

This case can be included in the scheme of general statistical models in the following manner: $\mathcal{E}=(\Omega, \underline{F}, F, P_\theta, P)$ where $\Omega = R^\infty$, $\underline{F}=\mathcal{B}(R^\infty)$, $F=(\underline{F}_n)_{n\geq 1}$, $\underline{F}_n=\mathcal{B}(R^n)$, $dP_\theta/d\Lambda=\prod_{i\geq 1}^{\infty} f(x_i,\theta)$, $d\Lambda/dP=\prod_{i\geq 1}^{\infty}\pi(x_i)$,

$\Lambda \sim P$, Λ is the Lebesgue measure. Then if $P_\theta^n = P_\theta|\underline{F}_n$, $P^n = P|\underline{F}^n$ are restrictions of the measures P_θ, P on σ-algebra $\underline{F}_n$, then $\rho_\theta^n = dP_\theta^n/dP^n = \prod_{i\geq 1}^{n} f(x_i,\theta)\pi(x_i) = \mathcal{E}_n(M_\theta)$, where $M_\theta = \{M_\theta(n), n\geq 1\}$ is a (F,P) martingale with $\Delta M_\theta(n) = M_\theta(n) - M_\theta(n-1) = f(x_n,\theta)\pi(x_n) - 1$, $\mathcal{E}(\cdot)$ is Dolean's exponential curve and $L_n(\dot{M}_\theta, M_\theta) = F(n,\theta) = \sum_{i\leq n} \dot{f}(x_i,\theta) f^{-1}(x_i,\theta)$ and the equation for the recursive estimator $\tilde{\theta}$ will be

$$\tilde{\theta}_n - \tilde{\theta}_{n-1} = \dot{f}(x_n, \tilde{\theta}_{n-1})\left[(1+n\tilde{I}_n(\tilde{\theta}_{n-1}))\, f(x_n,\tilde{\theta}_{n-1})\right]^{-1},$$

where $\tilde{I}_n(\theta) = \frac{1}{n}\sum_{i\leq n} (\dot{f}(x_i,\theta)/f(x_i,\theta))^2$.

2. Let the unknown parameter $\theta \in R^1$ be estimated on the basis of the observations of the diffusion type process

$$d\xi_t = a(t,\xi,\theta)\,dt + dW_t .$$

In this case the scheme of models will be: $\mathcal{E} = (\Omega, \underline{F}, F, P_\theta, P)$, where $\Omega = C_{[0,\infty)}$, $\underline{F} = \mathcal{B}(C_{[0,\infty)})$, $F = (\underline{F}_t)_{t\geq 0} = \mathcal{B}(C_{[0,t]})$, $P_\theta = P^\xi$, $P = P^w$. Let $P_\theta(t) \sim P(t), t\geq 0$, where $P_\theta(t) = P_\theta|\underline{F}_t$, $P(t) = P|\underline{F}_t$. Then $\rho_\theta(t) = dP_\theta(t)/dP(t) = \mathcal{E}_t(M_\theta)$ where $M_\theta = \{M_\theta(t) = \int_0^t a(s,x,\theta)\,dx_s, t\geq 0, x\in C\} \in \mathcal{M}(\underline{F}, P)$ and $F(t,\theta) = L_t(\dot{M}_\theta, M_\theta) = \int_0^t \dot{a}(s,x,\theta)(dx_s - a(s,x,\theta)\,ds)$, $x \in C_{[0,\infty)}$.

The equation for the recursive estimator has the following form

$$d\tilde{\theta}_t = \dot{a}(t,\xi,\tilde{\theta}_t)\left(1 - \int_0^t (\dot{a}(s,\xi,\theta))^2 ds\big|_{\theta=\tilde{\theta}_t}\right)^{-1}(d\xi_t - a(t,\xi,\tilde{\theta}_t)\,dt).$$

3. Let the unknown parameter θ be estimated on the basis of the observations of the point process N with the compensator $A(\cdot,\theta)$. We shall describe the scheme of the statistical models in this case. Let $\mathcal{E} = (\Omega, \underline{F}, F, P_\theta, P)$ where Ω is a space of piecewise constant functions $x = (x_t)_{t\geq 0}$ such that $x_0 = 0$, $x_t = x_{t-} + (0 \text{ or } 1)$, $\underline{F} = \sigma\{x : x_s, s\geq 0\}$, $\underline{F}_t = \sigma\{x : x_s, s\leq t\}$, $\tau_i(x) = \inf\{s\geq 0 : x_s = i\}$ assuming $\tau_i(x) = \infty$ if $\lim_{t\to\infty} x_t < i$. Evidently, $x_t = \sum_{i\geq 1} I_{\{\tau_i(x)\leq t\}}$ is a point process. Let P_θ (P) be a measure such that $\tau_\infty(x) = \infty$ P_θ- a.s. (P- a.s.) and w.r.t. the measure P_θ (P) the point process has a continuous compensator $A(t,\theta)$ ($A(t)$) and $dA(t,\theta)/dA(t) = Y(t,\theta) > 0$. Let $\rho_t(\theta) = dP_\theta(t)/dP(t) = \mathcal{E}_t(M_\theta)$, where $M_\theta = Y(\cdot,\theta)\cdot(N - A)$. Then $F(t,\theta) = L_t(\dot{M}_\theta, M_\theta) = \sum_{\tau_n\leq t} \dot{Y}(\tau_n,\theta)/Y(\tau_n,\theta) - \dot{Y}(\cdot,\theta)\circ A_t$. The equation for the recursive estimator has the form

$$d\tilde{\theta}_t = \frac{\dot{Y}(t,\tilde{\theta}_{t-})}{Y(t,\tilde{\theta}_{t-})} \; \frac{dN_t}{1+\tilde{I}(t,\tilde{\theta}_{t-})} - \frac{\dot{Y}(t,\tilde{\theta}_{t-})}{1+\tilde{I}(t-,\tilde{\theta}_{t-})}\, dA_t,$$

where $\tilde{I}(t,\theta) = \sum_{\tau_n \le t} (\dot{Y}(\tau_n,\theta)/Y(\tau_n,\theta))^2$.

c) Consider a particular case of the scheme of statistical models, which illustrates the method of the investigation of asymptotic properties of recursive estimators given in a). Let $M_\theta(t) = f(\theta) M(t)$ where $f(\theta)$ be a twice continuously differentiable function with $\dot{f}(\theta) \ne 0$, $\theta \in R^1$, $M \in \mathcal{M}^c$. We assume that $E_\theta \langle M \rangle_t \to \infty$ as $t \to \infty$, $P_\theta\text{-}\lim_{t\to\infty} \langle M \rangle_t / E_\theta \langle M \rangle_t = \text{const}$ and $\langle M \rangle_t / E_\theta \langle M \rangle_t \le C(\omega)$, $E_\theta C < \infty$. Under these conditions the MLE $\hat{\theta}$ is consistent, asymptotically normal and efficient (see 2°).

In this case the equation for the MLE has the form

$$d\hat{\theta}_t = \frac{1}{\dot{f}(\hat{\theta}_t)} \; \frac{dM(t) - f(\hat{\theta}_t) d\langle M \rangle_t}{\langle M \rangle_t} - \frac{\ddot{f}(\hat{\theta}_t)}{2[\dot{f}(\hat{\theta}_t)]^3} \; \frac{d\langle M \rangle_t}{\langle M \rangle_t}$$

and that for RE $\tilde{\theta}$ will be

$$d\tilde{\theta}_t = \frac{1}{\dot{f}(\tilde{\theta}_t)} \; \frac{dM(t) - f(\tilde{\theta}_t) d\langle M \rangle_t}{\langle M \rangle_t + 1}.$$

It can be easily shown that (using the stochastic Gronwall lemma) $E_\theta [\varphi_t^{-1}(\theta)(\tilde{\theta}_t - \hat{\theta}_t)]^2 \to 0$ as $t \to \infty$ where $\varphi_t^{-2}(\theta) = E_\theta \langle M \rangle_t$. Thus we conclude that RE $\tilde{\theta}$ has the same asymptotic properties as the MLE $\hat{\theta}$.

4°. Short review of Literature. Ito-Ventzel's formula was first obtained by A.D.Ventzel [2] and B.A.Rozovky [3] for the case of the Ito processes. Subsequently Ito-Ventzel's formula for semimartingales $F(x)$ having integral representation w.r.t. the semimartingale components was generalized by M.Bismut [4], H.Kunita [5], R.Mikulevichus [6] and E.I.Trofimov [7]. We have studied the case of the semimartingales $F(x)$ without the assumption of integral representation, which involves the necessity of consideration of the so called stochastic line integrals, the construction scheme for which given in 1° is due to M.G.Mania and R.J.Chitashvili [8].

The idea of the investigation of the problems of asymptotic statistics by LAN methods is due to L.LeCam [9] and was later developed in a fundamental book by I.A.Ibragimov and R.Z.Has'minsky [10].

By now, using the approach proposed in [9] and [10] both discrete and continuous models have been separately studied, as well as the semimartingale "scheme of observations". In particular, the above topics are developed in the works of the following authors: Yu.A.Kutoants [11] , Yu.I.Lin'kov [12] , A.F.Taraskin [13] , L.Vostrikova [14] , etc.

Recursive estimators with good asymptotic properties were constructed on the basis of the Robbins -Monro stochastic approximation, proposed by A.Albert and L.Gardner [15] and developed in the papers by Nevelson and Has'minsky [16] and others, where the discrete time case and that of continuous time were separately studied.

References

1. J.Jacod. Calcul stochastique et problemes de martingales. Lect.N. in Math. 714. Springer Verlag: 1978.

2. A.D.Ventzel. On equations of the theory of conditionally Markov processes. Teor.Verojat. i Primen.,1965,10,№ 2, pp.390-393.

3. B.L.Rozovsky. On Ito-Ventzel's formula. Vestnik Moscov.Univ., 1973, № 1, pp. 26-32 (in Russian).

4. J.M.Bismut. A generalized formula of Ito and some other properties of stochastic flows. Z.Wahrscheinlichkeitstheorie and Verw.Gebiete, 1981, 55, pp. 174-211.

5. H.Kunita. On the decomposition of solutions of stochastic differential equations.- Lect.Notes in Math., 1981, v.851, p.213-255.

6. R.Mikulevichus. On some properties of solutions of stochastic differential equations. Litovsk.Mat.Sb.,1983,XXIII,№ 4,pp.18-31.851 Springer Verlag: Berlin-Heidelberg-New York, 1981,p.213-255 (in Russian).

7. E.I.Trofimov. A theorem on the implicit function for semimartingales and its application. Uspehi Mat.Nauk,v.38,2,1983,pp.215-216 (in Russian).

8. R.J.Chitashvili, M.G.Mania. On optimal controls in the problem of locally absolutely continuous change of measures. Proc. of the IV Intern.Vilnius Conf. on Probab.Theory and Math.Stat.,Vilnius, 1985.

9. L.LeCam. Locally asymptotically normal families of distributions. Univ.Calif.Publs.Statist.,1960,3,p.37-98.

10. I.A.Ibragimov, R.Z.Has'minsky. Asymptotic theory of estimation. M.: Nauka, 1979 (in Russian).

11. Yu.A.Kutoyants. Estimation of parameter of a diffusion type process. Teor.Verojat. i Primen.,1978,23,pp.641-649(inRussian).

12. Yu.N.Lin'kov. Asymptotic properties of statistical estimators and criteria for Markov processes. In :Teor.Verojat. i Mat.Statist., Kiev: 1981, 25, p.76-91 (in Russian).

13. A.F.Taraskin. On the behaviour of the likelihood ratio of semimartingales. Teor.Verojat. i Priman., 1984,№ 3, pp.440-451.

14. L.Vostrikova. On a weak convergence of likelihood ratio processes of general statistical models. IV Intern.Vilnius Conf. on Probab. Theory and Math. Stat., Abstr. of Commun., v. 3, Vilnius, 1985, p. 331-332.

15. A.Albert, L.Gardner. Stochastic approximation and nonlinear regression. M.I.T.-Press, Cambrige, Massachusetts, 1967.

16. M.B.Nevelson, R.Z.Has'minsky. Stochastic approximation and recursive estimation., M., Nauka, 1972.

CONDITIONS FOR CONTIGUITY

F. Liese
Wilhelm-Pieck-Universität Rostock, Universitätsplatz 5
Rostock
DDR-2500

ABSTRACT

Conditions for contiguity of a sequence of distributions P_n with respect to Q_n are established in the present paper. P_n and Q_n are defined on a filtered space and the conditions are formulated in terms of functionals of Q_n and the predictable characteristics of the density processes.

RESULTS

Let $(\Omega_n, \mathcal{F}_n)$ be a sequence of measurable spaces and P_n, Q_n probability measures on $(\Omega_n, \mathcal{F}_n)$. The sequence P_n is said to be contiguous with respect to Q_n ($P_n \lhd Q_n$) if $\lim\limits_{n\to\infty} Q_n(A_n) = 0$ implies $\lim\limits_{n\to\infty} P_n(A_n) = 0$ for any sequence $A_n \in \mathcal{F}_n$. We denote by X_n the Radon-Nikodym derivative of P_n' with respect to Q_n where $P_n = P_n' + P_n''$, $P_n' \ll Q_n'$, $P_n'' \perp Q_n$, is the Lebesgue decomposition of P_n with respect to Q_n. The following criterion for contiguity is well known /7/.

Lemma 1: It holds $P_n \lhd Q_n$ iff $\lim\limits_{n\to\infty} P_n'(\Omega_n) = 1$ and

$$\limsup_{c\to\infty} (\limsup_{n\to\infty} P_n(X_n > c)) = 0.$$

Frequently it is difficult to calculate or to estimate $P_n(X_n > c)$ directly. Sometimes it is more convinient to deal with suitable functionals of X_n. Conditions for contiguity in terms of certain functionals of X_n will be established in Theorem 1. Suppose $\varphi_\alpha(x)$, $0 \le x < \infty$, $0 < \alpha < 1$, is a family of nonnegative measurable functions which fulfils the following conditions

A1) $\varphi_\alpha(x) \ge 0$ for every $0 \le x < \infty$, $0 < \alpha < 1$

A2) $\lim\limits_{\alpha\uparrow 1} \varphi_\alpha(x) = x$ uniformly on every finite interval

A3) $\lim\limits_{x\to\infty} \frac{1}{x}\varphi_\alpha(x) = 0$ for every fixed $\alpha, 0 < \alpha < 1$

A4) $\varphi_\alpha(x)$ is a concave function of x for every $0 < \alpha < 1$ and $x \ge 0$.

<u>Theorem 1:</u> Let P_n, Q_n be probability measures on $(\Omega_n, \mathcal{F}_n)$ and $\varphi_\alpha(x)$ a family which fulfils A1), ..., A4).
It holds $P_n \lhd Q_n$ iff

$$\liminf_{\alpha\uparrow 1} (\liminf_{n\to\infty} E_{Q_n} \varphi_\alpha(X_n)) = 1.$$

A typical family which fulfils A1), ..., A4) is $\varphi_\alpha(x) = x^\alpha$. In this case $E_{Q_n} \varphi_\alpha(X_n)$ is nothing else the Hellinger integral of order α. The statement of Theorem 1 in this special case have been already established in /1/, /3/, /4/. A further example of functions φ_α fulfilling A1), ..., A4) is

$$\varphi_\alpha(x) = (1-\alpha)^{-1}(1 - \exp - (1-\alpha)x).$$

We consider a sequence of stochastic bases $(\Omega_n, (\mathcal{F}_{n,t})_{t\geq o}, \mathcal{F}_n, Q_n)$ fulfilling the usual conditions.
We suppose that P_n are distributions on $\mathcal{F}_n$ and assume that $P_{n,t} << Q_{n,t}$ for every $n = 1,2,...$ and $0 \leq t < \infty$. If we are interested in conditions for $P_n \lhd Q_n$ then this assumption is not very restrictive in view of Lemma 1.
It was proved in /2/ that there exists a local martingale $X_{n,t}$ called density process so that $X_{n,t}$ is the Radon-Nikodym derivative of $P_{n,t}$ with respect to $Q_{n,t}$ where $P_{n,t}, Q_{n,t}$ are the restrictions of P,Q to $\mathcal{F}_t$. Denote by ν_n the compensator of the jump measure of $X_{n,t}$. Necessary and sufficient conditions for contiguity in terms of the predictable characteristics of $X_{n,t}$ formulated with respect to P_n can be found in /1/, /4/, /5/. In statistics P_n is a sequence of alternatives whereas Q_n corresponds to the null-hypotheses. Consequently, Q_n has in many cases a more simple structure than P_n. Therefore conditions for contiguity formulated with respect to Q_n are of interest. For this aim we generalize Theorem 1 to spaces equipped with a filtration.
As $\varphi_\alpha(X_{n,t})$ is a supermartingale $E_{Q_n}\varphi(X_{n,t})$ is non-increasing so that the limit for $t\to\infty$ exists. Put

$$\Phi_n(\alpha) = \lim_{t\to\infty} E_{Q_n} \varphi_\alpha(X_{n,t}).$$

<u>Theorem 2:</u> The sequence P_n is contiguous with respect to Q_n iff

$$\liminf_{\alpha\uparrow 1} (\liminf_{n\to\infty} \Phi_n(\alpha)) = 1.$$

Denote by C_2 the space of all real twice continuously differentiable functions on $\mathbb{R} = (-\infty,\infty)$. Let $(\Omega, (\mathcal{F}_t)_{t\geq o}, \mathcal{F}, Q)$ be a stochastic basis which fulfils the usual conditions. For every $\varphi \in \mathbb{C}_2$ and every local martingale X we introduce the process

$$V_t(\varphi, X) = \varphi(X_{t-} + u) - \varphi(X_{t-}) - \varphi'(X_{t-})u.$$

Let ν be the compensator of the jump measure of X. Then we set

$$(V(\varphi, X) * \nu)_t = \int_{(0,t]} \int_{\mathbb{R}_0} (\varphi(X_{s-} + u) - \varphi(X_{s-}) - \varphi'(X_{s-})u)\, \nu(ds, du)$$

where $\mathbb{R}_0 = \mathbb{R} \setminus \{0\}$. The process X_{t-} is also denoted by X_-.

<u>Theorem 3:</u> If $P_{n,t} << Q_{n,t}$ for every $n = 1,2,\ldots$ and $0 \leq t < \infty$ and $\varphi_\alpha \in \mathbb{C}_2$ is a family which satisfies the assumptions A1),...,A4) then $P_n \triangleleft Q_n$ iff the following conditions are fulfilled

$$P_{0,n} \triangleleft Q_{0,n}$$

$$\limsup_{\alpha \uparrow 1} (\limsup_{n \to \infty} E_{Q_n}(-\varphi_\alpha''(X_{n,-}) \circ \langle X_n^c \rangle_\infty) = 0$$

$$\limsup_{\alpha \uparrow 1} (\limsup_{n \to \infty} E_{Q_n}(-V(\varphi_\alpha, X_n) * \nu_n)_\infty) = 0$$

<u>Corollary:</u> It holds $P_n \triangleleft Q_n$ iff

$$P_{0,n} \triangleleft Q_{0,n}$$

$$\limsup_{\alpha \downarrow 0} (\limsup_{n \to \infty} E_{Q_n} \int_0^\infty \alpha e^{-\alpha X_{n,t-}} d\langle X_n^c \rangle_t) = 0$$

$$\limsup_{\alpha \downarrow 0} (\limsup_{n \to \infty} E_{Q_n} \int_0^\infty \int_{\mathbb{R}_0} e^{-\alpha X_{n,t-}} \left[u - \frac{1-e^{-\alpha u}}{\alpha}\right] \nu_n(dt, du) = 0.$$

Let $(\Omega, (\mathcal{F}_t)_{t \geq 0}, \mathcal{F}, Q)$ be a stochastic basis fulfilling the usual conditions and M be a continuous local martingale with $M_0 = 0$. Then the Doleans-Dade exponential

$$\mathcal{E}(M) = \exp\{M - \tfrac{1}{2}\langle M \rangle\}$$

is also a local martingale. A necessary and sufficient condition for uniform integrability of $X = \mathcal{E}(M)$ will be given in the next Proposition.

<u>Proposition:</u> $X = \mathcal{E}(M)$ is uniformly integrable iff

$$\lim_{\alpha \uparrow 1} (1-\alpha) E_Q(X^\alpha \circ \langle M \rangle)_\infty = 0.$$

<u>Remark:</u> The proof of Theorem 3 is based on the Doob-Meyer decomposition of $\varphi_\alpha(X_{n,t})$ where the non-increasing proces is calculated by Ito's formula. We remark that although x^α is not differentiable

for $x = 0$ the Doob-Meyer decomposition of $X_{n,t}$ leads to a statement analog to Corollary with x^{α} instead of $(1-\exp -(1-\alpha)x)/(1-\alpha)$. W furthermore note that the multiplicative decomposition of $X^{\alpha}_{n,t}$ in /1/ and Theorem 2 yield also sufficient conditions for contiguity.

PROOFS

Proof of Theorem 1: Put $a(\alpha,c) = \sup_{0 \le x \le c} |x - \varphi_{\alpha}(x)|$.

Then by A2) $\lim_{\alpha \uparrow 1} a(\alpha,c) = 0$ for every fixed $c \ge 0$.

$$|1-E_{Q_n} \varphi_{\alpha}(X_n)| \le |1-E_{Q_n} X_n| + E_{Q_n} |X_n - \varphi_{\alpha}(X_n)| I_{\{X_n \le c\}} + E_{Q_n} X_n I_{\{X_n > c\}}$$
$$+ E_{Q_n} \varphi_{\alpha}(X_n) I_{\{X_n > c\}}$$

holds for every $c \ge 0$. Note that $E_{Q_n} X_n = P_n'(\Omega_n)$ and

$E_{Q_n} X_n I_{\{X_n > c\}} = P_n(X_n > c)$. As φ_{α} is non-negative and concave we get

$$E_{Q_n} \varphi_{\alpha}(X_n) I_{\{X_n > c\}} \le E_{Q_n} \varphi_{\alpha}(X_n I_{\{X_n > c\}}) \le \varphi_{\alpha}(P_n(X_n > c))$$

and by $\qquad \varphi_{\alpha}(P_n(X_n > c)) \le P_n(X_n > c) + a(\alpha,1)$

(1) $|1-E_{Q_n} \varphi_{\alpha}(X_n)| \le |1-P_n'(\Omega_n)| + a(\alpha,c) + a(\alpha,1) + 2P_n(X_n > c)$.

Suppose now $P_n \triangleleft Q_n$. Then the first term tends to zero as $n \to \infty$ by Lemma 1. Using $a(\alpha,c) \xrightarrow[\alpha \uparrow 1]{} 0$ for every fixed $c > 0$ we get

$$\limsup_{\alpha \uparrow 1} (\limsup_{n \to \infty} |1-E_{Q_n} \varphi_{\alpha}(X_n)|) \le 2 \limsup_{n \to \infty} P_n(X_n > c).$$

Taking $c \to \infty$ we see from Lemma 1 that the term on the left hand side is zero and the necessity of the condition in Theorem 1 is proved. For showing the sufficiency we set

$$b(\alpha,c) = \sup_{x \ge c} \frac{1}{x} \varphi_{\alpha}(x).$$

Then by A3) $b(\alpha,c) \xrightarrow[c \to \infty]{} 0$ for every fixed $0 < \alpha < 1$.

The conditions A1) and A4) yield

$$E_{Q_n} \varphi_{\alpha}(X_n) \le \varphi_{\alpha}(P_n(X_n \le c)) + E_{Q_n} \varphi_{\alpha}(X_n) I_{\{X_n > c\}}$$
$$\le a(\alpha,1) + P_n(X_n \le c) + E_{P_n} \frac{1}{X_n} \varphi_{\alpha}(X_n) I_{\{X_n > c\}}$$
$$\le a(\alpha,1) + b(\alpha,c) + P_n(X_n \le c).$$

Taking on both sides $n \to \infty$ then $c \to \infty$ and finally $\alpha \uparrow 1$ we obtain

(2) $\liminf_{\alpha \uparrow 1} (\liminf_{n \to \infty} E_{Q_n} \varphi_{\alpha}(X_n) \le 1 - \limsup_{c \to \infty} (\limsup_{n \to \infty} P_n(X_n > c))$.

Furthermore by the concavity of φ_α and the definition of $a(\alpha,1)$

$$E_{Q_n} \varphi_\alpha(X_n) \le \varphi_\alpha(E_{Q_n} X_n) \le P_n^{\cdot}(\Omega_n) + a(\alpha,1)$$

and

(3) $\liminf_{\alpha \uparrow 1} (\liminf_{n\to\infty} E_{Q_n} \varphi_\alpha(X_n) \le \liminf_{n\to\infty} P_n^{\cdot}(\Omega_n).$

If the sequence P_n is not contiguous with respect to Q_n then by Lemma 1 at least one of the terms on the right hand side of (2) and (3), respectively, is less than one and the condition in Theorem 1 is not fufilled.

Let $(\Omega_n, \mathcal{F}_n)$ be a sequence of measurable spaces where each of them is equipped with a filtration $(\mathcal{F}_{n,t})_{0\le t<\infty}$ so that $\mathcal{F}_n = \sigma(\bigcup_{t\ge 0} \mathcal{F}_{n,t})$. Let P_n, Q_n be probability measures on $\mathcal{F}_n$. The following assertion is a direct consequence of the definition of contiguity and the fact that every set from $\mathcal{F}_n$ can be approximated by stes from $\bigcup_{t\ge 0} \mathcal{F}_{n,t}$ both with respect to P_n and Q_n.

Lemma 2: It holds $P_n \lhd Q_n$ iff for every sequence $0 \le t_n < \infty$

$$P_{n,t_n} \lhd Q_{n,t_n}.$$

Proof of Theorem 2: Let $0 \le t_n < \infty$ be any sequence and assume the condition in Theorem is fulfilled. Then by the monotonicity of $E_{Q_n} \varphi_\alpha(X_{n,t})$

$$\liminf_{\alpha\uparrow 1} (\liminf_{n\to\infty} E_{Q_n} \varphi_\alpha(X_{n,t_n})) \ge \liminf_{\alpha\uparrow 1} (\liminf_{n\to\infty} \Phi_n(\alpha)) = 1.$$

Hence by Theorem 1 $P_{n,t_n} \lhd Q_{n,t_n}$. As t_n was arbitrary Lemma 2 yields $P_n \lhd Q_n$.

For proving the converse implication we set

$$\Phi = \liminf_{\alpha\uparrow 1} (\liminf_{n\to\infty} \Phi_n(\alpha)).$$

There exist $\alpha_k \to 1$, $n_k \to \infty$ as $k \to \infty$ and $0 \le t_k < \infty$ so that

$$\Phi = \lim_{k\to\infty} E_{Q_{n_k}} \varphi_{\alpha_k}(X_{n_k,t_k}).$$

Inequality (1) yields

$$|1-E_{Q_{n_k}} \varphi_{\alpha_k}(X_{n_k,t_k})| \leq a(\alpha_k,c) + a(\alpha_k,1) + 2P_{n_k,t_k}(X_{n_k,t_k} > c)$$

as $P_{n,t} << Q_{n,t}$.

Suppose $P_n \lhd Q_n$. Lemma 3 yields $P_{n_k t_k} \lhd Q_{n_k,t_k}$.

The convergence to zero of the first and second term is a consequence of $\alpha_k \to 1, k \to \infty$. This means

$$\limsup_{k\to\infty} |1 - E_{Q_{n_k}} \varphi_{\alpha_k}(X_{n_k,t_k})| \leq 2 \limsup_{k\to\infty} P_{n_k,t_k}(X_{n_k,t_k} > c).$$

As $P_{n_k,t_k} \lhd Q_{n_k,t_k}$ the right hand term vanishes as $c \to \infty$ in accordance with Lemma 1. Thus $\Phi = 1$ is proved.

<u>Proof of Theorem 3:</u> Using a special version of Ito's formula (see /6/ Theorem 32.1) we obtain that

$$\varphi_\alpha(X_{n,t}) - \varphi_\alpha(X_{n,o}) - \tfrac{1}{2}\varphi_\alpha''(X_{n,-}) \circ \langle X_n^c \rangle_t - (V(\varphi_\alpha, X_n) * \nu_n)_t$$

is a local martingale which is zero for $t = 0$. Let T_m be a localizing sequence. As φ_α is concave there are real c,d so that $\varphi_\alpha(x) \leq c + dx$. By taking the expectation and the fact that $X_{n,t\wedge T_m}$, $m = 1,2,\ldots$ is uniformly integrable and $-\varphi_\alpha''(X_{n,-}) \circ \langle X_n^c \rangle$, $-V(\varphi_\alpha, X_n) * \nu_n$ are non-decreasing processes we get for $m \to \infty$

$$E_{Q_n} \varphi_\alpha(X_t) = E_{Q_n} \varphi_\alpha(X_{n,o}) + E_{Q_n} \tfrac{1}{2}\varphi_\alpha''(X_{n,-}) \circ \langle X_n^c \rangle_t + E_{Q_n} V(\varphi_\alpha, X_n) * \nu_n)_t.$$

By the monotone convergence theorem as $t \to \infty$

$$(4) \quad \Phi_n(\alpha) = E_{Q_n} \varphi_\alpha(X_{n,o}) + E_{Q_n} \tfrac{1}{2}\varphi_\alpha''(X_{n,-}) \circ \langle X_n^c \rangle_\infty + E_{Q_n} V(\varphi_\alpha, X_n) * \nu_n)_\infty.$$

The cocavity of φ_α and A2) imply

$$\limsup_{\alpha\uparrow 1} (\limsup_{n\to\infty} E_{Q_n} \varphi_\alpha(X_{n,o})) \leq 1.$$

Since $\varphi_\alpha'' \leq 0$ and $V(\varphi_\alpha, X_n) \leq 0$ the condition in Theorem 2 is fulfilled iff the second and third term in (4) tend to zero as $n \to \infty$ and then $\alpha \uparrow 1$ and simultaneously it holds

$$\liminf_{\alpha\uparrow 1} (\liminf_{n\to\infty} E_{Q_n} \varphi_\alpha(X_{n,o})) = 1.$$

This is by Theorem 1 equivalent to $P_{n,o} \lhd Q_{n,o}$.

The Corollary follows by setting $\varphi_\alpha(x) = (1-\alpha)^{-1}(1 - \exp\{(1-\alpha)x\})$.

Proof of Proposition: Put $T_k = \inf\{t: X_t \le \frac{1}{k}\}$ and $T = \lim_{k\to\infty} T_k$.

Let $\varphi_k \in \mathbb{C}_2$ be a function which coincides with x^α for $x \ge 1/k$. Applying Ito's formula to $\varphi_k(X)$ we get by the same arguments as in the preceding proof

$$E_Q X_{t\wedge T_k} = 1 - \frac{\alpha(1-\alpha)}{2} E_Q(X_{T_k\wedge t}^{-2} \circ \langle X\rangle)_{t\wedge T_k}.$$

AS $\langle X\rangle = X^2 \circ \langle M\rangle$ and $X^\alpha_{t\wedge T_k}$ is uniformly integrable we get

$$(5)\quad E_Q X^\alpha_{t\wedge T} = 1 - \frac{\alpha(1-\alpha)}{2} E_Q(X^\alpha \circ \langle M\rangle_{t\wedge T}).$$

Let $S_n \uparrow \infty$ be a localizing sequence for X. Put $(\Omega_n, (\mathcal{F}_{t,n})_{t\ge 0}, \mathcal{F}_n, Q_n) = (\Omega, (\mathcal{F}_{t\wedge T\wedge S_n})_{t\ge 0}, \mathcal{F}_{T\wedge S_n}, Q_{T\wedge S_n})$ and $dP_n = X_{T\wedge S_n} dQ_{T\wedge S_n}$. Then by Lemma 1 $P_n \vartriangleleft Q_n$ iff $X_{T\wedge S_n}$, $n = 1,2,\ldots$ is uniformly integrable. (5) implies

$$\lim_{n\to\infty} \Phi_n(\alpha) = 1 - \frac{\alpha(1-\alpha)}{2} E_Q(X^\alpha \circ \langle M\rangle_\infty)$$

as $X_t = 0$ for $t > T$. Hence the left hand side tends to one as $\alpha \uparrow 1$ iff the condition in the proposition is fulfilled. To end the proof we remark that $X_{t\wedge T}$ is uniformly integrable iff X_t has the same property since $X_t = 0$ for $t > T$.

REFERENCES

/1/ Jacod, J., Processus de Hellinger, absolue continuite, contiguite, Technical report, Universite de Rennes I (1983)

/2/ Kabanov, J. M., Liptser, R. S., Shiryayev, A. N., Absolute contunuity and singularity of locally absolutele continuous probability distributions I, II, Math. USSR Sbornik, Vol. 35, No. 5, 631 - 680 (1979) Vol. 36, No. 1, 31 - 58 (1980)

/3/ Liese, F., Admissible translations of Gaussian cylinder measures, Wiss. Ztsch. Friedrich-Schiller-Univ. Jena, Math.-Nat. R., 31. Jg. H. 4, 609 - 615 (1982)

/4/ Liese, F., Hellinger integrals of diffusion processes. Forsch. Erg. Friedrich-Schiller-Univ. Jena, No. 83/39 (1983) Statistics 17, 1, 63 - 78 (1986)

/5/ Liptser, R., Shiryayev, A. N., On the problem of predictable criteria for contiguity, Prob. theory and math. stat. (USSR-Japan symposium) Lect. Notes in Math. 1021, 346 - 418, Springer Verl. (1983)

/6/ Metivier, M., Seminartingales a course on stochastic processes Walter de Gruyter, Berlin New York (1982)

/7/ Roussas, G. G., Contiguity of probability measures Cambridge, Univ. Press 1972

STOCHASTIC CALCULUS ASSOCIATED WITH SKOROHOD'S INTEGRAL

D. NUALART
Facultat de Matemàtiques
Universitat de Barcelona
Gran Via 585
08007 BARCELONA SPAIN

and

E. PARDOUX
Mathématiques, case H
Université de Provence
13331 MARSEILLE CEDEX 3
FRANCE

Abstract : We study the stochastic integral defined by Skorohod [7] of a possibly anticipating integrand, and give conditions under which the resulting process, as a function of its upper limit, has a continuous modification. We formulate an extension of Itô's formula, and a Stratonovich version of our results.

1. Introduction

In [7], Skorohod constructed the stochastic integral of a possibly anticipating process having a certain kind of smoothness property, in the sense of the differential calculus on Wiener space. His integral has been further studied by Gaveau-Trauber [1] and Nualart-Zakaï [4]. We will give a new definition of Skorohod's integral, and study the properties of the resulting process, as a function of its upper limit. We next state a generalized Itô formula, study a Stratonovich version of Skorohod's integral, and the associated chain rule.

After our research was completed, we learned about the results of Sevljakov [6], Sekiguchi-Shiota [5] and Ustunel [8] concerning the generalized Itô formula, which are very close to part of our result, also the proofs differ significantly.

The present note presents most of our results in the particular case of one dimensional processes, without proofs. The complete results and proofs will appear in [3].

2. Skorohod's integral

Let $\Omega = C([0,1])$, $\mathbf{F}$ be the Borel field over Ω, P the standard Wiener measure, and $W_t(\omega) = \omega(t)$ be the canonical process. For any $t \in [0,1]$, we define $\mathbf{F}_t = \sigma\{W_s ; 0 \le s \le t\}$ and $\mathbf{F}^t = \sigma\{W_s - W_t ; t \le s \le 1\}$.

Skorohod's integral - see Skorohod [7], Gaveau-Trauber [1], Nualart-Zakaï [4] - can be defined as a closed linear unbounded operator δ from $L^2([0,1] \times \Omega)$ into $L^2(\Omega)$, which has the property that $\forall\, 0 \le s < t \le 1$, $\forall\, F \in L^2(\Omega)$ which is $\mathbf{F}_s \vee \mathbf{F}^t$ measurable, $F\, 1_{]s,t]} \in \operatorname{Dom} \delta$, and

$$\delta(F\, 1_{]s,t]}) = F(W_t - W_s) \tag{2.1}$$

Let now $u \in L^2([0,1] \times \Omega)$, and $\Pi^n = \{0 = t_{0,n} < t_{1,n} < \ldots < t_{n,n} = t\}$, $n \in N$, be a sequence of partitions of $[0,1]$ with $|\Pi^n| = \sup_k (t_{k+1,n} - t_{k,n}) \to 0$ as $n \to \infty$. We define the sequence of processes :

$$\tilde{u}^n_t = \sum_{k=0}^{n-1} \Big[\frac{1}{t_{k+1,n}-t_{k,n}} \int_{t_{k,n}}^{t_{k+1,n}} E(u_s / \mathbf{F}_{t_{k,n}} \vee \mathbf{F}^{t_{k+1,n}})\, ds \Big]\, 1_{]t_{k,n},t_{k+1,n}]}(t)$$

It is not hard to show that $\tilde{u}^n \to u$ in $L^2([0,1] \times \Omega)$. On the other hand, from the above facts, $\tilde{u}^n \in \operatorname{Dom} \delta$ and

$$\delta(\tilde{u}^n) = \sum_{k=0}^{n-1} \Big[\frac{1}{t_{k+1,n}-t_{k,n}} \int_{t_{k,n}}^{t_{k+1,n}} E(u_s / \mathbf{F}_{t_{k,n}} \vee \mathbf{F}^{t_{k+1,n}})\, ds \Big] (W_{t_{k+1,n}} - W_{t_{k,n}})$$

Since δ is closed, a sufficient condition for u to belong to Dom δ is that $\delta(\tilde{u}^n)$ converges in $L^2(\Omega)$ as $n \to \infty$. In that case, $\delta(u) = \lim_n \delta(\tilde{u}^n)$. It is easy to show that $\{\delta(\tilde{u}^n)\}$ converges in case where u_t is $\mathbf{F}_t$ adapted (resp. in case where u_t is $\mathbf{F}^t$ adapted). In this particular case, $\delta(u)$ coincides with the Itô forward integral (resp. with the Itô backward integral). We will now identify a third subset of Dom δ.

For that sake, let us first recall the notion of derivation (or gradient operator) on Wiener space. A random variable of the form $F = f(W_{t_1}, \ldots, W_{t_n})$, with $n \in \mathbf{N}$, $0 \le t_1 < \ldots < t_n \le 1$, $f \in C^\infty(\mathbf{R}^n)$ bounded as well as all its derivatives will be called a smooth functional, and the class of smooth functionals will be denoted by $\mathbf{S}$. For $F \in \mathbf{S}$, we define its derivative (or gradient) DF as the element of $L^2([0,1] \times \Omega)$ given by :

$$D_t F = \sum_{i=1}^{n} \frac{\partial f}{\partial x_i} (W_{t_1}, \ldots, W_{t_n}) \, 1_{[0,t_i]}(t)$$

The second derivative of $F \in \mathbf{S}$ is the element of $L^2([0,1]^2 \times \Omega)$ given by :

$$D_s D_t F = (D^2F)_{st} = \sum_{i,j=1}^{n} \frac{\partial^2 f}{\partial x_i \partial x_j} (W_{t_1}, \ldots, W_{t_n}) \, 1_{[0,t_i]}(t) \, 1_{[0,t_j]}(s)$$

More generally, $\forall$ $N \in \mathbf{N}$, we define the N-th derivative of $F \in \mathbf{S}$ as an element of $L^2([0,1]^N \times \Omega)$, or as an $L^2([0,1]^N)$ - valued random variable. We denote by $\|.\|_N$ the norm of $L^2([0,1]^N)$ and $\|.\|_p$ the norm of $L^2(\Omega)$.

For any $N \in \mathbf{N}$ and $p > 1$, we introduce the semi-norm on $\mathbf{S}$:

$$\|F\|_{p,N} = \|F\|_p + \| \, \|D^N F\|_N \|_p$$

and define the "Sobolev space" $\mathbf{D}_{p,N}$ as the completion of $\mathbf{S}$ with respect to $\|.\|_{p,N}$. $\mathbf{D}_{p,N}$ is a Banach space. Essentially usefull to us will be the space $\mathbf{D}_{2,1}$. Note that the derivation D is a closed linear operator from $\mathbf{D}_{2,1}$ into $L^2([0,1] \times \Omega)$.

We finally define $\mathbf{L}^2$ as the set of processes $u \in L^2([0,1] \times \Omega)$ which are such that $u_t \in \mathbf{D}_{2,1}$ for almost all $t \in [0,1]$, and there exists a measurable version of $D_s u_t$ such that:

$$E \int_0^1 \int_0^1 |D_s u_t|^2 \, ds \, dt < \infty$$

$\mathbf{L}^2$ is a Banach space with the norm :

$$\|u\| = (E \int_0^1 |u_t|^2 \, dt + E \int_0^1 \int_0^1 |D_s u_t|^2 \, ds \, dt)^{1/2}$$

We have :

<u>Proposition 2.1</u> : $\mathbf{L}^2 \subset \mathrm{Dom}\, \delta$, and for $u \in \mathbf{L}^2$,

(i) $E(\delta(u)) = 0$

(ii) $E(\delta(u)^2) = E \int_0^1 u_t^2 \, dt + E \int_0^1 \int_0^1 D_s u_t D_t u_s \, ds \, dt$

<u>Sketch of the proof</u> : It suffices to show that $E(\delta(\tilde{u}^n)\, \delta(\tilde{u}^m)) \to \chi$ as $n, m \to \infty$, where χ is the right side of (ii). Indeed, it will then follow that $\{\delta(\tilde{u}^n)\}$ is Cauchy sequence in $L^2(\Omega)$, which then converges to $\delta(u)$, and the two first moments converge. □

<u>Remark 2.2</u> : It follows from the definition of the derivative that if $F \in \mathbf{D}_{2,1}$ and F is $\mathbf{F}_t$ measurable (resp. $\mathbf{F}^t$ measurable), then $D_s F = 0$ for $s \in]t,1]$ (resp. for $s \in [0,t[$).

Therefore, formula (ii) reduces to the usual one in both cases where u_t is $\mathbf{F}_t$ adapted, or $\mathbf{F}^t$ adapted. □

We have explicited three subsets of Dom δ. For other examples of elements of Dom δ, see [3]. Note that Skorohod's integral does generalize both the forward and the backward Itô integrals. In the sequel, we restrict ourselves to integrands in $\mathbf{L}^2$. For $u \in \mathbf{L}^2$, we will write :

$$\delta(u) = \int_0^1 u_t \, dW_t$$

It is well known that one can define the stochastic integral of an $\mathbf{F}_t$-adapted process, provided $u \in L^2(0,1)$ a.s. (and not necessarily $u \in L^2([0,1] \times \Omega)$). Similarly, one can define a space $\mathbf{L}^2_{loc}$, and most of the results below can be "localized", see [3].

Let us finally state the fundamental result in Gaveau-Trauber [1], which says that δ is the adjoint of D :

Theorem 2.3 : Let $u \in$ Dom δ and $F \in \mathbf{D}_{2,1}$.

Then

$$E \int_0^1 u_t D_t F \, dt = E [F \, \delta(u)]. \ \square$$

Corollary 2.4 : Let $u \in \mathbf{L}^2$ and $G \in \mathbf{D}_{2,1}$. Suppose moreover that $u G \in$ Dom δ. Then :

$$\delta(uG) = G \, \delta(u) - \int_0^1 u_t D_t G \, dt. \ \square$$

Note that (2.1) is a particular case of Corollary 2.4 since, whenever F is $\mathbf{F}_s \vee \mathbf{F}^t$ measurable, $D_r F = 0$ for $r \in]s,t[$.

3. The Skorohod integral as a process

Let $u \in \mathbf{L}^2$, and $t \in [0,1]$. Then $u 1_{[0,t]} \in \mathbf{L}^2$, and we define the process $\{\int_0^t u_s \, dW_s, \ t \in [0,1]\}$ by :

$$\int_0^t u_s \, dW_s = \delta \, (u 1_{[0,t]})$$

This process is mean-square continuous and then measurable. It satisfies :

Proposition 3.1 : Let $u \in \mathbf{L}^2$, and $0 \leq s < t \leq 1$. Then :

$$\text{(i)} \ E \left(\int_s^t u_r \, dW_r \, / \, \mathbf{F}_s \vee \mathbf{F}^t\right) = 0$$

(ii) $E\left[\left(\int_s^t u_r\, dW_r\right)^2 / \mathbf{F}_s \vee \mathbf{F}^t\right] =$

$= E\left[\int_s^t u_r^2\, dr + \int_s^t \int_s^t D_\alpha u_r D_r u_\alpha\, dr\, d\alpha / \mathbf{F}_s \vee \mathbf{F}^t\right]$ □

We now give a sufficient condition for the existence of an a.s. continuous modification.

Theorem 3.2 : Let $u \in \mathbf{L}^2$. Then each one of the following conditions implies that the process $\{\int_0^t u_s\, dW_s,\ t \in [0,1]\}$ has an a.s. continuous modification :

(i) $\exists\, p > 1$ s.t. $\sup_{t \in [0,1]} E\left[\left(\int_0^1 |D_s u_t|^2\, ds\right)^p\right] < \infty$

(ii) $\exists\, p > 2$ s.t. $E \int_0^1 \left(\int_0^1 |D_s u_t|^2\, ds\right)^p dt < \infty$

Sketch of the proof : Define $v_t = u_t - E(u_t)$. It clearly suffices to show that $\{\int_0^t v_s\, dW_s\}$ possesses a continuous modification. Since $E(v_t) = 0$ and $D_s v_t = D_s u_t$, it follows from an L^q inequality for the operator δ (see Watanabe [9], or [3] - this inequality is a consequence of Meyer's inequalities for the equivalence of the Sobolev norms, see Meyer [2]) :

$$E\left(\left|\int_s^t v_r\, dW_r\right|^q\right) \le c_p\, E\left[\left(\int_s^t \int_0^1 |D_\alpha u_r|^2\, d\alpha\, dr\right)^{q/2}\right],\ q \ge 2.$$

It is then easy to conclude, using Kolmogorov's lemma, under either (i) or (ii). □

We next compute the "quadratic variation" of $\{\int_0^t u_s\, dW_s\}$, using the same sequence of partitions $\{\Pi^n\}$ as above.

Proposition 3.3 : Let $u \in \mathbf{L}^2$. Then :

$$\sum_{k=0}^{n-1} \left(\int_{t_{k,n}}^{t_{k+1,n}} u_s\, dW_s\right)^2 \to \int_0^1 u_s^2\, ds$$

in probability, as $n \to \infty$. □

4. The generalized Itô formula

Theorem 4.1 : Let $\Phi : \mathbf{R}^2 \to \mathbf{R}$ be a continuous function, such that the derivatives Φ'_x, Φ'_y, Φ''_{yx} and Φ''_{yy} exist and are continuous, and moreover let :

(i) $\left[\begin{array}{l} u \in \mathbf{L}^2 \text{ s.t. } u_t \in \mathbf{D}_{2,2} \text{ t a.e., and for some } p > 4 \\ \int_0^1 \int_0^1 |E(D_s u_t)|^p \, ds \, dt + E \int_0^1 \int_0^1 \int_0^1 |D_r D_s u_t|^p \, dr \, ds \, dt < \infty \end{array} \right.$

(ii) $\left[\begin{array}{l} (V_t,\ t \in [0,1]) \text{ be a continuous process with a.s. finite variation belonging to } \mathbf{L}^2, \text{ s.t.} \\ E \int_0^1 \int_0^1 (D_s V_t)^4 \, ds \, dt < \infty \text{ and the mapping } t \to D_s V_t \text{ is continuous with values in } L^4(\Omega), \text{ uniformly with respect to s.} \end{array} \right.$

Then for any $t \in [0,1]$, with the notation $U_t = \int_0^t u_s \, dW_s$, we have the following :

$$\Phi(V_t, U_t) = \Phi(V_0, 0) + \int_0^t \Phi'_x(V_s, U_s)\, dV_s +$$

$$+ \int_0^t \Phi'_y(V_s, U_s)\, u_s \, dW_s + \frac{1}{2} \int_0^t \Phi''_{yy}(V_s, U_s)\, u_s^2 \, ds +$$

$$+ \int_0^t [\Phi''_{yx}(V_s, U_s)\, D_s V_s + \Phi''_{yy}(V_s, U_s) \int_0^s D_s u_r \, dW_r]\, u_s \, ds$$

Sketch of the proof : First note that the Skorohod integral in the above formula makes sense, although the integrand does not belong to $\mathbf{L}^2$, but to $\mathbf{L}^2_{loc}$ - see [3]. Let us restrict our attention to the following simpler version of Itô's formula :

$$\Phi(U_t) = \Phi(0) + \int_0^t \Phi'(U_s)\, u_s \, dW_s + \frac{1}{2} \int_0^t \Phi''(U_s)\, u_s^2 \, ds + \int_0^t \Phi''(U_s)(\int_0^s D_s u_r \, dW_r)\, u_s \, ds$$

We use a sequence of partitions of $[0,t]$.

$$\begin{aligned} \Phi(U_t) - \Phi(0) &= \sum_k [\Phi(U_{t_{k+1}}) - \Phi(U_{t_k})] \\ &= \sum_k \Phi'(U_{t_k}) \int_{t_k}^{t_{k+1}} u_s dW_s + \\ &+ \frac{1}{2} \sum_k \Phi''(\overline{U}_k) (\int_{t_k}^{t_{k+1}} u_s dW_s)^2 \end{aligned}$$

where $\overline{U}_k$ is a random intermediate point between U_{t_k} and $U_{t_{k+1}}$. It is not hard to deduce from Proposition 3.3 the convergence :

$$\sum_k \Phi''(\overline{U}_k) (\int_{t_k}^{t_{k+1}} u_s dW_s)^2 \to \int_0^t \Phi''(U_s)\, u_s^2 \, ds$$

Using Corollary 2.4, we obtain :

$$\sum_k \Phi'(U_{t_k}) \int_{t_k}^{t_{k+1}} u_s dW_s = \sum_k \int_{t_k}^{t_{k+1}} \Phi'(U_{t_k}) u_s dW_s + $$

$$+ \sum_k \int_{t_k}^{t_{k+1}} D_s[\Phi'(U_{t_k})] u_s ds$$

It is then easy to conclude, once we have the fact (see [3]) that for $s > t_k$:

$$D_s[\Phi'(U_{t_k})] = \Phi''(U_{t_k}) \int_0^{t_k} D_s u_r dW_r . \square$$

Note that if both V_t and u_t are $\mathbf{F}_t$ adapted, then the new additionnal term vanishes, since $D_s u_r = 0$ for $s > r$ and $D_s V_s = 0$ (see the comment just after Theorem 5.4 below).

5. A generalized Stratonovich integral, and the associated chain rule

If u_t is a $\mathbf{F}_t$ semi-martingale, we can define its Stratonovich integral with respect to dW: $\int_0^1 u_t \circ dW_t$, and we have :

$$\int_0^1 u_t \circ dW_t = \int_0^1 u_t \, dW_t + \frac{1}{2} < u, W >_t \tag{5.1}$$

If v_t is a $\mathbf{F}^t$ "backward semi-martingale", we can define its backward Stratonovich integral, which we again note $\int_0^1 v_t \circ dW_t$, and we now have :

$$\int_0^1 v_t \circ dW_t = \int_0^1 v_t \, dW_t - \frac{1}{2} < v, W >_t \tag{5.2}$$

We will now generalize the Stratonovich integral, so as to include those two types of integrals.

Let $u \in L^2([0,1] \times \Omega)$, and (Π^n) be a sequence of partitions of $[0,1]$, with $|\Pi^n| \to 0$ as $n \to \infty$. Define :

$$\Delta_n(u) = \sum_{k=0}^{n-1} u_{k,n} \, (W_{t_{k+1,n}} - W_{t_{k,n}})$$

where

$$u_{k,n} = \frac{1}{t_{k+1,n}-t_{k,n}} \int_{t_{k,n}}^{t_{k+1,n}} u_s ds$$

Definition 5.1 : A process $u \in L^2([0,1] \times \Omega)$ will be said to be Stratonovich integrable whenever $\Delta_n(u)$ converges in probability as $n \to \infty$, and the limit does not depend on the particular sequence of partitions. In that case, we define the Stratonovich integral of u as being the limit of $\Delta_n(u)$:

$$\Delta(u) = \int_0^1 u_t \circ dW_t \qquad \square$$

Definition 5.2 : $\mathbf{L}^2{}_c$ will denote the subset of $\mathbf{L}^2$ consisting of those processes u s.t. for a given version of $D_s u_t$,

(i) $t \to D_s u_t$ is mean square continuous, uniformly with respect to s, for $t \neq s$.

(ii) $\underset{(s,t) \in [0,1]^2}{\text{ess sup}} E(|D_s u_t|^2) < \infty \qquad \square$

If $u \in \mathbf{L}^2{}_c$, we can define

$$D_t^+ u_t = \lim_{\substack{s \to t \\ s > t}} D_t u_s, \quad D_t^- u_t = \lim_{\substack{s \to t \\ s < t}} D_t u_s$$

as elements of $L^2([0,1] \times \Omega)$.

Theorem 5.3 : If $u \in \mathbf{L}^2{}_c$, then u is Stratonovich integrable, and

$$\int_0^1 u_t \circ dW_t = \int_0^1 u_t \, dW_t + \frac{1}{2} \int_0^1 [D_t^+ u_t + D_t^- u_t] \, dt \qquad (5.3)\square$$

In order to compare the Itô-Stratonovich correction term in (5.3) with those in (5.1) and (5.2), let us first state :

Theorem 5.4 : If $u \in \mathbf{L}^2{}_c$ and is mean-square continuous. Then :

$$\sum_{k=0}^{n-1} (u_{t_{k+1}} - u_{t_k})(W_{t_{k+1}} - W_{t_k}) \to \int_0^1 (D_t^+ u_t - D_t^- u_t) \, dt$$

in mean square, as $n \to \infty$. $\square$

Note that $D_s u_t$ is discontinuous accross the diagonal of $[0,1]^2$ iff the joint quadratic variation of u and W is non zero. If u has bounded variation, then we can assume that $D_s u_t$ is continuous (see condition (ii) in theorem 4.1) ; and if moreover u is $\mathbf{F}_t$ adpated, then $D_t u_t = 0$.

If $u \in \mathbf{L}^2_c$ and is a $\mathbf{F}_t$ semi-martingale, $D_t^- u_t = 0$. Then

$$\int_0^1 (D_t^+ u_t + D_t^- u_t)\, dt = \int_0^1 D_t^+ u_t\, dt = \langle u, W \rangle_1$$

If $u \in \mathbf{L}^2_c$ and is a $\mathbf{F}^t$ backward semi-martingale, $D_t^+ u_t = 0$. Then

$$\int_0^1 (D_t^+ u_t + D_t^- u_t)\, dt = \int_0^1 D_t^- u_t\, dt = -\langle u, W \rangle_1$$

Theorem 5.5 : Suppose that Φ, u, V satisfy the hypotheses of Theorem 4.1, and moreover that $u \in \mathbf{L}^2_c$, the process $(D_t^+ u_t + D_t^- u_t)$ belongs to $\mathbf{L}^2$ and

$$\sup_{t \in [0,1]} E \int_0^1 |D_t(D_s^+ u_s + D_s^- u_s)|^4\, ds < \infty$$

Then, with the notation $\mho_t = \int_s^t u_s \circ dW_s$,

$$\Phi(V_t, \mho_t) = \Phi(V_0, 0) + \int_0^t \Phi'_x (V_s, \mho_s)\, dV_s + \int_0^t \Phi'_y (V_s, \mho_s)\, u_s \circ dW_s \; \square$$

Theorem 5.5 is a consequence of theorem 4.1, by translating everything into Itô language (except that we need a version of theorem 4.1 with a two-dimensional bounded variation process). Note that all "additionnal terms" have disappeared in the generalized Stratonovich chain rule.

6. Bibliography

[1] B. Gaveau, P. Trauber : L'intégrale stochastique comme opérateur de divergence dans l'espace fonctionnel. J. Funct. Anal. 46, 230-238, 1982.

[2] P.A. Meyer : Tranformations de Riesz pour les lois gaussiennes. Séminaire de Probabilités XVIII. Lecture Notes in Math. 1059, 179-193, Springer-Verlag 1984.

[3] D. Nualart, E. Pardoux : Stochastic calculus with anticipating integrands, to appear.

[4] D. Nualart, M. Zakaï : Generalized stochastic integrals and the Malliavin Calculus, to appear.

[5] T. Sekiguchi, Y. Shiota : L^2-theory of non causal stochastic integrals. Math Rep. Toyama Univ. 8, 119-195, 1985.

[6] A. Sevljakov : The Itô formula for the extensed stochastic integral. Theory Prob. and Math. Statist. 22, 163-174, 1981.

[7] A. Skorohod : On a generalization of a stochastic integral. Theory of Prob. and Appl. XX, 219-233, 1975.

[8] A. Ustunel : La formule de changement de variable pour l'intégrale anticipante de Skorohod. Note CRAS, to appear.

[9] S. Watanabe : Lectures on stochastic differential equations and Malliavin Calculus. Tata Institute for Fund. Research. Springer Verlag, 1984.

ABSOLUTE CONTINUITY OF A SEMIMARTINGALE WITH RESPECT TO A CONTINUOUS INCREASING AND ADAPTED PROCESS

C. Stricker
Laboratoire de Mathématiques
Faculté des Sciences
16, Route de Gray
25030 Besançon Cedex - France

The aim of this paper is to study the absolute continuity of a semimartingale (X_t) w. r. t. a continuous increasing and adapted process (D_t). Such problems arise in filtering theory. In that case $dD_s = ds$.

Let $\left(\Omega, \mathcal{F}, P, (\mathcal{F}_t)_{0\leq t\leq 1}\right)$ be a filtered probability space satisfying the usual conditions. We suppose all processes are null at time 0 and they are indexed by $[0, 1]$ rather then all of $\mathbb{R}$. Assume (B_t) is a $(\mathcal{F}_t)$ Brownian motion and (h_s) is a predictable process such that $\int_0^1 |h_s|\, ds < +\infty$. It is well known that if for every $s \in [0, 1]$ $E[|h_s| \,|\, \mathcal{F}_s^X] < +\infty$ where $(\mathcal{F}_s^X)$ denotes the natural filtration of (X_t), then the process $\nu_t = X_t - \int_0^t E[h_s | \mathcal{F}_s^X] ds$ is a Brownian motion. One consequence of our main result will be that the condition $\int_0^1 |h_s| ds < +\infty$ implies the existence of a $(\mathcal{F}_t^X)$ predictable process $(\hat{h}_s)$ such that $\int_0^1 |\hat{h}_s| ds < +\infty$ and $\nu_t = X_t - \int_0^t \hat{h}_s\, ds$ is a $(\mathcal{F}_t^X)$ Brownian motion.

In the last section we improve the Lindquist-Picci semimartingale representation theorem ([4]).

1. ABSOLUTE CONTINUITY

Let $\left(\Omega, \mathcal{F}, P, (\mathcal{F}_t)_{0\leq t\leq 1}\right)$ be a filtered probability space satisfying the usual conditions. If (X_t) is a $(\mathcal{F}_t)$ adapted process, $\mathcal{F}_t^X = \bigcap_{u>t} \sigma(X_s, s\leq u) \cup \mathcal{N}$ where $\mathcal{N}$ is the family of P-negligible sets. Let $(D_t)_{0\leq t\leq 1}$ be a continuous increasing and adapted process. Suppose (X_t) as a right continuous semimartingale. The notation $(X_t) << (D_t)$ means that (X_t) satisfies the following condition :

1. 1. For every non negative bounded predictable process (H_t) such that $\int_0^1 H_s\, dD_s = 0$, $(H.X)_1 = 0$ where $(H.X)_t$ denotes the stochastic integral $\int_0^t H_s\, dX_s$.

Suppose (X_t) is a right continuous predictable process of bounded variation such that $(X_t) << (D_t)$. Set $dQ = k\left(1 + \int_0^1 |dX_s| + \int_0^1 |dD_s|\right)^{-1} dP$ where k is a constant such that $Q(\Omega) = 1$. Then $\mu(H) = E^Q[(1_H . X)_1]$ and $\nu(H) = E^Q[(1_H . D)_1]$ are measures on the predictable σ-field $\mathcal{P}$ such that $\mu << \nu$. Thus there exists a density (K_s) satisfying $d X_s = K_s \, d D_s$. So we get the following lemma [1].

1. 2. LEMMA. Suppose (X_t) is a right continuous predictable process of bounded variation. Then $(X_t) << (D_t)$ iff there exists a predictable process (K_s) such that $\int_0^1 |K_s| \, d D_s < +\infty$ and $d X_s = K_s \, d D_s$.

If (X_t) is a semimartingale and if $([X, X]_t)$ is locally integrable, $(\langle X, X\rangle_t)$ will be the dual predictable projection of $([X, X]_t)$.

1. 3. PROPOSITION. Suppose (X_t) is a right continuous semimartingale and $(X_t) << (D_t)$. Then :

i) the jumps of (X_t) are totally inaccessible,

ii) $([X, X]_t) << (D_t)$. If moreover $(\langle X, X\rangle_t)$ exists, then $(\langle X, X\rangle_t) << (D_t)$ and there exists a predictable process (m_s) such that $d\langle X, X\rangle_s = m_s \, dD_s$,

iii) if moreover (X_t) is a special semimartingale with canonical decomposition $X = M + A$, then $(M_t) << (D_t)$ and $(A_t) << (D_t)$,

iv) there exists a decomposition $X = M + A$ where (M_t) is a local martingale and (A_t) is a right continuous process of bounded variation such that $(M_t) << (D_t)$ and $(A_t) << (D_t)$.

PROOF. i) Let T be a predictable stopping time. Since (D_t) is continuous, $\left(1_{[T]} \cdot D\right)_1 = 0$. Therefore $0 = \left(1_{[T]} \cdot X\right)_1 = \Delta X_T$. Thus the jumps of (X_t) are totally inaccessible.

ii) Suppose H is a bounded nonnegative predictable process such that $(H. D)_1 = 0$. Then $\left(H^2. [X, X]\right)_1 = (H. X)_1 - 2\int_0^1 (H. X)_- dX = 0$. Therefore $([X, X]_t) << (D_t)$. If moreover $(\langle X, X\rangle_t)$ exists, then $([X, X]_t)$ is locally integrable (actually both conditions are equivalent). If T is a stopping time satisfying $E[X, X]_T < +\infty$, then for every bounded predictable process H, $E[(H^2. [X, X])_T] = E[(H^2. \langle X, X\rangle)_T]$. Hence $(\langle X, X\rangle_t) << (D_t)$.

iii) if (H_t) is a bounded predictable process, then the canonical decomposition of $H . X$ is $H. M + H. A$. Therefore, if $(X_t) << (D_t)$, then $(M_t) << (D_t)$ and $(A_t) << (D_t)$

iv) if $(X_t) << (D_t)$ and if (H_t) is a bounded nonnegative predictable process satisfying $(H. D)_1 = 0$, then $H. X = 0$. Thus $H. X$ does not jump and $H. A' = 0$ where $A'_t = \sum_{s \le t} \Delta X_s 1_{\{|\Delta X_s| \ge 1\}}$. Therefore $A' << D$. Since $X - A'$ is a special semimartingale, iii) implies iv).

1. 4. REMARK. There exist semimartingales which are not special and satisfy con-

dition 1.1. (see [8] for an example).

Suppose (X_t) is a right continuous semimartingale such that $([X,X]_t)$ is locally integrable. Then (X_t) is a special semimartingale with canonical decomposition $X=M+A$ where (A_t) is a right continuous predictable process of bounded variation and (M_t) is a local martingale. Moreover, since every right continuous predictable process is locally bounded (see [1]) and $([X,X]_t)$ is locally integrable, it follows that (M_t) is locally square integrable and $\langle M,M\rangle$ exists. Conversely, if X is a special semimartingale with canonical decomposition $X=M+A$ and $\langle M,M\rangle$ exists, then $[X,X]$ is locally integrable.

Let me introduce the following condition.

1.5. X is a right continuous special semimartingale such that its canonical decomposition $X=M+A$ satisfies : $\langle M,M\rangle$ exists, $d\langle M,M\rangle_s = m_s\, dD_s$ and $dA_s = a_s\, dD_s$ where (m_s) and (a_s) are predictable processes.

1.6. THEOREM. Suppose (X_t) is a right continuous semimartingale such that $[X,X]$ is locally integrable. Then (X_t) satisfies condition 1.5. iff $(X_t) \ll (D_t)$.

PROOF. Suppose (X_t) satisfies condition 1.5. Then $(A_t) \ll (D_t)$. If T is a stopping time such that $E[\langle M,M\rangle_T] < +\infty$, then for every bounded predictable process H $E[(H^2.\langle M,M\rangle)_T] = E[(H.M)_T^2]$. Therefore $(M_t) \ll (D_t)$ and $(X_t) \ll (D_t)$. Conversely suppose $(X_t) \ll (D_t)$. Then proposition 1.3 iii) implies that $(M_t) \ll (D_t)$ and $(A_t) \ll (D_t)$. Since $(\langle M,M\rangle_t)$ exists, we can apply proposition 1.3 ii) to (M_t). Therefore $(\langle M,M\rangle_t) \ll (D_t)$. By means of lemma 1.2 there exist two predictable processes (m_s) and (a_s) such that $d\langle M,M\rangle_s = m_s\, dD_s$ and $dA_s = a_s\, dD_s$.
It is easy to prove the following :

COROLLARY 1.7. Suppose (X_t) is a right continuous semimartingale satisfying condition 1.5. Then :

i) if $Q \ll P$ and if $[X,X]$ is locally integrable under Q, then X satisfies condition 1.5. under Q.

ii) if $(\mathcal{G}_t)$ is a subfiltration of $(\mathcal{F}_t)$ such that (X_t) is $(\mathcal{G}_t)$ adapted and $([X,X]_t)$ is locally integrable w.r.t. $(\mathcal{G}_t)$, then (X_t) satisfies condition 1.5. w.r.t. $(\mathcal{G}_t)$.

1.8. EXAMPLE. Let (B_t) be a Brownian motion w.r.t. $(\mathcal{F}_t)$. Suppose $X_t = B_t + \int_0^t h_s\, d_s$ where (h_s) is a $(\mathcal{F}_t)$ predictable process such that $\int_0^1 |h_s|\, ds < +\infty$. Then $(X_t) \ll (D_t)$ where $dD_s = ds$. Suppose $(\mathcal{G}_t)$ is a subfiltration of $(\mathcal{F}_t)$ such that (X_t) is $(\mathcal{G}_t)$ adapted. Then according to corollary 1.7 ii) there exists a $(\mathcal{G}_t)$ predictable process $(\hat{h}_s)$ such that $\int_0^1 |\hat{h}_s|\, ds < +\infty$ and $\nu_t = X_t - \int_0^t \hat{h}_s\, ds$ is a continuous local martingale w.r.t. $(\mathcal{G}_t)$. Since $[X,X]$ does not depend on the filtration, $[\nu,\nu]_t = [X,X]_t = t$. Therefore (ν_t) is a $(\mathcal{G}_t)$ Brownian motion.

Now we are going to give a Lebesgue decomposition of the semimartingale (X_t) w.r.t. (D_t).

1.9. PROPOSITION. Let (X_t) be a right continuous semimartingale. Then there exists a predictable set B such that $1_{B^c}.D=0$ and $X^1=1_B.X$ is absolutely continuous w.r.t. (D_t). If there exists another predictable set E such that $1_{E^c}.D=0$ and $Y^1=1_E.X$ is absolutely continuous w.r.t. (D_t), then $X^1=Y^1$.

PROOF. There exists a law Q which is equivalent to P such that $X_t=M_t+A_t$ where (M_t) is a square integrable martingale and (A_t) is a right continuous predictable process of integrable variation. Set $V_t=\langle M,M\rangle_t+\int_0^t|dA_s|$. $\mu(H)=E[(1_H.V)_1]$ and $\nu(H)=E[(1_H.D)_1]$ are measures on the predictable σ-field $\mathcal{P}$. According to the usual Lebesgue decomposition of a measure there exists a predictable set B such that $\big((1_B.V)_t\big)\ll(D_t)$ and $1_{B^c}.D=0$. Since $\big((1_B.V)_t\big)\ll(D_t)$, $\big((1_B.X)_t\big)\ll(D_t)$. Suppose there exists another predictable set E such that $1_{E^c}.D=0$ and $Y^1=1_E.X$ is absolutely continuous w.r.t. (D_t). Set $Y^2=1_{E^c}.X$ and $X^2=1_{B^c}.X$. Then $X^1-X^2=Y^2-Y^1$. Since $1_{E^c\cup B^c}.D=0$, we get $0=1_{E^c\cup B^c}.(X^1-X^2)$ $=1_{E^c\cup B^c}.(Y^1-Y^2)$. Therefore the decomposition of X is unique.

1.10. REMARK. The Lebesgue decomposition of (X_t) depends on $(\mathcal{F}_t)$. Let me give an example. Suppose (B_t) is a $(\mathcal{F}_t)$ Brownian motion and $dD_s=ds$. The process $X_t=2\sup_{s\le t}B_s-B_t$ is a Bessel process or order 3. The absolutely continuous part is (B_t) and the singular part is $(2\sup_{s\le t}B_s)$. If $(\mathcal{F}_t^X)$ denotes the natural filtration generated by (X_t), then it is well known that $X_t-\int_0^t X_s^{-1}ds$ is a $(\mathcal{F}_t^X)$ Brownian motion and $(X_t)\ll(D_t)$ w.r.t. $(\mathcal{F}_t^X)$. Therefore the Lebesgue decomposition of (X_t) depends on the filtration.

2. A SPECIAL CLASS OF SEMIMARTINGALES

The results of this section are not new but the proofs are different ([7]). For the sake of completeness we shall give all the details. Suppose Y is a <u>continuous</u> semimartingale such that its canonical decomposition $X=M+A$ satisfies the following condition :

2.1. $d\langle M,M\rangle_t=m_t\,dD_t$, $dA_t=a_t\,dD_t$ where (m_t) and (a_t) are predictable, (m_t) is locally bounded and $\int_0^1 a_s^2\,dD_s<+\infty$.

This condition implies :

2.2. $(H^n.Y)_1$ converges to 0 in L^0 whenever (H_t^n) is a sequence of predictable bounded step processes such that $(\int_0^1(H_s^n)^2\,dD_s)_n$ converges to 0 in L^0.

$\mathcal{S}$ denotes the set of bounded predictable step processes H satisfying $\int_0^1 H_s^2 dD_s \leq 1$. Then condition 2.2. implies :

2.3. $\{(H.Y)_1, H \in \mathcal{S}\}$ is bounded in L^0.

Let L(Y) be the set of all predictable processes H such that $\left(H^2.\langle M,M\rangle\right)_1 + \int_0^1 |H_s||dA_s| < +\infty$. Then $(H.Y)_1$ is well defined. It is easy to see that condition 2.1. implies.

2.4. The set of all predictable processes H such that $\int_0^1 H_s^2 dD_s < +\infty$ is included in L(Y).

The main result of this section is the following :

2.5. THEOREM. The four conditions above are equivalent.

Since 2.1. ⇒ 2.2. ⇒ 2.3., 2.1. ⇒ 2.4. and 2.4. ⇒ 2.2., it is sufficient to prove the implication 2.3. ⇒ 2.1. Let $\mathcal{S}'$ denote the set of bounded predictable processes H satisfying $\int_0^1 H_s^2 dD_s \leq 1$. According to the monotone class theorem $\mathcal{S}$ is dense in $\mathcal{S}'$. From now on we suppose condition 2.3. holds. Then $\{(H.Y)_1, H \in \mathcal{S}'\}$ is bounded in L^0 since $\mathcal{S}$ is dense in $\mathcal{S}'$. Therefore $(H.Y)_1 = 0$ whenever H is a bounded predictable process such that $\int_0^1 |H_s| dD_s = 0$. According to theorem 1.6. there exist two predictable processes (m_s) and (a_s) such that $d\langle M,M\rangle_s = m_s dD_s$ and $dA_s = a_s dD_s$.

2.6. LEMMA. $\int_0^1 |H_s m_s| dD_s < +\infty$ whenever H is a predictable process such that $\int_0^1 |H_s| dD_s < +\infty$.

PROOF. Suppose $H \in \mathcal{S}'$. Put $T_n = \inf\{t, |(H.Y)_t| \geq n\}$ and $K^n = (H.Y)^{T_n}$. Suppose $\varepsilon > 0$. Since $\{(H.Y)_1, H \in \mathcal{S}'\}$ is bounded in L^0, there exists an integer n such that for every $H \in \mathcal{S}'$ $P[|(H.Y)_1| \geq n] < \frac{\varepsilon}{2}$. Therefore :

$$P[|\int_0^1 (H.Y)_s H_s dY_s| \geq n^2] \leq P[|((K^n H).Y)_1| \geq n^2] + P[T_n < 1] \leq \frac{\varepsilon}{2} + \frac{\varepsilon}{2}$$

Hence $\{\int_0^1 (H.Y)_s H_s dY_s, H \in \mathcal{S}'\}$ is bounded in L^0. According to Ito's formula, $\{(H^2.[M,M])_1, H \in \mathcal{S}'\}$ is bounded in L^0. Therefore $\int_0^1 |K_s m_s| dD_s < +\infty$ whenever (K_s) is a predictable process such that $\int_0^1 |K_s| dD_s < +\infty$.

2.7. LEMMA. There exists a locally bounded version of (m_s).

PROOF. Put $G_n = \{|m| \geq n\}$, $b_n = \left(1_{G_n} . D\right)_1$ and $H_t = \sum_n n^{-2} b_n^{-1} 1_{G_n}(t)$. Since $(H.D)_1 < +\infty$, we get $\int_0^1 |H_s m_s| dD_s < +\infty$. Therefore for a.e. ω there exists $n(\omega)$ such that $b_{n(\omega)}(\omega) = 0$. Put $T_n = \inf\{t, \left(1_{G_n} . D\right)_t > 0\}$. Then $T_{n(\omega)}(\omega) = 1$, the process $\widetilde{m} = \sum_n (m_1 (n+1)) \vee (-n-1) 1_{]T_n, T_{n+1}]}$ is locally bounded and $\left(1_{\{m \neq \widetilde{m}\}} . D\right)_1 = 0$.

2. 8. LEMMA. Let $p>1$ and $\frac{1}{p}+\frac{1}{q}=1$. If the following set $\{\int_0^1 |K_s a_s| dD_s$, K is predictable and $\int_0^1 |K_s|^p dD_s \leq 1\}$ is bounded in L^0, then $\int_0^1 |a_s|^q dD_s <+\infty$.

PROOF. Put $H_s^n = |a_s^n|^{q-1}/(\int_0^1 |a_s^n|^q dD_s)^{1/p}$ where $a_s^n = a_s 1_{\{|a_s|\leq n\}}$. Then $\int_0^1 |H_s^n|^p dD_s \leq 1$ and the sequence $(\int_0^1 |H_s^n a_s| dD_s)_n$ is bounded in L^0. Now $\int_0^1 |H_s^n a_s| dD_s = \int_0^1 |a_s^n|^q dD_s$. Therefore $\int_0^1 |a_s|^q dD_s <+\infty$.

Now we are going to finish the proof of theorem 2. 5. Since condition 2. 3. holds, $d\langle M,M\rangle_s = m_s dD_s$ and $dA_s = a_s dD_s$. According to 2. 6. and 2. 7. there exists a locally bounded version of (m_s). Hence $\{(H.A)_1, H\in \mathcal{E}'\}$ is bounded in L^0. It follows from lemma 2. 8. that $\int_0^1 a_s^2 dD_s < +\infty$ and the four conditions of theorem 2. 5. are equivalent.

As in section 1 we get the following corollary :

2. 9. COROLLARY. Suppose (Y_t) is a continuous semimartingale satisfying one of the four equivalent conditions. If $Q<<P$, then (Y_t) satisfies the same conditions under the law Q. If $(\mathcal{G}_t)$ is a subfiltration of $(\mathcal{F}_t)$ and if (Y_t) is $(\mathcal{G}_t)$ adapted, then (Y_t) satisfies the same conditions w. r. t. $(\mathcal{G}_t)$.

2. 10. EXAMPLES.

i) Suppose (Y_t) is a $(\mathcal{F}_t)$ Brownian motion under the law P. If $Q<<P$, then under Q $Y_t = M_t + \int_0^t a_s\, ds$ where $\int_0^1 a_s^2\, ds <+\infty$ and (M_t) is a continuous local martingale. Since $t=[Y,Y]_t = [M,M]_t$, (M_t) is $(\mathcal{F}_t)$ a Brownian motion under the law Q.

ii) Suppose $Y_t = B_t + \int_0^t a_s\, ds$ where (B_t) is a $(\mathcal{F}_t)$ Brownian motion and $\int_0^1 a_s^2\, ds < +\infty$. Then for every subfiltration $(\mathcal{G}_t)$ such that (Y_t) is $(\mathcal{G}_t)$ adapted, there exists a $(\mathcal{G}_t)$ predictable process $(\hat{a}_s)$ such that $\int_0^1 \hat{a}_s^2\, ds <+\infty$ and $V_t = Y_t - \int_0^t \hat{a}_s\, ds$ is a $(\mathcal{G}_t)$ Brownian motion.

2. 11. REMARK. Suppose (Y_t) is a right continuous semimartingale such that $([Y,Y]_t)$ is locally integrable and $(Y_t) << (D_t)$. Then $Y_t = M_t + A_t$ where (M_t) is a local martingale, $d\langle M,M\rangle_s = m_s dD_s$, $dA_t = a_s dD_s$. If (m_s) is locally bounded and if $\int_0^1 a_s^2\, sD_s <+\infty$, then (Y_t) satisfies condition 2. 2. (which is equivalent to 2. 3. and 2. 4.). But the converse is not true. For instance suppose $\Omega = [0,\infty[$, $S(\omega)=\omega$, $D_s = s$ and $(\mathcal{F}_t)$ is the smallest right continuous filtration such that S is a stopping time. Then it is well known (see [1]) that the dual predictable projection of $1_{[S,+\infty[}(t)$ is $A_t = \int_0^{S\wedge t} \frac{dF(u)}{1-F(u^-)}$ where $F(u)=P[S\leq u]$. If $dP_1(\omega)=e^{-\omega}$, then $A_t = S_\wedge t$. Therefore $1_{[S,+\infty[}(t) = M_t + S_\wedge t$, $[M,M]_t = 1_{[S,+\infty[}(t)$ and $\langle M,M\rangle_t = S_\wedge t$. In that case

$m_t = a_t = 1_{[0,S]}(t)$. But if $dP_2(\omega) = \frac{k e^{-\omega}}{\sqrt{\omega}}$, then (m_t) is not locally bounded.

3. THE LINDQUIST-PICCI SEMIMARTINGALE REPRESENTATION THEOREM.

In this section we shall study the following condition :

3. 1. X_t is a special semimartingale and its canonical decomposition admits a representation of the form $X_t = M_t + \int_0^t Z_s \, dA_s$ where (M_t) is a local martingale and (A_t) is a predictable process of integrable variation.

Conditions of type 3. 1. appeared in [6], [4], [5], [2]. In the following theorem we shall give a necessary and sufficient condition for (X_t) to admit such a representation.

3. 2. THEOREM. Let (X_t) be a right continuous adapted process. (X_t) satisfies condition 3. 1. iff there exists an increasing sequence of stopping times (T_n) such that :

i) (T_n) converges stationary to 1,

ii) for every $t \in [0, 1]$ $X_{t \wedge T_n}$ is integrable,

iii) for every n the map μ^n defined by $\mu^n(H) = E[(H.X)_{T_n}]$ whenever H is a bounded predictable step function, can be extended to a measure on the predictable σ-field $\mathcal{P}$.

iv) for every n μ^n is absolutely continuous w. r. t. the measure ν defined on $\mathcal{P}$ by $\nu(H) = E[\int_0^1 H_s \, dA_s]$.

PROOF. Suppose (X_t) satisfies condition 3. 1. Then it is well known that (X_t) is locally in $\mathcal{H}^1$ and it is easy to see that (X_t) satisfies the four conditions of theorem 3. 2. Conversely if $\mu^n \ll \nu$, there exists a unique predictable process (B_t^n) of integrable variation such that $\mu^n(H) = E[(H.B)_1^n]$. Therefore $X^{T_n} - B^n$ is a martingale. Moreover $(B_t^n) \ll (A_t^n)$. Since (B_t^n) is unique and (T_n) is increasing, there exists a predictable process (B_t) of bounded variation such that $(X_t - B_t)$ is a local martingale and $(B_t) \ll (A_t)$.

According to theorem 3. 2. we get a result of Protter [5].

3. 3. COROLLARY. If (X_t) is a right continuous, adapted, and integrable process and if there exists an integrable random variable K such that for every t and h $|E[X_{t+h} - X_t | \mathcal{F}_t]| \le Kh$, then there exists a predictable process (Z_s) such that $E[\int_0^1 |Z_s| \, ds] < +\infty$ and $\left(X_t - \int_0^t Z_s \, ds\right)$ is a martingale.

PROOF. Let $A_t = \int_0^t E[K | \mathcal{F}_s] \, ds$ and $T_n = 1$ for every n. Then it is easy to see that $\mu + \nu$ is a positive additive set function defined on the set of bounded predictable step functions. Since $\mu + \nu \le 2\nu$, μ can be extended to a measure on $\mathcal{P}$ which is -

absolutely continuous w.r.t. ν. Then theorem 3.2. implies corollary 3.3.

Now we are going to study the case of a constant K which is not integrable. Let D be a denumerable dense subset of $[0,1]$. Put $\tilde{K}_t = \sup_{\substack{(h,s)\in D^2 \\ s\le t}} |E[X_{s+h}-X_s|\mathcal{F}_s]|$ and $K_t = \tilde{K}_{t+}$. Then (K_t) is a right continuous process. For $0\le s<t\le 1$ and $(s,t)\in D^2$ consider a partition $s=t_o<t_1<\dots<t_n=t$ where $t_i\in D$ for all i. Suppose α is a constant and T is a stopping time such that $K_{T-}\le\alpha$. Then for all $A\in\mathcal{F}_s$ $|E[1_A \sum_i (X_{t_{i+1}}-X_{t_i})\, 1_{\{t_i<T\}}]| \le \alpha(t-s)$.

Now the sum $N=\sum_i (X_{t_{i+1}}-X_{t_i})\, 1_{\{t_i<T\}}$ is equal to 0 if $T\le s$, $N=X_t-X_s$ if $T\ge t$, $N=X_{t_{i+1}}-X_s$ if $t_i<T\le t_{i+1}$. Therefore N converges to X_t-X_s if the mesh of the partition tends to 0. If (X_t) is of class D, then $|E[1_A(X_t^T-X_s^T)]|\le\alpha(t-s)$. Hence there exist a martingale (M_t) and a process (Z_t) such that $X_t^T=M_t+\int_o^t Z_s\,ds$. Moreover $E[\int_o^1 |Z_s|\,ds]<+\infty$. Since (K_t) is a right continuous increasing process, there exists a sequence (T_n) of stopping times such that (T_n) increases stationary to 1 and $K_{T_n-}\le n$ on $\{T_n>0\}$. So we can apply theorem 3.2. : there exist two processes (M_t) and (Z_t) such that $\left(1_{\{T_n>0\}} M_t^{T_n}\right)$ is a martingale for every n and $X_t = M_t + \int_o^t Z_s\,ds$.

3.5. REMARK. Suppose $\mathcal{H}$ is a Gaussian space and $E[X_t|\mathcal{F}_s]\in\mathcal{H}$ for every t and s. Then $\tilde{K}_t$ is integrable for every t and $X_t=M_t+\int_o^t Z_s\,ds$ where (M_t) is a martingale. Moreover $M_t\in\mathcal{H}$ and $Z_t\in\mathcal{H}$ for every t.

REFERENCES.

[1] Dellacherie, C. et Meyer, P.A. (1979). Probabilités et potentiel B, Théorie des martingales. Hermann, Paris.

[2] Ethier, S. and Kurtz, T. (1986). Markov processes : characterization and convergence. Wiley, New York.

[3] Lenglart, E., Lepingle, D. et Pratelli, M. (1980). Présentation unifiée de certaines inégalités de la théorie des martingales. Séminaire de Probabilités XV, 26-48. Lecture Notes in M. 721, Springer Verlag, Berlin.

[4] Lindquist, A. and Picci, G. (1985). Forward and backward semimartingale models for Gaussian processes with stationary increments. Stochastics 15 1-50.

[5] Protter, P. (1986). Reversing Gaussian semimartingales without Gauss. Preprint.

[6] Stricker, C. (1975). Une caractérisation des quasimartingales. Séminaire de Probabilités IX, 420-424. Lecture Notes in M. 465, Springer Verlag, Berlin.

[7] Stricker, C. (1984). Quelques remarques sur le problème de l'innovation. Proceedings of the CNET ENST Colloquium, 260-176. Lecture Notes in control and information sciences 61, Springer Verlag, Berlin.

[8] Stricker, C. (1985). Une remarque sur une certaine classe de semimartingales. Séminaire de Probabilités XIX, 218-221. Lecture Notes in M. 1123, Springer Verlag, Berlin.

TALKS DELIVERED ON THE CONFERENCE BUT NOT PUBLISHED IN THIS PROCEEDINGS

S.K. CHRISTENSEN

Linear stochastic differential equations on the dual of a nuclear space

N. CHRISTOPEIT

Last squares estimation in semimartingale regression models

M.H.A. DAVIS, J.B. RIBEIRO DO VAL

Optimality conditions for piecewise deterministic process

H.J. ENGELBERT, W. SCHMIDT

On absolute continuity and equivalence of diffusion processes with respect to the Wiener process

L. GIRAITIS

On Markov fields with finite number of values

B.I. GRIGELIONIS

Rigged Hilbert space valued semimartingales

A.A. GUCHTCHIN

Localisation for two parameter strong martingales

G. KALLIANPUR

Stochastic differential equations satisfied by inifinte dimensional processes with applications to neuronal behavior

P. KOTELENEZ

Necessary and sufficient conditions for limit theorems in reaction-diffusion systems

S.A. MOLCANOV

Magnetic fields in random flows

T. SHIGA

Stochastic differential equations for some measure valued diffusions

J. STOYANOV

Stochastic integrals and stochastic differential equations: counterexamples

Lecture Notes in Control and Information Sciences

Edited by M. Thoma

Vol. 43: Stochastic Differential Systems
Proceedings of the 2nd Bad Honnef Conference
of the SFB 72 of the DFG at the University of Bonn
June 28 – July 2, 1982
Edited by M. Kohlmann and N. Christopeit
XII, 377 pages. 1982.

Vol. 44: Analysis and Optimization of Systems
Proceedings of the Fifth International
Conference on Analysis and Optimization of Systems
Versailles. December 14–17, 1982
Edited by A. Bensoussan and J. L. Lions
XV, 987 pages, 1982

Vol. 45: M. Arató
Linear Stochastic Systems
with Constant Coefficients
A Statistical Approach
IX, 309 pages. 1982

Vol. 46: Time-Scale Modeling of Dynamic Networks
with Applications to Power Systems
Edited by J. H. Chow
X, 218 pages. 1982

Vol. 47: P. A. Ioannou, P. V. Kokotovic
Adaptive Systems with Reduced Models
V, 162 pages. 1983

Vol. 48: Yaakov Yavin
Feedback Strategies for Partially
Observable Stochastic Systems
VI, 233 pages, 1983

Vol. 49: Theory and Application of Random Fields
Proceedings of the IFIP-WG 7/1
Working Conference
hold under the joint auspices of the
Indian Statistical Institute
Bangalore, India, January 1982
Edited by G. Kallianpur
VI. 290 pages. 1983

Vol. 50: M. Papageorgiou
Applications of Automatic Control Concepts
to Traffic Flow Modeling and Control
IX, 186 pages. 1983

Vol. 51: Z. Nahorski, H.F. Ravn, R.V.V. Vidal
Optimization of Discrete Time Systems
The Upper Boundary Approach
V, 137 pages 1983

Vol. 52: A. L. Dontchev
Perturbations, Approximations and Sensitivity Analysis
of Optimal Control Systems
IV, 158 pages. 1983

Vol. 53: Liu Chen Hui
General Decoupling Theory of Multivariable
Process Control Systems
XI, 474 pages. 1983

Vol. 54: Control Theory for Distributed
Parameter Systems and Applications
Edited by F. Kappel, K. Kunisch,
W. Schappacher
VII, 245 pages. 1983.

Vol. 55: Ganti Prasada Rao
Piecewise Constant Orthogonal Functions
and Their Application to Systems and Control
VII, 254 pages. 1983.

Vol. 56: Dines Chandra Saha, Ganti Prasada Rao
Identification of Continuous
Dynamical Systems
The Poisson Moment Functional
(PMF) Approach
IX, 158 pages. 1983.

Vol. 57: T. Söderström, P. G. Stoica
Instrumental Variable Methods
for System Identification
VII, 243 pages. 1983.

Vol. 58: Mathematical Theory of
Networks and Systems
Proceedings of the MTNS-83 International
Symposium
Beer Sheva, Israel, June 20–24, 1983
Edited by P. A. Fuhrmann
X, 906 pages. 1984

Vol. 59: System Modelling and Optimization
Proceedings of the 11th IFIP Conference
Copenhagen, Denmark, July 25-29, 1983
Edited by P. Thoft-Christensen
IX, 892 pages. 1984

Vol. 60: Modelling and Performance
Evaluation Methodology
Proceedings of the International Seminar
Paris, France, January 24–26, 1983
Edited by F. Bacelli and G. Fayolle
VII, 655 pages. 1984

Vol. 61: Filtering and Control of Random
Processes
Proceedings of the E.N.S.T.-C.N.E.T. Colloquium
Paris, France, February 23–24, 1983
Edited by H. Korezlioglu, G. Mazziotto, and
J. Szpirglas
V, 325 pages. 1984

Lecture Notes in Control and Information Sciences

Edited by M. Thoma and A. Wyner

Vol. 62: Analysis and Optimization
of Systems
Proceedings of the Sixth International
Conference on Analysis and Optimization
of Systems
Nice, June 19–22, 1984
Edited by A. Bensoussan, J. L. Lions
XIX, 591 pages. 1984.

Vol. 63: Analysis and Optimization
of Systems
Proceedings of the Sixth International
Conference on Analysis and Optimization
of Systems
Nice, June 19–22, 1984
Edited by A. Bensoussan, J. L. Lions
XIX, 700 pages. 1984.

Vol. 64: Arunabha Bagchi
Stackelberg Differential Games
in Economic Models
VIII, 203 pages, 1984

Vol. 65: Yaakov Yavin
Numerical Studies
in Nonlinear Filtering
VIII, 273 pages, 1985.

Vol. 66: Systems and Optimization
Proceedings of the Twente Workshop
Enschede, The Netherlands, April 16–18, 1984
Edited by A. Bagchi, H. Th. Jongen
X, 206 pages, 1985.

Vol. 67: Real Time Control of Large Scale Systems
Proceedings of the First European Workshop
University of Patras, Greece, Juli 9–12, 1984
Edited by G. Schmidt, M. Singh, A. Titli,
S. Tzafestas
XI, 650 pages, 1985.

Vol. 68: T. Kaczorek
Two-Dimensional Linear Systems
IX, 397 pages, 1985.

Vol. 69: Stochastic Differential Systems –
Filtering and Control
Proceedings of the IFIP-WG 7/1 Working Conference
Marseille-Luminy, France, March 12-17, 1984
Edited by M. Metivier, E. Pardoux
X, 310 pages, 1985.

Vol. 70: Uncertainty and Control
Proceedings of a DFVLR International Colloquium
Bonn, Germany, March, 1985
Edited by J. Ackermann
IV, 236 pages, 1985.

Vol. 71: N. Baba
New Topics in Learning Automata
Theory and Applications
VII, 231 pages, 1985.

Vol. 72: A. Isidori
Nonlinear Control Systems:
An Introduction
VI, 297 pages, 1985.

Vol. 73: J. Zarzycki
Nonlinear Prediction
Ladder-Filters for Higher-Order
Stochastic Sequences
V, 132 pages, 1985.

Vol. 74: K. Ichikawa
Control System Design based on
Exact Model Matching Techniques
VII, 129 pages, 1985.

Vol. 75: Distributed Parameter
Systems
Proceedings of the 2nd International
Conference, Vorau, Austria 1984
Edited by F. Kappel, K. Kunisch,
W. Schappacher
VIII, 460 pages, 1985.

Vol. 76: Stochastic Programming
Edited by F. Archetti, G. Di Pillo,
M. Lucertini
V, 285 pages, 1986.

Vol. 77: Detection of
Abrupt Changes in Signals
and Dynamical Systems
Edited by M. Basseville,
A. Benveniste
X, 373 pages, 1986.

Vol. 78: Stochastic
Differential Systems
Proceedings of the 3rd Bad Honnef
Conference, June 3–7, 1985
Edited by N. Christopeit, K. Helmes,
M. Kohlmann
V, 372 pages, 1986.

Vol. 79: Signal
Processing for Control
Edited by K. Godfrey, P. Jones
XVIII, 413 pages, 1986.

Vol. 80: Artificial Intelligence
and Man-Machine Systems
Edited by H. Winter
IV, 211 pages, 1986.

Lecture Notes in Control and Information Sciences

Edited by M. Thoma and A. Wyner